AF094684

MANY-BODY GREEN'S FUNCTIONS FOR TIME-DEPENDENT PROBLEMS

Quantum many-body systems are a central feature of condensed matter physics and are relevant to important, modern research areas such as ultrafast light–matter interactions and quantum information. This book covers the contour Green's function formalism in detail – an approach that could be successfully applied to solve the quantum many-body and time-dependent problems present within such systems. Divided into three parts, the text provides a structured overview of the relevant theoretical and practical tools, with specific focus on the Schwinger–Keldysh formalism. Part I introduces the mathematical frameworks that make use of functions in normal-phase states. Part II covers fermionic superfluid phases with a discussion of topics such as the BCS–BEC crossover and superconducting systems. Part III deals with the application of the Schwinger–Keldysh formalism to various topics of experimental interest. Graduate students and researchers will benefit from this book's comprehensive treatment of the subject matter and its novel arrangement of topics.

GIANCARLO CALVANESE STRINATI is Emeritus Professor of Physics at the University of Camerino, and his research is focused on condensed matter physics and ultracold atoms. He earned his PhD at the University of Chicago in 1977 with support from the Fulbright Program, before spending a year as Humboldt Fellow at the Max Planck Institute for Solid State Research. He joined the faculty of the Sapienza University of Rome – initially as Assistant Professor and then as Associate Professor – and later worked at the Scuola Normale Superiore in Pisa. He has been Fellow of the American Physical Society since 2010.

"Professor Strinati provides us with a pleasant and expert overview of the various standard techniques used in handling the many-body problem, mainly in fermionic systems. The book covers not only the classical Green's functions methods appropriate for equilibrium situations but also the remarkable extension to out-of-equilibrium situations with the Keldysh formalism. It also addresses the famous Kadanoff and Baym approach. As a physical example for application, Professor Strinati has chosen the fascinating phenomenon of the BEC–BCS crossover in ultracold fermionic atoms, a field in which he has made very important contributions. I highly recommend this remarkable book."

Roland Combescot, École Normale Supérieure, Paris

"Echoing the pedagogical style of Enrico Fermi, Professor Strinati guides the reader step-by-step through the nonequilibrium Green's function formalism. The text stands out for its mathematical completeness, offering rigorous derivations that expose theoretical details frequently omitted elsewhere. Bridging this foundational clarity with modern applications in ultracold gases, it serves as a valuable tutorial for students and an essential reference for researchers."

Hui Zhai, Institute for Advanced Study, Tsinghua University

"Since Feynman introduced diagrammatic methods in quantum field theory, the techniques have undergone extensive development and now underpin a wide range of applications. These methods have required even further expansion to meet the challenges of describing nonequilibrium phenomena and nonlinear processes. This book is devoted to a key breakthrough in this field – the Keldysh formalism or nonequilibrium Green's functions. Professor Giancarlo Calvanese Strinati, a world-renowned expert in condensed-matter theory, first presents the foundations and subtleties of the method and then applies it to a broad set of modern problems – from high-temperature superconductivity and ultracold gases to strongly correlated systems, mesoscopic physics, and nanoscale electronics.

I am confident that this monograph will help new generations of researchers master modern diagrammatic techniques, appreciate their elegance, and use them to drive future advances in understanding complex quantum systems."

Andrey Varlamov, CNR-SPIN, Rome

MANY-BODY GREEN'S FUNCTIONS FOR TIME-DEPENDENT PROBLEMS

GIANCARLO CALVANESE STRINATI
University of Camerino

CAMBRIDGE
UNIVERSITY PRESS

CAMBRIDGE
UNIVERSITY PRESS

Shaftesbury Road, Cambridge CB2 8EA, United Kingdom

One Liberty Plaza, 20th Floor, New York, NY 10006, USA

477 Williamstown Road, Port Melbourne, VIC 3207, Australia

314–321, 3rd Floor, Plot 3, Splendor Forum, Jasola District Centre, New Delhi – 110025, India

103 Penang Road, #05–06/07, Visioncrest Commercial, Singapore 238467

Cambridge University Press is part of Cambridge University Press & Assessment, a department of the University of Cambridge.

We share the University's mission to contribute to society through the pursuit of education, learning and research at the highest international levels of excellence.

www.cambridge.org
Information on this title: www.cambridge.org/9781009411547

DOI: 10.1017/9781009411509

© Giancarlo Calvanese Strinati 2026

This publication is in copyright. Subject to statutory exception and to the provisions of relevant collective licensing agreements, no reproduction of any part may take place without the written permission of Cambridge University Press & Assessment.

When citing this work, please include a reference to the DOI 10.1017/9781009411509

First published 2026

A catalogue record for this publication is available from the British Library

A Cataloging-in-Publication data record for this book is available from the Library of Congress

ISBN 978-1-009-41154-7 Hardback

Cambridge University Press & Assessment has no responsibility for the persistence or accuracy of URLs for external or third-party internet websites referred to in this publication and does not guarantee that any content on such websites is, or will remain, accurate or appropriate.

For EU product safety concerns, contact us at Calle de José Abascal, 56, 1°, 28003 Madrid, Spain, or email eugpsr@cambridge.org

To my parents, and to my wife Claudia
and sons Emilio, Adriano, and Marcello

Contents

Author's Note

Before delving into the full body of this book, I would like to say a few words about the general perspective through which it has been organized. In particular, I would like to reveal how my personal admiration for Impressionist art has influenced not only the choice of this book's cover, where the photograph, by capturing the presence of "dramatic swirling clouds over a quiet suburban neighborhood," gives the visual impression of a strongly nonequilibrium situation here due to a natural phenomenon that has inspired many painters, including Vincent Van Gogh in his famous artwork "The Starry Night", but also the choice of *A Guided Tour* as a possible subtitle, which is meant to be "virtually" attached to the title of this book, as well as the number of chapters (36 in total, excluding the Introduction) that form this book.

My personal admiration for Impressionist art has led me to visit several museums in Paris over the years, especially the Musée d'Orsay (located at the Gare d'Orsay in Paris) that houses the largest collection of Impressionist art in the world, with more than 480 paintings on display. As I enjoyed visiting the museum several times independently, without the guidance of a museum docent, each time I was invariably overwhelmed by the abundance of the artworks on display, however without grasping to my full satisfaction what ought to have been an underlying common thread.

Until one day, around the middle of February 2011, during a physics workshop held to mark the retirement of an eminent colleague, the participants were taken to the Musée d'Orsay during the free afternoon. On that occasion, the participants were thoughtfully divided into small groups, and each group was entrusted to a highly professional guide. After a short initial briefing, each guide led their assigned group of visitors through the museum but stopping only in front of a limited selection of Impressionist paintings. For my group, the guide selected only 36 paintings (which, by the way, is just the number of chapters in this book), emphasizing the connections between each painting and the next as we moved along. I must confess that I learned more about the deeper spirit of Impressionist art during that "guided tour" than in all of my independent visits to the Musée d'Orsay.

It is in the light of this artistic perspective that I planned to offer to readers, who are interested in the subject of *Many-Body Green's Functions for Time-Dependent Problems*, a "guided tour" through a focused number of selected and extensively analyzed topics in this field. This virtual tour, on the one hand, is unavoidably conditioned by personal preferences but, on the other hand, aims at providing a logical excursus through the essential developments of and implications for the treatment of time-dependent many-body problems in terms of appropriate Green's functions, thereby seeking to frame them within the most possible general landscape that is amenable to future developments. The expectation is that the readers will appreciate these efforts.

Part I

Normal Phase

1
Introduction

Heard melodies are sweet, but those unheard are sweeter
"Ode on a Grecian Urn" by John Keats

This book is meant to provide an advanced and detailed, yet pedagogical, account of the theoretical formalism used to describe quantum many-body systems that depart from equilibrium under fairly general conditions. It deals specifically with the *contour Green's function formalism*, which represents a general and versatile framework that can be applied to finite and extended quantum many-body systems, whether in equilibrium or in nonequilibrium, either at zero or finite temperature. The seed for this formalism was originally introduced by Schwinger [1] and subsequently extended by Keldysh [2], while an independent yet related approach was simultaneously developed by Baym and Kadanoff [3–5]. Among the advantages of the contour Green's function formalism is the fact that it allows one to derive many-body approximations consistent with conservation laws, even in the presence of time-dependent perturbations.

1.1 Relevance Time-Dependent Phenomena in Quantum Many-Body Systems

Quite generally, many-body Green's function methods aim at reducing the detailed information contained in the wave functions, by addressing directly dynamical quantities that are more closely related to experiments. In addition, they allow one to use physical insight when solving many-body problems, also by relying on the fact that they are "modular" in nature such that relevant refinements could be readily added when needed. These motivations are also reflected in more conventional Green's function methods other than the contour Green's function formalism discussed here, such as the zero-temperature method used to describe systems at most weakly perturbed out of equilibrium at low temperatures or the Matsubara method for systems in equilibrium at any temperature [6, 7]. Actually, the contour Green's function formalism encompasses and generalizes both these conventional methods, by maintaining, on the one hand, the same diagrammatic formal structure of the zero-temperature method, but obtaining, on the other hand, the explicit expressions of diagrammatic terms via a Wick's theorem similar to that of the Matsubara method. It is possibly for these reasons that the contour Green's function (or Schwinger–Keldysh) formalism is sometimes referred to as the "theory of everything" [8].

There are a number of excellent books describing the zero-temperature and Matsubara methods, which deal essentially with all aspects related to condensed-matter problems [6, 7, 9–11] or deal extensively with the Landau theory of Fermi liquids [12] and superconductivity [13]. The reader is advised to consult at least some of these books, which will help in grasping the generic aspects of the many-body Green's function methods (such as their connection to physical quantities) and in being introduced to the general form of the associated diagrammatic structure(s). Confident in this approach of "standing on the shoulders of giants," the discussion of these generic and well-known aspects will not be replicated here. Rather, we shall concentrate on the novel and specific aspects of the contour Green's function formalism, although sometimes these will unavoidably overlap with those of the zero-temperature and Matsubara methods.

It is worth mentioning in this context that, thus far, there are only a limited number of books available that deal with the nonequilibrium contour Green's function approach beginning from the late noughties [14–17], although, as already mentioned, the origin of this method goes back to the early sixties [1–5]. Among the reasons for this wide delay, one should mention that (i) while the zero-temperature and Matsubara methods often lead to analytic or semi-analytic results for which only limited numerical effort may be sufficient (especially when they originate from expansions in terms of small parameters or when recourse to self-consistency is not strictly required), implementing the (Schwinger–Keldysh) contour Green's function formalism unavoidably requires considerable numerical effort and (ii) only quite recently, experiments have developed the capability of following the time evolution of transient and/or metastable out-of-equilibrium dynamics of condensed-matter systems (as well as of ultracold gases).

Specifically, on the technological side, the interest in ultrafast excitation and relaxation phenomena has been prompted by the needs of nano-electronics and the attainability of ultrashort coherent light sources (generated, for instance, from electron storage rings), with pulses at the level of femtosecond (10^{-15}s) and even sub-femtosecond in ultrafast pump–probe spectroscopies. The importance of this new branch of physics was also recognized by the 2023 Nobel Prize in Physics, awarded to Pierre Agostini, Ferenc Krausz, and Anne L'Huillier "for experimental methods that generate attosecond pulses (10^{-18}s) of light for the study of electron dynamics in matter." This is because, although pioneering research in ultrafast laser science has opened up investigations of electronic motion in atoms, molecules, and materials, truly capturing the actual motion of electrons requires surpassing the performance limitations of even state-of-the-art lasers and accessing the attosecond regime. For instance, this approach has made it possible to quantify the timescale of the photoelectric effect of electrons in neon atoms, revealing that ionization from the 2p subshell occurs approximately 21 attoseconds longer than the ionization from the 2s subshell [18, 19].

On the theoretical side, fundamental questions arise about the behavior of matter on these ultrashort timescales, where structural dynamics and the formation of many-particle correlations acquire special significance. All of this should then be described within a nonequilibrium fashion, without any a priori assumption about the statistical distribution of particles out of equilibrium, the separation of timescales, the weakness of external perturbations, and the degree of spatial inhomogeneities. These are precisely the tasks that the contour Green's function (or Schwinger–Keldysh) formalism is optimally suited for.

1.2 Purpose, Contents, and Style of This Book

This book focuses on the general and broad task of equipping readers with a practical working knowledge on how to use the tools of the contour many-body Green's functions for time-dependent problems in all their details. Its main scope will then be to highlight the universality and versatility of the contour Schwinger–Keldysh formalism to a whole class of physical phenomena. To this end, it provides a self-contained introduction to the topic, together with a considerable amount of detailed derivations, which makes the text accessible to graduate students with minimal training in Green's function methods. At the same time, this book possesses a distinct degree of originality and presents material that is not commonly found in other books or review articles on the subject.

This book is divided into three parts. The main introductory aspects of the Schwinger–Keldysh formalism are presented in Part I, where bosons and fermions are dealt with on equal footing in the "normal" phase (in the sense that no condensate is considered). Part II deals with the "superfluid" phase in fermions, which are also allowed to undergo the BCS–BEC crossover [20]. In this way, emphasis will be given to the extension of the contour Schwinger–Keldysh formalism to superconductors driven out of equilibrium and, more generally, to fermionic superfluids (like ultracold Fermi gases) with their vastly different characteristic timescale compared to electron systems. These systems have recently acquired notable importance since time-dependent effects may lead to transient and/or metastable out-of-equilibrium dynamics with the emergence of some kind of long-range order [21–25]. Part III is devoted to the discussion of a number of specific applications of the nonequilibrium Schwinger–Keldysh formalism to various topics, including driven open quantum systems (also in the superfluid phase and with emphasis on dissipation), and its relation to the Lindblad equation, which is gaining increasing importance in the literature on nonequilibrium processes. In addition, although this task would be more pertinent to review articles of topical interest, Part III also covers at a qualitative level anticipated applications of the Schwinger–Keldysh formalism to various topics of current and increasing experimental interest, such as pump and probe photoemission, metastable photoinduced superconductivity, and dynamics induced by quenches and ramps in closed quantum systems. State-of-the-art numerical methods will also be discussed in Part III.

Throughout, this book is planned to proceed at an even pace while providing all necessary details, without "sweeping under the rug" the theoretical subtleties that will be encountered along the way and rather addressing them in full depth, with the intention of making this book a future reference manual for practitioners and even specialists in the field. To this end, the many formal derivations presented in this book will be fully detailed, self-contained, and rigorous, but still reader-friendly. In this respect, the presentation style of this book has admittedly been influenced by the classical review paper by Enrico Fermi on the "Quantum Theory of Radiation" [26], in which Fermi "takes the readers by the hand" and guides them through difficult topics with clarity and logic, providing detailed derivations throughout.

It is unavoidable that the contents of this book partially overlap with those of other books on the same subject, at least in some aspects. In this context, three books in particular deserve mention. The Schwinger–Keldysh formalism is developed elaborately in the book by Rammer [14], which expands on a previous review article [27] and places particular emphasis on the transport theory of metals in terms of kinetic equations such as the Boltzmann equation and its generalizations (although it also briefly mentions the extension of the formalism to superconductors). The book by Stefanucci and Van Leeuwen [15] is closer in spirit to this book and deals in a rather comprehensive way with the nonequilibrium Schwinger–Keldysh approach, although it is restricted to the normal phase, with no mention of the superfluid phase, and, more generally, places a different emphasis and relevance on various topics. Finally, one should mention the pioneering approach on nonequilibrium Green's functions by Kadanoff and Baym [5] (to be utilized in Part II of this book), which rests on the assumption of analytic continuation from imaginary to real times even in the presence of a time-dependent perturbation but makes no explicit link to the real-time Schwinger–Keldysh formalism. Although this book takes a different perspective that distinguishes it from the previous ones, the reader is strongly encouraged to consult those in order to gain a broader understanding of the topic.[1]

1.3 Target Audience

The book will be of interest to researchers in condensed matter physics, quantum physics and technologies, and ultracold Fermi gases undergoing the BCS–BEC crossover. It is especially meant to be a pedagogical text for independent reading on time-dependent problems in quantum many-body systems, with a particular focus on superconductors, in such a way that even advanced researchers can consult it as needed.

In addition, one of the purposes of this book is to provide a ready-to-use manual, which instructors could easily use for their courses on this topic. This book is intended for readers at the graduate and postgraduate levels. Accordingly, it may serve, on the one hand, as a text for a one- or two-semester graduate course and, on the other hand, as a useful reference for a broad audience of physicists dealing with physical problems where the use of the contour Schwinger–Keldysh formalism is relevant.

[1] In this way, one may avoid confronting with the Latin phrase "hominem unius libri timeo" ("I fear the man of a single book.") attributed to St Thomas Aquinas.

2

The Schrödinger and Heisenberg Representations: Time-Evolution Operator

In this chapter, the Hamiltonian operator is introduced, which is associated with the many-particle system that will be considered throughout, including its time-dependent part. An expression is also derived for the corresponding time-evolution operator, which depends only on the Hamiltonian and not on the initial preparation of the system before the time-dependent part begins to act. The connection between the Schrödinger and Heisenberg representations is finally discussed. In what follows, the language of second quantization will be utilized for its convenience in a many-body context.

2.1 Generic Time-Dependent Hamiltonian for a Quantum Many-Particle System, Either Fermions or Bosons

We consider a many-particle system made of identical particles (either fermions or bosons) with mass m, described by a time-dependent Hamiltonian $\mathcal{H}(t)$ of the form:

$$\mathcal{H}(t) = H + H^{\text{ext}}(t), \tag{2.1}$$

where $H = H_0 + H^{\text{int}}$ is the system Hamiltonian that contains a one-body (H_0) and a two-body (H^{int}) part, and $H^{\text{ext}}(t)$ is a time-dependent perturbation associated with the action of an external potential $V_{\text{ext}}(\mathbf{r}, t)$. These Hamiltonians can be conveniently expressed in terms of the field operator $\psi_\sigma(\mathbf{r})$ at spatial position $\mathbf{r}$ and with spin label σ, where $\sigma = 0$ for bosons and $\sigma = \pm 1$ for fermions, which satisfy the commutation (anti-commutation) relations [7]:

$$\left[\psi_\sigma(\mathbf{r}), \psi^\dagger_{\sigma'}(\mathbf{r}')\right]_\pm = \psi_\sigma(\mathbf{r})\psi^\dagger_{\sigma'}(\mathbf{r}') \pm \psi^\dagger_{\sigma'}(\mathbf{r}')\psi_\sigma(\mathbf{r}) = \delta_{\sigma\sigma'}\, \delta(\mathbf{r} - \mathbf{r}'), \tag{2.2}$$

where the plus sign holds for fermions and the minus sign for bosons. One obtains [7]:

$$H_0 = \sum_\sigma \int d\mathbf{r}\, \psi^\dagger_\sigma(\mathbf{r}) h_0(\mathbf{r}) \psi_\sigma(\mathbf{r}), \tag{2.3}$$

$$H^{\rm int} = \frac{1}{2}\sum_{\sigma,\sigma'}\int d\mathbf{r}d\mathbf{r}'\,\psi^\dagger_\sigma(\mathbf{r})\psi^\dagger_{\sigma'}(\mathbf{r}')v(\mathbf{r}-\mathbf{r}')\psi_{\sigma'}(\mathbf{r}')\psi_\sigma(\mathbf{r}), \tag{2.4}$$

$$H^{\rm ext}(t) = \sum_\sigma\int d\mathbf{r}\,\psi^\dagger_\sigma(\mathbf{r})V_{\rm ext}(\mathbf{r},t)\psi_\sigma(\mathbf{r})\,. \tag{2.5}$$

In Eq. (2.3), $h_0(\mathbf{r}) = -\frac{\hbar^2}{2m}\nabla^2 + V(\mathbf{r})$ is a single-particle Hamiltonian that includes a time-independent potential $V(\mathbf{r})$ acting on the particles, while, in Eq. (2.4), $v(\mathbf{r}-\mathbf{r}')$ is the interparticle (two-body) interaction. Depending on the experimental conditions, the interparticle interaction may as well depend on time, as it is the case for ultracold gases for which $v(\mathbf{r}-\mathbf{r}',t)$ can be subject to a sudden quench. In addition, the external potential $V_{\rm ext}$ may also depend on spin, as in the presence of a magnetic field.

2.2 Time-Evolution Operator $U(t,t_0)$ Obtained from the Schrödinger Representation and Its Properties

We now consider the state $|\Psi(t)\rangle$ of the many-body system with time-dependent Hamiltonian $\mathcal{H}(t)$, which evolves in time according to the Schrödinger equation

$$i\hbar\frac{d}{dt}|\Psi(t)\rangle = \mathcal{H}(t)|\Psi(t)\rangle, \tag{2.6}$$

with initial condition $|\Psi(t_0)\rangle$ at time t_0. However, instead of describing the system dynamics in terms of the state $|\Psi(t)\rangle$, it can be convenient to introduce a *time-evolution operator* $U(t,t_0)$ that connects states at different times

$$|\Psi(t)\rangle = U(t,t_0)|\Psi(t_0)\rangle, \tag{2.7}$$

where either $t > t_0$ or $t < t_0$. We will prove that:

$$U(t,t_0) = \begin{cases} \mathcal{T}\left\{e^{-\frac{i}{\hbar}\int_{t_0}^{t}dt'\mathcal{H}(t')}\right\} & \text{for } t \geq t_0 \\ \tilde{\mathcal{T}}\left\{e^{\frac{i}{\hbar}\int_{t}^{t_0}dt'\mathcal{H}(t')}\right\} & \text{for } t \leq t_0, \end{cases} \tag{2.8}$$

where $\mathcal{T}$ is the *time-ordering operator* and $\tilde{\mathcal{T}}$ is the *anti-time-ordering operator*. By definition, the time-ordering operator $\mathcal{T}$ takes any product of operators at different times and changes the order so that each operator has only operators at later times to its left and operators at earlier times to its right, while the anti-time-ordering operator $\tilde{\mathcal{T}}$ does just the opposite. When $\mathcal{H}(t) \to H$ does not depend on time, the evolution operator reduces to the simple form $U(t,t_0) = e^{-\frac{i}{\hbar}H(t-t_0)}$. Let's consider the two cases of Eq. (2.8) separately.

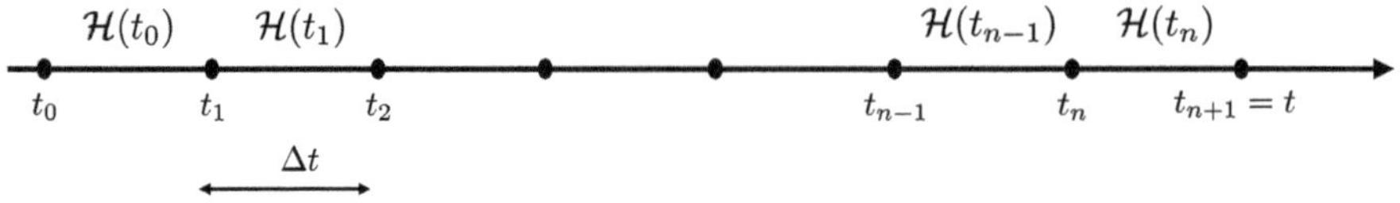

Figure 2.1 The time interval (t, t_0) is partitioned into $n+1$ subintervals, each of which has a length $\Delta t = (t - t_0)/(n+1)$. The Hamiltonian operator $\mathcal{H}(t_k)$ at time $t_k = t_0 + k\Delta t$ with $k = (0, 1, \cdots, n)$ evolves the system within the interval (t_k, t_{k+1}).

CASE $U(t, t_0)$ with $t > t_0$

We divide the time interval (t_0, t) into $n+1$ subintervals of equal length Δt, such that $t - t_0 = (n+1)\Delta t$ (cf. Fig. 2.1). If Δt is sufficiently small, we can consider $\mathcal{H}(t)$ to be approximately constant in each of these subintervals. This implies that within, say, the time subinterval (t_{k-1}, t_k)

$$|\Psi(t_k)\rangle = U(t_k, t_{k-1})|\Psi(t_{k-1})\rangle \simeq e^{-\frac{i}{\hbar}\mathcal{H}(t_{k-1})\Delta t}|\Psi(t_{k-1})\rangle \tag{2.9}$$

since the evolution from t_{k-1} to t_k occurs via the constant Hamiltonian $\mathcal{H}(t_{k-1})$. This expression can be extended to the whole interval (t_0, t), by writing (cf. Fig. 2.1)

$$\begin{aligned} |\Psi(t)\rangle &= |\Psi(t_{n+1})\rangle \simeq e^{-\frac{i}{\hbar}\mathcal{H}(t_n)\Delta t}|\Psi(t_n)\rangle \\ &\simeq e^{-\frac{i}{\hbar}\mathcal{H}(t_n)\Delta t}\, e^{-\frac{i}{\hbar}\mathcal{H}(t_{n-1})\Delta t}|\Psi(t_{n-1})\rangle \\ &\simeq e^{-\frac{i}{\hbar}\mathcal{H}(t_n)\Delta t}\, e^{-\frac{i}{\hbar}\mathcal{H}(t_{n-1})\Delta t} \cdots e^{-\frac{i}{\hbar}\mathcal{H}(t_0)\Delta t}|\Psi(t_0)\rangle, \end{aligned} \tag{2.10}$$

where the approximation is expected to improve with the number $n+1$ of subintervals. Note at this point that the (exponential) operators appearing in the expression (2.10) are ordered in order of increasing time, that is, with $t_n > t_{n-1} > \cdots > t_1 > t_0$. Nothing thus changes if we act on this string of operators by the time-ordering operator $\mathcal{T}$, such that

$$\begin{aligned} |\Psi(t_{n+1})\rangle &\simeq \mathcal{T}\Big\{e^{-\frac{i}{\hbar}\mathcal{H}(t_n)\Delta t}\, e^{-\frac{i}{\hbar}\mathcal{H}(t_{n-1})\Delta t} \cdots e^{-\frac{i}{\hbar}\mathcal{H}(t_0)\Delta t}\Big\}\, |\Psi(t_0)\rangle \\ &= \mathcal{T}\Big\{e^{-\frac{i}{\hbar}\sum_{k=0}^{n}\mathcal{H}(t_k)\Delta t}\Big\}\, |\Psi(t_0)\rangle \\ &= \lim_{n\to\infty} \mathcal{T}\Big\{e^{-\frac{i}{\hbar}\int_{t_0}^{t} dt' \mathcal{H}(t')}\Big\}\, |\Psi(t_0)\rangle\,, \end{aligned} \tag{2.11}$$

where we have used the properties that the operators under the $\mathcal{T}$ sign can be treated as commuting operators and that $e^A e^B = e^{A+B}$ for commuting operators A and B. This is an important feature because the operators $\mathcal{H}(t_k)$ at different times t_k do not necessarily commute with each other. Comparison of the result (2.11) with

Eq. (2.7) then identifies the evolution operator $U(t, t_0)$ with the upper expression in Eq. (2.8).

CASE $U(t_0, t)$ with $t > t_0$

Quite generally, the evolution operator obeys the *group property*

$$U(t_3, t_2)\, U(t_2, t_1) = U(t_3, t_1), \tag{2.12}$$

from which it follows that $U(t_1, t_2)\, U(t_2, t_1) = U(t_1, t_1) = \mathbb{1}$ is the unity operator $\mathbb{1}$. We thus obtain:

$$\lim_{n\to\infty} e^{-\frac{i}{\hbar}\mathcal{H}(t_n)\Delta t}\, e^{-\frac{i}{\hbar}\mathcal{H}(t_{n-1})\Delta t} \cdots e^{-\frac{i}{\hbar}\mathcal{H}(t_0)\Delta t}\, U(t_0, t) = \mathbb{1}\,, \tag{2.13}$$

that is to say

$$\begin{aligned} U(t_0, t) &= \lim_{n\to\infty} e^{\frac{i}{\hbar}\mathcal{H}(t_0)\Delta t} \cdots e^{\frac{i}{\hbar}\mathcal{H}(t_{n-1})\Delta t}\, e^{\frac{i}{\hbar}\mathcal{H}(t_n)\Delta t} \\ &= \tilde{\mathcal{T}}\left\{ e^{\frac{i}{\hbar}\int_{t_0}^{t} dt'\mathcal{H}(t')} \right\} \end{aligned} \tag{2.14}$$

where $\tilde{\mathcal{T}}$ is the anti-time-ordering operator of Eq. (2.8). Note also that

$$\begin{aligned} U^\dagger(t, t_0) &= \lim_{n\to\infty} \left[e^{-\frac{i}{\hbar}\mathcal{H}(t_n)\Delta t}\, e^{-\frac{i}{\hbar}\mathcal{H}(t_{n-1})\Delta t} \cdots e^{-\frac{i}{\hbar}\mathcal{H}(t_0)\Delta t} \right]^\dagger \\ &= \lim_{n\to\infty} e^{\frac{i}{\hbar}\mathcal{H}(t_0)\Delta t} \cdots e^{\frac{i}{\hbar}\mathcal{H}(t_{n-1})\Delta t}\, e^{\frac{i}{\hbar}\mathcal{H}(t_n)\Delta t} \\ &= U(t_0, t)\,, \end{aligned} \tag{2.15}$$

which implies that the evolution operator is unitary since

$$\mathbb{1} = U(t, t_0)\, U(t_0, t) = U(t, t_0)\, U^\dagger(t, t_0)\,. \tag{2.16}$$

Remark: An alternative route to U(t, t$_0$)

An equivalent way to arrive at the result (2.8) is as follows. A combination of Eqs. (2.6) and (2.7) yields

$$i\hbar\frac{d}{dt}|\Psi(t)\rangle = i\hbar\frac{d}{dt}U(t, t_0)|\Psi(t_0)\rangle \underset{\substack{\uparrow \\ \text{[Eq.(2.6)]}}}{=} \mathcal{H}(t)|\Psi(t)\rangle = \mathcal{H}(t)U(t, t_0)|\Psi(t_0)\rangle\,, \tag{2.17}$$

from which it follows that

$$i\hbar\frac{d}{dt}U(t, t_0) = \mathcal{H}(t)\, U(t, t_0) \tag{2.18}$$

since the state $|\Psi(t_0)\rangle$ is arbitrary. The differential equation (2.18) can then be integrated from t_0 to t to give

$$i\hbar \int_{t_0}^{t} dt' \frac{d}{dt'} U(t', t_0) = i\hbar \left(U(t, t_0) - \mathbb{1} \right) \underset{[\text{Eq.}(2.18)]}{\underset{\uparrow}{=}} \int_{t_0}^{t} dt' \mathcal{H}(t') U(t', t_0) \,, \tag{2.19}$$

that is,

$$U(t, t_0) = \mathbb{1} - \frac{i}{\hbar} \int_{t_0}^{t} dt' \mathcal{H}(t') U(t', t_0) \,. \tag{2.20}$$

This expression can be iterated as follows:

$$\begin{aligned} U(t, t_0) &= \mathbb{1} - \frac{i}{\hbar} \int_{t_0}^{t} dt' \mathcal{H}(t') \left(\mathbb{1} - \frac{i}{\hbar} \int_{t_0}^{t'} dt'' \mathcal{H}(t'') U(t'', t_0) \right) \\ &= \mathbb{1} - \frac{i}{\hbar} \int_{t_0}^{t} dt' \mathcal{H}(t') + \left(\frac{i}{\hbar} \right)^2 \int_{t_0}^{t} dt' \mathcal{H}(t') \int_{t_0}^{t'} dt'' \mathcal{H}(t'') + \cdots \\ &= \mathbb{1} + \sum_{k=1}^{\infty} \left(-\frac{i}{\hbar} \right)^k \int_{t_0}^{t} dt_1 \int_{t_0}^{t_1} dt_2 \cdots \int_{t_0}^{t_{k-1}} dt_k \, \mathcal{H}(t_1) \mathcal{H}(t_2) \cdots \mathcal{H}(t_k), \end{aligned} \tag{2.21}$$

where the Hamiltonian operators are ordered in order of increasing time. Nothing then happens if we act on this string of operators by the time-ordering operator $\mathcal{T}$, thereby writing

$$U(t, t_0) = \mathbb{1} + \sum_{k=1}^{\infty} \left(-\frac{i}{\hbar} \right)^k \int_{t_0}^{t} dt_1 \int_{t_0}^{t_1} dt_2 \cdots \int_{t_0}^{t_{k-1}} dt_k \, \mathcal{T}\{\mathcal{H}(t_1) \mathcal{H}(t_2) \cdots \mathcal{H}(t_k)\}. \tag{2.22}$$

We note at this point that the $k!$ permutations of the dummy time variables $\{t_j; j = (1, \cdots, k)\}$, with the simultaneous modification of the upper limits of the integrals, which identify a specific sector of the k-dimensional hypercube, give formally the same contribution to the k-dimensional integral of Eq. (2.22) since the Hamiltonian operators under the $\mathcal{T}$-sign can be treated as commuting operators. In this way, we obtain

$$\begin{aligned} U(t, t_0) &= \mathbb{1} + \sum_{k=1}^{\infty} \left(-\frac{i}{\hbar} \right)^k \frac{1}{k!} \int_{t_0}^{t} dt_1 \int_{t_0}^{t} dt_2 \cdots \int_{t_0}^{t} dt_k \, \mathcal{T} \left\{ \mathcal{H}(t_1) \mathcal{H}(t_2) \cdots \mathcal{H}(t_k) \right\} \\ &= \mathcal{T} \left\{ e^{-\frac{i}{\hbar} \int_{t_0}^{t} dt' \mathcal{H}(t')} \right\} \end{aligned} \tag{2.23}$$

which again coincides with the upper expression in Eq. (2.8).

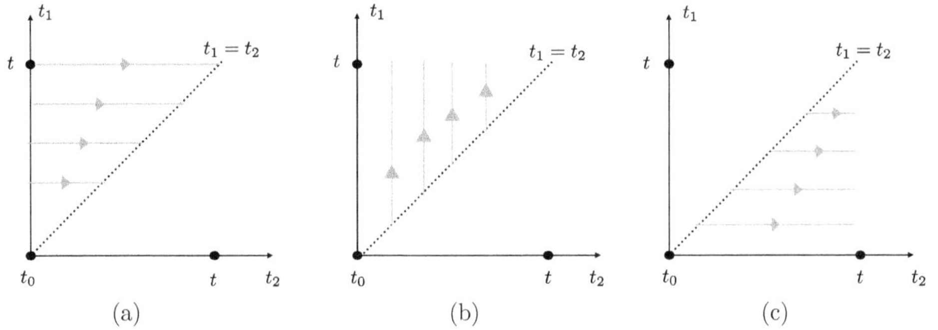

Figure 2.2 Order of integration in the $t_1 - t_2$ plane, corresponding to (a) the left term, (b) the middle term, and (c) the right term in the upper line of Eq. (2.24).

As a simple example of the above argument, consider the case with $k = 2$ for which the corresponding integral in Eq. (2.21) can be manipulated as follows:

$$\int_{t_0}^{t} dt_1 \int_{t_0}^{t_1} dt_2 \, \mathcal{H}(t_1)\mathcal{H}(t_2) = \int_{t_0}^{t} dt_2 \int_{t_2}^{t} dt_1 \, \mathcal{H}(t_1)\mathcal{H}(t_2) = \int_{t_0}^{t} dt_1 \int_{t_1}^{t} dt_2 \, \mathcal{H}(t_2)\mathcal{H}(t_1)$$

$$= \frac{1}{2} \int_{t_0}^{t} dt_1 \int_{t_0}^{t} dt_2 \, \left[\theta(t_1 - t_2)\mathcal{H}(t_1)\mathcal{H}(t_2) + \theta(t_2 - t_1)\mathcal{H}(t_2)\mathcal{H}(t_1) \right] . \tag{2.24}$$

Note that in the upper line of the expression (2.24), the left term corresponds to Fig. 2.2(a), the middle term to Fig. 2.2(b), and the right term to Fig. 2.2(c). The lower line of the expression (2.24) gives the desired result in terms of the time-ordering operator for this simple case.

Throughout the book, the time t_0 (sometimes in the literature referred to as *reference time* and indicated alternatively by t_r) will play a special role because it will represent the time at which the system is known to be prepared in a specific state $|\Psi(t_0)\rangle = |\Psi_0\rangle$ (or, more generally, in a given ensemble average) just before the time-dependent perturbation contained in $\mathcal{H}(t)$ begins to act on the system.

2.3 Operators in the Heisenberg Representation

In Section 2.2, the time-evolution operator $U(t, t_0)$ was meant to apply to a generic state of the physical system. It is sometimes convenient to transfer the time evolution from the states to the operators that are of relevance for describing the physical phenomena of interest. Accordingly, this will transform the time evolution of an observable from the Schrödinger to the Heisenberg representation.

Consider an operator $A(t)$ that might itself depend explicitly on time. The *Heisenberg representation* for this operator is defined by:

$$A_{\mathrm{H}}(t) = U^{\dagger}(t, t_0)\, A(t)\, U(t, t_0) \underset{\substack{\uparrow \\ \text{[Eq.(2.15)]}}}{=} U(t_0, t)\, A(t)\, U(t, t_0), \tag{2.25}$$

where the suffix H stands for Heisenberg and the time-evolution operator $U(t, t_0)$ is given by the expression (2.8). The definition (2.25) stems from the expression of the dynamical quantum average of the operator $A(t)$ over the state of the system $|\Psi(t)\rangle$ at time t, which reads:

$$\begin{aligned} \langle\Psi(t)|A(t)|\Psi(t)\rangle &\underset{\substack{\uparrow \\ \text{[Eq.(2.7)]}}}{=} \langle\Psi(t_0)|U^{\dagger}(t, t_0)A(t)U(t, t_0)|\Psi(t_0)\rangle \\ &\underset{\substack{\uparrow \\ \text{[Eq.(2.15)]}}}{=} \langle\Psi(t_0)|U(t_0, t)A(t)U(t, t_0)|\Psi(t_0)\rangle \underset{\substack{\uparrow \\ \text{[Eq.(2.25)]}}}{=} \langle\Psi(t_0)|A_{\mathrm{H}}(t)|\Psi(t_0)\rangle\,. \end{aligned} \tag{2.26}$$

Note that, in the definition (2.25), we have kept explicit reference to the time t_0 at which the time-dependent perturbation begins to act on the system. Recall also that the time-evolution operator in Eq. (2.25) depends on the *total* system Hamiltonian $\mathcal{H}(t)$ given by Eq. (2.1).

The operator $A_{\mathrm{H}}(t)$ in the Heisenberg representation (2.25) obeys a simple *equation of motion*, which is obtained from Eq. (2.18) and its adjoint

$$-i\frac{d}{dt}U^{\dagger}(t, t_0) = U^{\dagger}(t, t_0)\mathcal{H}(t) \tag{2.27}$$

that holds for the Hermitian operator $\mathcal{H}(t)$. [From now on, the reduced Planck constant $\hbar$ will be set equal to unity for convenience since we shall only deal with quantum phenomena throughout.] The equation of motion for $A_{\mathrm{H}}(t)$ then reads

$$i\frac{d}{dt}A_{\mathrm{H}}(t) = i\frac{d}{dt}\left(U^{\dagger}(t, t_0)A(t)U(t, t_0)\right) = [A_{\mathrm{H}}(t), \mathcal{H}_{\mathrm{H}}(t)] + i\frac{\partial}{\partial t}A_{\mathrm{H}}(t) \tag{2.28}$$

with the notation

$$\begin{aligned} \mathcal{H}_{\mathrm{H}}(t) &= U^{\dagger}(t, t_0)\mathcal{H}(t)U(t, t_0) \\ \frac{\partial}{\partial t}A_{\mathrm{H}}(t) &= U^{\dagger}(t, t_0)\left(\frac{d}{dt}A(t)\right)U(t, t_0)\,, \end{aligned} \tag{2.29}$$

where only the derivative with respect to the explicit time dependence of the operator $A(t)$ has to be taken. Note that when $\mathcal{H}(t) \to H$ does not depend on time, $U(t, t_0) = e^{-\frac{i}{\hbar}H(t-t_0)}$ commutes with H such that $\mathcal{H}_{\mathrm{H}}(t) \to H$, and the equation of motion (2.28) acquires the more familiar form of elementary Quantum Mechanics [28].

The definition (2.25) of the Heisenberg representation applies not only to operators associated with observable quantities but also to all operators including the field operators $\psi_\sigma(\mathbf{r})$ in terms of which the quantum many-body problem is conveniently formulated.

Although the Schrödinger and Heisenberg representations for the quantum average of an operator (corresponding to the first and last terms of Eq. (2.26)) are totally equivalent to each other, an advantage of the Heisenberg representation is that the quantum average of the operator $A(t)$ is taken over the state $|\Psi(t_0)\rangle$, specifying the *initial condition* of the system at time t_0 when the external agent begins to act, while the subsequent time evolution enters only the operator $A_\mathrm{H}(t)$.

3

Splitting the Hamiltonian: Heisenberg and Interaction Pictures

In both the Schrödinger and Heisenberg representations of Chapter 2, the time-evolution operator $U(t, t_0)$ depends on the full Hamiltonian $\mathcal{H}(t) = H + H^{\text{ext}}(t)$ of Eq. (2.1), where in addition $H = H_0 + H^{\text{int}}$. Out of the full time dependence due to $\mathcal{H}(t)$, it could be useful to isolate at the outset the time dependences due to *either* H *or* H_0. These two cases correspond to what are usually referred to as the Heisenberg and interaction pictures, respectively, which we now consider separately. In particular, the two cases give rise to perturbation theories in terms of the external perturbation $H^{\text{ext}}(t)$ and of the interaction part H^{int} of the system Hamiltonian, respectively.

3.1 The Time-Evolution Operator $V(t, t_0)$ in the Heisenberg Picture

A state in the Heisenberg (h) picture is defined as follows:

$$|\Psi_{\text{h}}(t)\rangle = e^{iHt}|\Psi(t)\rangle \,, \tag{3.1}$$

where $|\Psi(t)\rangle$ evolves in time according to the Schrödinger equation (2.6), and at this stage, there is no mention of the reference time t_0 when the external perturbation begins to act. Accordingly, the time evolution of $|\Psi_{\text{h}}(t)\rangle$ is given by:

$$i\frac{d}{dt}|\Psi_{\text{h}}(t)\rangle = e^{iHt}H^{\text{ext}}(t)|\Psi(t)\rangle = H_{\text{h}}^{\text{ext}}(t)|\Psi_{\text{h}}(t)\rangle, \tag{3.2}$$

where

$$H_{\text{h}}^{\text{ext}}(t) = e^{iHt}H^{\text{ext}}(t)\, e^{-iHt} \,. \tag{3.3}$$

At this point, we may define the time-evolution operator $V(t, t_0)$ in the Heisenberg picture, such that

$$|\Psi_{\text{h}}(t)\rangle = V(t, t_0)|\Psi_{\text{h}}(t_0)\rangle \,. \tag{3.4}$$

Comparison with the treatment of Section 2.2 leads us to identify ($t \geq t_0$)

$$V(t, t_0) = \mathcal{T}\left\{e^{-i\int_{t_0}^{t} dt' H_{\mathrm{h}}^{\mathrm{ext}}(t')}\right\} \tag{3.5}$$

which evolves in time according to $H_{\mathrm{h}}^{\mathrm{ext}}(t)$ given by Eq. (3.3). In reverse, we can write

$$|\Psi(t)\rangle = e^{-iHt}|\Psi_{\mathrm{h}}(t)\rangle = e^{-iHt}V(t, t_0)\, e^{iHt_0}|\Psi(t_0)\rangle\,, \tag{3.6}$$

from which (cf. Eq. (2.7))

$$U(t, t_0) = e^{-iHt}V(t, t_0)\, e^{iHt_0} \tag{3.7}$$

or else

$$V(t, t_0) = e^{iHt}U(t, t_0)\, e^{-iHt_0}\,, \tag{3.8}$$

such that

$$V^{\dagger}(t, t_0) = e^{iHt_0}U^{\dagger}(t, t_0)\, e^{-iHt} \underset{\substack{\uparrow \\ [\mathrm{Eq.}(2.15)]}}{=} e^{iHt_0}U(t_0, t)\, e^{-iHt} = V(t_0, t)\,. \tag{3.9}$$

In addition, the operator $A_{\mathrm{H}}(t)$ given by Eq. (2.25) can be rewritten in the form

$$A_{\mathrm{H}}(t) = e^{-iHt_0}\left(V^{\dagger}(t, t_0)A_{\mathrm{h}}(t)V(t, t_0)\right) e^{iHt_0} = \tilde{V}^{\dagger}(t, t_0)\, \tilde{A}_{\mathrm{h}}(t)\, \tilde{V}(t, t_0), \tag{3.10}$$

where we have defined

$$A_{\mathrm{h}}(t) = e^{iHt}A(t)\, e^{-iHt} \quad , \quad \tilde{A}_{\mathrm{h}}(t) = e^{iH(t-t_0)}A(t)\, e^{-iH(t-t_0)} \tag{3.11}$$

in line with Eq. (3.3), as well as

$$\tilde{V}(t, t_0) = e^{-iHt_0}V(t, t_0)\, e^{iHt_0} \underset{\substack{\uparrow \\ [\mathrm{Eq.}(3.8)]}}{=} e^{iH(t-t_0)}U(t, t_0) \underset{\substack{\uparrow \\ [\mathrm{Eq.}(3.5)]}}{=} \mathcal{T}\left\{e^{-i\int_{t_0}^{t} dt' \tilde{H}_{\mathrm{h}}^{\mathrm{ext}}(t')}\right\} \tag{3.12}$$

with $\tilde{H}_{\mathrm{h}}^{\mathrm{ext}}(t) = e^{iH(t-t_0)}H^{\mathrm{ext}}(t)e^{-iH(t-t_0)}$ in line with the definitions adopted in Ref. [14]. These results will be utilized in Chapter 18.

3.2 The Time-Evolution Operator $W(t, t_0)$ in the Interaction Picture

A state in the interaction (I) picture is defined as follows:

$$|\Psi_{\mathrm{I}}(t)\rangle = e^{iH_0 t}|\Psi(t)\rangle\,, \tag{3.13}$$

where $|\Psi(t)\rangle$ again evolves in time according to the Schrödinger equation (2.6), and H_0 is the non-interacting part (2.3) of the system Hamiltonian. We can now proceed as in Eq. (3.2) and write

$$i\frac{d}{dt}|\Psi_{\rm I}(t)\rangle = e^{iH_0t}\left(H^{\rm int}+H^{\rm ext}(t)\right)|\Psi(t)\rangle = \left(H_{\rm I}^{\rm int}(t)+H_{\rm I}^{\rm ext}(t)\right)|\Psi_{\rm I}(t)\rangle, \qquad (3.14)$$

where

$$H_{\rm I}^{\rm int}(t) = e^{iH_0t}H^{\rm int}\,e^{-iH_0t}, \qquad (3.15)$$

$$H_{\rm I}^{\rm ext}(t) = e^{iH_0t}H^{\rm ext}(t)\,e^{-iH_0t}, \qquad (3.16)$$

in analogy to Eq. (3.3).

The time-evolution operator $W(t,t_0)$ in the interaction picture can now be defined, such that

$$|\Psi_{\rm I}(t)\rangle = W(t,t_0)|\Psi_{\rm I}(t_0)\rangle, \qquad (3.17)$$

where (for $t > t_0$)

$$W(t,t_0) = \mathcal{T}\left\{e^{-i\int_{t_0}^{t}dt'\left(H_{\rm I}^{\rm int}(t')+H_{\rm I}^{\rm ext}(t')\right)}\right\} \qquad (3.18)$$

in analogy to Eqs. (2.8) and (3.5). Also, in this case, we may proceed in reverse and write

$$|\Psi(t)\rangle = e^{-iH_0t}|\Psi_{\rm I}(t)\rangle = e^{-iH_0t}W(t,t_0)\,e^{iH_0t_0}|\Psi(t_0)\rangle\,, \qquad (3.19)$$

such that

$$U(t,t_0) = e^{-iH_0t}W(t,t_0)\,e^{iH_0t_0} \qquad (3.20)$$

or else

$$W(t,t_0) = e^{iH_0t}U(t,t_0)\,e^{-iH_0t_0}\,, \qquad (3.21)$$

such that

$$W^\dagger(t,t_0) = e^{iH_0t_0}U^\dagger(t,t_0)\,e^{-iH_0t} \underset{\uparrow \atop [\text{Eq.(2.15)}]}{=} e^{iH_0t_0}U(t_0,t)\,e^{-iH_0t} = W(t_0,t)\,. \qquad (3.22)$$

Finally, the operator $A_{\rm H}(t)$ given by Eq. (2.25) can be rewritten in the form

$$A_{\rm H}(t) = e^{-iH_0t_0}\left(W^\dagger(t,t_0)A_{\rm I}(t)W(t,t_0)\right)e^{iH_0t_0} = \tilde{W}^\dagger(t,t_0)\,\tilde{A}_{\rm I}(t)\,\tilde{W}(t,t_0), \qquad (3.23)$$

where we have defined

$$A_{\rm I}(t) = e^{iH_0t}A(t)\,e^{-iH_0t} \quad , \quad \tilde{A}_{\rm I}(t) = e^{iH_0(t-t_0)}A(t)\,e^{-iH_0(t-t_0)}, \qquad (3.24)$$

in line with Eqs. (3.15) and (3.16), as well as

$$\tilde{W}(t,t_0) = e^{-iH_0t_0}W(t,t_0)\,e^{iH_0t_0} \underset{\uparrow \atop [\text{Eq.(3.21)}]}{=} e^{iH_0(t-t_0)}U(t,t_0) \underset{\uparrow \atop [\text{Eq.(3.18)}]}{=} \mathcal{T}\left\{e^{-i\int_{t_0}^{t}dt'\left(\tilde{H}_{\rm I}^{\rm int}(t')+\tilde{H}_{\rm I}^{\rm ext}(t')\right)}\right\} \qquad (3.25)$$

with $\tilde{H}_{\mathrm{I}}^{\mathrm{int}}(t) = e^{iH_0(t-t_0)} H^{\mathrm{int}} e^{-iH_0(t-t_0)}$ and $\tilde{H}_{\mathrm{I}}^{\mathrm{ext}}(t) = e^{iH_0(t-t_0)} H^{\mathrm{ext}}(t) e^{-iH_0(t-t_0)}$ in line with the definitions adopted in Ref. [14]. These results will be utilized in Chapters 8 and 10 of Part I and in Chapters 35 and 37 of Part III.

Remark: A first connection between real and imaginary times

From the relation (3.20) between the full time-evolution operator $U(t, t_0)$ and the time-evolution operator $W(t, t_0)$ in the interaction picture and from the expression (3.18) for $W(t, t_0)$, one can readily derive a fundamental relation in the context of the equilibrium Green's functions at finite temperature T, from which a perturbation expansion of the grand partition function of the interacting many-particle system can also be obtained [7].

To this end, one has to rotate the real-time t into the imaginary time $-i\tau$, where τ extends to the limited interval $0 \le \tau \le \beta = (k_B T)^{-1}$ (k_B being the Boltzmann constant) and set $t_0 = 0$ since t_0 is an irrelevant parameter in stationary (equilibrium) situations. In addition, at thermodynamic equilibrium, it is convenient to replace the Hamiltonian H by the grand-canonical Hamiltonian $K = H - \mu N$, where μ is the chemical potential and N is the total number of particles in the system, and correspondingly replace the noninteracting Hamiltonian H_0 by grand-canonical Hamiltonian $K_0 = H_0 - \mu N$ [7]. Accordingly, one obtains the desired thermodynamic relation from Eqs. (3.18) and (3.20)

$$U(t) \to U(\tau) = e^{-K\tau} = e^{-K_0\tau}\, \mathcal{T}_\tau \left\{ e^{-\int_0^\tau d\tau' H_{\mathrm{I}}^{\mathrm{int}}(\tau')} \right\} , \tag{3.26}$$

where $\mathcal{T}_\tau$ is the imaginary-time-ordering operator, which orders the operators according to their values of τ, with the smallest at the right, and

$$H_{\mathrm{I}}^{\mathrm{int}}(t) = e^{K_0\tau} H^{\mathrm{int}} e^{-K_0\tau} . \tag{3.27}$$

Note that the integration over the variable τ' on the right-hand side of Eq. (3.26) can be represented as an integration along the negative imaginary axis in the complex t-plane, as depicted in Fig. 3.1. This "vertical track" will play an important role in what

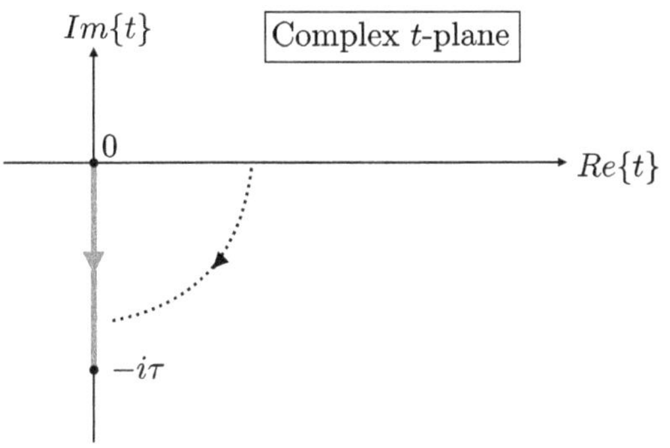

Figure 3.1 Integration over the interval from $t = 0$ to $t = -i\tau$ along the negative imaginary axis in the complex t-plane.

follows when the initial configuration of the system is taken at thermal equilibrium before the time-dependent perturbation begins to act. Note also that the result (3.26) could alternatively be obtained using Feynman's technique for the disentangling of operators [29].

4

Time-Dependent Quantum and Ensemble Averages: Initial Preparation of the System

In this chapter, we consider the initial preparation of the many-particle system, whose control is achieved before the reference time t_0 introduced in Chapters 2 and 3. After t_0, the system is let to evolve in time according to the full Hamiltonian $\mathcal{H}(t)$. The initial control can be either *full* or *partial*, where, by full control, we mean that at time t_0 the system is prepared in a definite ("pure") quantum state $|\Psi(t_0)\rangle$ (for instance, in the ground state $|\Psi_0\rangle$), while, by partial control, we mean that initially the system is only known to be in a "mixture" of states $\{|\chi_n\rangle\}$ with probabilities $\{w_n\}$ (whereby the information on the phases of the superposition is lost). We treat these two cases separately.

FULL CONTROL: Time-dependent quantum averages

Like in Section 2.3, let $A(t)$ be an operator that depends parametrically on time (this dependence may even be formal). Like in Eq. (2.7), let $|\Psi(t)\rangle$ be the state that develops in time for $t > t_0$ out of the state $|\Psi(t_0)\rangle$, in which the system was initially prepared at time t_0. The expectation value of the operator $A(t)$ over the state $|\Psi(t_0)\rangle$ is then simply given by (cf. also Eq. (2.26)):

$$\langle\Psi(t)|A(t)|\Psi(t)\rangle = \langle\Psi(t_0)|U^{\dagger}(t, t_0)A(t)U(t, t_0)|\Psi(t_0)\rangle \,. \tag{4.1}$$

PARTIAL CONTROL: Time-dependent ensemble averages

At the initial time t_0, the ensemble average of the operator $A(t)$ is given by

$$\langle A(t_0)\rangle = \sum_n w_n \langle\chi_n|A(t_0)|\chi_n\rangle, \tag{4.2}$$

where $\langle\chi_n|\chi_n\rangle = 1$ and $\sum_n w_n = 1$ (extension to continuum states is straightforward). This expression can be cast in a more meaningful form by introducing the *density matrix operator* ρ [30], defined by

$$\rho = \sum_n |\chi_n\rangle w_n \langle \chi_n| \,, \tag{4.3}$$

such that

$$\begin{aligned}
\langle A(t_0)\rangle &= \sum_n w_n \langle \chi_n | A(t_0) | \chi_n\rangle = \sum_n w_n \sum_k \langle \chi_n|\phi_k\rangle\langle\phi_k|A(t_0)|\chi_n\rangle \\
&= \sum_k \sum_n \langle\phi_k|A(t_0)|\chi_n\rangle w_n \langle\chi_n|\phi_k\rangle = \sum_k \langle\phi_k|A(t_0)\rho|\phi_k\rangle \\
&= \mathrm{Tr}\{A(t_0)\rho\} = \mathrm{Tr}\{\rho A(t_0)\}.
\end{aligned} \tag{4.4}$$

In this expression, $\{|\phi_k\rangle\}$ is a complete set of states, such that $\sum_k |\phi_k\rangle\langle\phi_k| = 1$, and Tr stands for the trace operation. We note the following properties of the operator ρ:

(i) $\rho = \rho^\dagger$ is a Hermitian operator (as w_n are real numbers).
(ii) The trace of ρ is given by

$$\begin{aligned}
\mathrm{Tr}\{\rho\} &= \sum_k \langle\phi_k|\rho|\phi_k\rangle = \sum_n w_n \sum_k \langle\phi_k|\chi_n\rangle\langle\chi_n|\phi_k\rangle \\
&= \sum_n w_n \langle\chi_n| \left(\sum_k |\phi_k\rangle\langle\phi_k| \right) |\chi_n\rangle = \sum_n w_n = 1 \,.
\end{aligned} \tag{4.5}$$

(iii) For any state $|\Psi\rangle$,

$$\langle\Psi|\rho|\Psi\rangle = \sum_n w_n \langle\Psi|\chi_n\rangle\langle\chi_n|\Psi\rangle = \sum_n w_n \,|\langle\chi_n|\Psi\rangle|^2 \geq 0 \,. \tag{4.6}$$

(iv) The states $\{|\phi_k\rangle\}$ can be taken to be the eigenvectors of the operator ρ with eigenvalues $\{\rho_k\}$, such that

$$\rho|\phi_k\rangle = |\phi_k\rangle \rho_k \,. \tag{4.7}$$

Accordingly, the operator ρ can be rewritten in the form

$$\rho = \sum_k |\phi_k\rangle \rho_k \langle\phi_k|, \tag{4.8}$$

where the sum is now over the complete set $\{|\phi_k\rangle\}$. Note that the difference between the expressions (4.8) and (4.3) of the density matrix operator ρ stems from the fact that the states $\{|\chi_n\rangle\}$, although normalized, may not be orthogonal to each other and may not form a complete set.

(v) The eigenvalues $\{\rho_k\}$ are positive semidefinite for each k since by replacing $|\Psi\rangle \to |\phi_k\rangle$ in Eq. (4.6), we obtain

$$\rho_k = \langle\phi_k|\rho|\phi_k\rangle = \sum_n w_n \langle\phi_k|\chi_n\rangle\langle\chi_n|\phi_k\rangle = \sum_n w_n \, |\langle\chi_n|\phi_k\rangle|^2 \geq 0\,. \tag{4.9}$$

In addition, $\rho_k \leq 1$ owing to the property (4.5).

(vi) The most general expression for ρ_k, which incorporates the above properties, is

$$\rho_k = \frac{e^{-x_k}}{\sum_{k'} e^{-x_{k'}}}, \tag{4.10}$$

where $x_k \geq 0$ for all k. We can then write $x_k = \beta E_k^{\mathrm{M}}$, where β is a positive constant (with dimension of the inverse of an energy), and construct an operator H^{M} (with the suffix M standing for "mixture"), such that

$$H^{\mathrm{M}} = \sum_k |\phi_k\rangle E_k^{\mathrm{M}} \langle\phi_k|\,. \tag{4.11}$$

The density matrix operator (4.8) then takes the form

$$\rho = \frac{\sum_k |\phi_k\rangle e^{-\beta E_k^{\mathrm{M}}} \langle\phi_k|}{\sum_{k'} e^{-\beta E_{k'}^{\mathrm{M}}}} = \frac{e^{-\beta H^{\mathrm{M}}}}{Z}, \tag{4.12}$$

where

$$Z = \sum_k e^{-\beta E_k^{\mathrm{M}}} = \mathrm{Tr}\{e^{-\beta H^{\mathrm{M}}}\}\,. \tag{4.13}$$

In particular, when the N-particle system described by the Hamiltonian H is initially in *thermodynamic equilibrium* at temperature T and chemical potential μ, one knows from quantum statistical mechanics that $H^{\mathrm{M}} \leftrightarrow H - \mu N$ and $\beta = (k_B T)^{-1}$ in the expressions (4.12) and (4.13) [31].

When one considers the time dependence of an operator $A(t)$, starting from an initial configuration at time t_0 described by Eq. (4.2) with $|\chi_n\rangle = |\chi_n(t_0)\rangle$ for all n, it is reasonable to assume that it is given by the expression

$$\begin{aligned} \langle A(t)\rangle &= \sum_n w_n \langle\chi_n(t)|A(t)|\chi_n(t)\rangle = \sum_n w_n \langle\chi_n|U^\dagger(t,t_0)A(t)U(t,t_0)|\chi_n\rangle \\ &= \mathrm{Tr}\{\rho\, U^\dagger(t,t_0)A(t)U(t,t_0)\} = \mathrm{Tr}\{U(t,t_0)\rho\, U^\dagger(t,t_0)A(t)\}, \end{aligned} \tag{4.14}$$

where use has been made of the cyclic property of the trace. Note that in the second line of Eq. (4.14), the expression on the left-hand side can be readily obtained from Eq. (4.4), with the replacement $A(t_0) \to U^\dagger(t,t_0)A(t)U(t,t_0)$, while in the expression on the right-hand side one can identify $U(t,t_0)\rho\, U^\dagger(t,t_0)$ as the time-dependent density matrix operator $\rho(t)$. We shall return to this representation for the time dependence of the density matrix operator in Part III when specifically addressing the Lindblad equation.

The physical assumption underlying the result (4.14) is that, when the time-dependent part $H^{\text{ext}}(t)$ of the Hamiltonian $\mathcal{H}(t)$ in Eq. (2.1) starts to act on the system at time t_0, there is no longer any influence on the system from the "machinery," which has served to prepare the system at time t_0 (like the reservoir with which the system has been in contact for sufficiently long time). It is then as though for times $t > t_0$ the system would be disconnected from the reservoir, which had duly prepared it in equilibrium with temperature T and chemical potential μ.

5

Quantum Averages over the Ground State and the Gell–Mann–Low Theorem

In this chapter, we consider the expectation value of an operator A (or of products of two operators A and B) over the *ground state* $|\Psi_0\rangle$ *of the interacting system* with Hamiltonian H, when the external time-dependent perturbation $H^{\text{ext}}(t)$ in Eq. (2.1) is switched off. These limitations apply to systems in equilibrium at zero temperature, which include important cases like insulators and semiconductors [32], as well as Fermi liquids [6, 12], for which the energy gap and the Fermi energy are, respectively, much larger than the available thermal energy $k_B T$. There are also two additional reasons to consider this case here, although in the following we shall mainly be interested in out-of-equilibrium situations to be dealt with in terms of the Schwinger–Keldysh formalism. The first reason is that the diagram techniques leading to perturbation expansions in terms of the interaction part H^{int} of the Hamiltonian H are formally similar in the zero-temperature and Schwinger–Keldysh formalisms. The second reason is to emphasize that the present formalism for ground-state averages at zero temperature relies specifically on an "adiabatic assumption," which cannot be applied as it is when excited states are involved in the ensemble averages.

5.1 Adiabatic Turning on of the Interaction

When $H^{\text{ext}}(t) = 0$, the time t_0 loses its original meaning of the reference time at which $H^{\text{ext}}(t)$ begins to act. In this case, we can conveniently set $t_0 = 0$ that lies in the time range when the physical system is subject to measurements. In addition, we assume that the interaction part H^{int} of the system Hamiltonian is first *adiabatically turned on* and then *adiabatically switched off*, meaning that we replace

$$H^{\text{int}} \longrightarrow H^{\text{int}}_{\epsilon} = e^{-|\epsilon|t} H^{\text{int}}, \tag{5.1}$$

where $0 < \epsilon \ll 1$ (in units of an inverse time), with the understanding that the limit $\epsilon \longrightarrow 0$ will eventually be taken at the end of the calculation. By this assumption, one is supposing that at time $t \longrightarrow -\infty$ in the far past no interaction is acting between particles and that afterwards the interaction is turned on infinitely slowly. As a consequence, at time $t \to -\infty$, the ground state $|\Psi_0\rangle$ of the interacting system reduces to the ground state $|\Phi_0\rangle$ of the noninteracting system, such that

$$|\Psi_{\rm I}(t \to -\infty)\rangle = |\Phi_0\rangle \tag{5.2}$$

according to Eq. (3.13); while at time $t = 0$, when the interaction is fully turned on

$$|\Psi_{\rm I}(t=0)\rangle = W_\epsilon(0,-\infty)|\Phi_0\rangle = |\Psi_0\rangle \tag{5.3}$$

according to Eq. (3.17). The last equality implies the crucial assumption that the perturbation (5.1) develops so slowly in time that the system remains always in the (non-degenerate) ground state of the instantaneous Hamiltonian (apart from a phase factor, which is irrelevant at this point but to which we shall return later on).

With these premises, we now consider the expectation value of an operator $A(t)$ (which may depend explicitly on time) over the ground state $|\Psi_0\rangle$ of the interacting system. We obtain:

$$\begin{aligned}\langle A(t)\rangle &\underset{\substack{\uparrow\\ \text{[Eq.(4.1) with } t_0=0]}}{=} \langle\Psi_0|U_\epsilon^\dagger(t,0)A(t)U_\epsilon(t,0)|\Psi_0\rangle \underset{\substack{\uparrow\\ \text{[Eqs.(3.20) and (3.24) with } t_0=0]}}{=} \langle\Psi_0|W_\epsilon^\dagger(t,0)A_{\rm I}(t)W_\epsilon(t,0)|\Psi_0\rangle \\ &\underset{\substack{\uparrow\\ \text{[Eqs.(5.3) and (3.18)]}}}{=} \langle\Phi_0|W_\epsilon^\dagger(t,-\infty)A_{\rm I}(t)W_\epsilon(t,-\infty)|\Phi_0\rangle\,. \end{aligned} \tag{5.4}$$

At this point, we can make use of the unitarity and group properties of the time-evolution operator W in the interaction picture (cf. Eq. (3.18)) and rewrite in the last term of Eq. (5.4):

$$\begin{aligned} W_\epsilon^\dagger(t,-\infty) &= W_\epsilon(-\infty,t) = W_\epsilon(-\infty,+\infty)\,W_\epsilon(+\infty,t) \\ &= W_\epsilon^\dagger(+\infty,-\infty)\,W_\epsilon(+\infty,t)\,. \end{aligned} \tag{5.5}$$

In this way, the expression (5.4) becomes:

$$\langle A(t)\rangle = \langle\Phi_0|W_\epsilon^\dagger(+\infty,-\infty)\,W_\epsilon(+\infty,t)A_{\rm I}(t)W_\epsilon(t,-\infty)|\Phi_0\rangle\,. \tag{5.6}$$

Two additional steps are still required to bring this expression to a form that is directly amenable to perturbative calculations.

5.2 Step 1: Adiabatic Evolution from $t = -\infty$ to $t = +\infty$

The first step addresses the action of the operator $W_\epsilon^\dagger(+\infty,-\infty)$ on the left of the expression (5.6). Recall that, according to Eq. (5.1), the interaction part of the Hamiltonian is assumed to be adiabatically switched on "in the past" between $t = -\infty$ and $t = 0$ and then to be adiabatically switched off "in the future" between $t = 0$ and $t = +\infty$. For a *non-degenerate* ground state $|\Phi_0\rangle$ of the noninteracting system, the adiabatic assumption implies that the effect of $W_\epsilon(+\infty,-\infty)$ on

$|\Phi_0\rangle$ is to return it back to the state $|\Phi_0\rangle$ *apart from a phase factor*, that is to say,

$$W_\epsilon(+\infty, -\infty)|\Phi_0\rangle = |\Phi_0\rangle\, e^{i\phi_\epsilon} . \tag{5.7}$$

By projecting both sides of this expression onto $\langle\Phi_0|$ and taking into account that $\langle\Phi_0|\Phi_0\rangle = 1$, we then obtain

$$\langle\Phi_0|W_\epsilon(+\infty, -\infty)|\Phi_0\rangle = \langle\Phi_0|\Phi_0\rangle\, e^{i\phi_\epsilon} = e^{i\phi_\epsilon} , \tag{5.8}$$

from which

$$\begin{aligned}(W_\epsilon(+\infty, -\infty)|\Phi_0\rangle)^\dagger &= \langle\Phi_0|W_\epsilon^\dagger(+\infty, -\infty) = e^{-i\phi_\epsilon}\langle\Phi_0| \\ &= \frac{\langle\Phi_0|}{\langle\Phi_0|W_\epsilon(+\infty, -\infty)|\Phi_0\rangle} .\end{aligned} \tag{5.9}$$

Entering this result into the expression (5.6), we obtain eventually the desired result

$$\langle A(t)\rangle = \frac{\langle\Phi_0|W_\epsilon(+\infty, t)A_\mathrm{I}(t)W_\epsilon(t, -\infty)|\Phi_0\rangle}{\langle\Phi_0|W_\epsilon(+\infty, -\infty)|\Phi_0\rangle} \tag{5.10}$$

known as the *Gell–Mann–Low theorem*.

It should be mentioned that in the limit $\epsilon \to 0$, the phase ϕ_ϵ of Eq. (5.7) diverges like ϵ^{-1} [33]. The ratio in Eq. (5.10) is, however, well defined in the limit $\epsilon \to 0$, such that this limit can safely be taken [7]. In addition, it should be emphasized that the result (5.10) is valid only for an average over the ground state of the system, which is nondegenerate since it corresponds to the minimum of the energy. Accordingly, under the action of an infinitely slow perturbation, this state cannot make transitions into other (excited) states, as it was assumed in Eq. (5.7). However, this is not the case when the average is taken over an excited state of the system, which being degenerate can make transitions to other states during the adiabatic evolution [6].

5.3 Step 2: Folding Theorem for Time-Ordered Products

The second step aims at folding the two time-ordered products in the numerator of Eq. (5.10) into a single one, by writing therein

$$\mathcal{T}\left\{e^{-i\int_{-\infty}^{+\infty}dt' H_\mathrm{I}^\mathrm{int}(t')} A_\mathrm{I}(t)\right\} = W_\epsilon(+\infty, t)A_\mathrm{I}(t)W_\epsilon(t, -\infty) . \tag{5.11}$$

Proof of Eq. (5.11)

We begin by exploiting the presence of the time t associated with the operator $A_\mathrm{I}(t)$ and split the integration from $-\infty$ to $+\infty$ on the left-hand side of Eq. (5.11) into a first integration from $-\infty$ to t and a second integration from t to $+\infty$. We do this in each term originating from the expansion of the exponential operator therein, namely,

$$\mathcal{T}\left\{e^{-i\int_{-\infty}^{+\infty}dt' H_{\mathrm{I}}^{\mathrm{int}}(t')}A_{\mathrm{I}}(t)\right\} = \sum_{n=0}^{\infty}\frac{(-i)^n}{n!}\int_{-\infty}^{+\infty}dt_1\int_{-\infty}^{+\infty}dt_2\cdots\int_{-\infty}^{+\infty}dt_n \times\mathcal{T}\left\{H_{\mathrm{I}}^{\mathrm{int}}(t_1)\,H_{\mathrm{I}}^{\mathrm{int}}(t_2)\cdots H_{\mathrm{I}}^{\mathrm{int}}(t_n)A_{\mathrm{I}}(t)\right\} \tag{5.12}$$

where we write

$$\int_{-\infty}^{+\infty}dt_1\int_{-\infty}^{+\infty}dt_2\cdots\int_{-\infty}^{+\infty}dt_n = \left[\int_{-\infty}^{t}dt_1+\int_{t}^{+\infty}dt_1\right]\left[\int_{-\infty}^{t}dt_2+\int_{t}^{+\infty}dt_2\right]\cdots\left[\int_{-\infty}^{t}dt_n+\int_{t}^{+\infty}dt_n\right]. \tag{5.13}$$

To make the notation more compact, we identify

$$\int_{-\infty}^{t}dt_i \longleftrightarrow x_i \quad \text{and} \quad \int_{t}^{+\infty}dt_i \longleftrightarrow y_i \tag{5.14}$$

with $i = (1, \ldots, n)$, such that the expression (5.13) becomes:

$$\prod_{i=1}^{n}(x_i+y_i) = \sum_{m=0}^{n}\frac{n!}{m!(n-m)!}\,\overbrace{y_1\cdots y_m}^{m\text{ times}}\;\overbrace{x_{m+1}\cdots x_n}^{n-m\text{ times}}\,. \tag{5.15}$$

Note that we have here utilized the binomial coefficients of the binomial theorem for grouping terms with an equal number of x-factors (and thus also of y-factors) because in the present case the variables can be renamed (as they correspond to different dummy integration labelings). We further introduce the variable $k = n - m$ on the right-hand side of Eq. (5.15), such that

$$\prod_{i=1}^{n}(x_i+y_i) = \sum_{m=0}^{n}\sum_{k=0}^{n}\delta_{n,k+m}\frac{n!}{m!\,k!}\,y_1\cdots y_m\;x_1\cdots x_k, \tag{5.16}$$

where we have again renamed the dummy integration variables associated with the x-terms. When this expression is eventually summed over the variable n as in Eq. (5.12) with the appropriate coefficients appearing therein, we obtain eventually the result:

$$\begin{aligned}\sum_{n=0}^{\infty}\frac{(-i)^n}{n!}\prod_{i=1}^{n}(x_i+y_i) &= \sum_{n=0}^{\infty}\frac{(-i)^n}{n!}\sum_{m=0}^{n}\sum_{k=0}^{n}\delta_{n,k+m}\frac{n!}{m!\,k!}\,y_1\cdots y_m\;x_1\cdots x_k\\ &= \sum_{m=0}^{\infty}\frac{(-i)^m}{m!}\,y_1\cdots y_m\sum_{k=0}^{\infty}\frac{(-i)^k}{k!}x_1\cdots x_k\,.\end{aligned} \tag{5.17}$$

At this point, we can go back to the identification (5.14) and take into account the presence of the time-ordered product in Eq. (5.12) to rewrite this equation in the desired form:

$$\begin{aligned}\mathcal{T}\left\{e^{-i\int_{-\infty}^{+\infty}dt' H_{\mathrm{I}}^{\mathrm{int}}(t')}A_{\mathrm{I}}(t)\right\} &= \mathcal{T}\left\{e^{-i\int_{t}^{+\infty}dt' H_{\mathrm{I}}^{\mathrm{int}}(t')}\right\} A_{\mathrm{I}}(t)\mathcal{T}\left\{e^{-i\int_{-\infty}^{t}dt'' H_{\mathrm{I}}^{\mathrm{int}}(t'')}\right\}\\ &= W_{\epsilon}(+\infty, t)\, A_{\mathrm{I}}(t)\, W_{\epsilon}(t, -\infty)\,. \qquad \text{[QED]}\end{aligned} \tag{5.18}$$

With the result (5.11), the expression (5.10) can be cast in its final form (where one is allowed to take $\epsilon \longrightarrow 0$):

$$\langle A(t)\rangle = \frac{\langle \Phi_0|\mathcal{T}\left\{e^{-i\int_{-\infty}^{+\infty}dt' H_{\mathrm{I}}^{\mathrm{int}}(t')}A_{\mathrm{I}}(t)\right\}|\Phi_0\rangle}{\langle \Phi_0|\mathcal{T}\left\{e^{-i\int_{-\infty}^{+\infty}dt' H_{\mathrm{I}}^{\mathrm{int}}(t')}\right\}|\Phi_0\rangle}\,. \tag{5.19}$$

Remark: Extension to products of operators

The above result can be extended to the case of more than one operator. Let's consider the case of two operators $A(t)$ and $B(t)$, which are ordered by the time-ordered operator in the form $\langle \Psi_0|\mathcal{T}\,\{A_{\mathrm{H}}(t_a)B_{\mathrm{H}}(t_b)\}\,|\Psi_0\rangle$. Suppose for definiteness that $t_a > t_b$. Following the steps that have led to Eq. (5.6), we now obtain:

$$\begin{aligned}&\langle \Psi_0|A_{\mathrm{H}}(t_a)B_{\mathrm{H}}(t_b)|\Psi_0\rangle\\ &\underset{\substack{\uparrow\\ [\text{Eq.(3.23) with } t_0=0]}}{=} \langle \Psi_0|W_{\epsilon}^{\dagger}(t_a,0)A_{\mathrm{I}}(t_a)W_{\epsilon}(t_a,0)W_{\epsilon}^{\dagger}(t_b,0)B_{\mathrm{I}}(t_b)W_{\epsilon}(t_b,0)|\Psi_0\rangle\\ &\underset{\substack{\uparrow\\ [\text{Eq.(5.3)}]}}{=} \langle \Phi_0|W_{\epsilon}^{\dagger}(t_a,-\infty)A_{\mathrm{I}}(t_a)W_{\epsilon}(t_a,t_b)B_{\mathrm{I}}(t_b)W_{\epsilon}(t_b,-\infty)|\Phi_0\rangle\\ &\underset{\substack{\uparrow\\ [\text{Eq.(5.5)}]}}{=} \langle \Phi_0|W_{\epsilon}^{\dagger}(+\infty,-\infty)W_{\epsilon}(+\infty,t_a)A_{\mathrm{I}}(t_a)W_{\epsilon}(t_a,t_b)B_{\mathrm{I}}(t_b)W_{\epsilon}(t_b,-\infty)|\Phi_0\rangle\\ &\underset{\substack{\uparrow\\ [\text{Eq.(5.9)}]}}{=} \frac{\langle \Phi_0|W_{\epsilon}(+\infty,t_a)A_{\mathrm{I}}(t_a)W_{\epsilon}(t_a,t_b)B_{\mathrm{I}}(t_b)W_{\epsilon}(t_b,-\infty)|\Phi_0\rangle}{\langle \Phi_0|W_{\epsilon}(+\infty,-\infty)|\Phi_0\rangle}\,,\end{aligned} \tag{5.20}$$

where the limit $\epsilon \longrightarrow 0$ can safely be taken. In analogy to the result (5.11), we can now prove the operator identity

$$\begin{aligned}&\mathcal{T}\left\{e^{-i\int_{-\infty}^{+\infty}dt' H_{\mathrm{I}}^{\mathrm{int}}(t')}A_{\mathrm{I}}(t_a)B_{\mathrm{I}}(t_b)\right\}\\ &= \begin{cases} W_{\epsilon}(+\infty,t_a)A_{\mathrm{I}}(t_a)W_{\epsilon}(t_a,t_b)B_{\mathrm{I}}(t_b)W_{\epsilon}(t_b,-\infty) & \text{for } t_a > t_b\\[2ex] W_{\epsilon}(+\infty,t_b)B_{\mathrm{I}}(t_b)W_{\epsilon}(t_b,t_a)A_{\mathrm{I}}(t_a)W_{\epsilon}(t_a,-\infty) & \text{for } t_b > t_a\,,\end{cases}\end{aligned} \tag{5.21}$$

with the further provision that in the lower line on the right-hand side there should also appear a ± sign, depending on whether A and B are both bosonic (+ sign) of fermionic (− sign) operators.

Proof of Eq. (5.21)

In analogy to what we did in Eq. (5.13), we now split the integration from $-\infty$ to $+\infty$ on the left-hand side of Eq. (5.21) into three intervals, namely, when $t_a > t_b$ first from $-\infty$ to t_b, then from t_b to t_a, and finally from t_a to $+\infty$ (while t_a and t_b have to be interchanged in the limits of integration when $t_b > t_a$). We thus write in the place of Eq. (5.13) when $t_a > t_b$:

$$\int_{-\infty}^{+\infty} dt_1 \int_{-\infty}^{+\infty} dt_2 \cdots \int_{-\infty}^{+\infty} dt_n = \left[\int_{-\infty}^{t_b} dt_1 + \int_{t_b}^{t_a} dt_1 + \int_{t_a}^{+\infty} dt_1\right] \cdots \left[\int_{-\infty}^{t_b} dt_n + \int_{t_b}^{t_a} dt_n + \int_{t_a}^{+\infty} dt_n\right] . \quad (5.22)$$

This leads us to complement the compact notation of Eq. (5.14), in the form

$$\int_{-\infty}^{t_b} dt_i \longleftrightarrow x_i \quad , \quad \int_{t_b}^{t_a} dt_i \longleftrightarrow y_i \quad , \quad \int_{t_a}^{+\infty} dt_i \longleftrightarrow z_i \quad (5.23)$$

with $i = (1, \cdots, n)$, such that:

$$\prod_{i=1}^{n} (x_i + y_i + z_i) = \prod_{i=1}^{n} (x_i + w_i) = \sum_{m=0}^{n} \sum_{k=0}^{n} \delta_{n,k+m} \frac{n!}{m!\,k!}\, w_1 \cdots w_m \; x_1 \cdots x_k \, , \quad (5.24)$$

where we have set $w_i = y_i + z_i$ and utilized the result (5.16) (as well as exploited again the freedom of renaming the variables). Proceeding now as in Eq. (5.17), we obtain

$$\sum_{n=0}^{\infty} \frac{(-i)^n}{n!} \prod_{i=1}^{n} (x_i + w_i) = \sum_{m=0}^{\infty} \frac{(-i)^m}{m!} w_1 \cdots w_m \sum_{k=0}^{\infty} \frac{(-i)^k}{k!} x_1 \cdots x_k \, , \quad (5.25)$$

where we can further write

$$\begin{aligned} \sum_{m=0}^{\infty} \frac{(-i)^m}{m!} w_1 \cdots w_m &= \sum_{m=0}^{\infty} \frac{(-i)^m}{m!} \prod_{i=1}^{m} (y_i + z_i) \\ &= \sum_{\mu=0}^{\infty} \frac{(-i)^\mu}{\mu!} z_1 \cdots z_\mu \sum_{\nu=0}^{\infty} \frac{(-i)^\nu}{\nu!} y_1 \cdots y_\nu \, . \end{aligned} \quad (5.26)$$

Taking into account the presence of the time-ordered product on the left-hand side of Eq. (5.21), the above results imply that Eq. (5.21) is satisfied. [QED]

In Chapter 6, we will show how the result (5.11) can be obtained in a simpler fashion in terms of the Schwinger–Keldysh formalism, by exploiting directly the properties of the time-ordered operator without recourse to combinatorics as we did here. In addition, since the Schwinger–Keldysh formalism avoids recourse to the adiabatic evolution, it can be applied to averages over any state of the system (and not necessarily over the ground state), as well as in cases when a time-dependent perturbation is active.

Nevertheless, as we have already mentioned, one of the main reasons to give here an extended mention to the zero-temperature approach is that *at a formal level* its diagrammatics [6, 7] is analogous to that of the Schwinger–Keldysh formalism, to be discussed in Chapter 10.

6

The Contour Idea for Time-Dependent Averages: Forward and Backward Branches

In this chapter, we introduce the contour Schwinger–Keldysh method for time-dependent averages, in light of its relevance for non-equilibrium processes [1, 2, 34]. This is because a key feature of this approach is that it leaves open the possibility that no state of a system in the future can be identified with any of its states in the past.

We illustrate in detail this method with reference to the time-dependent quantum averages introduced in Chapter 4, whereby for definiteness the system is initially prepared at the reference time t_0 in a definite quantum state $|\Psi(t_0)\rangle$, and then, it is let to evolve with the full Hamiltonian $\mathcal{H}(t)$ of Eq. (2.1). [Extension to ensemble averages is straightforward and will be considered in Chapter 8.] The operator $A(t)$ we consider may or may not depend explicitly on time. In any case, we shall keep its (although formal) dependence on time for future convenience.

In Chapter 3, we have seen that the time-dependent average of the operator $A(t)$, over the state that evolves in time for $t > t_0$ out of an initial state $|\Psi(t_0)\rangle$, can be expressed in three equivalent ways since

$$A_{\mathrm{H}}(t) \underset{\substack{\uparrow \\ \text{[Eq.(2.25)]}}}{=} U^{\dagger}(t,t_0)A(t)U(t,t_0) \underset{\substack{\uparrow \\ \text{[Eq.(3.10)]}}}{=} e^{-iHt_0}\left(V^{\dagger}(t,t_0)A_{\mathrm{h}}(t)V(t,t_0)\right)e^{iHt_0}$$

$$\underset{\substack{\uparrow \\ \text{[Eq.(3.23)]}}}{=} e^{-iH_0t_0}\left(W^{\dagger}(t,t_0)A_{\mathrm{I}}(t)W(t,t_0)\right)e^{iH_0t_0}\,. \qquad (6.1)$$

Here, U develops in time with the full Hamiltonian $\mathcal{H}(t)$ (cf. Eq. (2.8)), V with $H_{\mathrm{h}}^{\mathrm{ext}}(t)$ only (cf. Eq. (3.5)), and W with $H_{\mathrm{I}}^{\mathrm{int}}(t) + H_{\mathrm{I}}^{\mathrm{ext}}(t)$ (cf. Eq. (3.18)). Note also that, even though $A(t)$ may not depend explicitly on time, its related versions $A_{\mathrm{h}}(t)$ and $A_{\mathrm{I}}(t)$ acquire a time dependence from their definitions (3.11) and (3.24), respectively. In addition, owing to the unitary property of the evolution operator(s) (cf. Eqs. (2.15), (3.9), and (3.22)), the equalities (6.1) can be cast in the alternative forms

$$A_{\mathrm{H}}(t) = U(t_0,t)A(t)U(t,t_0) = e^{-iHt_0}\Big(V(t_0,t)A_{\mathrm{h}}(t)V(t,t_0)\Big)e^{iHt_0}$$
$$= e^{-iH_0t_0}\Big(W(t_0,t)A_{\mathrm{I}}(t)W(t,t_0)\Big)e^{iH_0t_0}\,, \tag{6.2}$$

which can be of use in different contexts. For definiteness, in the following, we shall adopt the form in terms of U on the left-hand side of Eq. (6.2) to introduce the contour method.

Consider now the quantum average over the initial state $|\Psi(t_0)\rangle$ (cf. Eq. (4.1))

$$\langle A(t)\rangle = \langle\Psi(t_0)|U(t_0,t)A(t)U(t,t_0)|\Psi(t_0)\rangle\,, \tag{6.3}$$

where according to Eq. (2.8) for $t > t_0$

$$U(t,t_0) = \mathcal{T}\Big\{e^{-i\int_{t_0}^{t}dt'\mathcal{H}(t')}\Big\} \tag{6.4}$$

and

$$U(t_0,t) = \tilde{\mathcal{T}}\Big\{e^{-i\int_{t}^{t_0}dt'\mathcal{H}(t')}\Big\} \tag{6.5}$$

where the time-ordering operator $\mathcal{T}$ orders products of operators with *later* time arguments to the left, while the anti-time-ordering operator $\tilde{\mathcal{T}}$ orders products of operators with *earlier* time argument to the left. With reference to Eqs. (2.10) for $U(t,t_0)$ and (2.14) for $U(t_0,t)$ of Chapter 2, we end up with the operator identity (cf. Fig. 6.1):

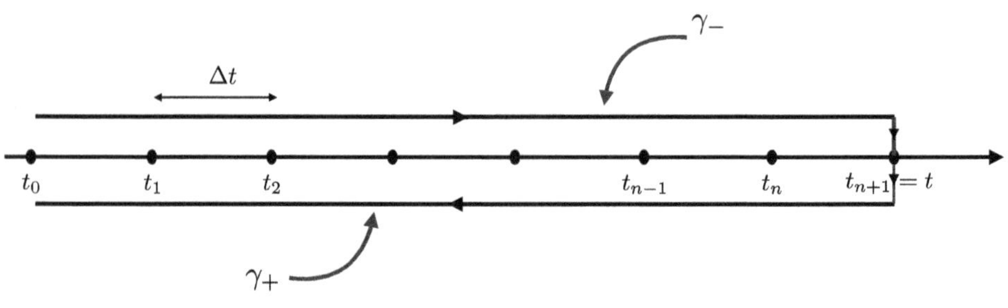

Figure 6.1 The chronological (forward) branch γ_- running forward from t_0 to t and the anti-chronological (backward) branch γ_+ running backward from t to t_0 (for $t > t_0$) are lumped together into the contour $\gamma = \gamma_- + \gamma_+$.

$$
\begin{aligned}
U(t_0, t)A(t)U(t, t_0) &= \lim_{n\to\infty} e^{i\mathcal{H}(t_0)\Delta t} \cdots e^{i\mathcal{H}(t_{n-1})\Delta t}\, e^{i\mathcal{H}(t_n)\Delta t}\, A(t) \\
&\quad \times\ e^{-i\mathcal{H}(t_n)\Delta t}\, e^{-i\mathcal{H}(t_{n-1})\Delta t} \cdots e^{-i\mathcal{H}(t_0)\Delta t} \\
&= \mathcal{T}_\gamma\Big\{ e^{-i\int_{\gamma_+} dz\, \mathcal{H}(z)} A(t)\, e^{-i\int_{\gamma_-} dz\, \mathcal{H}(z)} \Big\} \\
&= \mathcal{T}_\gamma\Big\{ e^{-i\int_{\gamma} dz\, \mathcal{H}(z)} A(t) \Big\}\ . \qquad (6.6)
\end{aligned}
$$

In the above expression, we have:

(i) Introduced the *contour time-ordering operator* $\mathcal{T}_\gamma$, which orders products of operators according to the position of their time arguments along the contour γ, where an earlier contour time places an operator to the right. The operator $\mathcal{T}_\gamma$ thus recognizes whether the operators belong either to the *chronological* (γ_-) or the *anti-chronological* (γ_+) parts of the product, with $\gamma = \gamma_- + \gamma_+$ (cf. Fig. 6.1). [In connection with future applications, the contour γ may be imagined as lying in the complex time plane.]

(ii) Considered that the operators $\mathcal{H}(t)$ and $A(t)$ are the same on the two (forward γ_- and backward γ_+) branches of γ for given t. All operators associated with physical quantities have this property. However, sometimes one may need to introduce operators that differ on the two branches γ_- and $\gamma_+$of the contour γ (like later on in Chapter 17).

(iii) Left the time label t in the operator $A(t)$ occurring in the last line, even though the operator A may not depend explicitly on time, in order to mark the point along the contour γ where this operator was originally placed by the evolution operator in the quantum average (6.3). [This subtlety will be especially useful when field operators will be considered in the place of A.]

(iv) Used the property that operators inside the $\mathcal{T}_\gamma$-sign can be treated as commuting operators, such that it is possible to write

$$
\int_{\gamma_+} dz\, \mathcal{H}(z) + \int_{\gamma_-} dz\, \mathcal{H}(z) = \int_{\gamma=\gamma_-+\gamma_+} dz\, \mathcal{H}(z) \qquad (6.7)
$$

in the exponent of the exponential operator.

Entering the operator identity (6.6) into Eq. (6.3), we obtain eventually

$$
\begin{aligned}
\langle A(t)\rangle &= \langle \Psi(t_0)|\mathcal{T}_\gamma\Big\{ e^{-i\int_\gamma d\bar{z}\, \mathcal{H}(\bar{z})} A(z)\Big\}\, |\Psi(t_0)\rangle \\
&= \langle \Psi_{\mathrm{h}}(t_0)|\mathcal{T}_\gamma\Big\{ e^{-i\int_\gamma d\bar{z}\, H_{\mathrm{h}}^{\mathrm{ext}}(\bar{z})} A_{\mathrm{h}}(z)\Big\}\, |\Psi_{\mathrm{h}}(t_0)\rangle \\
&= \langle \Psi_{\mathrm{I}}(t_0)|\mathcal{T}_\gamma\Big\{ e^{-i\int_\gamma d\bar{z}\, (H_{\mathrm{I}}^{\mathrm{int}}(\bar{z}) + H_{\mathrm{I}}^{\mathrm{ext}}(\bar{z}))} A_{\mathrm{I}}(z)\Big\}\, |\Psi_{\mathrm{I}}(t_0)\rangle\,, \qquad (6.8)
\end{aligned}
$$

where we have adapted the result (6.6) to each expression entering Eq. (6.2). Note that, for making the notation internally consistent, we have replaced $t \longrightarrow z$ in the time argument of the operator A.

We emphasize that the *key feature* of the result (6.6) is that a *single* contour time-ordering operator $\mathcal{T}_\gamma$ along the composite contour $\gamma = \gamma_- + \gamma_+$ encompasses both the time-ordering operator $\mathcal{T}$ along the forward branch γ_- and the anti-time-ordering operator $\tilde{\mathcal{T}}$ along the backward branch γ_+.

Remark: The extended contour

The contour γ of Fig. 6.1, which runs from t_0 to t and then back to t_0, depends explicitly on the time t at which the operator A is considered. As this feature may appear unpleasant, it would be desirable to introduce a "universal" contour that would not depend on the measuring time t (although it should still depend on the *reference time* t_0 at which the system was initially prepared in the state $|\Psi(t_0)\rangle\rangle$.

This can be achieved by considering the unitary property (2.16), in the form

$$U(t, +\infty)\, U(+\infty, t) = \mathbb{1}\,, \tag{6.9}$$

such that in the expression (6.3), we can replace the operator on the left side by

$$U(t_0, t) = U(t_0, t)\, U(t, +\infty)\, U(+\infty, t) = U(t_0, +\infty)\, U(+\infty, t) \tag{6.10}$$

where the group property (2.12) has been used. In this way, the contour $\gamma_- + \gamma_+$ of Fig. 6.1 can be replaced by the *extended oriented contour* $\gamma = \gamma_- + \tilde{\gamma} + \gamma_+$ shown in Fig. 6.2, which runs forward from t_0 to $+\infty$ and then backward from $+\infty$ to t_0 without running into the measuring time t. For notational convenience, in the following, we shall refer to γ meaning either $\gamma_- + \gamma_+$ or $\gamma_- + \tilde{\gamma} + \gamma_+$.

Remark: An alternative way to prove Eq. (5.11)

At the end of Chapter 5, we had anticipated that we could have proven the result (5.11) without recourse to the combinatorics, which is required for splitting the time-ordered product on the left-hand side into the two time-ordered products on the right-hand side of that expression.

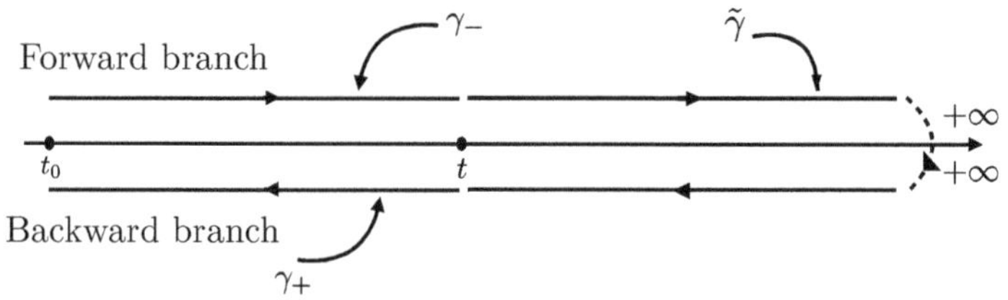

Figure 6.2 The "universal" extended contour $\gamma = \gamma_- + \tilde{\gamma} + \gamma_+$, running forward from t_0 to $+\infty$ and backward from $+\infty$ to t_0 without intercepting the measuring time t.

This conclusion can be reached by considering again the way Eq. (6.6) was proven, where only the elementary properties of the time-ordered operator were utilized without the need for combinatorics. To this end, it is enough to establish the formal correspondence

$$\int_{-\infty}^{t} dt' \longleftrightarrow \int_{\gamma_-} dz \quad , \quad \int_{t}^{+\infty} dt' \longleftrightarrow \int_{\gamma_+} dz \, , \tag{6.11}$$

where the left sides of these correspondences refer to Eq. (5.11) and the right sides to Eq. (6.6). [QED]

For dealing with more than one operator at the same time, it is convenient to introduce at the outset the contour time-ordering operator via the result (6.6), as it is done in Chapter 7.

7

Closed-Time-Path Green's Functions

In this chapter, we consider products of operators in the Heisenberg representation and express them in terms of the contour time-ordering operator, as we did in Chapter 6 for a single operator. For the time being, we average this product of operators over an initial state $|\Psi(t_0)\rangle$ at the reference time t_0. Since the relative order in which the operators enter the quantum average matters, this order will have to be specified in detail. This procedure will also lead us to consider the single- and two-particle Green's functions, where the product contains, respectively, two and four field operators, which satisfy the relations (2.2), and in terms of which the Hamiltonian (2.1) has conveniently been expressed at the outset. These operators will then be at the core of the diagrammatic many-particle theory to be developed in what follows.

7.1 Averaging Products of Operators over a Pure State along a Closed Oriented Contour

To describe the aforementioned procedure in detail, we initially consider two Bose-like operators $A(t_a)$ and $B(t_b)$ in the Heisenberg representation, respectively, at times t_a and t_b. Two cases need to be examined.

CASE $t_a > t_b$

In this case, we may define the "greater" correlation function $F^{>}_{AB}(t_a, t_b)$

$$\begin{aligned} F^{>}_{AB}(t_a, t_b) &= \langle\Psi(t_0)|A_{\mathrm{H}}(t_a)B_{\mathrm{H}}(t_b)|\Psi(t_0)\rangle \\ &\underset{[\mathrm{Eq.}(2.25)]}{\overset{\uparrow}{=}} \langle\Psi(t_0)|U(t_0, t_a)A(t_a)U(t_a, t_0)U(t_0, t_b)B(t_b)U(t_b, t_0)|\Psi(t_0)\rangle \\ &\underset{[\mathrm{Eq.}(6.6)]}{\overset{\uparrow}{=}} \langle\Psi(t_0)|\mathcal{T}_{\gamma_a}\left\{e^{-i\int_{\gamma_a} dz\, \mathcal{H}(z)} A(t_a)\right\} \mathcal{T}_{\gamma_b}\left\{e^{-i\int_{\gamma_b} dz\, \mathcal{H}(z)} B(t_b)\right\} |\Psi(t_0)\rangle \\ &= \langle\Psi(t_0)|\mathcal{T}_{\gamma_a+\gamma_b}\left\{e^{-i\int_{\gamma_a+\gamma_b} dz\, \mathcal{H}(z)} A(t_a)B(t_b)\right\} |\Psi(t_0)\rangle, \qquad (7.1) \end{aligned}$$

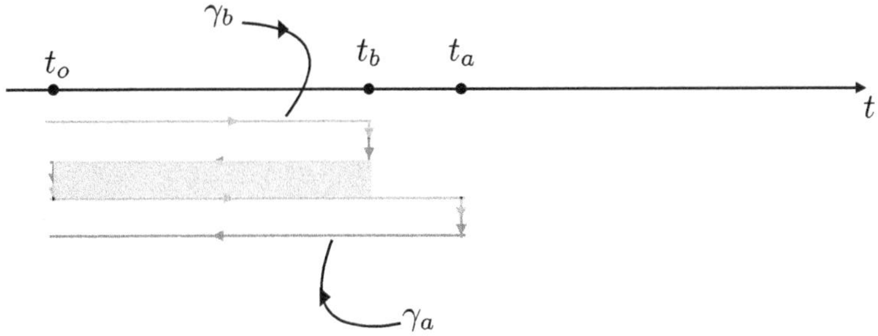

Figure 7.1 The two separate contours γ_a and γ_b combine into the total contour $\gamma_a + \gamma_b$ when $t_a > t_b$.

where (cf. Fig. 7.1):

(i) The contour γ_a starts at t_0, passes through t_a, and returns to t_0;
(ii) The contour γ_b starts at t_0, passes through t_b, and returns to t_0;
(iii) The combined contour $\gamma_a + \gamma_b$ starts at t_0, stretches to $\max\{t_a, t_b\} = t_a > t_b$, and returns to t_0.

This is because the contribution from the "closed loop" (which surrounds anti-clockwise the shaded part in Fig. 7.1) yields

$$\mathcal{T}_{\square}\left\{e^{-i\int_{\square} dz\, \mathcal{H}(z)}\right\} = U(t_b, t_0)\, U(t_0, t_b) = \mathbb{1}\,, \tag{7.2}$$

to the extent that both operators $A(t_a)$ at time t_a and $B(t_b)$ at time t_b are not involved in the closed loop. In addition, similar to what was done in Chapter 6, the contour $\gamma_a + \gamma_b$ can be stretched from $\max\{t_a, t_b\} = t_a > t_b$ all the way to $+\infty$, thus recovering the extended oriented contour γ of Fig. 6.2.

CASE $t_b > t_a$

In this case, we may define the "lesser" correlation function $F^{<}_{AB}(t_a, t_b)$

$$\begin{aligned} F^{<}_{AB}(t_a, t_b) &= \langle \Psi(t_0)| B_{\mathrm{H}}(t_b) A_{\mathrm{H}}(t_a) |\Psi(t_0)\rangle \\ &= \langle \Psi(t_0)| \mathcal{T}_{\gamma_b+\gamma_a}\left\{e^{-i\int_{\gamma_b+\gamma_a} dz\, \mathcal{H}(z)} B(t_b) A(t_a)\right\} |\Psi(t_0)\rangle\,, \end{aligned} \tag{7.3}$$

which corresponds to the situation shown in Fig. 7.2. Once again, the contour $\gamma_a + \gamma_b$ can be replaced by the extended oriented contour γ of Fig. 6.2.

In both cases (7.1) and (7.3), we have managed to bring the operators under the same contour time-ordering operator. What can further be done is to let the time variables of the operators A and B to run freely along the extended oriented contour γ. Let τ_a and τ_b be the corresponding time variables along the extended oriented

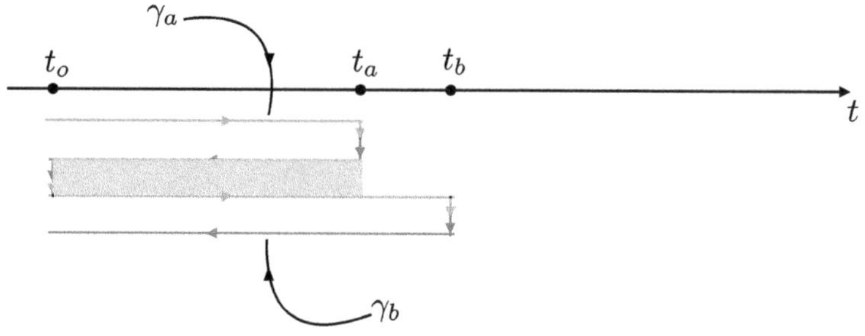

Figure 7.2 The two separate contours γ_b and γ_a combine into the total contour $\gamma_b + \gamma_a$ when $t_b > t_a$.

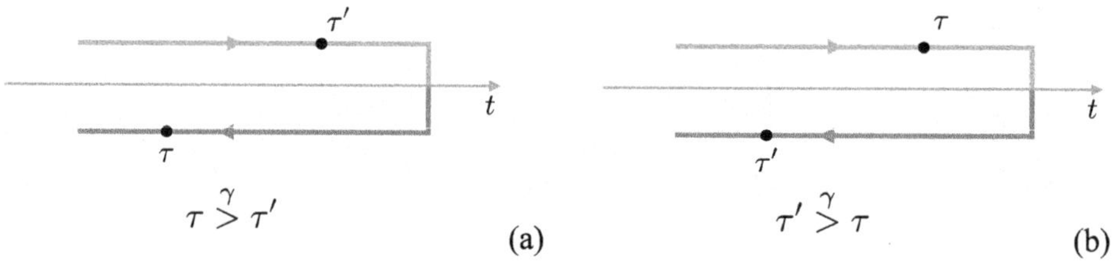

Figure 7.3 Two possible (a) and (b) positions of the variables τ and τ' along the extended oriented contour γ, which determine the values of the function (7.4).

contour γ and introduce the *contour Heaviside unit step function*

$$\theta(\tau, \tau') = \begin{cases} 1 & \text{if } \tau \text{ is later than } \tau' \\ 0 & \text{if } \tau \text{ is earlier than } \tau' \end{cases} \tag{7.4}$$

along the contour γ (cf. Fig. 7.3). Quite generally, we can define the following correlation function of the operators $A(\tau_a)$ and $B(\tau_b)$, where τ_a and τ_b lie along the extended oriented contour γ:

$$\begin{aligned} &F_{AB}(\tau_a, \tau_b) \\ &= \theta(\tau_a, \tau_b)\, \langle \Psi(t_0)| T_\gamma \left\{ e^{-i \int_\gamma dz\, \mathcal{H}(z)} A(\tau_a) \right\} T_\gamma \left\{ e^{-i \int_\gamma dz\, \mathcal{H}(z)} B(\tau_b) \right\} |\Psi(t_0)\rangle \\ &\quad + \theta(\tau_b, \tau_a)\, \langle \Psi(t_0)| T_\gamma \left\{ e^{-i \int_\gamma dz\, \mathcal{H}(z)} B(\tau_b) \right\} T_\gamma \left\{ e^{-i \int_\gamma dz\, \mathcal{H}(z)} A(\tau_a) \right\} |\Psi(t_0)\rangle\, . \end{aligned} \tag{7.5}$$

Here, several cases can be distinguished when associating the time variable τ along the contour γ with the time variable t along the ordinary time axis.

CASE 1: Both τ_a and τ_b on the "forward" branch of γ

In this case, the expression (7.5) reduces to (cf. Fig. 7.4 (a)):

$$F_{AB}(\tau_a, \tau_b) \underset{\substack{\uparrow \\ \text{[Eq.(6.6)]}}}{=} \theta(t_a - t_b)\, \langle \Psi(t_0)|A_{\rm H}(t_a)\, B_{\rm H}(t_b)|\Psi(t_0)\rangle$$
$$+ \theta(t_b - t_a)\langle \Psi(t_0)|B_{\rm H}(t_b)\, A_{\rm H}(t_a)|\Psi(t_0)\rangle, \quad (7.6)$$

which coincides with the conventional *chronological* correlation function.

CASE 2: τ_a on the "forward" and τ_b in the "backward" branch of γ

In this case, $\tau_b > \tau_a$ by definition and the expression (7.5) reduces to (cf. Fig. 7.4 (b)):

$$F_{AB}(\tau_a, \tau_b) = \langle \Psi(t_0)|B_{\rm H}(t_b)\, A_{\rm H}(t_a)|\Psi(t_0)\rangle \underset{\substack{\uparrow \\ \text{[Eq.(7.3)]}}}{=} F^{<}_{AB}(t_a, t_b) \quad (7.7)$$

no matter what are the relative values of t_a and t_b along the ordinary time axis.

CASE 3: τ_a on the "backward" and τ_b on the "forward" branch of γ

In this case, $\tau_a > \tau_b$ by definition and the expression (7.5) reduces to (cf. Fig. 7.4 (c)):

$$F_{AB}(\tau_a, \tau_b) = \langle \Psi(t_0)|A_{\rm H}(t_a)\, B_{\rm H}(t_b)|\Psi(t_0)\rangle \underset{\substack{\uparrow \\ \text{[Eq.(7.1)]}}}{=} F^{>}_{AB}(t_a, t_b) \quad (7.8)$$

again no matter what are the relative values of t_a and t_b along the ordinary time axis.

CASE 4: Both τ_a and τ_b on the "backward" branch of γ

In this case, the expression (7.5) reduces to (cf. Fig. 7.4 (d)):

$$F_{AB}(\tau_a, \tau_b) = \theta(t_b - t_a)\, \langle \Psi(t_0)|A_{\rm H}(t_a)\, B_{\rm H}(t_b)|\Psi(t_0)\rangle$$
$$+ \theta(t_a - t_b)\langle \Psi(t_0)|B_{\rm H}(t_b)\, A_{\rm H}(t_a)|\Psi(t_0)\rangle, \quad (7.9)$$

which coincides with the conventional *anti-chronological* correlation function.

Note, finally, that the expression (7.5) can be rewritten in the compact form:

$$F_{AB}(\tau_a, \tau_b) = \langle \Psi(t_0)|\mathcal{T}_\gamma \left\{ e^{-i \int_\gamma dz\, \mathcal{H}(z)} A(\tau_a)\, B(\tau_b) \right\} |\Psi(t_0)\rangle, \quad (7.10)$$

where the operators $A(\tau_a)$ and $B(\tau_b)$ are ordered by the contour time-ordering operator $\mathcal{T}_\gamma$ along the extended oriented contour γ in order of increasing time, such that the operator with the earliest time on the contour γ is placed to the right.

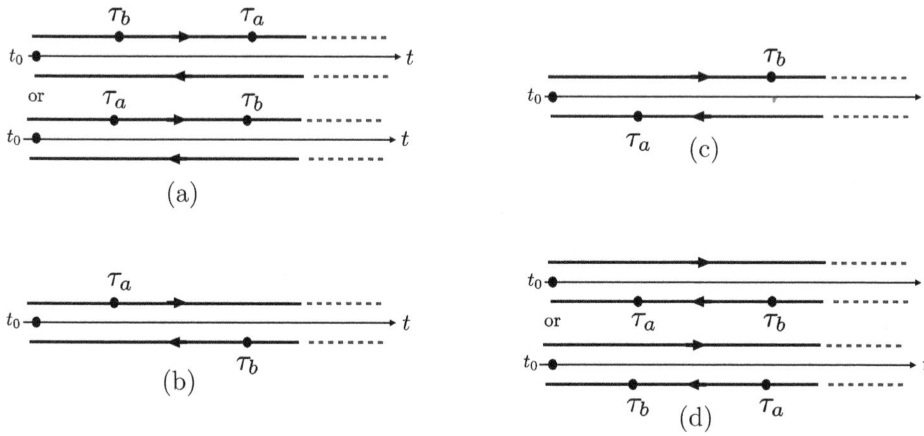

Figure 7.4 Four possible [from (a) to (d)] ways of associating the time variables τ_a and τ_b along the contour γ in the expression (7.5) with the time variables t_a and t_b along the ordinary time axis.

Reference to the quantities introduced above will be taken over and further expanded in Chapter 13.

7.2 Single-Particle and Two-Particle Green's Functions Defined on the Contour γ

Thus far, we have considered $A(\tau_a)$ and $B(\tau_b)$ to be Bose-like operators, such that no minus sign appears when these operators are interchanged within the contour time-ordering like in the expression (7.10). In the following, we shall mostly be interested in the *single-particle Green's function*, whereby

$$\begin{cases} A(t_a) & \longleftrightarrow & \psi_{\sigma_a}(\mathbf{r}_a, t_a) = \psi_{\mathrm{H}}(\mathbf{x}_a, t_a) \\ B(t_b) & \longleftrightarrow & \psi^{\dagger}_{\sigma_b}(\mathbf{r}_b, t_b) = \psi^{\dagger}_{\mathrm{H}}(\mathbf{x}_b, t_b) \, , \end{cases} \tag{7.11}$$

which are field operators with spin σ at the spatial position $\mathbf{r}$ in the Heisenberg representation (2.25). [For convenience, here and in the following, the notation $\mathbf{x} = (\mathbf{r}, \sigma)$ signifies the set of space and spin variables.] When these operators refer to a fermionic system, a minus sign appears for any interchange within the time-ordered product. We then define the *contour single-particle Green's function* (that is to say, with the time arguments running on the extended oriented contour γ), in the form:

$$\begin{aligned} i\, G(\mathbf{x}_a\tau_a, \mathbf{x}_b\tau_b) &= \langle \Psi(t_0) | \mathcal{T}_\gamma \left\{ \psi_{\mathrm{H}}(\mathbf{x}_a, \tau_a)\, \psi^{\dagger}_{\mathrm{H}}(\mathbf{x}_b, \tau_b) \right\} | \Psi(t_0) \rangle \\ &= \langle \Psi(t_0) | \mathcal{T}_\gamma \left\{ e^{-i \int_\gamma dz\, \mathcal{H}(z)} \psi(\mathbf{x}_a, \tau_a)\, \psi^{\dagger}(\mathbf{x}_b, \tau_b) \right\} | \Psi(t_0) \rangle \, . \end{aligned} \tag{7.12}$$

Like we did in Chapter 6, in the last line of Eq. (7.12), we have kept the contour time argument τ in the field operators, even though they do not bear any explicit time dependence. This is because we need to specify their positions along the contour, thus rendering unambiguous the action of the contour time-ordering operator $\mathcal{T}_\gamma$. Accordingly, once the operators are properly ordered, we can omit the time arguments when there is no explicit time dependence. On the other hand, an explicit time dependence of the field operators would result when expressing them within the Heisenberg picture (cf. Section 3.1) or within the interaction picture (cf. Section 3.2).

The definition (7.12) can be extended to the *contour two-particle Green's function*, in the form:

$$i^2\, G_2(a, b; a', b') = \langle \Psi(t_0) | \mathcal{T}_\gamma \left\{ \psi_{\mathrm{H}}(a)\, \psi_{\mathrm{H}}(b)\, \psi^\dagger_{\mathrm{H}}(b')\, \psi^\dagger_{\mathrm{H}}(a') \right\} | \Psi(t_0) \rangle , \qquad (7.13)$$

where we have further simplified the notation by letting $(\mathbf{r}_a, \sigma_a, \tau_a) \rightarrow a$, and so on. The function (7.13) will be useful in what follows, for instance, when determining the equation of motion of the single-particle Green's function (7.12) or when limiting to consider the linear response of the system to the action of an external perturbation.

7.3 The Contour Heaviside Unit Step Function versus the Contour Dirac Delta Function

Along with the contour Heaviside unit step function (7.4), we can define its derivative such that

$$\delta(\tau, \tau') = \frac{d}{d\tau}\, \theta(\tau, \tau') = -\frac{d}{d\tau'}\, \theta(\tau, \tau') \,. \qquad (7.14)$$

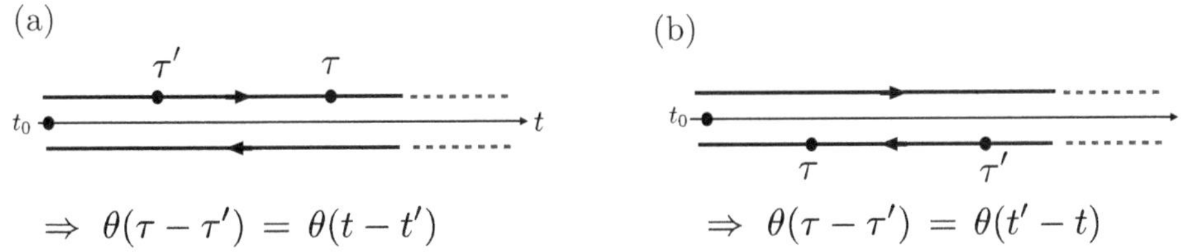

Figure 7.5 Panels (a) and (b) show the relation between the Heaviside unit step functions for variables (τ, τ') along the contour γ and variables (t, t') along the real-time axis.

Two cases are possible:

(i) Both τ and τ' lie on the "forward" (chronological) branch (cf. Fig. 7.5 (a)), such that $\theta(\tau - \tau') = \theta(t - t')$ and

$$\frac{d}{d\tau}\,\theta(\tau, \tau') = \frac{d}{dt}\,\theta(t, t') = \delta(t, t') \tag{7.15}$$

is the standard Dirac delta function with real-time arguments.

(ii) Both τ and τ' lie on the "backward" (anti-chronological) branch (cf. Fig. 7.5 (b)), such that $\theta(\tau - \tau') = \theta(t' - t)$ and

$$\frac{d}{d\tau}\,\theta(\tau, \tau') = \frac{d}{dt}\,\theta(t', t) = -\delta(t, t')\,. \tag{7.16}$$

On the other hand, when τ and τ' lie on different branches, $\frac{d}{d\tau}\,\theta(\tau, \tau') = 0$ by construction.

Accordingly, we can define the *contour Dirac delta function* $\delta(z, z')$ (with $z \leftrightarrow \tau$ and $z' \leftrightarrow \tau'$) such that

$$\int_{z_i}^{z_f} dz'\;\delta(z, z')\,A(z') = A(z) \tag{7.17}$$

for any function (or operator) $A(z)$ defined along the contour γ from z_i to z_f. This property will be utilized, for instance, in Chapters 10 and 11 to introduce the time-dependent interparticle interaction $v(a, a') = v(\mathbf{r}_a - \mathbf{r}_{a'})\,\delta(z_a, z_{a'})$ along the contour γ.

8

Dynamics for a Correlated Initial State and Various Kinds of Contours in the Complex Time Plane

In this chapter, we extend the treatment of the previous Chapters 6 and 7 when the system is initially prepared (at time t_0 and before the time dependence sets in) not in a definite quantum state $|\Psi(t_0)\rangle$ but rather in an ensemble average. This will lead us to add a "vertical track" like that of Fig. 3.1 to the oriented contour of Fig. 6.1 or to its extended version of Fig. 6.2 (both referred to simply as γ in the following). Alternative formalisms will be considered in this context, depending on the way the vertical track is dealt with.

8.1 Averaging over a Mixed State along a Closed Oriented Contour: The Vertical Track and the Ensuing Total Contour

The results obtained in Chapters 6 and 7 relied mostly on the operator identity (6.6) and did not depend on the specific initial state $|\Psi(t_0)\rangle$ in which the quantum system was prepared at the reference time t_0. As a consequence, when the system is instead prepared not in a single "pure" state but in an incoherent superposition of states as described by a density matrix, the averages of the form $\langle\Psi(t_0)|\cdots|\Psi(t_0)\rangle$ considered in Chapters 6 and 7 are replaced by

$$\langle\Psi(t_0)|\cdots|\Psi(t_0)\rangle \quad \longrightarrow \quad \mathrm{Tr}\{\rho\cdots\} \tag{8.1}$$

like in Eq. (4.14), where the dots refer to a string of operators under the $\mathcal{T}_\gamma$-sign like in Eq. (7.10). In addition, similar to what we did at the end of Chapter 3, we can rewrite the density matrix operator ρ in Eq. (8.1) by exploiting the trivial identity

$$e^{-\beta H^{\mathrm{M}}} = e^{-i\int_{t_0}^{t_0-i\beta} dz\, H^{\mathrm{M}}} \tag{8.2}$$

irrespective of t_0, where H^{M} is generically given in Eq. (4.11). Accordingly, we may refer to the segment γ^{M} from t_0 to $t_0 - i\beta$ in the complex time plane as the *vertical track* (sometimes also named the "appendix contour"). Recall also from

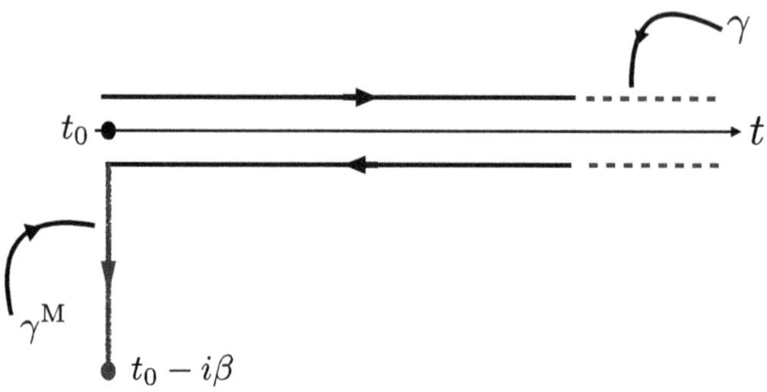

Figure 8.1 The vertical track γ^{M} combines with the contour γ to yield the total contour $\gamma_{\mathrm{T}} = \gamma + \gamma^{\mathrm{M}}$.

Chapter 4 that, for an N-particle system initially at thermal equilibrium at temperature T with an external reservoir, $H^{\mathrm{M}} \leftrightarrow H - \mu N$ and $\beta = (k_B T)^{-1}$, where μ is the chemical potential.

Since any point on γ^{M} is "later" than the points on the extended oriented contour γ (cf. Fig. 8.1), we can rewrite the numerator of Eq. (8.1) in the form

$$\begin{aligned} \mathrm{Tr}\{e^{-\beta H^{\mathrm{M}}}\, \mathcal{T}_{\gamma}\{e^{-i\int_{\gamma} dz\, \mathcal{H}(z)} \cdots\}\} &\underset{[\text{Eq.(8.2)}]}{\overset{}{=}} \mathrm{Tr}\{e^{-i\int_{t_0}^{t_0-i\beta} dz\, H^{\mathrm{M}}}\, \mathcal{T}_{\gamma}\{e^{-i\int_{\gamma} dz\, \mathcal{H}(z)} \cdots\}\} \\ &= \mathrm{Tr}\{\mathcal{T}_{\gamma+\gamma^{\mathrm{M}}}\{e^{-i\int_{\gamma+\gamma^{\mathrm{M}}} dz\, \mathcal{H}(z)} \cdots\}\}, \end{aligned} \tag{8.3}$$

where we have:

(i) added the vertical track γ^{M} to the original extended oriented contour γ and obtained in this way the *total contour* $\gamma_{\mathrm{T}} = \gamma + \gamma^{\mathrm{M}}$ in the complex time plane and

(ii) identified $\mathcal{H}(z) \to H^{\mathrm{M}}$ for any z on γ^{M}.

In addition, by exploiting the property (7.2), the denominator of Eq. (8.1) can also be rewritten in the form

$$Z \underset{[\text{Eq.(4.13)}]}{=} \mathrm{Tr}\{e^{-\beta H^{\mathrm{M}}}\} = \mathrm{Tr}\{e^{-i\int_{\gamma^{\mathrm{M}}} dz\, \mathcal{H}(z)}\} = \mathrm{Tr}\{\mathcal{T}_{\gamma_{\mathrm{T}}}\{e^{-i\int_{\gamma_{\mathrm{T}}} dz\, \mathcal{H}(z)}\}\}, \tag{8.4}$$

such that Eq. (8.1) becomes eventually

$$\mathrm{Tr}\{\rho \cdots\} = \frac{\mathrm{Tr}\{\mathcal{T}_{\gamma_{\mathrm{T}}}\{e^{-i\int_{\gamma_{\mathrm{T}}} dz\, \mathcal{H}(z)} \cdots\}\}}{\mathrm{Tr}\{\mathcal{T}_{\gamma_{\mathrm{T}}}\{e^{-i\int_{\gamma_{\mathrm{T}}} dz\, \mathcal{H}(z)}\}\}}, \tag{8.5}$$

where the dots refer again to a string of operators under the $\mathcal{T}_{\gamma_{\mathrm{T}}}$-sign.

Remark: Averaging over a mixed state in the interaction picture
When working in the interaction picture of Section 3.2, we can use the last line of Eq. (6.8) and the result (3.26) to write in the place of Eq. (8.5):

$$\mathrm{Tr}\{\rho\cdots\} = \frac{\mathrm{Tr}\{e^{-\beta H_0^{\mathrm{M}}}\mathcal{T}_{\gamma_\mathrm{T}}\{e^{-i\int_{\gamma_\mathrm{T}}dz\,(H_\mathrm{I}^{\mathrm{int}}(z)+H_\mathrm{I}^{\mathrm{ext}}(z))}\cdots\}\}}{\mathrm{Tr}\{e^{-\beta H_0^{\mathrm{M}}}\mathcal{T}_{\gamma_\mathrm{T}}\{e^{-i\int_{\gamma_\mathrm{T}}dz\,(H_\mathrm{I}^{\mathrm{int}}(z)+H_\mathrm{I}^{\mathrm{ext}}(z))}\}\}}, \tag{8.6}$$

where $H = H_0 + H^{\mathrm{int}}$, with the understanding that H^{ext} vanishes along the vertical track γ^{M}.

Remark: Matsubara formalism
For an operator A^{M} taken along the vertical track γ^{M}, in the expression in the numerator of Eq. (8.5), we can use the identities (7.2) and (8.4) in reverse, such that in this case Eq. (8.5) becomes

$$\mathrm{Tr}\{\rho A^{\mathrm{M}}\} = \frac{\mathrm{Tr}\{\mathcal{T}_{\gamma_\mathrm{T}}\{e^{-i\int_{\gamma_\mathrm{T}}dz\,\mathcal{H}(z)}A^{\mathrm{M}}\}\}}{\mathrm{Tr}\{\mathcal{T}_{\gamma_\mathrm{T}}\{e^{-i\int_{\gamma_\mathrm{T}}dz\,\mathcal{H}(z)}\}\}} = \frac{\mathrm{Tr}\{e^{-\beta H^{\mathrm{M}}}A^{\mathrm{M}}\}}{\mathrm{Tr}\{e^{-\beta H^{\mathrm{M}}}\}}, \tag{8.7}$$

which coincides with a standard thermal average at equilibrium. Quite generally, when more than one operator along the vertical track γ^{M} is considered, from the expression (8.5) one can construct the Green's functions within the so-called *Matsubara formalism*, which accounts for equilibrium properties [35]. In particular, when using the interaction picture within the Matsubara formalism, a result related to Eq. (8.6) can readily be obtained. We shall return to this topic in Section 15.1.

8.2 Konstantinov–Perel versus Keldysh formalisms

The expression (8.5) in principle represents an "exact" result for average values of time-ordered operators, like the contour single- and two-particle Green's functions. The contour γ_T utilized in Eq. (8.5) and depicted in Fig. 8.1 corresponds to the so-called *Konstantinov–Perel formalism* [36]. (Recall in this context that, if the time arguments of the operators corresponding to the dots in Eq. (8.5) extend up to, say, t_{max}, then there is no reason to continue the contour γ_T past t_{max}.) Accordingly, the Konstantinov–Perel formalism entails full control of the initial system configuration at the reference time t_0, which for a many-particle system at thermal equilibrium would in practice require facing the nontrivial task of solving for the Matsubara single-particle Green's function in a fully self-consistent fashion, as discussed in Section 15.1.

An approximate version of the above formalism is provided by the *Keldysh formalism* [2], which can be obtained from the Konstantinov–Perel formalism on the basis of the "adiabatic approximation" for the density matrix. In addition, a further simplified version of the Keldysh approach can be obtained on physical grounds from the general expression (8.6), as discussed in Section 8.3.

Let's first consider the adiabatic approximation for the density matrix. Similar to what we have done in Chapter 5, when the interaction was adiabatically turned on, we assume that the interacting part H^{int} of the system Hamiltonian is switched on very slowly from $t = -\infty$ to t_0, where t_0 is again the time at which the system is perturbed by an external agent via $H^{\text{ext}}(t)$. [1] We thus write:

$$\mathcal{H}(t) = \begin{cases} H_0 + e^{-\eta|t-t_0|}\, H^{\text{int}} & \text{for } t < t_0 \\ H_0 + H^{\text{int}} + H^{\text{ext}}(t) & \text{for } t > t_0, \end{cases} \tag{8.8}$$

where the parameter η should be sufficiently small for the adiabatic process to represent a convenient tool to drive the system to the desired target configuration. A further discussion on the Keldysh's switch-on process can be found in Ref. [37].

According to the adiabatic assumption, we now make the vertical track γ^{M} to recede from t_0 to $t = -\infty$, such that $H^{\text{M}} \to H_0^{\text{M}} = H_0 - \mu N$ on this *displaced vertical track*. As a consequence, $\rho \to \rho_0$ in the displaced vertical track. The expression (8.5) then becomes:

$$\text{Tr}\{\rho \cdots\} \xrightarrow{\text{adiabatic assumption}} \frac{\text{Tr}\{e^{-\beta H_0^{\text{M}}} \mathcal{T}_{\gamma_{\text{K}}} \{e^{-i\int_{\gamma_{\text{K}}} dz\, \mathcal{H}(z)} \cdots\}\}}{\text{Tr}\{e^{-\beta H_0^{\text{M}}}\}} \tag{8.9}$$

where γ_{K} is the *Keldysh contour* depicted in Fig. 8.2(a). In other words, with reference to Fig. 8.2(b), in the present context, the adiabatic assumption replaces the vertical track γ^{M} attached to t_0 by the displaced vertical track γ_0^{M} attached to $t = -\infty$ plus the horizontal lines running forward from $t = -\infty$ to t_0 and then backward from t_0 to $t = -\infty$, with the understanding that in these horizontal lines the upper line of Eq. (8.8) applies since $t < t_0$. As a further step, we may replace the operators inside the $\mathcal{T}_{\gamma_{\text{K}}}$ ordering in the numerator of Eq. (8.5) by their version in the interaction picture, according to the last line of Eq. (6.8) where now $\gamma \to \gamma_{\text{K}}$.

The adiabatic switch-on procedure of the interparticle interaction needed in the Keldysh formalism has been well tested in practice [38], when it was seen to require $t = -\infty$ as the actual starting point of the procedure to ensure that no additional energy (over and above that related to the interaction) is transferred to the system.

[1] Note, however, that in the Keldysh formalism, there is no need for a subsequent adiabatic switching off of the interaction in the far future, as it is the case in the zero-temperature formalism of Chapter 5.

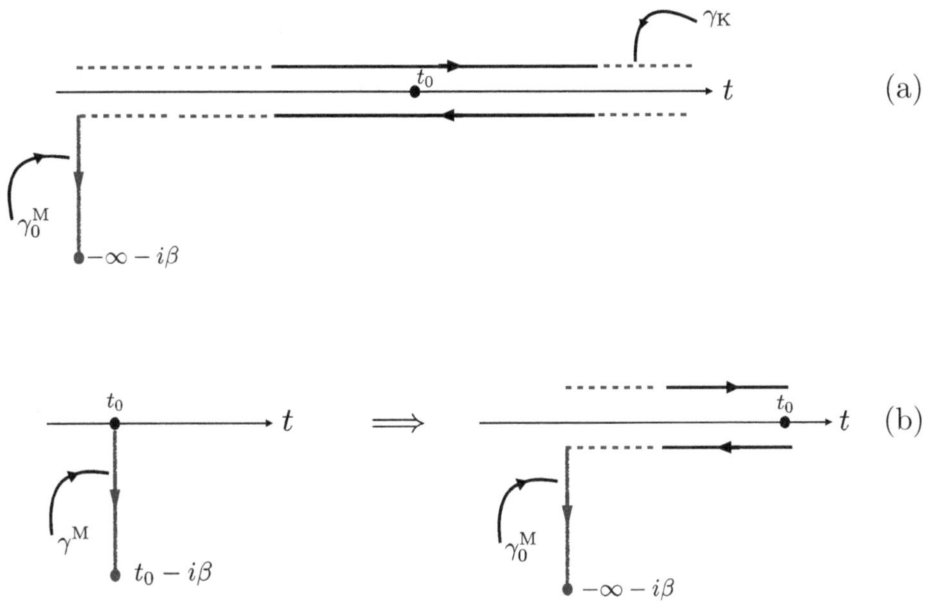

Figure 8.2 (a) The displaced vertical track γ_0^M is attached in the far past to the Keldysh contour γ_K, which runs from $-\infty$ to $+\infty$ and then back $-\infty$ passing through t_0. (b) Replacement of the vertical track γ^M by the displaced vertical track γ_0^M plus a closed contour ranging from $t = -\infty$ to t_0, in which the upper line of Eq. (8.8) applies.

8.3 Transient Phenomena versus the Adiabatic Assumption

Under appropriate physical circumstances, one may also arrive at *a simplified version* of the Keldysh result (8.9), starting from the general expression (8.6) in the interaction picture by reasoning on physical grounds as follows [14].

Suppose that the time variables occurring in the operators corresponding to the dots in Eq. (8.6) are sufficiently later that t_0 and consider a characteristic time δt associated with the collision time of the system due to the interparticle interaction. If the difference between the time variables in those operators and t_0 is much larger than δt, then one may let $t_0 \to -\infty$ for all practical purposes, in such a way that any *memory* about the initial preparation of the system is effectively *lost*. In other words, the interparticle interaction has the physical effect of an "intrinsic damping," which makes the system to lose memory of its initial preparation. As a consequence, as fas as one is not interested in *transient phenomena*, which arise just after the external agent has begun to act on the system (and, accordingly, one considers the difference between the time variables of the relevant operators and t_0 much larger than the characteristic time δt due to collisions), one may disregard the effects of the interparticle interaction in the vertical track in the contour γ_T of

Eq. (8.6) and replace γ_T by the Keldysh contour γ_K of Fig. 8.2(a). Under these circumstances, the adiabatic switching on of the interparticle interaction (like that in Eq. (8.8), which was meant to build up a specific initial configuration) is also not strictly required. Accordingly, for all practical purposes, one may consider the expression (8.6) with $\gamma_T \to \gamma_K$ to coincide with Eq. (8.9), with no adiabatic building up of the interaction (once the operators inside the $\mathcal{T}_{\gamma_T}$-ordering therein are replaced by their interaction counterparts).

Or else, one may arrive at this "simplified version" of the Keldysh approach [14] by (i) ignoring the above adiabatic building up of the interaction from $-\infty$ to t_0; (ii) replacing it by a *sudden quench* of the system Hamiltonian that occurs in the "remote" past (say, at $-\infty$) from the noninteracting H_0 to the fully interacting H; (iii) keeping H_0 in the vertical track, which is now placed at $-\infty$; (iv) considering the effects that this quench has on the system only at times t in the "close" past such that, again, all transient effects due to the sudden quench have eventually died out; and (v) switching on the time-dependent external perturbation only afterward at a later time t_0. At a formal level, this approximate approach corresponds to that adopted in the early review of Ref. [39], where it was assumed that the initial configuration specified at an initial time t_0 is non-correlated, and as such, it is represented by the noninteracting density matrix ρ_0. These features simplify the application of the Wick's decomposition, to be discussed in Chapter 9.

In addition, a practical case for which the Keldysh formalism is applicable without the use of the adiabatic switching of the interaction is that of the nonequilibrium steady state that is established in a dissipative system coupled to an external heat bath. In this case, the initial correlations are expected to disappear in the long-time limit since the large number of degrees of freedom of the heat bath affect the long-time dynamics and wipe out the information of the initial state and the initial transient dynamics [40].

On the other hand, if one is specifically interested in transient effects, which occur for times close to t_0 at which the time dependence is activated, this approximate version of the Keldysh approach cannot be utilized. In this case, it is probably better to resort back to the Konstantinov–Perel formalism in which the initial correlations are supposed to be relevant as they built in the initial configuration of the system.

9

Perturbation Theory: Wick's Theorem for Strings of Operators Ordered along a Contour

In this chapter, we consider a general form of the Wick's theorem, which leads to a perturbation expansion of the (contour) single- and two-particle Green's functions, which are expressed in terms of the contour time-ordering operator.

Generally, the calculation of the single- and two-particle Green's functions for a many-particle quantum system cannot be done in a closed form, and approximations are unavoidably required. In the context of a diagrammatic approach, there are mainly two procedures to perform the relevant approximations. The first procedure is due to Feynman, which relies on a perturbative expansion in the coupling constant of the interparticle interaction (which, in the presence of a time-dependent interaction with an external agent, should in principle be supplemented by a perturbative expansion also in this interaction [14]). A second procedure is due to Schwinger, which considers the exact integral equations satisfied by the single- and two-particle Green's functions and adopts (even "conserving") approximations on the kernels of these integral equations. In this chapter, we consider the perturbative Feynman procedure and postpone consideration to the Schwinger procedure to Chapter 17. Nevertheless, we anticipate that the results of the perturbative Feynman procedure discussed in this chapter will be useful also when considering the Schwinger procedure to the extent that the choice of the approximate kernels of the integral equations therein relies on the topological structure of those kernels, which can mostly be drawn from the experience nurtured on the basis of the perturbative expansion.

The strategy for proving the Wick's theorem considered here is similar to that adopted within the Matsubara formalism for the Green's functions at finite temperature [6, 7] and relies on form $e^{-\beta H_0^{\mathrm{M}}}$ of the statistical operator that occurs in the interaction picture. It thus contrasts the proof of the Wick's theorem in the zero-temperature formalism, which relies instead on averaging over a "vacuum" state that is annihilated by destruction operators [6, 7].

9.1 A Few Preliminary Properties

The averages we plan to calculate via a perturbative expansion have either the general form of Eq. (8.6) with the contour γ_{T} shown in Fig. 8.1 or the form of Eq. (8.9) with the Keldysh contour γ_{K} shown in Fig. 8.2(a), once supplemented by the last line of Eq. (6.8) applied also to more than one operators. In both cases, transient phenomena can be included. Otherwise, the form corresponding to the simplified version of the Keldysh approach discussed in Section 8.3, which keeps the Keldysh contour but excludes transient effects by construction. The following arguments do not depend on which kind of contour is utilized. We shall then indicate the generic contour utilized below by C.

When expanding the exponential in Eq. (8.6) or in Eq. (8.9), one ends up with a string of field operators in the interaction picture (cf. Eq. (3.24)), which are ordered by the contour time-ordering operator $\mathcal{T}_{\mathrm{C}}$ along the contour C:

$$\frac{\mathrm{Tr}\{e^{-\beta H_0^{\mathrm{M}}}\mathcal{T}_{\mathrm{C}}\{c_{q_n}(\tau_n)\,c_{q_{n-1}}(\tau_{n-1})\cdots c_{q_2}(\tau_2)\,c_{q_1}(\tau_1)\}}{\mathrm{Tr}\{e^{-\beta H_0^{\mathrm{M}}}\}}, \tag{9.1}$$

where $c_{q_i}(\tau_i)$ with $i=(1,2,\cdots,n)$ denotes *either* a creation *or* a destruction operator at time τ_i along the contour C, with the suffix q_i to be shortly specified. For definiteness, we assume that in Eq. (9.1) the contour-time labeling already corresponds to the contour-ordered one, such that

$$\tau_n \overset{C}{>} \tau_{n-1} \overset{C}{>} \cdots \overset{C}{>} \tau_2 \overset{C}{>} \tau_1\,. \tag{9.2}$$

In addition, we take H_0^{M} of the form

$$H_0^{\mathrm{M}} = \sum_q \xi_q\, a_q^{\dagger} a_q, \tag{9.3}$$

where a_q ($a_q^{\dagger}$) are now destruction (creation) operators with the index q distinguishing also between the spin components and $\xi_q = \epsilon_q$ in the canonical ensemble and $\xi_q = \epsilon_q - \mu$ in the grand-canonical ensemble with chemical potential μ.

With these premises, we consider the following elementary properties involving the operators c_{q_i}, which will serve to prove the Wick's theorem.

Property # 1:

The first property reads:

$$\boxed{c_q\, e^{-\beta H_0^{\mathrm{M}}} = e^{\lambda_q \beta \xi_q}\, e^{-\beta H_0^{\mathrm{M}}}\, c_q} \tag{9.4}$$

with $\beta = (k_B T)^{-1}$ and where $\lambda_q = +1$ ($\lambda_q = -1$) if c_q is a creation (destruction) operator.

Proof of Eq. (9.4):

From the identity $[a, a^\dagger a] = a$ for both Bose and Fermi cases, it follows that $a\, e^{-\nu\, a^\dagger a} = e^{-\nu\, a^\dagger a}\, a\, e^{-\nu}$ as well as $e^{-\nu\, a^\dagger a}\, a^\dagger = a^\dagger\, e^{-\nu\, a^\dagger a}\, e^{-\nu}$ for real ν. The result (9.4) then follows with the choice $\nu = \beta \xi_q$.

[QED]

Property # 2:

The second property reads:

$$\langle [c_q, A]_\mp \rangle_0 = \langle c_q A \rangle_0 \left(1 \mp e^{\lambda_q \beta \xi_q}\right), \tag{9.5}$$

where A is an arbitrary operator and the minus (plus) sign refers to a bosonic (fermionic) operator c_q (with the further provision on λ_q that holds for Eq. (9.4)), and $\langle \cdots \rangle_0$ stands for

$$\langle \cdots \rangle_0 = \frac{\mathrm{Tr}\{e^{-\beta H_0^{\mathrm{M}}} \cdots\}}{\mathrm{Tr}\{e^{-\beta H_0^{\mathrm{M}}}\}} . \tag{9.6}$$

Proof of Eq. (9.5):

In the Bose case, we obtain:

$$\begin{aligned} \langle [c_q, A]_- \rangle_0 &= \frac{\mathrm{Tr}\{[e^{-\beta H_0^{\mathrm{M}}}, c_q]_- A\}}{\mathrm{Tr}\{e^{-\beta H_0^{\mathrm{M}}}\}} \underset{[\mathrm{Eq.}(9.4)]}{\overset{\uparrow}{=}} \left(1 - e^{\lambda_q \beta \xi_q}\right) \frac{\mathrm{Tr}\{e^{-\beta H_0^{\mathrm{M}}} c_q A\}}{\mathrm{Tr}\{e^{-\beta H_0^{\mathrm{M}}}\}}, \\ &= \left(1 - e^{\lambda_q \beta \xi_q}\right) \langle c_q A \rangle_0 \end{aligned} \tag{9.7}$$

where the cyclic property of the trace has been used. Similarly, in the Fermi case, we obtain:

$$\langle [c_q, A]_+ \rangle_0 = \frac{\mathrm{Tr}\{[e^{-\beta H_0^{\mathrm{M}}}, c_q]_+ A\}}{\mathrm{Tr}\{e^{-\beta H_0^{\mathrm{M}}}\}} \underset{[\mathrm{Eq.}(9.4)]}{\overset{\uparrow}{=}} \left(1 + e^{\lambda_q \beta \xi_q}\right) \langle c_q A \rangle_0 . \tag{9.8}$$

[QED]

In addition, in view of future manipulations, it is convenient to introduce the notation $s = -1$ for bosons and $s = +1$ for fermions and rewrite Eq. (9.5) in the form:

$$\langle [c_q, A]_s \rangle_0 = \langle c_q A \rangle_0 \left(1 + s\, e^{\lambda_q \beta \xi_q}\right) . \tag{9.9}$$

Remark: Averaging over states with a definite number of particles
From the property (9.5), the following results follow:

(i) When $A = \mathbb{1}$, for bosons, it follows that $[c_q, A]_- = 0$ and thus $\langle c_q \rangle_0 = 0$, while for fermions, $[c_q, A]_+ = 2c_q$, from which $2\langle c_q \rangle_0 = \left(1 + e^{\lambda_q \beta \xi_q}\right) \langle c_q \rangle_0$ and thus also $\langle c_q \rangle_0 = 0$.
(ii) When $A = a_q = c_q$, for both bosons and fermions, it follows that $[c_q, A]_\mp = [a_q, a_q]_\mp = 0$ and thus also $\langle a_q\, a_q \rangle_0 = 0$.

This is because the states we are considering for averaging contain a definite number N of particles. In Section 9.3, we will consider a key ingredient for generalizing Wick's theorem to superfluid Bose and Fermi systems by considering "anomalous" averages for which the relevant states do not contain a definite number of particles.

Property # 3:

The third property reads:

$$\boxed{[c_q^{\mathrm{I}}(\tau), c_{q'}^{\mathrm{I}}(\tau')]_\mp = \delta_{q,q'} \left(1 \mp e^{\lambda_q \beta \xi_q}\right) \langle c_q^{\mathrm{I}}(\tau)\, c_{q'}^{\mathrm{I}}(\tau') \rangle_0,} \tag{9.10}$$

where the operator $c_q^{\mathrm{I}}(\tau)$ is in the interaction picture at time τ along the contour C, the minus (plus) sign refers to bosonic (fermionic) operators, and $\delta_{q,q'}$ is the Kronecker delta function.

Proof of Eq. (9.10):

For an operator a_q in the interaction picture (cf. Eq. (3.24), where we leave t_0 for reference), we have

$$a_q^{\mathrm{I}}(t) \underset{\substack{\uparrow \\ \text{[Eq.(3.24)]}}}{=} e^{iH_0(t-t_0)} a_q e^{-iH_0(t-t_0)} \underset{\substack{\uparrow \\ \text{[Eq.(9.3)]}}}{=} e^{i\epsilon_q a_q^\dagger a_q (t-t_0)} a_q e^{-i\epsilon_q a_q^\dagger a_q (t-t_0)} = a_q\, e^{-i\epsilon_q (t-t_0)}, \tag{9.11}$$

where the identities utilized to prove Eq. (9.4) have been used, from which it follows that $a_q^{\mathrm{I}}(t)^\dagger = a_q^\dagger\, e^{i\epsilon_q (t-t_0)}$. These results can be rewritten in the compact form

$$c_q^{\mathrm{I}}(t) = c_q\, e^{i\lambda_q \epsilon_q (t-t_0)}, \tag{9.12}$$

where again $\lambda_q = +1$ if c_q is a creation operator and $\lambda_q = -1$ if c_q is a destruction operator (and with the further understanding that $\epsilon_q \to \xi_q$ when $t \to -i\tau$ along the vertical track – cf. Eqs. (3.26) and (3.27)).

In the Bose case, the result (9.12) enables us to express the commutator between $c_q^{\mathrm{I}}(\tau)$ and $c_{q'}^{\mathrm{I}}(\tau')$ as follows:

$$[c_q^{\mathrm{I}}(\tau), c_{q'}^{\mathrm{I}}(\tau')]_- = [c_q, c_{q'}]_-\, e^{i\lambda_q \epsilon_q (\tau - t_0)}\; e^{i\lambda_{q'} \epsilon_{q'} (\tau' - t_0)}, \tag{9.13}$$

where $[c_q, c_{q'}]_-$ is a c-number such that $[c_q, c_{q'}]_- = \langle [c_q, c_{q'}]_- \rangle_0$ with the notation (9.6). At this point, we make use of the property (9.5) with $A \rightarrow c_{q'}$, to obtain:

$$
\begin{aligned}
[c^{\mathrm{I}}_q(\tau), c^{\mathrm{I}}_{q'}(\tau')]_- &= \langle [c_q, c_{q'}]_- \rangle_0 \, e^{i\lambda_q \epsilon_q (\tau - t_0)} \, e^{i\lambda_{q'} \epsilon_{q'} (\tau' - t_0)} \\
&= \delta_{q,q'} \, \langle c_q c_{q'} \rangle_0 \left(1 - e^{\lambda_q \beta \xi_q}\right) e^{i\lambda_q \epsilon_q (\tau - t_0)} \, e^{i\lambda_{q'} \epsilon_{q'} (\tau' - t_0)} \\
&\underset{[\text{Eq.}(9.12)]}{\underset{\uparrow}{=}} \delta_{q,q'} \left(1 - e^{\lambda_q \beta \xi_q}\right) \langle c^{\mathrm{I}}_q(\tau) \, c^{\mathrm{I}}_{q'}(\tau') \rangle_0 \, .
\end{aligned}
\tag{9.14}
$$

In the Fermi case, we obtain in a similar way the expression for the anti-commutator between $c^{\mathrm{I}}_q(\tau)$ and $c^{\mathrm{I}}_{q'}(\tau')$:

$$
\begin{aligned}
[c^{\mathrm{I}}_q(\tau), c^{\mathrm{I}}_{q'}(\tau')]_+ &= \langle [c_q, c_{q'}]_+ \rangle_0 \, e^{i\lambda_q \epsilon_q (\tau - t_0)} \, e^{i\lambda_{q'} \epsilon_{q'} (\tau' - t_0)} \\
&= \delta_{q,q'} \, \langle c_q c_{q'} \rangle_0 \left(1 + e^{\lambda_q \beta \xi_q}\right) e^{i\lambda_q \epsilon_q (\tau - t_0)} \, e^{i\lambda_{q'} \epsilon_{q'} (\tau' - t_0)} \\
&\underset{[\text{Eq.}(9.12)]}{\underset{\uparrow}{=}} \delta_{q,q'} \left(1 + e^{\lambda_q \beta \xi_q}\right) \langle c^{\mathrm{I}}_q(\tau) \, c^{\mathrm{I}}_{q'}(\tau') \rangle_0 \, .
\end{aligned}
\tag{9.15}
$$

The result (9.10) is thus proved for both bosons and fermions. [QED]

In addition, similar to Eq. (9.9), with the notation $s = -1$ for bosons and $s = +1$ for fermions, Eq. (9.10) can be rewritten in the form:

$$
[c^{\mathrm{I}}_q(\tau), c^{\mathrm{I}}_{q'}(\tau')]_s = \delta_{q,q'} \left(1 + s \, e^{\lambda_q \beta \xi_q}\right) \langle c^{\mathrm{I}}_q(\tau) \, c^{\mathrm{I}}_{q'}(\tau') \rangle_0 \, . \tag{9.16}
$$

In the following, we shall further drop the suffix "I" and leave it implicit in Eq. (9.16).

In practice, most of the commutators and anti-commutators in the expressions (9.16) vanish, and the only relevant ones are for

$$
\begin{cases}
c_q \leftrightarrow a_q \quad \text{and} \quad c_{q'} \leftrightarrow a^{\dagger}_{q'} \, , \\
c_q \leftrightarrow a^{\dagger}_q \quad \text{and} \quad c_{q'} \leftrightarrow a_{q'} \, .
\end{cases}
\tag{9.17}
$$

9.2 Wick's Theorem for Time Ordering along a Generic Contour, for Both Bosons and Fermions

We are now in a position to prove Wick's theorem for the string of operators (9.1) in the interaction picture along the contour C. The proof proceeds along the following steps.

Step # 1 Rewrite the expression (9.1) in the form

$$
\langle \mathcal{T}_{\mathrm{C}} \{ c_{q_{2N}}(\tau_{2N}) \, c_{q_{2N-1}}(\tau_{2N-1}) \cdots c_{q_2}(\tau_2) \, c_{q_1}(\tau_1) \} \rangle_0 \, , \tag{9.18}
$$

where we have taken an even ($2N$) number of operators because we are considering here the "normal" phase with no anomalous averages.

Step # 2 Like in Eq. (9.2), assume that the contour-time labeling already corresponds to the contour-ordered one, such that we can remove the contour time-ordering operator $\mathcal{T}_C$ from Eq. (9.18), which becomes:

$$\langle c_{q_{2N}}(\tau_{2N})\, c_{q_{2N-1}}(\tau_{2N-1}) \cdots c_{q_2}(\tau_2)\, c_{q_1}(\tau_1)\rangle_0 \,. \tag{9.19}$$

Step # 3 In Eq. (9.19), identify the string of operators $c_{q_{2N-1}}(\tau_{2N-1}) \cdots c_{q_2}(\tau_2)$ $c_{q_1}(\tau_1)$ with the operator A of Eq. (9.9) and write accordingly

$$\begin{aligned} &\langle c_{q_{2N}}(\tau_{2N})\; \overbrace{c_{q_{2N-1}}(\tau_{2N-1}) \cdots c_{q_2}(\tau_2)\, c_{q_1}(\tau_1)}^{A}\; \rangle_0 \\ &\qquad \longrightarrow \langle c_{q_{2N}}(\tau_{2N})\, A\rangle_0 \underset{[\text{Eq.}(9.9)]}{\underset{\uparrow}{=}} \frac{\langle [c_{q_{2N}}(\tau_{2N}), A]_s\rangle_0}{1 + s\, e^{\lambda_{q_{2N}}\beta}\, \xi_{q_{2N}}}\,, \end{aligned} \tag{9.20}$$

where we treat the bosonic case ($s = -1$) and the fermionic case ($s = +1$) on the same footing.

Step # 4 Manipulate the numerator on the right-hand side of the expression (9.20) as follows:

$$[c_{q_{2N}}(\tau_{2N}), A]_s = c_{q_{2N}}(\tau_{2N})A + sAc_{q_{2N}}(\tau_{2N}), \tag{9.21}$$

as well as

$$c_{q_{2N}}(\tau_{2N})\, c_{q_{2N-1}}(\tau_{2N-1}) = [c_{q_{2N}}(\tau_{2N}), c_{q_{2N-1}}(\tau_{2N-1})]_s - s c_{q_{2N-1}}(\tau_{2N-1}) c_{q_{2N}}(\tau_{2N})\,, \tag{9.22}$$

from which we obtain the operator identity:

$$\begin{aligned} [c_{q_{2N}}(\tau_{2N}), A]_s \quad &\longrightarrow \quad c_{q_{2N}}(\tau_{2N})\, c_{q_{2N-1}}(\tau_{2N-1})\, c_{q_{2N-2}}(\tau_{2N-2}) \cdots c_{q_2}(\tau_2)\, c_{q_1}(\tau_1) \\ &+ \quad s\, c_{q_{2N-1}}(\tau_{2N-1})\, c_{q_{2N-2}}(\tau_{2N-2}) \cdots c_{q_2}(\tau_2)\, c_{q_1}(\tau_1)\, c_{q_{2N}}(\tau_{2N}) \\ &= \quad [c_{q_{2N}}(\tau_{2N}), c_{q_{2N-1}}(\tau_{2N-1})]_s\, c_{q_{2N-2}}(\tau_{2N-2}) \cdots c_{q_2}(\tau_2)\, c_{q_1}(\tau_1) \\ &- \quad s\, c_{q_{2N-1}}(\tau_{2N-1})\, c_{q_{2N}}(\tau_{2N})\, c_{q_{2N-2}}(\tau_{2N-2}) \cdots c_{q_2}(\tau_2)\, c_{q_1}(\tau_1) \\ &+ \quad s c_{q_{2N-1}}(\tau_{2N-1})\, c_{q_{2N-2}}(\tau_{2N-2}) \cdots c_{q_2}(\tau_2)\, c_{q_1}(\tau_1)\, c_{q_{2N}}(\tau_{2N})\,. \end{aligned} \tag{9.23}$$

Step # 5 Take the average of the operator identity (9.23) according to Eq. (9.6) and use the property (9.16), where recall that the factor $[c_{q_{2N}}(\tau_{2N}), c_{q_{2N-1}}(\tau_{2N-1})]_s$ is a c-number and can thus be taken outside the average:

$$
\begin{aligned}
&\langle [c_{q_{2N}}(\tau_{2N}), A]_s \rangle_0 \\
&\longrightarrow \delta_{q_{2N},q_{2N-1}} \left(1 + s\, e^{\lambda_{q_{2N}}\beta\, \xi_{q_{2N}}}\right) \langle c_{q_{2N}}(\tau_{2N}) c_{q_{2N-1}}(\tau_{2N-1}) \rangle_0 \\
&\quad \times \langle c_{q_{2N-2}}(\tau_{2N-2}) \cdots c_{q_2}(\tau_2)\, c_{q_1}(\tau_1) \rangle_0 \\
&\quad - s\, \langle c_{q_{2N-1}}(\tau_{2N-1}) \left([c_{q_{2N}}(\tau_{2N}), c_{q_{2N-2}}(\tau_{2N-2})]_s - s c_{q_{2N-2}}(\tau_{2N-2}) c_{q_{2N}}(\tau_{2N}) \right) \\
&\quad \times c_{q_{2N-3}}(\tau_{2N-3}) \cdots c_{q_2}(\tau_2)\, c_{q_1}(\tau_1) \rangle_0 \\
&\quad + s\, \langle c_{q_{2N-1}}(\tau_{2N-1})\, c_{q_{2N-2}}(\tau_{2N-2}) \cdots c_{q_2}(\tau_2)\, c_{q_1}(\tau_1)\, c_{q_{2N}}(\tau_{2N}) \rangle_0 \\
&= \boxed{\delta_{q_{2N},q_{2N-1}} \left(1 + s\, e^{\lambda_{q_{2N}}\beta\, \xi_{q_{2N}}}\right) \langle c_{q_{2N}}(\tau_{2N}) c_{q_{2N-1}}(\tau_{2N-1}) \rangle_0} \\
&\quad \times \langle c_{q_{2N-2}}(\tau_{2N-2}) \cdots c_{q_2}(\tau_2)\, c_{q_1}(\tau_1) \rangle_0 \\
&\quad - s \boxed{\delta_{q_{2N},q_{2N-2}} \left(1 + s\, e^{\lambda_{q_{2N}}\beta\, \xi_{q_{2N}}}\right) \langle c_{q_{2N}}(\tau_{2N}) c_{q_{2N-2}}(\tau_{2N-2}) \rangle_0} \\
&\quad \times \langle c_{q_{2N-1}}(\tau_{2N-1})\, c_{q_{2N-3}}(\tau_{2N-3}) \cdots c_{q_2}(\tau_2)\, c_{q_1}(\tau_1) \rangle_0 \\
&\quad + (-s)^2 \langle c_{q_{2N-1}}(\tau_{2N-1})\, c_{q_{2N-2}}(\tau_{2N-2})\, \overbrace{c_{q_{2N}}(\tau_{2N})\, c_{q_{2N-3}}(\tau_{2N-3})} \\
&\qquad \cdots c_{q_2}(\tau_2)\, c_{q_1}(\tau_1) \rangle_0 \\
&\quad + s\, \langle c_{q_{2N-1}}(\tau_{2N-1})\, c_{q_{2N-2}}(\tau_{2N-2}) \cdots c_{q_2}(\tau_2)\, c_{q_1}(\tau_1)\, c_{q_{2N}}(\tau_{2N}) \rangle_0,
\end{aligned}
\tag{9.24}
$$

where the property (9.16) has once again been used to arrive to the last expression.

Step # 6 Continue transferring the operator $c_{q_{2N}}(\tau_{2N})$ to the right with the use of Eq. (9.22), starting from the factor $\overbrace{c_{q_{2N}}(\tau_{2N})\, c_{q_{2N-3}}(\tau_{2N-3})}$ given earlier. At each jump of the operator $c_{q_{2N}}(\tau_{2N})$ to the right, a term like those encapsulated in boxes in Eq. (9.24) is generated, until eventually the last term on the right-hand side of Eq. (9.24) is cancelled since

$$
(-s)^{2N-1} = (-s)^{2N}(-s)^{-1} = (-s)^{-1} = -s\,. \tag{9.25}
$$

Entering the result (9.24) into Eq. (9.20), the factor $\left(1 + s\, e^{\lambda_{q_{2N}}\beta\, \xi_{q_{2N}}}\right)$ cancels out, and we are left with the result:

$$
\begin{aligned}
&\langle c_{q_{2N}}(\tau_{2N})\, c_{q_{2N-1}}(\tau_{2N-1}) \cdots c_{q_2}(\tau_2)\, c_{q_1}(\tau_1) \rangle_0 \\
&\quad = \delta_{q_{2N},q_{2N-1}} \langle c_{q_{2N}}(\tau_{2N}) c_{q_{2N-1}}(\tau_{2N-1}) \rangle_0 \\
&\qquad \times \langle c_{q_{2N-2}}(\tau_{2N-2})\, c_{q_{2N-3}}(\tau_{2N-3}) \cdots c_{q_2}(\tau_2)\, c_{q_1}(\tau_1) \rangle_0 \\
&\qquad - s\, \delta_{q_{2N},q_{2N-2}} \langle c_{q_{2N}}(\tau_{2N}) c_{q_{2N-2}}(\tau_{2N-2}) \rangle_0 \\
&\qquad \times \langle c_{q_{2N-1}}(\tau_{2N-1})\, c_{q_{2N-3}}(\tau_{2N-3}) \cdots c_{q_2}(\tau_2)\, c_{q_1}(\tau_1) \rangle_0 \\
&\qquad + (-s)^2\, \delta_{q_{2N},q_{2N-3}} \langle c_{q_{2N}}(\tau_{2N}) c_{q_{2N-3}}(\tau_{2N-3}) \rangle_0
\end{aligned}
$$

$$
\begin{aligned}
&\times \langle c_{q_{2N-1}}(\tau_{2N-1})\, c_{q_{2N-2}}(\tau_{2N-2}) \cdots c_{q_2}(\tau_2)\, c_{q_1}(\tau_1)\rangle_0 \\
&+ \cdots \\
&+ (-s)^{2N-2}\, \delta_{q_{2N},q_1}\, \langle c_{q_{2N}}(\tau_{2N}) c_{q_1}(\tau_1)\rangle_0 \\
&\times \langle c_{q_{2N-1}}(\tau_{2N-1})\, c_{q_{2N-2}}(\tau_{2N-2}) \cdots c_{q_2}(\tau_2)\rangle_0,
\end{aligned} \tag{9.26}
$$

where again $(-s) = +1$ for bosons and $(-s) = -1$ for fermions.

Step # 7 Wick's theorem is eventually proved by induction, by applying the same sequence of steps that has led us to the result (9.26) to each of the remaining averages of strings of $2N - 2$ operators contained therein, namely,

$$
\left\{
\begin{array}{l}
\langle c_{q_{2N-2}}(\tau_{2N-2})\, c_{q_{2N-3}}(\tau_{2N-3}) \cdots c_{q_2}(\tau_2)\, c_{q_1}(\tau_1)\rangle_0 \\
\langle c_{q_{2N-1}}(\tau_{2N-1})\, c_{q_{2N-3}}(\tau_{2N-3}) \cdots c_{q_2}(\tau_2)\, c_{q_1}(\tau_1)\rangle_0 \\
\langle c_{q_{2N-1}}(\tau_{2N-1})\, c_{q_{2N-2}}(\tau_{2N-2}) \cdots c_{q_2}(\tau_2)\, c_{q_1}(\tau_1)\rangle_0 \\
\cdots \\
\langle c_{q_{2N-1}}(\tau_{2N-1})\, c_{q_{2N-2}}(\tau_{2N-2}) \cdots c_{q_3}(\tau_3)\, c_{q_2}(\tau_2)\rangle_0 \, .
\end{array}
\right. \tag{9.27}
$$

The end result is that *the original average* (9.18) *factorizes into products of pairwise contractions* $\langle \mathcal{T}_C\{c_q(\tau)\, c_{q'}(\tau')\}\rangle_0$, with an appropriate sign due to the transpositions that are required in the Fermi case to bring nearby in the string the operators to be contracted (while in the Bose case, no sign is required). For instance, with four Fermi field operators, we get:

$$
\begin{aligned}
&\langle \mathcal{T}_C\, \{c_{q_4}(\tau_4)\, c_{q_3}(\tau_3)\, c_{q_2}(\tau_2)\, c_{q_1}(\tau_1)\}\rangle_0 \\
&\quad = \langle \mathcal{T}_C\, \{c_{q_4}(\tau_4)\, c_{q_3}(\tau_3)\}\rangle_0\, \langle \mathcal{T}_C\, \{c_{q_2}(\tau_2)\, c_{q_1}(\tau_1)\}\rangle_0 \\
&\quad - \langle \mathcal{T}_C\, \{c_{q_4}(\tau_4)\, c_{q_2}(\tau_2)\}\rangle_0\, \langle \mathcal{T}_C\, \{c_{q_3}(\tau_3)\, c_{q_1}(\tau_1)\}\rangle_0 \\
&\quad + \langle \mathcal{T}_C\, \{c_{q_4}(\tau_4)\, c_{q_1}(\tau_1)\}\rangle_0\, \langle \mathcal{T}_C\, \{c_{q_3}(\tau_3)\, c_{q_2}(\tau_2)\}\rangle_0 \, .
\end{aligned} \tag{9.28}
$$

Step # 8 Only a few pairwise contractions $\langle \mathcal{T}_C\{c_q(\tau)\, c_{q'}(\tau')\}\rangle_0$ are actually nonvanishing owing to the result (9.16), as anticipated in Eq. (9.17). For instance, in the Fermi case, the nonvanishing contraction $\langle \mathcal{T}_C\{a_q(\tau)\, a^\dagger_{q'}(\tau')\}\rangle_0$ takes the following form when either $\tau \overset{C}{>} \tau'$ or $\tau' \overset{C}{>} \tau'$ along the contour C:

$\tau \overset{C}{>} \tau'$

$$
\langle \mathcal{T}_C\{a_q(\tau)\, a^\dagger_{q'}(\tau')\}\rangle_0 = \langle a_q(\tau)\, a^\dagger_{q'}(\tau')\rangle_0 \underset{\substack{\uparrow \\ \text{[Eq.(9.16)]}}}{=} \delta_{q,q'}\, \frac{\langle [a_q(\tau), a^\dagger_{q'}(\tau')]_+\rangle_0}{1 + e^{-\beta\xi_q}}
$$

$$
\underset{\substack{\uparrow \\ \text{[Eq.(9.11)]}}}{=} \delta_{q,q'} \frac{e^{\beta\xi_q}}{1 + e^{\beta\xi_q}}\, e^{-i\epsilon_q(\tau-\tau')} \langle [a_q, a^\dagger_{q'}]_+\rangle_0 = \delta_{q,q'}\, \left(1 - f_F(\xi_q)\right) e^{-i\epsilon_q(\tau-\tau')}, \tag{9.29}
$$

where $f_F(x) = (e^{\beta x} + 1)^{-1}$ is the Fermi–Dirac distribution function.

$$\boxed{\tau' \overset{C}{>} \tau} \quad \langle \mathcal{T}_C\{a_q(\tau)\, a^\dagger_{q'}(\tau')\}\rangle_0 = -\,\langle a^\dagger_{q'}(\tau')\, a_q(\tau)\rangle_0 \underset{\substack{\uparrow \\ \text{[Eq.(9.16)]}}}{=} -\,\delta_{q,q'}\,\frac{\langle [a^\dagger_{q'}(\tau'), a_q(\tau)]_+\rangle_0}{1+e^{\beta\xi_q}}$$

$$\underset{\substack{\uparrow \\ \text{[Eq.(9.11)]}}}{=} -\,\delta_{q,q'}\,\frac{1}{1+e^{\beta\xi_q}}\,e^{-i\epsilon_q(\tau-\tau')}\langle [a^\dagger_{q'}, a_q]_+\rangle_0 = -\,\delta_{q,q'}\, f_F(\xi_q)\, e^{-i\epsilon_q(\tau-\tau')}\,. \quad (9.30)$$

9.3 Extension to Superfluid Bose and Fermi Systems within the η-Ensemble

Previously, we made a remark about the extension of the Wick's theorem to superfluid systems, for which the relevant states do not contain a definite number of particles, and "anomalous" averages have to be taken into account besides the "normal" averages like (9.29) and (9.30). Here, we sketch a procedure that shows how this situation is dealt with in practice within the so-called *η-ensemble* originally introduced in Ref. [41] (see also Ref. [42]). In Part II, the anomalous averages briefly discussed here will be fully included in the diagrammatic structure of the contour Green's functions for superfluid fermions, while for condensed bosons, these averages will be of use for the BEC limit of the BCS–BEC crossover, as outlined in Chapter 26. Here, we consider Bose and Fermi systems separately, and in both cases, we restrict to one single-particle mode for simplicity.

Bose system:

Within the η-ensemble, the density matrix operator that we consider has the form

$$\rho_\eta = e^{-\beta\left(\xi_q a^\dagger_q a_q - \eta^*_q a_q - \eta_q a^\dagger_q\right)} \quad (9.31)$$

for a given mode q, where the "external" (complex) field η_q acts to give more weight to a particular phase of the order parameter. To simplify the notation, we let $\beta\,\xi_q \to \nu$ and $\beta\,\eta_q \to \lambda$ and drop the suffix q throughout, such that

$$\beta\left(\xi_q a^\dagger_q a_q - \eta^*_q a_q - \eta_q a^\dagger_q\right) \to \nu a^\dagger a - \lambda^* a - \lambda a^\dagger \equiv \gamma = \nu\alpha^\dagger\alpha - \frac{|\lambda|^2}{\nu}, \quad (9.32)$$

where we have introduced the "shifted" operator $\alpha = a - \frac{\lambda}{\nu}$ for which $[\alpha, \alpha^\dagger]_- = [a, a\dagger]_- = 1$. Similar to what we did when proving the property (9.4), we can now write

$$\begin{cases} \alpha\, e^{-\gamma} \;= e^{-\gamma}\,\alpha\, e^{-\nu} \\ \alpha^\dagger\, e^{-\gamma} = e^{-\gamma}\,\alpha^\dagger\, e^{\nu}\,. \end{cases} \quad (9.33)$$

In this way, we obtain in analogy to Eq. (9.7):

$$\langle [\alpha, A]_-] \rangle_\eta = \frac{\mathrm{Tr}\{[e^{-\gamma}, \alpha]_- A\}}{\mathrm{Tr}\{e^{-\gamma}\}} = \frac{\mathrm{Tr}\{(1 - e^{-\nu})\, e^{-\gamma} \alpha A\}}{\mathrm{Tr}\{e^{-\gamma}\}} = (1 - e^{-\nu})\ \langle \alpha A \rangle_\eta , \tag{9.34}$$

where A is an arbitrary operator. In particular, if we take $A = \mathbb{1}$, the result (9.34) implies that $\langle \alpha \rangle_\eta = 0$, that is, $\langle a \rangle_\eta = \frac{\lambda}{\nu} \rightarrow \frac{\eta_q}{\xi_q}$ according to the definitions introduced in Eq. (9.32). This result shows that the phase of the *order parameter* $\langle a \rangle_\eta$ is fixed by the complex field η_q, a result that is at the essence of the phenomenon of "broken symmetry" for the case of superfluidity [42]. Note also that within the η-ensemble, the emergence of the "anomalous" average $\langle a \rangle_\eta$ is due to the fact that the density matrix operator (9.31) does not commute with the number operator $a^\dagger a$.

In practice, for a superfluid Bose system, one assumes at the outset that $\langle a \rangle_\eta \neq 0$ and determines its value in a self-consistent way within the chosen approximate theory. In the homogeneous case, this assumption affects only the lowest-energy state with $q = 0$, as originally considered in the seminal work by Bogoliubov [43].

Fermi system:

In the case of superfluid fermions, the spin $\sigma = (\uparrow, \downarrow)$ plays an essential role and should then be taken into account. Accordingly, the density matrix operator now reads

$$\rho_\eta = e^{-\beta\left(\xi a_\uparrow^\dagger a_\uparrow + \xi a_\downarrow^\dagger a_\downarrow - \eta^* a_\downarrow a_\uparrow - \eta a_\uparrow^\dagger a_\downarrow^\dagger\right)} , \tag{9.35}$$

where, for simplicity, reference to a specific q mode has been dropped. Like for the Bose system, let $\beta\,\xi = \nu$ and $\beta\,\eta = \lambda$ and rewrite the operator in the exponent (9.35) in the following compact way:

$$\gamma \equiv \nu a_\uparrow^\dagger a_\uparrow + \nu a_\downarrow^\dagger a_\downarrow - \lambda^* a_\downarrow a_\uparrow - \lambda a_\uparrow^\dagger a_\downarrow^\dagger = \left(a_\uparrow^\dagger, a_\downarrow\right) \begin{pmatrix} \nu & -\lambda \\ -\lambda^* & -\nu \end{pmatrix} \begin{pmatrix} a_\uparrow \\ a_\downarrow^\dagger \end{pmatrix} + \nu\,. \tag{9.36}$$

The quadratic form in Eq. (9.36) can be diagonalized by the unitary matrix $U = \begin{pmatrix} u & v \\ -v & u \end{pmatrix}$, such that $U^\dagger \begin{pmatrix} \nu & -\lambda \\ -\lambda & -\nu \end{pmatrix} U = \begin{pmatrix} \epsilon & 0 \\ 0 & -\epsilon \end{pmatrix}$ (with λ real without loss of generality), where

$$\begin{cases} \begin{pmatrix} u \\ -v \end{pmatrix} & \text{is the eigenvector with energy } \epsilon \\[2ex] \begin{pmatrix} v \\ u \end{pmatrix} & \text{is the eigenvector with energy } -\epsilon \end{cases} \tag{9.37}$$

with $\epsilon = \sqrt{\nu^2 + \lambda^2}$, $u = \frac{1}{\sqrt{2}} \sqrt{1 + \frac{\nu}{\epsilon}}$, and $v = \frac{1}{\sqrt{2}} \sqrt{1 - \frac{\nu}{\epsilon}}$, such that $u^2 + v^2 = 1$. We thus obtain:

$$\gamma = \left(a_\uparrow^\dagger, a_\downarrow\right) U \begin{pmatrix} \epsilon & 0 \\ 0 & -\epsilon \end{pmatrix} U^\dagger \begin{pmatrix} a_\uparrow \\ a_\downarrow^\dagger \end{pmatrix} + \nu \, . \tag{9.38}$$

By introducing the new fermionic (Bogoliubov–Valatin) operators [44, 45]

$$\begin{pmatrix} \alpha \\ \beta^\dagger \end{pmatrix} = U^\dagger \begin{pmatrix} a_\uparrow \\ a_\downarrow^\dagger \end{pmatrix} = \begin{pmatrix} u & -v \\ v & u \end{pmatrix} \begin{pmatrix} a_\uparrow \\ a_\downarrow^\dagger \end{pmatrix} = \begin{cases} u a_\uparrow - v a_\downarrow^\dagger \\ \\ v a_\uparrow + u a_\downarrow^\dagger \end{cases} , \tag{9.39}$$

the expression (9.38) becomes

$$\gamma = \left(\alpha^\dagger, \beta\right) \begin{pmatrix} \epsilon & 0 \\ 0 & -\epsilon \end{pmatrix} \begin{pmatrix} \alpha \\ \beta^\dagger \end{pmatrix} + \nu = \epsilon \, \alpha^\dagger \alpha - \epsilon \, \beta \beta^\dagger + \nu \, . \tag{9.40}$$

In addition, from the Bogoliubov–Valatin transformation (9.39) and the original anti-commutation relations of the operators a_σ and $a_\sigma^\dagger$, it follows that the operators α and β satisfy the following anti-commutation relations

$$[\alpha, \alpha^\dagger]_+ = 1 \quad , \quad [\beta, \beta^\dagger]_+ = 1 \quad , \quad [\alpha, \beta]_+ = 0 \quad , \quad [\alpha, \beta^\dagger]_+ = 0 \quad , \tag{9.41}$$

such that the quadratic expression (9.40) becomes eventually

$$\gamma = \epsilon \left(\alpha^\dagger \alpha + \beta^\dagger \beta\right) + \nu - \epsilon \, . \tag{9.42}$$

The transformation (9.39) can further be inverted, to give

$$\begin{pmatrix} a_\uparrow \\ a_\downarrow^\dagger \end{pmatrix} = U \begin{pmatrix} \alpha \\ \beta^\dagger \end{pmatrix} = \begin{pmatrix} u & v \\ -v & u \end{pmatrix} \begin{pmatrix} \alpha \\ \beta^\dagger \end{pmatrix} = \begin{cases} u\alpha + v\beta^\dagger \\ \\ -v\alpha + u\beta^\dagger \end{cases} , \tag{9.43}$$

such that the "anomalous" average $\langle a_\uparrow a_\downarrow \rangle_\eta$ within the η-ensemble we are considering reads:

$$\langle a_\uparrow a_\downarrow \rangle_\eta = -uv \, \langle \alpha \alpha^\dagger \rangle_\eta + u^2 \, \langle \alpha \beta \rangle_\eta - v^2 \, \langle \beta^\dagger \alpha^\dagger \rangle_\eta + uv \, \langle \beta^\dagger \beta \rangle_\eta \, . \tag{9.44}$$

To calculate these averages, we take advantage of the anti-commutator relations (9.41) to obtain $\alpha \, e^{-\gamma} = e^{-\gamma} \alpha \, e^{-\epsilon}$ and $\beta^\dagger \, e^{-\gamma} = e^{-\gamma} \beta^\dagger \, e^{\epsilon}$. For a generic operator A, we can thus write

$$\langle [\alpha, A]_+ \rangle_\eta = \frac{\mathrm{Tr}\{e^{-\gamma} \alpha A + \alpha e^{-\gamma} A\}}{\mathrm{Tr}\{e^{-\gamma}\}} = (1 + e^{-\epsilon}) \, \langle \alpha A \rangle_\eta \tag{9.45}$$

as well as

$$\langle [\beta^\dagger, A]_+ \rangle_\eta = \frac{\mathrm{Tr}\{e^{-\gamma} \beta^\dagger A + \beta^\dagger e^{-\gamma} A\}}{\mathrm{Tr}\{e^{-\gamma}\}} = (1 + e^{\epsilon}) \, \langle \beta^\dagger A \rangle_\eta \, , \tag{9.46}$$

which generalizes the property (9.5) of the normal phase to the superfluid phase. We then take in Eq. (9.45)

$$\begin{cases} A = \alpha^{\dagger} & \Rightarrow \quad \langle \alpha \alpha^{\dagger} \rangle_{\eta} = 1 - f_F(\epsilon/\beta) \\ A = \beta & \Rightarrow \quad \langle \alpha \beta \rangle_{\eta} = 0 \end{cases} \tag{9.47}$$

and in Eq. (9.46)

$$\begin{cases} A = \alpha^{\dagger} & \Rightarrow \quad \langle \beta^{\dagger} \alpha^{\dagger} \rangle_{\eta} = 0 \\ A = \beta & \Rightarrow \quad \langle \beta^{\dagger} \beta \rangle_{\eta} = f_F(\epsilon/\beta)\,, \end{cases} \tag{9.48}$$

where the anti-commutation relations (9.41) have been utilized. In conclusion, the anomalous average (9.44) becomes

$$\langle a_{\uparrow} a_{\downarrow} \rangle_{\eta} = -uv\,(1 - f_F(\epsilon/\beta)) + uv f_F(\epsilon/\beta) = -uv\,(1 - 2 f_F(\epsilon/\beta))\;, \tag{9.49}$$

which is nonvanishing as long that $\eta \neq 0$ in Eq. (9.35), such that $uv \neq 0$.

Similar to a superfluid Bose system, it proves convenient also for a superfluid Fermi system to assume at the outset that $\langle a_{\uparrow} a_{\downarrow} \rangle \neq 0$ and then determine its value in a self-consistent way within the chosen approximate theory, as first proposed by Gor'kov [46].

10

Non-equilibrium Diagrammatics: Feynman Rules

In this chapter, we consider the application of the Wick's theorem (or decomposition) discussed in Chapter 9 to the contour single-particle Green's function (7.12) when the average over a pure state therein is more generally replaced by that over an incoherent superposition of states as in Eq. (8.1), which may also refer to a thermal average. To this end, this average has to be represented in the interaction picture following the prescription (8.6), where in both numerator and denominator, the exponential operator is expanded as in Eq. (5.12). In addition, as in Chapter 9, there is no need to specify at the outset which kind of contour is being used, so that we shall indicate the generic contour by C. This procedure will end up in a set of *Feynman diagrammatic rules*, which are reported schematically in the present chapter. Apart from the difference in the time contours, these rules are quite similar to those occurring in the zero-temperature formalism [6, 7], which we refer to for their extended proof.

For definiteness, we shall explicitly consider the effect of the interaction part $H_{\mathrm{I}}^{\mathrm{int}}$ of the Hamiltonian in Eq. (8.6) on the contour single-particle Green's function (7.12), while the effect of the external perturbation $H_{\mathrm{I}}^{\mathrm{ext}}$ will briefly be considered along similar lines at the end of this chapter. In addition, the diagrammatic rules can be readily extended to the contour two-particle Green's function (7.13) as well. Finally, in the following, we shall limit to consider the normal phase, while, in Part II, the Feynman rules will be extended to superfluid Fermi systems in line with the discussion made in Section 9.3.

Rule # 1: Cancellation of disconnected diagrams

The arguments of the contour single-particle (7.12) and two-particle (7.13) Green's functions correspond to "external" variables, which are not to be integrated over in space and time (as well in spin) when the exponential operator in the numerator of Eq. (8.6)) is expanded as in Eq. (5.12). Once proceeding with the Wick's decomposition, subunits appear (referred to as *disconnected diagrams*) that are not connected to the parts containing these external variables. On the other hand, the related expansion of the denominator of Eq. (8.6)), where by construction no external variables are present, results only in disconnected diagrams. The point is that the disconnected diagrams of the denominator cancel exactly those occurring in the numerator of Eq. (8.6). The proof of this statement proceeds along similar lines to those in the zero-temperature formalism [7] and is based on a partitioning similar to that used in Section 5.3 to prove the unfolding of a single time-ordered product into two time-ordered products.

Remark: Disconnected diagrams – To be or not to be

Sometimes, in the literature, one encounters the statement that nonequilibrium perturbation theory has a simpler structure than the standard equilibrium theory to the extent that there is no need to cancel the disconnected diagrams in the first place [14, 39]. This statement relies on the property (7.2), once applied in reverse (i.e., from right to left) to Eq. (8.4) and then transferred to the denominator of Eq. (8.6). However, this statement is justified only if one assumes that the initial system configuration is non-correlated, such that the integration contour γ_{T} in the denominator of Eq. (8.6) excludes the appendix contour γ^{M} and thus coincides with the contour γ of Fig. 8.1. On the other hand, if the initial configuration of the system is correlated, one cannot simply rely on the property (7.2) to invoke the absence of disconnected diagrams but has instead to follow the standard cancellation route as recalled earlier [15].

Rule # 2: Topologically distinct connected diagrams

Referring to the interaction part $H_{\mathrm{I}}^{\mathrm{int}}$ of the Hamiltonian in Eq. (8.6), we utilize the Wick's decomposition to draw all topologically connected diagrams, with n interaction terms and $2n+1$ non-interacting Green's functions $G^0(x, y)$ directed from the second (y) to the first (x) argument. Here, x signifies the set of space ($\mathbf{r}$), spin (σ), and time (either τ or z) variables, while the subscript "I" standing for the interaction picture has been omitted in G^0 for simplicity. To an interaction line, there corresponds a factor $v(\mathbf{r}-\mathbf{r}')\,\delta(z, z')$, where $v(\mathbf{r}-\mathbf{r}')$ is the interparticle interaction[1] entering H^{int} (cf. Eq. (2.4)), and $\delta(z, z')$ is the contour Dirac delta function of Eq. (7.17).

[1] More generally, we may consider the interparticle interaction $v(\mathbf{r}-\mathbf{r}', z)$ to have an explicit time dependence, as discussed in Section 11.3.

Rule # 3: Overall factor for each *n*th order term

To compute $G(x, y)$, assign to each diagrammatic term, a factor $(-i)\,(-i)^n\,(i)^{2n+1} = i^n$, where:

(a) the first factor $(-i)$ comes from the definition of $G(x, y)$;
(b) the second factor $(-i)^n$ comes from the n interaction terms; and
(c) the third factor $(i)^{2n+1}$ comes from the presence of $2n + 1$ noninteracting Green's functions G^0 in the given diagram.

Rule # 4: Green's functions with equal time variables

A Green's function with the same contour-time arguments must be interpreted with the second argument to be infinitesimally later than the first, that is, $G^0(\mathbf{r}\sigma z, \mathbf{r}'\sigma' z^+)$. This is because this kind of Green's function originates by expanding the time-evolution operator in the numerator of Eq. (8.6) in terms of the interaction part of the Hamiltonian in the interaction picture, namely,

$$H_{\mathrm{I}}^{\mathrm{int}}(z) = \frac{1}{2} \sum_{\sigma,\sigma'} \int d\mathbf{r} d\mathbf{r}'\, \psi_{\mathrm{I}}^{\dagger}(\mathbf{r}\sigma z^+)\, \psi_{\mathrm{I}}^{\dagger}(\mathbf{r}'\sigma' z^+)\, v(\mathbf{r} - \mathbf{r}')\, \psi_{\mathrm{I}}(\mathbf{r}'\sigma' z)\, \psi_{\mathrm{I}}(\mathbf{r}\sigma z), \tag{10.1}$$

where, to comply with the time ordering in Eq. (8.6), z^+ is meant to be infinitesimally later than z along the contour (for definiteness, we are considered fermions with spin σ). For instance,

$$\langle \mathcal{T}_C\{\psi_{\mathrm{I}}(\mathbf{r}'\sigma' z)\, \psi_{\mathrm{I}}^{\dagger}(\mathbf{r}\sigma z^+)\}\rangle_0 \longleftrightarrow G_0(\mathbf{r}'\sigma' z, \mathbf{r}\sigma z^+)\,. \tag{10.2}$$

Rule # 5: Overall sign of each diagram

The overall sign of each diagram is determined by the sign inherited from the application of Wick's theorem to the string of operators within the time-ordered product. Consider, for instance, the two diagrams arising at the first order in the interaction depicted in Fig. 10.1 (again for the fermion case), where the corresponding arrangements of the field operators are reported, and where we have introduced the compact notation 1, 2, $\cdots$ to signify the set of space, spin, and time variables. These first-order contributions are known as (a) Fock and (b) Hartree diagrams, respectively.

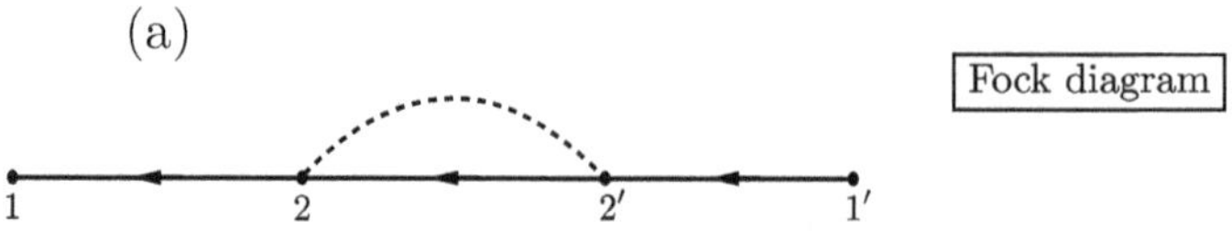

$$\langle \mathcal{T}_C\{\psi_I(1)\,\psi_I^\dagger(1')\}\rangle_0 \;\longrightarrow\; \langle \mathcal{T}_C\{\psi_I(1)\,\psi_I^\dagger(2)\,\psi_I^\dagger(2')\,\overbrace{\psi_I(2')\,\psi_I(2)\,\psi_I^\dagger(1')}\}\rangle_0 \;\Longrightarrow\; +\ \text{sign}$$

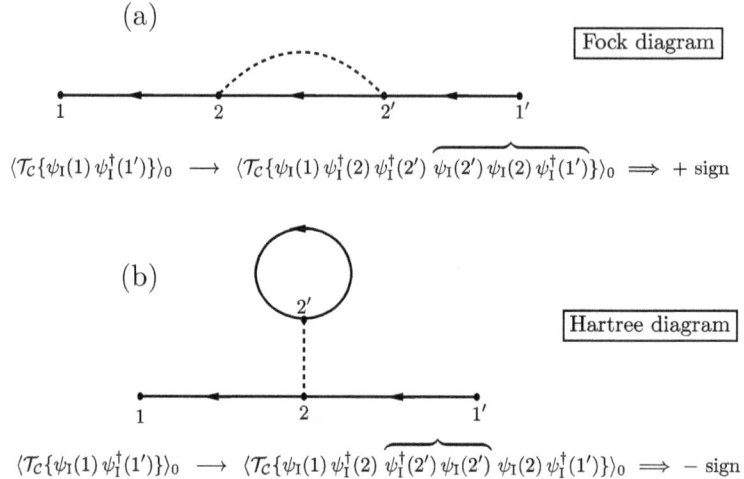

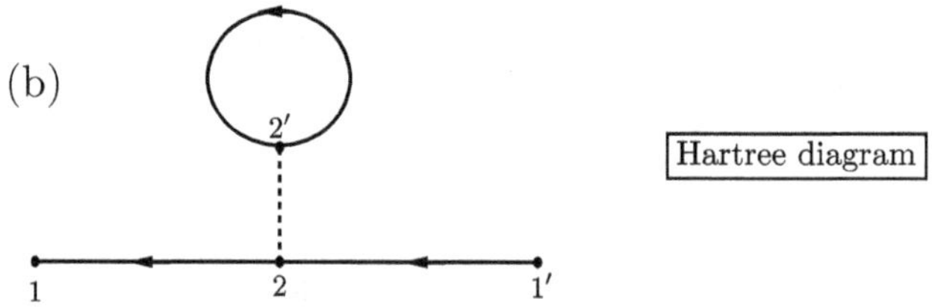

$$\langle \mathcal{T}_C\{\psi_I(1)\,\psi_I^\dagger(1')\}\rangle_0 \;\longrightarrow\; \langle \mathcal{T}_C\{\psi_I(1)\,\psi_I^\dagger(2)\,\overbrace{\psi_I^\dagger(2')\,\psi_I(2')}\,\psi_I(2)\,\psi_I^\dagger(1')\}\rangle_0 \;\Longrightarrow\; -\ \text{sign}$$

Figure 10.1 (a) Fock and (b) Hartree diagrams, where the different signs refer to those of the fermion case.

More generally, for fermions, each time a fermion line closes on itself, the term acquires an extra minus sign, in accordance with the following rearrangement:

$$\begin{aligned}
&\langle \mathcal{T}_C\{\psi_I^\dagger(1)\,\psi_I(1)\,\psi_I^\dagger(2)\,\psi_I(2)\ \ldots\ \psi_I^\dagger(N-1)\,\psi_I(N-1)\,\psi_I^\dagger(N)\,\psi_I(N)\}\rangle_0 \\
=\;& \langle \mathcal{T}_C\{\underbrace{\psi_I^\dagger(1)\ \overbrace{\psi_I(1)\,\psi_I^\dagger(N)}\ \overbrace{\psi_I(N)\,\psi_I^\dagger(N-1)}\ \psi_I(N-1)\ \ldots\ \psi_I^\dagger(2)\,\psi_I(2)}\}\rangle_0 \\
=-\;& \langle \mathcal{T}_C\{\underbrace{\psi_I(1)\,\psi_I^\dagger(N)}\ \underbrace{\psi_I(N)\,\psi_I^\dagger(N-1)}\ \ldots\ \underbrace{\psi_I(3)\,\psi_I^\dagger(2)}\ \underbrace{\psi_I(2)\,\psi_I^\dagger(1)}\}\rangle_0\,.
\end{aligned} \tag{10.3}$$

This is because each closed loop corresponds to a cycle of interactions, which begins at the left or right side of some vertex, connects it to one side of another vertex, then connects it further to some side of yet another vertex, and so on, until it finally returns to the original side of the original vertex (as shown schematically in Fig. 10.2).

Rule # 6: Integration over all internal variables

For each internal vertex of a diagram, integrate over the space variable, sum over the spin variable, and integrate the time variable over the contour C (where again the collection of these variables is indicated by the short-hand notation $1, 2, \cdots$). The last prescription (about the time integration) is actually the only difference

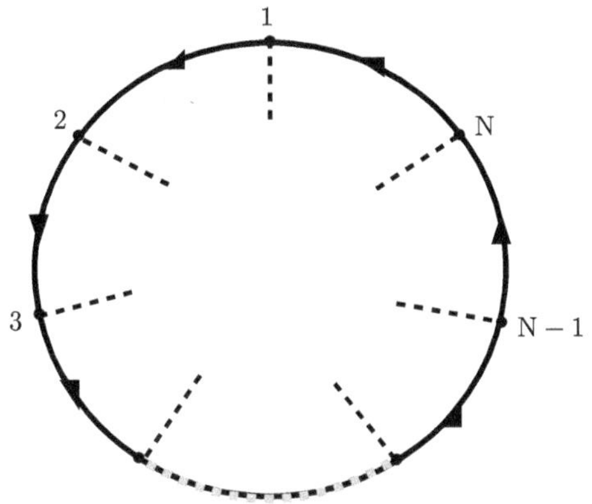

Figure 10.2 A typical closed loop occurring in the fermionic diagrammatic structure. Here and in the diagrams of Fig. 10.1, full lines correspond to the non-interacting single-particle Green's function G^0 and dashed lines to the interparticle interaction v.

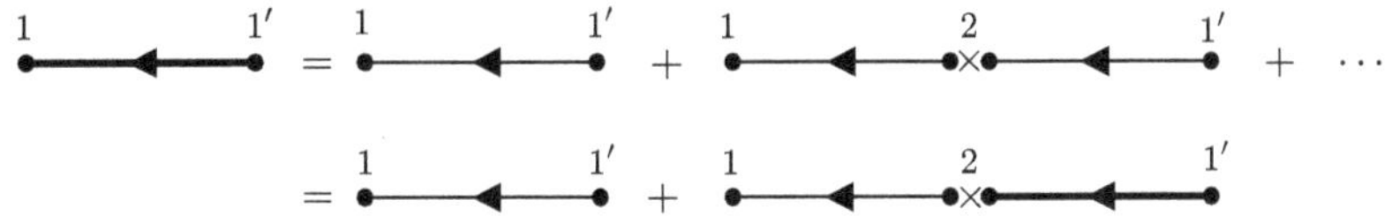

Figure 10.3 Diagrammatic series due to an external perturbation. Thin and thick lines correspond to the bare (G_0) and dressed ($\bar{G}$) single-particle Green's functions, respectively, with the arrows pointing from their second to first argument, and crosses represent the external time-dependent potential V_{ext}.

with respect to the zero-temperature formalism [6, 7], where all time integrations run instead from $-\infty$ to $+\infty$, rather than along the contour C, as in the present case.

Remark: Effects of an external perturbation

The effects of the time-dependent external perturbation on the diagrammatic structure of the contour single-particle (as well of the two-particle) Green's function can be readily taken into account because $H^{\text{ext}}(z)$ given by the expression (2.5) is a one-body operator. For the single-particle Green's function, one obtains the diagrammatic series shown schematically in Fig. 10.3, which can be summed to yield the following integral equation:

$$\bar{G}(1, 1') = G_0(1, 1') + \int_C d2\, G_0(1, 2)\, V_{\text{ext}}(2)\, \bar{G}(2, 1') \,. \tag{10.4}$$

The structure of this equation anticipates the full non-equilibrium Dyson integral equation considered in Chapter 11 (cf. Eq. (11.40)), in which G_0 is meant to already contain the effects of the external time-dependent potential V_{ext} (cf. Eq. (11.32)).

In this respect, Eq. (10.4) can be regarded as a pre-summation made in the full Dyson equation (11.40), in line with the general arguments discussed in Section 20.1. Finally, we remark that, although one may as well approximate G by its successive terms in powers of V_{ext} (thus giving rise to the so-called linear- and nonlinear response theories – cf. Chapter 18), an advantage of the Schwinger–Keldysh approach resides in considering the effects of V_{ext} to *all* orders, as in Eqs. (10.4) or (11.40).

11

Non-equilibrium Dyson Equations

In this chapter, we consider again some of the items discussed in the Chapters 2, 6, and 7 and cast them in a more formal way so as to adapt them for future developments. In this way, we will arrive at the integro-differential form of the Dyson equation for the contour single-particle Green's function (as well as to its integral counterpart), which will play an important role in throughout the book for capturing the dynamical evolution of the physical system under consideration.

The system Hamiltonian is again given by Eqs. (2.1)–(2.5), where it is now convenient to lump the one-body terms (2.3) and (2.5) into a single term, by defining

$$h(\mathbf{r}, t) = h_0(\mathbf{r}) + V_{\text{ext}}(\mathbf{r}, t). \tag{11.1}$$

Below, we shall indicate by C a generic contour in the time domain with a forward and a backward branch and by z the time variable running along it, whereby z_i and z_f are the initial and final time variables along the contour (for instance, with reference to Fig. 8.1, we may take C = γ, where $z_i = t_0$). In the context of the formal considerations of this chapter, there is no need to attach a vertical track to the contour C. Recall also that all operators (as well as their time derivatives) considered here take the same value at corresponding times along the forward and backward branches of the contour.

11.1 Evolution Operator and Heisenberg Representation along the Contour C

In analogy with Eq. (2.8), the *contour evolution operator* $U(z_2, z_1)$, for z_1 and z_2 belonging to the contour C, is defined by [15]

$$U(z_2, z_1) = \begin{cases} \mathcal{T}_{\mathrm{C}}\left\{e^{-i\int_{z_1}^{z_2} dz'\mathcal{H}(z')}\right\} & \text{for } z_2 \overset{C}{>} z_1, \\ \tilde{\mathcal{T}}_{\mathrm{C}}\left\{e^{i\int_{z_2}^{z_1} dz'\mathcal{H}(z')}\right\} & \text{for } z_2 \overset{C}{<} z_1, \end{cases} \tag{11.2}$$

where $\mathcal{T}_{\mathrm{C}}$ is the contour time-ordering operator and $\tilde{\mathcal{T}}_{\mathrm{C}}$ is the contour anti-time-ordering operator along C. The operator $U(z_2, z_1)$ has the following properties:

(i) $U(z_1, z_1) = \mathbb{1}$ (cf. Eq. (2.16));
(ii) $U(z_2, z_3)U(z_3, z_1) = U(z_2, z_1)$ (cf. Eq. (2.12));
(iii) $i\frac{d}{dz}U(z, z_0) = \mathcal{T}_{\mathrm{C}}\left\{\mathcal{H}(z)e^{-i\int_{z_0}^{z}dz'\mathcal{H}(z')}\right\} = \mathcal{H}(z)\,U(z, z_0)$ for $z \overset{C}{>} z_0$ (cf. Eq. (2.18));

and

(iv) $i\frac{d}{dz}U(z_0, z) = -\tilde{\mathcal{T}}_{\mathrm{C}}\left\{e^{i\int_{z_0}^{z}dz'\mathcal{H}(z')}\,\mathcal{H}(z)\right\} = -U(z_0, z)\,\mathcal{H}(z)$ again for $z \overset{C}{>} z_0$.

Given an operator $A(z)$ along the contour C, its *contour Heisenberg representation* is defined in analogy to the expression on the right-hand side of Eq. (2.25)

$$A_{\mathrm{H}}(z) = U(z_i, z)A(z)U(z, z_i) \tag{11.3}$$

with $z \overset{C}{>} z_i$, where the definition applies also when the time z lies on the vertical track [6, 7]. Its equation of motion is then given by (cf. Eq. (2.28))

$$i\frac{d}{dz}A_{\mathrm{H}}(z) = i\frac{d}{dz}\left(U(z_i, z)A(z)U(z, z_i)\right) = [A_{\mathrm{H}}(z), \mathcal{H}_{\mathrm{H}}(z)] + i\frac{\partial}{\partial z}A_{\mathrm{H}}(z) \tag{11.4}$$

with the notation

$$\mathcal{H}_{\mathrm{H}}(z) = U(z_i, z)\,\mathcal{H}(z)\,U(z, z_i), \tag{11.5}$$

$$\frac{\partial}{\partial z}A_{\mathrm{H}}(z) = U(z_i, z)\left(\frac{d}{dz}A(z)\right)\,U(z, z_i)\,, \tag{11.6}$$

where only the derivative with respect to the explicit time dependence of the operator $A(z)$ has to be taken. We may also rewrite the Heaviside unit step function on the contour C in the form (cf. Eq. (7.4))

$$\theta(z_1, z_2) = \begin{cases} 1 & \text{for } z_1 \overset{C}{>} z_2\,, \\ 0 & \text{for } z_1 \overset{C}{<} z_2\,, \end{cases} \tag{11.7}$$

such that the contour Dirac delta function is obtained by (cf. Eqs. (7.15) and (7.16)):

$$\delta(z_1, z_2) = \frac{d}{dz_1}\theta(z_1, z_2) = -\frac{d}{dz_1}\theta(z_2, z_1)\,. \tag{11.8}$$

11.2 Operator Correlators on the Contour and Their Equations of Motion

With these premises, we next consider two operators $A_{\mathrm{H}}(z)$ and $B_{\mathrm{H}}(z)$ in the Heisenberg representation (11.3) and form their contour-ordered product (cf. Eq. (7.10))

$$\mathcal{T}_C\{A_H(z_1)\,B_H(z_2)\} = A_H(z_1)\,B_H(z_2)\,\theta(z_1,z_2) \pm B_H(z_2)\,A_H(z_1)\,\theta(z_2,z_1)\,, \quad (11.9)$$

where the + sign holds for bosonic operators and the − sign for fermionic operators. With the help of Eq. (11.8), we then obtain the following operator identities

$$\frac{d}{dz_1}\mathcal{T}_C\,\{A_H(z_1)\,B_H(z_2)\} = \mathcal{T}_C\left\{\left(\frac{d}{dz_1}A_H(z_1)\right)\,B_H(z_2)\right\} + \delta(z_1,z_2)\,\left[A_H(z_1),B_H(z_2)\right]_{\mp} \quad (11.10)$$

and

$$\frac{d}{dz_2}\mathcal{T}_C\,\{A_H(z_1)\,B_H(z_2)\} = \mathcal{T}_C\left\{A_H(z_1)\,\left(\frac{d}{dz_2}B_H(z_2)\right)\right\} - \delta(z_1,z_2)\,\left[A_H(z_1),B_H(z_2)\right]_{\mp}\,, \quad (11.11)$$

where the upper sign refers to bosons and the lower sign to fermions. In both cases, the contour time derivatives of the operators in the Heisenberg representation is given by Eq. (11.4).

We can then consider the *correlator* between the above two operators (cf. Eq. (7.10))

$$F_{AB}(z_1,z_2) = \langle \mathcal{T}_C\,\{A_H(z_1)\,B_H(z_2)\}\rangle = \frac{\mathrm{Tr}\left\{e^{-\beta H^M}\mathcal{T}_C\,\{A_H(z_1)\,B_H(z_2)\}\right\}}{\mathrm{Tr}\left\{e^{-\beta H^M}\right\}}\,. \quad (11.12)$$

[Here, we could as well represent $e^{-\beta H^M} = e^{-i\int_{z_i}^{z_f} dz H^M} = U(z_f,z_i)$ in line with Eq. (11.2), where $z_i = t_0$ and $z_f = t_0 - i\beta$, and accordingly replace the contour C by the extended contour γ_T that includes the vertical track γ^M (cf. Fig. 8.1)). However, we ignore this replacement since it is not necessary for the following formal considerations.] Owing to the operator identities (11.10) and (11.11), the correlator (11.12) satisfies the *equations of motions*:

$$\frac{d}{dz_1}F_{AB}(z_1,z_2) = \left\langle\frac{d}{dz_1}\mathcal{T}_C\,\{A_H(z_1)\,B_H(z_2)\}\right\rangle = \left\langle\mathcal{T}_C\left\{\left(\frac{d}{dz_1}A_H(z_1)\right)\,B_H(z_2)\right\}\right\rangle$$
$$+\,\delta(z_1,z_2)\,\langle[A_H(z_1),B_H(z_2)]_{\mp}\rangle \quad (11.13)$$

and

$$\frac{d}{dz_2}F_{AB}(z_1,z_2) = \left\langle\frac{d}{dz_2}\mathcal{T}_C\,\{A_H(z_1)\,B_H(z_2)\}\right\rangle = \left\langle\mathcal{T}_C\left\{A_H(z_1)\,\left(\frac{d}{dz_2}B_H(z_2)\right)\right\}\right\rangle$$
$$-\,\delta(z_1,z_2)\,\langle[A_H(z_1),B_H(z_2)]_{\mp}\rangle\,. \quad (11.14)$$

11.3 Equations of Motion for the Single-Particle Green's Function

Of special interest in the context of time-dependent quantum many-body theory is the contour single-particle Green's function (7.12), which is now extended as in Eq. (8.1) to include a thermal average for the initial preparation, such that, in Eq. (11.12), we identify

$$\begin{cases} A_{\mathrm{H}}(z_1) & \longrightarrow \quad \psi_{\mathrm{H}}(\mathbf{r}_1,\sigma_1,z_1)=\psi_{\mathrm{H}}(1) \\ B_{\mathrm{H}}(z_2) & \longrightarrow \quad \psi^{\dagger}_{\mathrm{H}}(\mathbf{r}_2,\sigma_2,z_2)=\psi^{\dagger}_{\mathrm{H}}(2) \\ F_{AB}(z_1,z_2) & \longrightarrow \quad i\,G(1,2)=\left\langle \mathcal{T}_{\mathrm{C}}\left\{\psi_{\mathrm{H}}(1)\,\psi^{\dagger}_{\mathrm{H}}(2)\right\}\right\rangle , \end{cases} \tag{11.15}$$

with the short-hand notation $1 \leftrightarrow (\mathbf{r}_1,\sigma_1,z_1)$ and so on. In this case, the last term on the right-hand side of Eqs. (11.13) and (11.14) reduces to:

$$\begin{aligned} &\delta(z_1,z_2)\;[A_{\mathrm{H}}(z_1),B_{\mathrm{H}}(z_2)]_{\mp} \longrightarrow \delta(z_1,z_2)\;\left[\psi_{\mathrm{H}}(1),\psi^{\dagger}_{\mathrm{H}}(2)\right]_{\mp} \\ = \;& \delta(z_1,z_2)\,U(z_i,z_1)\left(\psi(\mathbf{r}_1,\sigma_1)\,\psi^{\dagger}(\mathbf{r}_2,\sigma_2)\mp\psi^{\dagger}(\mathbf{r}_2,\sigma_2)\,\psi(\mathbf{r}_1,\sigma_1)\right)U(z_1,z_i) \\ = \;& \delta(z_1,z_2)\,\delta(\mathbf{r}_1-\mathbf{r}_2)\,\delta_{\sigma_1,\sigma_2}\equiv\delta(1,2), \end{aligned} \tag{11.16}$$

where the properties (i) and (ii) of the contour evolution operator $U(z_1,z_2)$ given in Section 11.1 and the commutation (anti-commutation) relations (2.2) for bosons (fermions) have been utilized. In addition, the equation of motion (11.4) becomes

$$\begin{aligned} i\frac{d}{dz}A_{\mathrm{H}}(z) \quad &\longrightarrow \quad i\frac{d}{dz}\psi_{\mathrm{H}}(\mathbf{r},\sigma,z)=[\psi_{\mathrm{H}}(\mathbf{r},\sigma,z),\mathcal{H}_{\mathrm{H}}(z)] \\ &= \quad U(z_i,z)\;[\psi(\mathbf{r},\sigma),\mathcal{H}(z)]\;U(z,z_i), \end{aligned} \tag{11.17}$$

where Eqs. (11.3) and (11.5) have been utilized, and the last term on the right-hand side of Eq. (11.4)) vanishes because the field operator $\psi(\mathbf{r},\sigma)$ does not contain any explicit time dependence. Since the total Hamiltonian $\mathcal{H}(z)=H_0+H^{\mathrm{ext}}(z)+H^{\mathrm{int}}$ can itself be expressed in terms of the field operators as in Eqs. (2.3)–(2.5), the commutator in Eq. (11.17) can be readily calculated in terms of the commutation (anti-commutation) relations (2.2), to give

$$[\psi(\mathbf{r},\sigma),\mathcal{H}(z)]=h(\mathbf{r},z)\psi(\mathbf{r},\sigma)+\sum_{\sigma'}\int d\mathbf{r}'\psi^{\dagger}(\mathbf{r}',\sigma')\psi(\mathbf{r}',\sigma')\,v(\mathbf{r}'-\mathbf{r})\,\psi(\mathbf{r},\sigma), \tag{11.18}$$

which holds for both bosons and fermions. With this result, Eq. (11.17) becomes eventually

$$i\frac{d}{dz}\psi_{\rm H}(\mathbf{r},\sigma,z) = h(\mathbf{r},z)\,\psi_{\rm H}(\mathbf{r},\sigma,z) + \sum_{\sigma'}\int d\mathbf{r}'\int_{\rm C} dz'\,\psi^{\dagger}_{\rm H}(\mathbf{r}',\sigma',z')\,\psi_{\rm H}(\mathbf{r}',\sigma',z')\,v(\mathbf{r}'z',\mathbf{r}z)\,\psi_{\rm H}(\mathbf{r},\sigma,z)\,, \tag{11.19}$$

where, in the second term on the right-hand side, we have inserted an integration over the contour C by relying on the property (7.17) of the contour Dirac delta function and defined

$$v(\mathbf{r}z,\mathbf{r}'z') = \delta(z,z')\,v(\mathbf{r}-\mathbf{r}') \tag{11.20}$$

for z and z' on the contour C.

Remark: Interparticle interaction that depends explicitly on time

The definition (11.20) can be extended to the case when the interparticle interaction $v(\mathbf{r}-\mathbf{r}',z)$ depends explicitly on time, thus writing

$$v(\mathbf{r}z,\mathbf{r}'z') = \delta(z,z')\,v(\mathbf{r}-\mathbf{r}',z)\,, \tag{11.21}$$

which will be formally considered to hold true henceforth. In this way, the cases are automatically included of an adiabatic switching on of the interaction, as needed in the Keldysh formalism (cf. Section 8.2), and of the time dependence, which can be experimentally induced and controlled in ultracold gases (cf. Section 32.6).

The equation of motion for the operator $\psi^{\dagger}_{\rm H}(\mathbf{r},\sigma,z)$ is similarly obtained. With the short-hand notation $1 \leftrightarrow (\mathbf{r}_1,\sigma_1,z_1)$ and so on, these equations of motion eventually acquire the compact form:

$$\begin{cases} i\frac{d}{dz_1}\psi_{\rm H}(1) = h(1)\,\psi_{\rm H}(1) + \int d3\,\psi^{\dagger}_{\rm H}(3)\,\psi_{\rm H}(3)\,v(3,1)\,\psi_{\rm H}(1) \\ -i\frac{d}{dz_2}\psi^{\dagger}_{\rm H}(2) = h(2)\,\psi^{\dagger}_{\rm H}(2) + \psi^{\dagger}_{\rm H}(2)\int d3\,\psi^{\dagger}_{\rm H}(3)\,\psi_{\rm H}(3)\,v(3,2) \end{cases} \tag{11.22}$$

with the provision that the time component of the integrals runs over the contour C (although not explicitly indicated). These equations are formally identical to those occurring in the zero-temperature formalism [32], where, however, the time component of the integrals runs over the real axis from $-\infty$ to $+\infty$.

The results (11.22) can now be used in conjunction with Eqs. (11.13) and (11.14) to obtain the *equations of motion for the single-particle Green's function* on the contour of Eq. (11.15). These equations read:

$$\left[i\frac{d}{dz_1} - h(1)\right] G(1,2) \pm i\int d3\,v(1,3)\,G_2(1,3^{+};2,3^{++}) = \delta(1,2), \tag{11.23}$$

$$\left[-i\frac{d}{dz_2} - h(2)\right] G(1,2) \pm i\int d3\,v(2,3)\,G_2(1,3^{--};2,3^{-}) = \delta(1,2), \tag{11.24}$$

where G_2 is the two-particle Green's function on the contour defined in Eq. (7.13), $3^{\pm}$ implies that the time variable z_3 is augmented (diminished) by a positive infinitesimal along the contour C, and the upper (lower) sign refers to fermions (bosons).

11.4 The Contour Self-Energy Operator and the Differential Forms of the Dyson Equation

The presence of the two-particle Green's function G_2 in the equations (11.23) and (11.24) signals that, if we would go ahead and derive also the equations of motion for G_2, a whole hierarchy of equations of motion involving still higher-order Green's functions would be generated in this way. It is customary to avoid running into this hierarchy of equations and introduce instead the *self-energy operator* Σ as follows. This can be done in two different ways, by acting on either Eq. (11.23) or Eq. (11.24):

(i) The second term on the left-hand side of Eq. (11.23) can be rewritten in the form

$$\pm i \int d3\, v(1,3)\, G_2(1,3^+;2,3^{++}) = - \int d3\, \Sigma(1,3)\, G(3,2)\,, \tag{11.25}$$

such that Eq. (11.23) becomes

$$\left[i\frac{d}{dz_1} - h(1) \right] G(1,2) - \int d3\, \Sigma(1,3)\, G(3,2) = \delta(1,2)\,; \tag{11.26}$$

(ii) The second term on the left-hand side of Eq. (11.24) can be rewritten in the alternative form

$$\pm i \int d3\, v(2,3)\, G_2(1,3^{--};2,3^-) = - \int d3\, G(1,3)\, \bar{\Sigma}(3,2), \tag{11.27}$$

where in principle $\bar{\Sigma}$ differs from Σ, such that Eq. (11.24) becomes

$$\left[-i\frac{d}{dz_2} - h(2) \right] G(1,2) - \int d3\, G(1,3)\, \bar{\Sigma}(3,2) = \delta(1,2)\,. \tag{11.28}$$

We shall prove in Section 11.5 below that $\bar{\Sigma} = \Sigma$. With this result, Eqs. (11.26) and (11.28) are known as the *differential* (actually, integro-differential) *forms of the Dyson equation*.

Note how, on the right-hand side of Eq. (11.25), the self-energy Σ appears at the left of the single-particle Green's function G, while, on the right-hand side of Eq. (11.27), it appears at the right. This asymmeyry is due to the difference in the arguments of the interparticle interaction v entering the left-hand side of the expressions (11.25) and (11.27), in line with the reasoning followed in

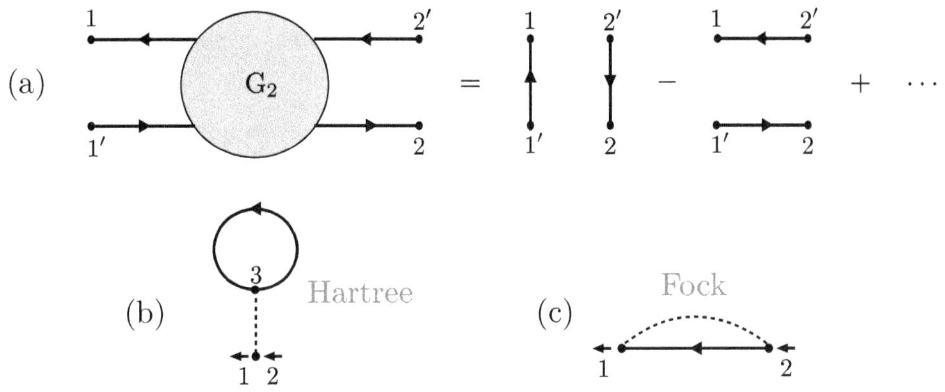

Figure 11.1 (a) Lowest-order diagrammatic contributions to the two-particle Green's function, which give rise, respectively, to (b) the Hartree term $\Sigma_{\mathrm{H}}(1,2)$ and (c) the Fock term $\Sigma_{\mathrm{F}}(1,2)$ of the single-particle self-energy (only the fermionic case is considered here).

Section 31.1 of Part II (cf. Fig. 31.1 therein), where Dyson equations like (11.26) and (11.28) are obtained in terms of the diagrammatic structure of Σ (which shows directly that $\bar{\Sigma} = \Sigma$).

Remark: Hartree and Fock self-energies

The two-particle Green's function on the contour (7.13) can be represented in diagrammatic terms according to the procedure described in Chapter 9. In particular, its lowest-order contributions originate from the Wick's factorization given in Eq. (9.28), whereby one identifies $\psi(1) \leftrightarrow c_{q_4}(\tau_4), \psi(2) \leftrightarrow c_{q_3}(\tau_3), \psi^{\dagger}(2') \leftrightarrow c_{q_2}(\tau_2), \psi^{\dagger}(1') \leftrightarrow c_{q_1}(\tau_1)$ and disregards the first term on the right-hand side of Eq. (9.28) because only the normal phase is considered here. In the fermionic case, we obtain accordingly

$$G_2(1,2;1',2') = G(1,1')\,G(2,2') \;-\; G(1,2')\,G(2,1') \;+\; \cdots \tag{11.29}$$

as depicted in Fig. 11.1(a). Substituting the expansion (11.29) into the left-hand side of Eq. (11.25) yields two distinct terms of the single-particle self-energy, namely, the Hartree term (cf. Fig. 11.1(b))

$$\Sigma_{\mathrm{H}}(1,2) = \delta(1,2)\,(-i)\int d3\; v(1,3)\,G(3,3^{+}) \tag{11.30}$$

and the Fock term (cf. Fig. 11.1(c))

$$\Sigma_{\mathrm{F}}(1,2) = i\,v(1,2)\,G(1,2^{+})\,, \tag{11.31}$$

respectively. By a similar token, substituting the expansion (11.29) into the left-hand side of Eq. (11.27), one readily verifies that $\bar{\Sigma}_{\mathrm{H}}(1,2) = \Sigma_{\mathrm{H}}(1,2)$, as given by the expression (11.30), as well as that $\bar{\Sigma}_{\mathrm{F}}(1,2) = \Sigma_{\mathrm{F}}(1,2)$, as given by the expression

(11.31). In conclusion, we have proved that $\bar{\Sigma}_{\mathrm{HF}} = \Sigma_{\mathrm{HF}}$, at least within the Hartree–Fock (HF) approximation. The general proof for the exact quantities is provided in Section 11.5.

11.5 The Non-interacting Single-Particle Green's Function and the Integral Form of the Dyson Equation

In the following, it proves useful to introduce the *non-interacting single-particle Green's function* $G_0(1,2)$, which by definition satisfies Eq. (11.26) with $\Sigma = 0$ and Eq. (11.28) with $\bar{\Sigma} = 0$:

$$\left[i\frac{d}{dz_1} - h(1) \right] G_0(1,2) = \left[-i\frac{d}{dz_2} - h(2) \right] G_0(1,2) = \delta(1,2) \,. \tag{11.32}$$

As anticipated at the end of Chapter 10, note that G_0 here contains the effects of the time-dependent external potential V_{ext}, which is included in the definition (11.1) of the single-particle Hamiltonian h.

This implies that the corresponding *inverse* Green's function is given by

$$G_0^{-1}(1,2) = \left[i\frac{d}{dz_1} - h(1) \right] \delta(1,2) = \left[-i\frac{d}{dz_2} - h(2) \right] \delta(1,2) \,. \tag{11.33}$$

Proof of Eq. (11.33)

The proof proceeds by assuming that Eq. (11.33) holds true and then looking at its internal consistency:

(i) Apply G_0^{-1} to G_0 from the left

$$\begin{aligned} \int d3\, G_0^{-1}(1,3)\, G_0(3,2) &= \int d3 \left[i\frac{d}{dz_1} - h(1) \right] \delta(1,3)\, G_0(3,2) \\ &= \left[i\frac{d}{dz_1} - h(1) \right] G_0(1,2) = \delta(1,2) \,; \end{aligned} \tag{11.34}$$

(ii) Apply G_0^{-1} to G_0 from the right

$$\begin{aligned} \int d3\, G_0(1,3)\, G_0^{-1}(3,2) &= \int d3\, G_0(1,3) \left[i\frac{d}{dz_3} - h(3) \right] \delta(3,2) \\ &\underset{\substack{\uparrow \\ \text{[by parts]}}}{=} \int d3\, \delta(3,2) \left[-i\frac{d}{dz_3} - h(3) \right] G_0(1,3) \\ &= \int d3\, \delta(3,2)\, \delta(1,3) = \delta(1,2) \,. \end{aligned} \tag{11.35}$$

[QED]

We are now in a position to prove that $\bar{\Sigma} = \Sigma$ (for which we follow Appendix B of Ref. [32]).

Proof that $\bar{\Sigma} = \Sigma$

To this end, we begin by rewriting in Eq. (11.26)

$$\left[i\frac{d}{dz_1} - h(1) \right] G(1,2) = \int d3\, G_0^{-1}(1,3)\, G(3,2) \tag{11.36}$$

as well as in Eq. (11.28)

$$\left[-i\frac{d}{dz_2} - h(2) \right] G(1,2) = \int d3\, G(1,3)\, G_0^{-1}(3,2)\,. \tag{11.37}$$

In this way, Eqs. (11.26) and (11.28) become

$$\int d3\ \left[G_0^{-1}(1,3) - \Sigma(1,3) \right]\ G(3,2) = \delta(1,2), \tag{11.38}$$

$$\int d3\, G(1,3)\ \left[G_0^{-1}(3,2) - \bar{\Sigma}(3,2) \right] = \delta(1,2)\,, \tag{11.39}$$

which imply that $G_0^{-1}(1,3) - \Sigma(1,3) = G^{-1}(1,3)$ is the inverse of $G(1,3)$ "from the left" and that $G_0^{-1}(3,2) - \bar{\Sigma}(3,2) = G^{-1}(3,2)$ is the inverse of $G(3,2)$ "from the right." To the extent that the inverse of $G(1,2)$ is unique, it then follows that $\bar{\Sigma} = \Sigma$. [QED]

The abovementioned result for $G^{-1}(1,2)$ also enables us to transform the differential form (11.26) of the Dyson equation into a corresponding *integral form of the Dyson equation*. This is achieved by multiplying the expression $G^{-1}(1,2) = G_0^{-1}(1,2) - \Sigma(1,2)$ by $G_0(3,1)$ from the left and by $G(2,4)$ from the right and then integrating over 1 and 2, yielding

$$G(1,2) = G_0(1,2) + \int d34\, G_0(1,3)\, \Sigma(3,4)\, G(4,2)\,. \tag{11.40}$$

Alternatively, we could multiply the expression $G^{-1}(1,2) = G_0^{-1}(1,2) - \Sigma(1,2)$ by $G(3,1)$ from the left and by $G_0(2,4)$ from the right, then integrate over 1 and 2, to obtain

$$G(1,2) = G_0(1,2) + \int d34\, G(1,3)\, \Sigma(3,4)\, G_0(4,2)\,. \tag{11.41}$$

The two equations (11.40) and (11.41) are represented graphically in Figs. 11.2(a) and 11.2(b), respectively. They are equivalent to each other, as it can be shown by expanding them formally in powers of the self-energy Σ. Recall that, by the present formulation, *all* time integrals, either in the integro-differential form (11.26) or in

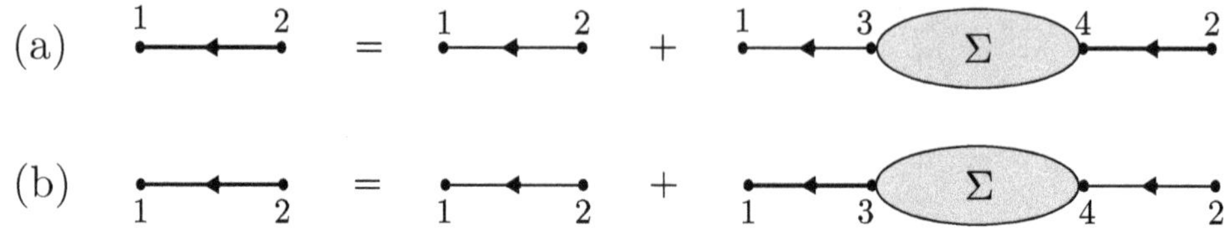

Figure 11.2 Graphical representation of the integral form of the Dyson equation, corresponding to (a) Eq. (11.40) and (b) Eq. (11.41). Thick and thin lines represent full (G) and bare (G_0) single-particle Green's functions, respectively, where the arrows point from the second to the first argument.

the integral form (11.40) of the Dyson equation, run over the contour C as specified at the outset.

Remark: The emergence of the "skeleton diagrams"

Explicit expressions for the self-energy Σ can be obtained from the diagrammatic perturbation theory on the basis of the integral equation (11.40), whereby the self-energy is identified with the sum of all one-particle *irreducible* diagrams that cannot be separated into two parts by cutting single G_0 lines. The diagrammatic contributions identified in this way can be transformed into "skeleton diagrams," by in turn identifying the single-particle lines within a given diagrammatic contribution to Σ with full single-particle Green's functions G instead of the bares G_0 (this is actually what was done in Fig. 11.1 where thick lines correspond to G).

Remark: "Conserving approximations" for the self-energy

Of particular importance, and especially in transport phenomena like in the context of the Landau theory of Fermi liquids [12], are the so-called conserving approximations for the self-energy, whose choice makes the approximate theory to satisfy the same conservation laws of the exact theory. The restrictions for a given approximation to be conserving were first formulated by Baym and Kadanoff [3] and then suitably cast in a diagrammatic context by Baym [4]. A concise summary of how the fulfillment of conservation laws can be formally enforced in approximate forms of the self-energy can be found in Ref. [32], whose zero-temperature formulation can be extended unchanged as well to the present more general case. We shall return to this formulation in Chapter 17, where, similar to Ref. [32], the diagrammatic structure of the Green's functions will be obtained through functional derivatives.

12

Kubo–Martin–Schwinger Boundary Conditions

It was shown in Chapter 11 that the contour single-particle Green's function $G(1,2)$ obeys two distinct integro-differential Dyson equations, given by Eq. (11.26), where the differential operator acts on the space and time variables 1, and by Eq. (11.28) (with $\bar{\Sigma} = \Sigma$), where the differential operator acts instead on the space and time variables 2. In nonequilibrium situations, these two equations contain in principle different information because, between the time z_1 of the first argument and the time z_2 of the second argument of $G(1,2)$, the external time-dependent potential $V_{\text{ext}}(\mathbf{r}, z)$ of Eq. (11.1) (or, when applicable, the time-dependent interparticle interaction $v(\mathbf{r}z, \mathbf{r}'z')$ of Eq. (11.21)) has in general evolved. This implies that, when solving the equations (11.26) and (11.28) of first order in time, two distinct boundary conditions (one for z_1 and the other for z_2) need to be specified.

In this chapter, we consider the boundary conditions on the time variables z_1 and z_2, which naturally emerge when the contour single-particle Green's function is quite generally represented in the form (cf. Eq. (7.12))

$$G(\mathbf{r}_1\sigma_1 z_1, \mathbf{r}_2\sigma_2 z_2) = -\,i\,\frac{\text{Tr}\left\{\mathcal{T}_C\left\{e^{-i\int_C dz\mathcal{H}(z)}\,\psi(\mathbf{r}_1\sigma_1 z_1)\,\psi^\dagger(\mathbf{r}_2\sigma_2 z_2)\right\}\right\}}{\text{Tr}\left\{\mathcal{T}_C\left\{e^{-i\int_C dz\mathcal{H}(z)}\right\}\right\}}\,. \tag{12.1}$$

In principle, C is here a *generic* contour in the complex-time z-plane on which the time-ordering operator $\mathcal{T}_C$ acts, starting at z_i and ending at z_f (provided, of course, that one can specify the dynamics contained in $\mathcal{H}(z)$ along C). In practice, commonly used shapes for C of physical interest will be considered. On the other hand, the boundary conditions on the spatial variables $\mathbf{r}_1$ and $\mathbf{r}_2$ of $G(1,2)$ will depend on the particular physical problem one is considering and need thus not be specified at the outset in the following considerations.

12.1 Boundary Conditions for the Single-Particle Green's Function on a Generic Contour C

The (Kubo–Martin–Schwinger) boundary conditions on the time variables z_1 and z_2 read

$$G(\mathbf{r}_1\sigma_1 z_i, \mathbf{r}_2\sigma_2 z_2) = \pm\, G(\mathbf{r}_1\sigma_1 z_f, \mathbf{r}_2\sigma_2 z_2), \tag{12.2}$$

$$G(\mathbf{r}_1\sigma_1 z_1, \mathbf{r}_i\sigma_2 z_i) = \pm\, G(\mathbf{r}_1\sigma_1 z_1, \mathbf{r}_2\sigma_2 z_f), \tag{12.3}$$

where the upper (lower) sign refers to bosons (fermions). The proof of Eqs. (12.2) and (12.2) is based on the cyclic invariance of the trace and on the definition of the time-ordering operator $\mathcal{T}_C$ along C.

Proof of Eq. (12.2)

In terms of the representation (12.1), we write

$$\begin{aligned}
&\mathrm{Tr}\left\{\mathcal{T}_C\left\{e^{-i\int_C dz \mathcal{H}(z)}\right\}\right\} i\, G(\mathbf{r}_1\sigma_1 z_f, \mathbf{r}_2\sigma_2 z_2) \\
&= \mathrm{Tr}\left\{\mathcal{T}_C\left\{e^{-i\int_C dz \mathcal{H}(z)}\, \psi(\mathbf{r}_1\sigma_1 z_f)\, \psi^\dagger(\mathbf{r}_2\sigma_2 z_2)\right\}\right\} \\
&= \mathrm{Tr}\left\{\psi(\mathbf{r}_1\sigma_1)\, \mathcal{T}_C\left\{e^{-i\int_C dz \mathcal{H}(z)}\, \psi^\dagger(\mathbf{r}_2\sigma_2 z_2)\right\}\right\} \\
&= \mathrm{Tr}\left\{\mathcal{T}_C\left\{e^{-i\int_C dz \mathcal{H}(z)}\, \psi^\dagger(\mathbf{r}_2\sigma_2 z_2)\right\} \psi(\mathbf{r}_1\sigma_1)\right\} \\
&= \mathrm{Tr}\left\{\mathcal{T}_C\left\{e^{-i\int_C dz \mathcal{H}(z)}\, \psi^\dagger(\mathbf{r}_2\sigma_2 z_2)\, \psi(\mathbf{r}_1\sigma_1 z_i)\right\}\right\} \\
&= \pm\, \mathrm{Tr}\left\{\mathcal{T}_C\left\{e^{-i\int_C dz \mathcal{H}(z)}\, \psi(\mathbf{r}_1\sigma_1 z_i)\, \psi^\dagger(\mathbf{r}_2\sigma_2 z_2)\right\}\right\},
\end{aligned} \tag{12.4}$$

where:

(i) in the first line, the definition of $G(\mathbf{r}_1\sigma_1 z_f, \mathbf{r}_2\sigma_2 z_2)$ has been utilized;
(ii) in the second line, the operator $\psi(\mathbf{r}_1\sigma_1 z_f)$ at the latest time z_f on the contour C has been brought to the left of the time-ordering operator $\mathcal{T}_C$ and accordingly spoiled of its fictitious time dependence (cf. the discussion of point (iii) after Eq. (6.6));
(iii) in the third line, the cyclic invariance of the trace has been exploited;
(iv) in the fourth line, the operator $\psi(\mathbf{r}_1\sigma_1)$ now standing at the right of the time-ordering operator $\mathcal{T}_C$ has been endowed of the earliest time z_i on the contour C and then brought inside $\mathcal{T}_C$; and
(v) in the fifth line, the bosonic (fermionic) nature of the field operators has been taken into account, with a minus sign appearing when interchanging fermionic operators within the time-ordered product.

The result (12.2) then follows. [QED]

Proof of Eq. (12.3)

Again in terms of the representation (12.1), we write

$$
\begin{aligned}
&\mathrm{Tr}\left\{\mathcal{T}_C\left\{e^{-i\int_C dz\mathcal{H}(z)}\right\}\right\} i\,G(\mathbf{r}_1\sigma_1 z_1, \mathbf{r}_2\sigma_2 z_f) \\
&= \mathrm{Tr}\left\{\mathcal{T}_C\left\{e^{-i\int_C dz\mathcal{H}(z)}\,\psi(\mathbf{r}_1\sigma_1 z_1)\,\psi^\dagger(\mathbf{r}_2\sigma_2 z_f)\right\}\right\} \\
&= \pm\,\mathrm{Tr}\left\{\mathcal{T}_C\left\{e^{-i\int_C dz\mathcal{H}(z)}\,\psi^\dagger(\mathbf{r}_2\sigma_2 z_f)\,\psi(\mathbf{r}_1\sigma_1 z_1)\right\}\right\} \\
&= \pm\,\mathrm{Tr}\left\{\psi^\dagger(\mathbf{r}_2\sigma_2)\,\mathcal{T}_C\left\{e^{-i\int_C dz\mathcal{H}(z)}\,\psi(\mathbf{r}_1\sigma_1 z_1)\right\}\right\} \\
&= \pm\,\mathrm{Tr}\left\{\mathcal{T}_C\left\{e^{-i\int_C dz\mathcal{H}(z)}\,\psi(\mathbf{r}_1\sigma_1 z_1)\right\}\psi^\dagger(\mathbf{r}_2\sigma_2)\right\} \\
&= \pm\,\mathrm{Tr}\left\{\mathcal{T}_C\left\{e^{-i\int_C dz\mathcal{H}(z)}\,\psi(\mathbf{r}_1\sigma_1 z_1)\,\psi^\dagger(\mathbf{r}_2\sigma_2, z_i)\right\}\right\},
\end{aligned}
\tag{12.5}
$$

where:

(i) in the first line, the definition of $G(\mathbf{r}_1\sigma_1 z_1, \mathbf{r}_2\sigma_2 z_f)$ has been utilized;
(ii) in the second line, the field operators have been interchanged within the time-ordering operator, with a minus sign occurring for fermionic operators;
(iii) in the third line, the operator $\psi^\dagger(\mathbf{r}_2\sigma_2 z_f)$ at the latest time z_f on the contour C has been brought to the left of the time-ordering operator $\mathcal{T}_C$ and accordingly spoiled of its fictitious time dependence (cf. the discussion of point (iii) after Eq. (6.6));
(iv) in the fourth line, the cyclic invariance of the trace has been exploited;
(v) in the fifth line, the operator $\psi^\dagger(\mathbf{r}_2\sigma_2)$ now standing at the right of the time-ordering operator $\mathcal{T}_C$ has been endowed of the earliest time z_i on the contour C and then brought inside $\mathcal{T}_C$.

The result (12.3) then follows. [QED]

Properties similar to Eqs. (12.2) and (12.3) hold also for higher-order contour Green's functions (and, in particular, for the contour two-particle Green's function (7.13)).

12.2 Particular Contours of Interest

The commonly used shapes of physical interest for C are as follows.

Matsubara contour

The simplest contour of interest when restricting to equilibrium situations is the vertical track γ^{M}, whereby $z_i = 0$ and $z_f = -i\beta$ with $\beta = (k_B T)^{-1}$ the inverse

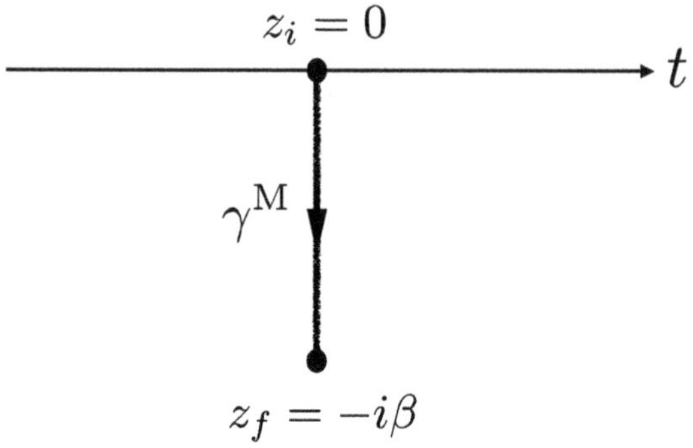

Figure 12.1 The Matsubara contour runs along the imaginary time axis from $z_i = 0$ to $z_f = -i\beta$.

temperature (cf. Fig. 12.1). In this case, both variables z_1 and z_2 in the expression (12.1) lie on γ^{M}. Accordingly, the Heisenberg representation (11.3) of the field operator on γ^{M} reads

$$\psi(\mathbf{r}_1\sigma_1 z_1) = U(z_i, z_1)\,\psi(\mathbf{r}_1\sigma_1)\,U(z_1, z_i) = e^{\tau_1 H^{\mathrm{M}}}\,\psi(\mathbf{r}_1\sigma_1)\,e^{-\tau_1 H^{\mathrm{M}}}, \qquad (12.6)$$

where we have set $z_1 - z_i = -i\tau_1$ and $H^{\mathrm{M}} = H - \mu N$ on the vertical track γ^{M} (cf. Section 8.1), and similarly $\psi^\dagger(\mathbf{r}_1\sigma_1 z_1) = U(z_i, z_1)\,\psi^\dagger(\mathbf{r}_1\sigma_1)\,U(z_1, z_i) = e^{\tau_1 H^{\mathrm{M}}}\,\psi^\dagger(\mathbf{r}_1\sigma_1)\,e^{-\tau_1 H^{\mathrm{M}}}$ for the adjoint field operator.

The single-particle (now referred to as "temperature" (or Matsubara)) Green's function is then homogeneous in the imaginary time τ and, due to the periodic (antiperiodic) boundary conditions (12.2) and (12.3) for boson (fermions), can be expanded in Fourier series

$$\begin{aligned} G(\mathbf{r}_1\sigma_1 z_1, \mathbf{r}_2\sigma_2 z_2) &\to G(\mathbf{r}_1\sigma_1, \mathbf{r}_2\sigma_2; \tau_1 - \tau_2) \\ &= \frac{i}{\beta}\sum_n e^{-i\omega_n(\tau_1-\tau_2)}\, G(\mathbf{r}_1\sigma_1, \mathbf{r}_2\sigma_2; \omega_n) \end{aligned} \qquad (12.7)$$

for $0 \le (\tau_1 - \tau_2) \le \beta$, where the Matsubara frequencies [6, 7, 35]

$$\omega_n = \begin{cases} 2n\,\frac{\pi}{\beta} & \text{for bosons} \\ (2n+1)\,\frac{\pi}{\beta} & \text{for fermions} \end{cases} \qquad (12.8)$$

(n integer) are different for bosons and fermions.

Konstantinov–Perel contour

In nonequilibrium situations, it is often the case that the initial preparation of the system at time $z_i = t_0$ corresponds to a thermal equilibrium described by the density

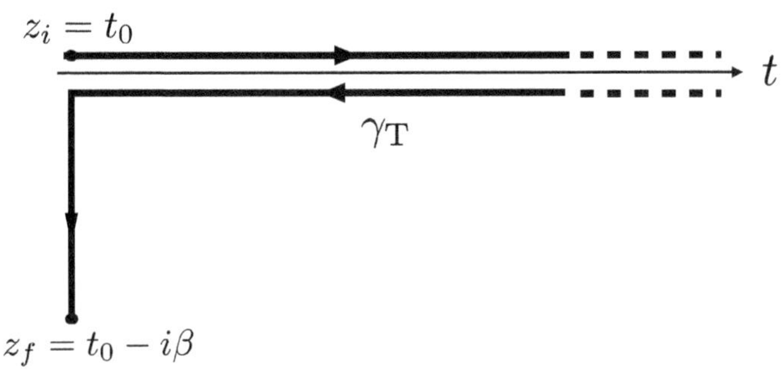

Figure 12.2 The Konstantinov–Perel contour coincides with the total contour γ_T of Fig. 8.1.

matrix $\rho = e^{-\beta H^M}/\mathrm{Tr}\{e^{-\beta H^M}\}$. The contour C in Eq. (12.1) thus becomes the total contour γ_T of Fig. 8.1, which includes the vertical track such that $z_f = t_0 - i\beta$ (cf. Fig. 12.2).

In this case, the information about the initial thermal equilibrium contained in the vertical part (or track) of γ_T is suitably transferred to the horizontal (forward and backward) branches of γ_T, in which the dynamical (time-dependent) perturbation of interest is acting, by considering the "mixed" configurations where one of the variables z_1 and z_2 of the contour single-particle Green's function belongs to the vertical track and the other to one of the horizontal branches of γ_T. We shall return to this important point in Chapter 15 when discussing the Kadanoff–Baym equations.

Keldysh contour

In this case, both z_i and z_f recede to $-\infty$ and the initial configuration reduces to the thermal equilibrium of a noninteracting system (cf. Fig. 12.3 and Eq. (8.9)). Accordingly, the boundary conditions (12.2) and (12.3) do not provide any new relevant information.

Kadanoff–Baym contour

In this case, $z_i = 0$ and $z_f = -i\beta$ like for the Matsubara contour (cf. Fig. 12.4), with the difference, however, that an external perturbation that depends on the imaginary time is now acting on this vertical contour [5]. Analytic continuation from imaginary to real times eventually restores the relevant physical conditions [5]. In Part II, we shall utilize this "Kadanoff–Baym" device for formal considerations.

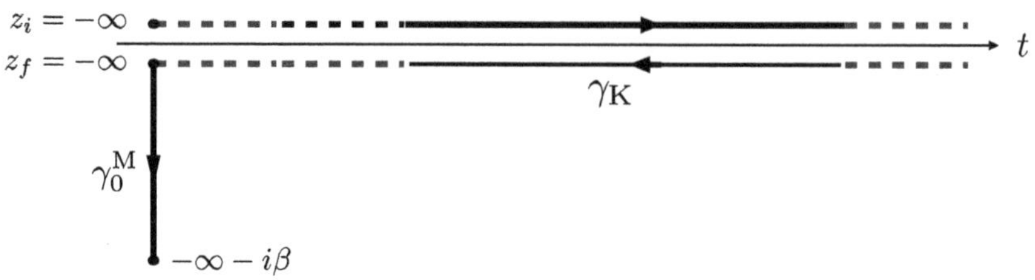

Figure 12.3 The Keldysh contour with the vertical track γ^M receding back in the far past.

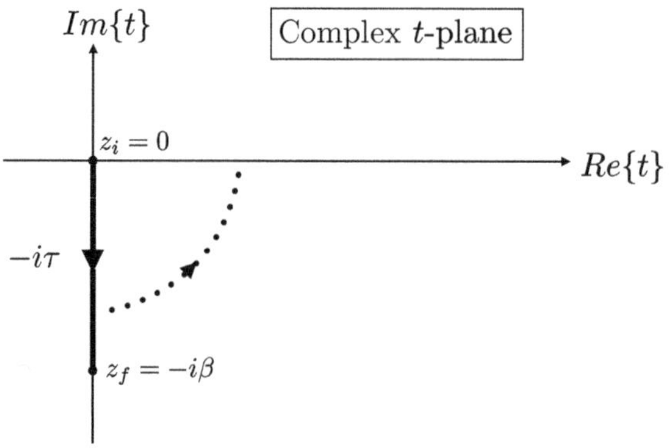

Figure 12.4 The Kadanoff–Baym contour formally coincides with the Matsubara contour.

Remark: Boundary conditions for the integral form of the Dyson equation

The Kubo–Martin–Schwinger boundary conditions (12.2) and (12.3), respectively, for the left (z_1) and right (z_2) time variables of the contour single-particle Green's function $G(1, 2)$ can be readily applied to the integral form of the Dyson equation (11.40) and (11.41) directly in terms of its non-interacting counterpart $G_0(1, 2)$. This is because in Eq. (11.40) the left time variable z_1 involved in the boundary condition (12.2) enters only in $G_0(1, 2)$ and $G_0(1, 3)$, while in Eq. (11.41) the right time variable z_2 involved in the boundary condition (12.3) enters only in $G_0(1, 2)$ and $G_0(4, 2)$.

13

Converting Contour-Time to Real-Time Arguments

The nonequilibrium diagrammatics of Chapter 10 and the Dyson equations of Chapter 11 contain integrations over the time variables that run over the generic contour C considered in Chapter 9, which comprises the total contour γ_T of Fig. 8.1 and the Keldysh contour γ_K of Fig. 8.2(a). In both cases, the time variables run first forward and then backward along the ordinary time axis, which is, in practice, an inconvenient procedure when calculating the time integrals. It is thus required to convert these time integrals into ordinary time integrals. To this end, it is first necessary to single out all possible combinations of the pair of time variables in the contour single-particle Green's function (12.1) (where, for simplicity, we lump the space $\mathbf{r}$ and spin σ variables into the single variable $\mathbf{x} = (\mathbf{r}, \sigma)$). This will be done in this chapter, by considering quite generally the total contour γ_T and splitting it into three branches, as shown in Fig. 13.1 (while for the Keldysh contour γ_K, two branches are sufficient). The properties ensuing from this splitting will be considered in detail.

13.1 The Single-Particle Green's Function in a 3 × 3 Matrix Form

Let $G(\mathbf{x}z, \mathbf{x}'z')$ be the contour single-particle Green's function, where the time variables z and z' lie on the contour of Fig. 13.1. Because this contour is made up of three branches (namely, the forward branch C_1, the backward branch C_2, and the vertical track C_3), the Green's function has 3×3 components

$$G_{ij}(\mathbf{x}z, \mathbf{x}'z') \quad \text{with } z \in C_i \text{ and } z' \in C_j, \tag{13.1}$$

where $(i, j) = (1, 2, 3)$. In particular, when $(i, j) = (1, 2)$, these are standard real times, while $(i, j) = 3$ corresponds to an imaginary time. We can thus conventionally represent these components in a 3×3 matrix form

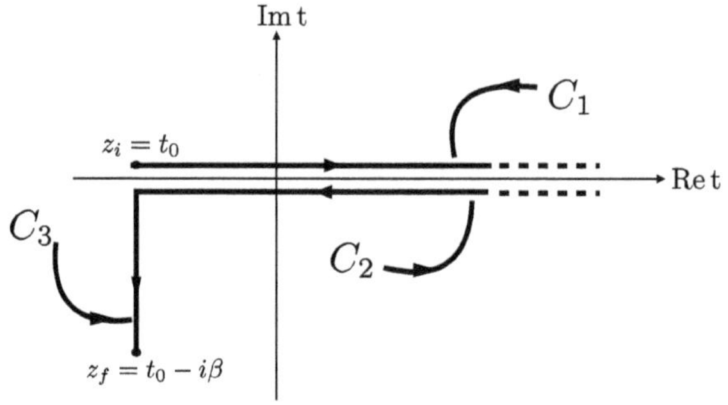

Figure 13.1 The total contour γ_T of Fig. 8.1 is split into three branches such that $\gamma_T = C_1 + C_2 + C_3$, where C_1 runs from t_0 to $+\infty$, C_2 from $+\infty$ back to t_0, and C_3 from t_0 to $t_0 - i\beta$ parallel to the negative imaginary time axis (t_0 being as usual the reference time).

$$\hat{G} = \begin{bmatrix} G_{11} & G_{12} & G_{13} \\ G_{21} & G_{22} & G_{23} \\ G_{31} & G_{32} & G_{33} \end{bmatrix}, \tag{13.2}$$

where the space and spin coordinates **x** are implicit. We next consider the various components of this matrix separately (Figs. 13.2–13.11).

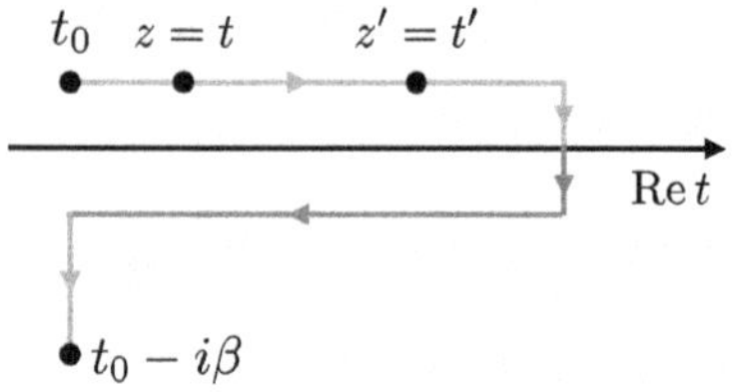

Figure 13.2 Graphical view of Eq. (13.3).

Component 11

$$G_{11}(\mathbf{x}t, \mathbf{x}'t') = -i \left\langle \mathcal{T}\left\{\psi_H(\mathbf{x}t)\, \psi_H^\dagger(\mathbf{x}'t')\right\}\right\rangle, \tag{13.3}$$

where $\mathcal{T}$ is the time-ordering operator introduced in Chapter 2.

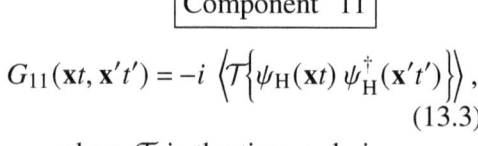

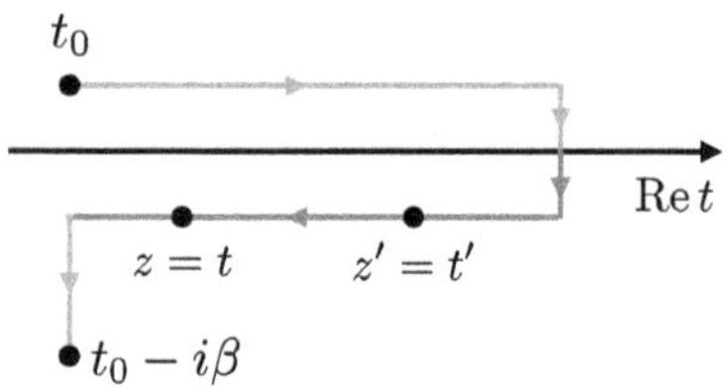

Figure 13.3 Graphical view of Eq. (13.4).

Component 22

$$G_{22}(\mathbf{x}t, \mathbf{x}'t') = -i \left\langle \tilde{\mathcal{T}}\left\{\psi_H(\mathbf{x}t)\, \psi_H^\dagger(\mathbf{x}'t')\right\}\right\rangle, \tag{13.4}$$

where $\tilde{\mathcal{T}}$ is the anti-time-ordering operator introduced in Chapter 2.

Component 12

$$G_{12}(\mathbf{x}t, \mathbf{x}'t') = \mp i \langle \psi_{\mathrm{H}}^{\dagger}(\mathbf{x}'t')\, \psi_{\mathrm{H}}(\mathbf{x}t) \rangle$$
$$\equiv G^{<}(\mathbf{x}t, \mathbf{x}'t') \qquad (13.5)$$

with the minus (plus) sign for bosons (fermions), which is called the *lesser Green's function* (with reference to the way the operators occur in the thermal average and not to the relative values of the real times t and t').

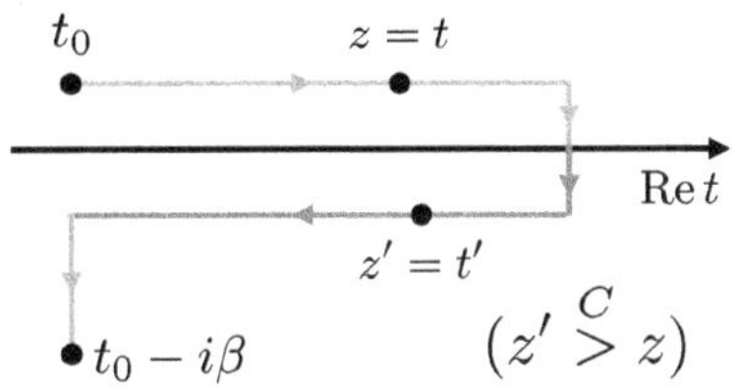

Figure 13.4 Graphical view of Eq. (13.5).

Component 21

$$G_{21}(\mathbf{x}t, \mathbf{x}'t') = -i \langle \psi_{\mathrm{H}}(\mathbf{x}t)\, \psi_{\mathrm{H}}^{\dagger}(\mathbf{x}'t') \rangle$$
$$\equiv G^{>}(\mathbf{x}t, \mathbf{x}'t'), \qquad (13.6)$$

which is called the *greater Green's function* (again with reference to the way the operators occur in the thermal average).

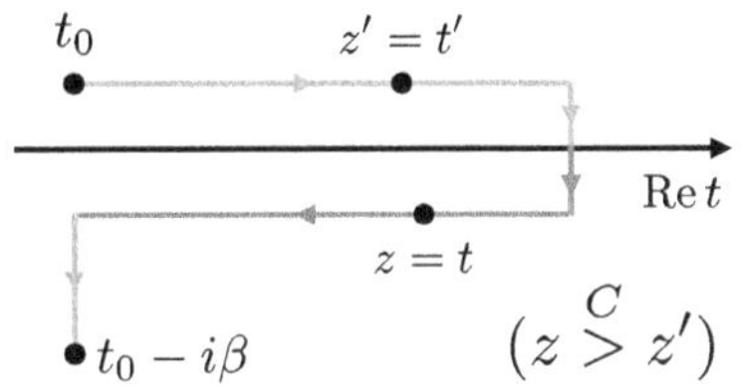

Figure 13.5 Graphical view of Eq. (13.6).

Component 13

$$G_{13}(\mathbf{x}t, \mathbf{x}'t_0 - i\tau')$$
$$= \mp i \langle \psi_{\mathrm{H}}^{\dagger}(\mathbf{x}'t_0 - i\tau')\, \psi_{\mathrm{H}}(\mathbf{x}t) \rangle$$
$$\equiv G^{\rceil}(\mathbf{x}t, \mathbf{x}'\tau') \qquad (13.7)$$

with the minus (plus) sign for bosons (fermions). This is called the *right Keldysh component* (where "right" refers to the position of the vertical segment of the hook $\rceil$ with respect to the horizontal segment). Recall that $\psi_{\mathrm{H}}^{\dagger}(\mathbf{x}'t_0 - i\tau')$ with time on the vertical segment is given by Eq. (12.6).

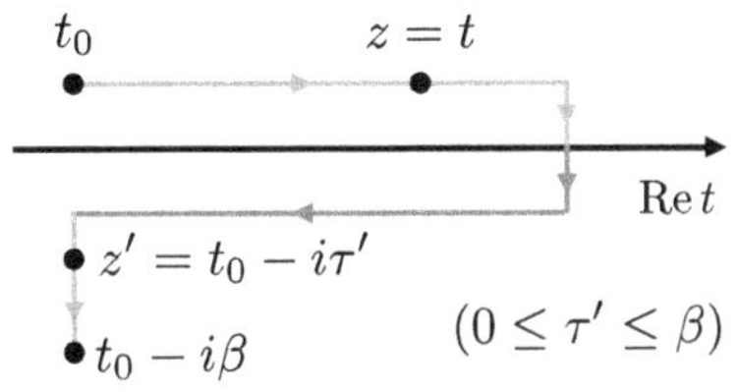

Figure 13.6 Graphical view of Eq. (13.7).

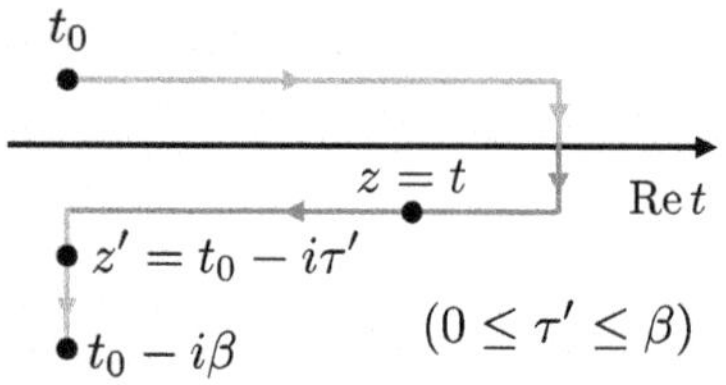

Figure 13.7 Graphical view of Eq. (13.8).

Component 23

$$\begin{aligned} G_{23}(\mathbf{x}t, \mathbf{x}'t_0 - i\tau') &= \mp i \langle \psi_{\mathrm{H}}^{\dagger}(\mathbf{x}'t_0 - i\tau')\, \psi_{\mathrm{H}}(\mathbf{x}t)\rangle \\ &= G_{13}(\mathbf{x}t, \mathbf{x}'t_0 - i\tau') \\ &\equiv G^{\rceil}(\mathbf{x}t, \mathbf{x}'\tau'). \end{aligned} \tag{13.8}$$

This is because, owing to the property (7.2), the time $z = t$ can be moved from the backward branch C_2 to the forward branch C_1, as shown in Fig. 13.8.

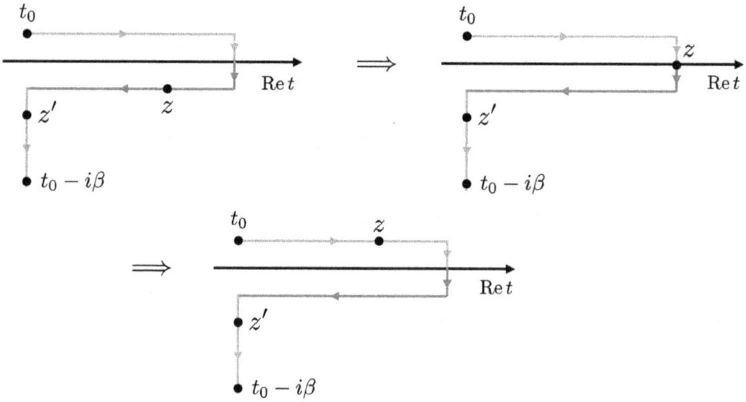

Figure 13.8 The way how the "closed-loop" property (7.2) works for the expression (13.8).

Component 31

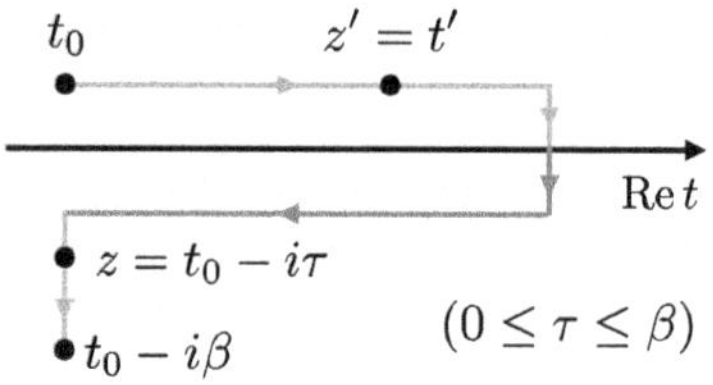

Figure 13.9 Graphical view of Eq. (13.9).

$$\begin{aligned} G_{31}(\mathbf{x}t_0 - i\tau, \mathbf{x}'t') &= -i \langle \psi_{\mathrm{H}}(\mathbf{x}t_0 - i\tau)\, \psi_{\mathrm{H}}^{\dagger}(\mathbf{x}'t')\rangle \\ &\equiv G^{\lceil}(\mathbf{x}\tau, \mathbf{x}'t'), \end{aligned} \tag{13.9}$$

which is called the *left Keldysh component* (where "left" refers to the position of the vertical segment of the hook $\lceil$ with respect to the horizontal segment). Recall that $\psi_{\mathrm{H}}(\mathbf{x}t_0 - i\tau)$ with time on the vertical segment is given by Eq. (12.6).

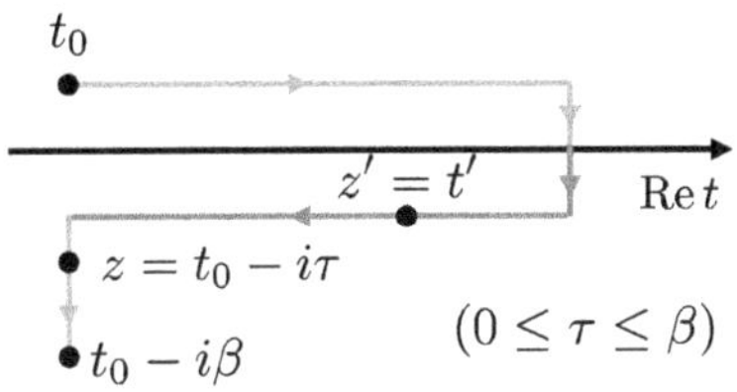

Figure 13.10 Graphical view of Eq. (13.10).

Component 32

$$\begin{aligned} &G_{32}(\mathbf{x}t_0 - i\tau, \mathbf{x}'t') \\ &\quad = -i\,\langle\, \psi_{\mathrm{H}}(\mathbf{x}t_0 - i\tau)\, \psi^{\dagger}_{\mathrm{H}}(\mathbf{x}'t')\rangle \\ &\quad = G_{31}(\mathbf{x}t_0 - i\tau, \mathbf{x}'t') \\ &\quad \equiv G^{\lceil}(\mathbf{x}\tau, \mathbf{x}'t')\,. \end{aligned} \tag{13.10}$$

Again, this is because the time $z' = t'$ can be moved with no harm from the backward branch C_2 to the forward branch C_1, similar to what done in Fig. 13.8.

Component 33

$$\begin{aligned} &G_{33}(\mathbf{x}t_0 - i\tau, \mathbf{x}'t_0 - i\tau') \\ &\quad = -i\,\left\langle \mathcal{T}_\tau \left\{ \psi_{\mathrm{H}}(\mathbf{x}t_0 - i\tau)\psi^{\dagger}_{\mathrm{H}}(\mathbf{x}'t_0 - i\tau')\right\}\right\rangle, \end{aligned} \tag{13.11}$$

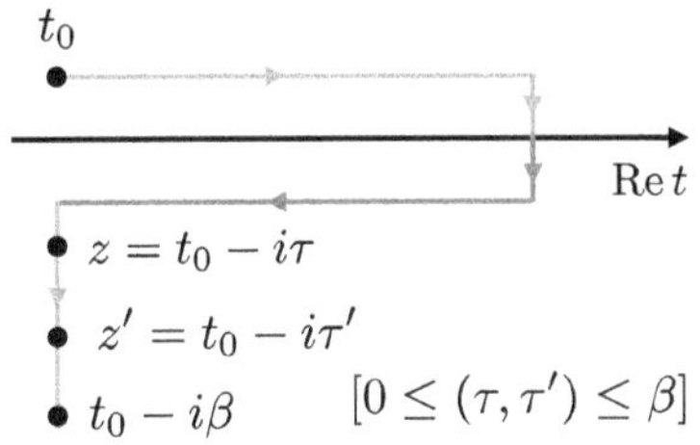

Figure 13.11 Graphical view of Eq. (13.11).

where $\mathcal{T}_\tau$ is the time-ordering operator along the imaginary time axis [7] (Figs. 13.9–13.11). As mentioned in Section 12.2, when both z and z' belong to the vertical track C_3, one speaks of "temperature" (or *Matsubara*) Green's function. In addition, a standard notation for thermodynamic equilibrium [7] would identify this function with

$$\begin{aligned} &G^{\mathrm{M}}_{\mathrm{FW}}(\mathbf{x}\tau, \mathbf{x}'\tau') \\ &\quad = -i\, G_{33}(\mathbf{x}t_0 - i\tau, \mathbf{x}'t_0 - i\tau')\,. \end{aligned} \tag{13.12}$$

Note that the component 33 plays a separate and special role in the theory because it contains the grand-canonical Hamiltonian H^{M} and is invariant with respect to a common shift of τ and τ' (i.e., it is a function of the difference $\tau - \tau'$ as in Eq. (12.7)).

13.2 Redundancy Conditions: Advanced, Retarded, and Keldysh Green's Functions

The possibility of moving one of the time variables from the forward branch C_1 to the backward branch C_2 (or viceversa) under appropriate circumstances (as in Fig. 13.8) enables us to reveal additional *redundancy* between the components 11, 12, 21, and 22 of the matrix $\hat{G}$ of Eq. (13.2). In this way, we will be able to eliminate three out of the nine components of the matrix (13.2), thus ending up with six independent physical Green's functions. The proof proceeds as indicated schematically in Fig. 13.12.

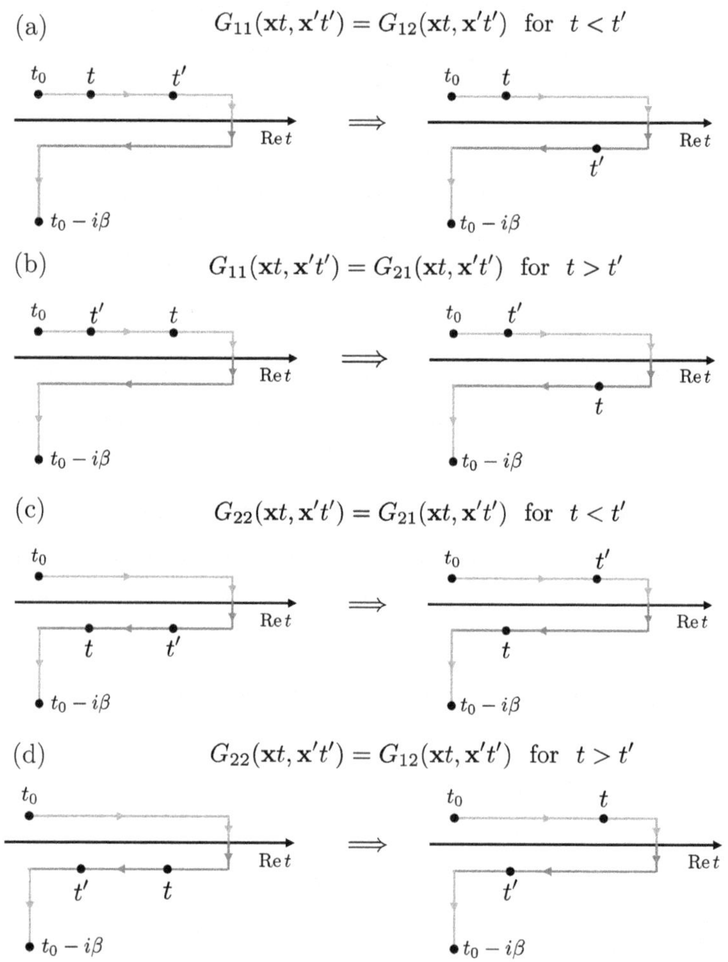

Figure 13.12 (a–d) Redundancy between the components 11, 12, 21, and 22 of the matrix $\hat{G}$ of Eq. (13.2).

Note that the equal sign that appears in (a) and (d) is consistent with the convention that, for equal times, $\psi^\dagger$ occurs at the left of ψ (cf. the "rule # 4" of Chapter 10). The four relations (a)–(d) can be further summarized, by writing

$$G_{11}(\mathbf{x}t, \mathbf{x}'t') + G_{22}(\mathbf{x}t, \mathbf{x}'t') = \begin{cases} G_{12}(\mathbf{x}t, \mathbf{x}'t') + G_{21}(\mathbf{x}t, \mathbf{x}'t') & \text{for } t < t' , \\ \\ G_{21}(\mathbf{x}t, \mathbf{x}'t') + G_{12}(\mathbf{x}t, \mathbf{x}'t') & \text{for } t > t' , \end{cases} \tag{13.13}$$

where violation of this relation for $t = t'$ is negligible over the whole time evolution. As anticipated, the properties (13.8), (13.10), and (13.13) allow us to eliminate three out of the nine components of the matrix (13.2), thus ending up with six independent components.

In particular, the four components (G_{11}, G_{12}, G_{21}, and G_{22}), which are linearly dependent according to Eq. (13.13), can be manipulated as follows:

$$\begin{aligned} \begin{bmatrix} G_{11} & G_{12} \\ G_{21} & G_{22} \end{bmatrix} &\longrightarrow \begin{bmatrix} 1 & 0 \\ 0 & -1 \end{bmatrix} \begin{bmatrix} G_{11} & G_{12} \\ G_{21} & G_{22} \end{bmatrix} = \begin{bmatrix} G_{11} & G_{12} \\ -G_{21} & -G_{22} \end{bmatrix} \\ &\longrightarrow \frac{1}{\sqrt{2}} \begin{bmatrix} 1 & -1 \\ 1 & 1 \end{bmatrix} \begin{bmatrix} G_{11} & G_{12} \\ -G_{21} & -G_{22} \end{bmatrix} \frac{1}{\sqrt{2}} \begin{bmatrix} 1 & 1 \\ -1 & 1 \end{bmatrix} \\ &= \frac{1}{2} \begin{bmatrix} G_{11} - G_{12} + G_{21} - G_{22} & , \; G_{11} + G_{12} + G_{21} + G_{22} \\ G_{11} - G_{12} - G_{21} + G_{22} & , \; G_{11} + G_{12} - G_{21} - G_{22} \end{bmatrix} . \end{aligned} \tag{13.14}$$

Repeated use of the property (13.13) then yields the following results for the elements of the matrix on the right-hand side of Eq. (13.14) (where first the time variables and then also the space–spin variables are restored):

Element 11 $\Longrightarrow$ The *retarded Green's function* G^{R}

$$\begin{aligned} &\frac{1}{2} \left(G_{11}(t,t') - G_{12}(t,t') + G_{21}(t,t') - G_{22}(t,t') \right) \underset{[\text{Eq.}(13.13)]}{\overset{\uparrow}{=}} G_{11}(t,t') - G_{12}(t,t') \\ \longrightarrow \;& \theta(t-t') \left(-i \, \langle \psi_{\mathrm{H}}(\mathbf{x}t) \, \psi^\dagger_{\mathrm{H}}(\mathbf{x}'t') \rangle \pm i \, \langle \psi^\dagger_{\mathrm{H}}(\mathbf{x}'t') \, \psi_{\mathrm{H}}(\mathbf{x}t) \rangle \right) \\ + \;& \theta(t'-t) \left(\mp i \, \langle \psi^\dagger_{\mathrm{H}}(\mathbf{x}'t') \, \psi_{\mathrm{H}}(\mathbf{x}t) \rangle \pm i \, \langle \psi^\dagger_{\mathrm{H}}(\mathbf{x}'t') \, \psi_{\mathrm{H}}(\mathbf{x}t) \rangle \right) \\ = \;& -i \, \theta(t-t') \left\langle \left[\psi_{\mathrm{H}}(\mathbf{x}t), \psi^\dagger_{\mathrm{H}}(\mathbf{x}'t') \right]_{\mp} \right\rangle \equiv G^{\mathrm{R}}(\mathbf{x}t, \mathbf{x}'t'), \end{aligned} \tag{13.15}$$

where the upper (lower) sign refers to bosons (fermions). The expression (13.15) is referred to as the *retarded Green's function*.

Element 12 $\Longrightarrow$ The *Keldysh Green's function* G^{K}

$$\frac{1}{2}\left(G_{11}(t,t')+G_{12}(t,t')+G_{21}(t,t')+G_{22}(t,t')\right) \underset{\substack{\uparrow \\ [\text{Eq.}(13.13)]}}{=} G_{12}(t,t')+G_{21}(t,t')$$

$$\begin{aligned} \longrightarrow \quad & \mp i\,\langle\psi_{\mathrm{H}}^{\dagger}(\mathbf{x}'t')\,\psi_{\mathrm{H}}(\mathbf{x}t)\rangle - i\,\langle\psi_{\mathrm{H}}(\mathbf{x}t)\,\psi_{\mathrm{H}}^{\dagger}(\mathbf{x}'t')\rangle \\ = \quad & -i\left\langle\left[\psi_{\mathrm{H}}(\mathbf{x}t),\psi_{\mathrm{H}}^{\dagger}(\mathbf{x}'t')\right]_{\pm}\right\rangle \equiv G^{\mathrm{K}}(\mathbf{x}t,\mathbf{x}'t'), \end{aligned} \tag{13.16}$$

where again the upper (lower) sign refers to bosons (fermions). The expression (13.16) is referred to as the *Keldysh Green's function.*

Element 22 $\Longrightarrow$ The *advanced Green's function* G^{A}

$$\frac{1}{2}\left(G_{11}(t,t')+G_{12}(t,t')-G_{21}(t,t')-G_{22}(t,t')\right) \underset{\substack{\uparrow \\ [\text{Eq.}(13.13)]}}{=} G_{11}(t,t')-G_{21}(t,t')$$

$$\begin{aligned} \longrightarrow \quad & \theta(t-t')\left(-i\,\langle\psi_{\mathrm{H}}(\mathbf{x}t)\,\psi_{\mathrm{H}}^{\dagger}(\mathbf{x}'t')\rangle + i\,\langle\psi_{\mathrm{H}}(\mathbf{x}t)\,\psi_{\mathrm{H}}^{\dagger}(\mathbf{x}'t')\rangle\right) \\ & +\theta(t'-t)\left(\mp i\,\langle\psi_{\mathrm{H}}^{\dagger}(\mathbf{x}'t')\,\psi_{\mathrm{H}}(\mathbf{x}t)\rangle + i\,\langle\psi_{\mathrm{H}}(\mathbf{x}t)\,\psi_{\mathrm{H}}^{\dagger}(\mathbf{x}'t')\rangle\right) \\ = \quad & i\,\theta(t'-t)\left\langle\left[\psi_{\mathrm{H}}(\mathbf{x}t),\psi_{\mathrm{H}}^{\dagger}(\mathbf{x}'t')\right]_{\mp}\right\rangle \equiv G^{\mathrm{A}}(\mathbf{x}t,\mathbf{x}'t'), \end{aligned} \tag{13.17}$$

where once again the upper (lower) sign refers to bosons (fermions). The expression (13.17) is referred to as the *advanced Green's function.*

Finally, the element 21 of the matrix on the right-hand side of Eq. (13.14) vanishes owing to the property (13.13). In this way, we have shown that we can replace the four linearly dependent functions (G_{11}, G_{12}, G_{21}, and G_{22}) by the three independent functions (G^{R}, G^{K}, and G^{A}).

Remark: Lesser $G^{<}$ and greater $G^{>}$ functions also expressed in terms of ($G^{\mathrm{R}}, G^{\mathrm{K}}$, and G^{A})

From the definition (13.16) of the Keldysh Green's function, we can write:

$$G^{\mathrm{K}}(\mathbf{x}t,\mathbf{x}'t') = -\,i\left(\langle\psi_{\mathrm{H}}(\mathbf{x}t)\,\psi_{\mathrm{H}}^{\dagger}(\mathbf{x}'t')\rangle \pm \langle\psi_{\mathrm{H}}^{\dagger}(\mathbf{x}'t')\,\psi_{\mathrm{H}}(\mathbf{x}t)\rangle\right) \underset{\substack{\uparrow \\ [\text{Eqs.}(13.5)\text{ and }(13.6)]}}{=} G^{>}(\mathbf{x}t,\mathbf{x}'t') + G^{<}(\mathbf{x}t,\mathbf{x}'t')\,. \tag{13.18}$$

From the definitions (13.15) and (13.17), we can instead write:

$$
\begin{aligned}
G^{\mathrm{R}}(\mathbf{x}t, \mathbf{x}'t') - G^{\mathrm{A}}(\mathbf{x}t, \mathbf{x}'t') &= -\, i\Big(\theta(t-t') \Big\langle \Big[\psi_{\mathrm{H}}(\mathbf{x}t), \psi^{\dagger}_{\mathrm{H}}(\mathbf{x}'t')\Big]_{\mp}\Big\rangle \\
&\quad + \theta(t'-t) \Big\langle \Big[\psi_{\mathrm{H}}(\mathbf{x}t), \psi^{\dagger}_{\mathrm{H}}(\mathbf{x}'t')\Big]_{\mp}\Big\rangle\Big) \\
&= -\, i \,\langle \psi_{\mathrm{H}}(\mathbf{x}t)\, \psi^{\dagger}_{\mathrm{H}}(\mathbf{x}'t')\rangle \;\pm\; i\, \langle \psi^{\dagger}_{\mathrm{H}}(\mathbf{x}'t')\, \psi_{\mathrm{H}}(\mathbf{x}t)\rangle \\
&\underset{\uparrow \atop \text{[Eqs.(13.5) and (13.6)]}}{=} G^{>}(\mathbf{x}t, \mathbf{x}'t') - G^{<}(\mathbf{x}t, \mathbf{x}'t')
\end{aligned}
\tag{13.19}
$$

since $\theta(t-t') + \theta(t'-t) = 1$. [The relation (13.19) is sometimes referred to as the *spectral identity*.] Adding and subtracting the results (13.18) and (13.19), we obtain eventually the desired relations:

$$
2\, G^{>}(\mathbf{x}t, \mathbf{x}'t') = G^{\mathrm{K}}(\mathbf{x}t, \mathbf{x}'t') + G^{\mathrm{R}}(\mathbf{x}t, \mathbf{x}'t') - G^{\mathrm{A}}(\mathbf{x}t, \mathbf{x}'t') \tag{13.20}
$$

$$
2\, G^{<}(\mathbf{x}t, \mathbf{x}'t') = G^{\mathrm{K}}(\mathbf{x}t, \mathbf{x}'t') - G^{\mathrm{R}}(\mathbf{x}t, \mathbf{x}'t') + G^{\mathrm{A}}(\mathbf{x}t, \mathbf{x}'t') \,. \tag{13.21}
$$

13.3 Symmetry Conditions

In addition to the redundancy conditions (13.8), (13.10), and (13.13), there are also a few *symmetry conditions* to be taken into account. They are as follows:

Symmetry property # 1

Consider the lesser Green's function $G^{<}(\mathbf{x}t, \mathbf{x}'t') = \mp\, i \,\langle\, \psi^{\dagger}_{\mathrm{H}}(\mathbf{x}'t')\, \psi_{\mathrm{H}}(\mathbf{x}t)\rangle$ defined in Eq. (13.5) and take its complex conjugate, yielding:

$$
G^{<}(\mathbf{x}t, \mathbf{x}'t')^{*} = \pm\, i \,\langle\, \psi^{\dagger}_{\mathrm{H}}(\mathbf{x}t)\, \psi_{\mathrm{H}}(\mathbf{x}'t')\rangle = -\, G^{<}(\mathbf{x}'t', \mathbf{x}t) \,. \tag{13.22}
$$

Symmetry property # 2

Consider the greater Green's function $G^{>}(\mathbf{x}t, \mathbf{x}'t') = -\, i \,\langle\, \psi_{\mathrm{H}}(\mathbf{x}t)\, \psi^{\dagger}_{\mathrm{H}}(\mathbf{x}'t')\rangle$ defined in Eq. (13.6) and take its complex conjugate, yielding:

$$
G^{>}(\mathbf{x}t, \mathbf{x}'t')^{*} = i \,\langle\, \psi_{\mathrm{H}}(\mathbf{x}'t')\, \psi^{\dagger}_{\mathrm{H}}(\mathbf{x}t)\rangle = -\, G^{>}(\mathbf{x}'t', \mathbf{x}t) \,. \tag{13.23}
$$

Symmetry property # 3

Consider the Keldysh Green's function $G^{\mathrm{K}}(\mathbf{x}t, \mathbf{x}'t') = G^{>}(\mathbf{x}t, \mathbf{x}'t') + G^{<}(\mathbf{x}t, \mathbf{x}'t')$ given by Eq. (13.18), take its complex conjugate, and take into account Eqs. (13.22) and (13.23), yielding:

$$
G^{\mathrm{K}}(\mathbf{x}t, \mathbf{x}'t')^{*} = -\, G^{\mathrm{K}}(\mathbf{x}'t', \mathbf{x}t) \,. \tag{13.24}
$$

Symmetry property # 4

Consider the retarded Green's function $G^{\mathrm{R}}(\mathbf{x}t, \mathbf{x}'t') = -i\,\theta(t-t')$ $\left\langle \left[\psi_{\mathrm{H}}(\mathbf{x}t), \psi^{\dagger}_{\mathrm{H}}(\mathbf{x}'t')\right]_{\mp} \right\rangle$ given by Eq. (13.15), take its complex conjugate, and take into account Eq. (13.17), yielding:

$$G^{\mathrm{R}}(\mathbf{x}t, \mathbf{x}'t')^{*} = i\,\theta(t-t') \left\langle \left[\psi_{\mathrm{H}}(\mathbf{x}'t'), \psi^{\dagger}_{\mathrm{H}}(\mathbf{x}t)\right]_{\mp} \right\rangle = G^{\mathrm{A}}(\mathbf{x}'t', \mathbf{x}t)\,. \tag{13.25}$$

Symmetry property # 5

Consider the right Keldysh component given by Eq. (13.7) (where Eq. (12.6) is also utilized) $G^{\rceil}(\mathbf{x}t, \mathbf{x}'\tau') = \mp i\,\langle\, \psi^{\dagger}_{\mathrm{H}}(\mathbf{x}'t_0 - i\tau')\, \psi_{\mathrm{H}}(\mathbf{x}t)\rangle = \mp i\,\langle\, e^{\tau' H^{\mathrm{M}}} \psi^{\dagger}(\mathbf{x}')$ $e^{-\tau' H^{\mathrm{M}}}\, \psi_{\mathrm{H}}(\mathbf{x}t)\rangle$, and take its complex conjugate, yielding:

$$G^{\rceil}(\mathbf{x}t, \mathbf{x}'\tau')^{*} = \pm i\,\langle\, \psi^{\dagger}_{\mathrm{H}}(\mathbf{x}t)\, e^{-\tau' H^{\mathrm{M}}} \psi(\mathbf{x}')\, e^{\tau' H^{\mathrm{M}}} \rangle = \pm i\,\langle\, \psi^{\dagger}_{\mathrm{H}}(\mathbf{x}t)\, \psi(\mathbf{x}'t_0 + i\tau'))\rangle\,. \tag{13.26}$$

This result has to be compared with the left Keldysh component (13.9) with displaced argument, namely,

$$\begin{aligned} G^{\lceil}(\mathbf{x}'\beta - \tau', \mathbf{x}t) &= -i\,\langle\, e^{(\beta-\tau')H^{\mathrm{M}}} \psi(\mathbf{x}')\, e^{-(\beta-\tau')H^{\mathrm{M}}}\, \psi^{\dagger}_{\mathrm{H}}(\mathbf{x}t)\rangle \\ &= -i\,\langle\, \psi^{\dagger}_{\mathrm{H}}(\mathbf{x}t)\, e^{-\tau' H^{\mathrm{M}}} \psi(\mathbf{x}')\, e^{\tau' H^{\mathrm{M}}} \rangle \\ &= -i\,\langle\, \psi^{\dagger}_{\mathrm{H}}(\mathbf{x}t)\, \psi(\mathbf{x}'t_0 + i\tau')\rangle, \end{aligned} \tag{13.27}$$

where the cyclic property of the trace has been used. Comparing Eq. (13.27) with Eq. (13.26), we obtain eventually:

$$G^{\rceil}(\mathbf{x}t, \mathbf{x}'\tau')^{*} = \mp\, G^{\lceil}(\mathbf{x}'\beta - \tau', \mathbf{x}t)\,. \tag{13.28}$$

Note that, to derive the properties (13.22)–(13.28), we have used the fact that the operator U defined by Eq. (11.2) is unitary for real times. Note also that, owing to the symmetry properties # 4 and # 5, the six independent functions ($G^{\mathrm{R}}, G^{\mathrm{K}}, G^{\mathrm{A}}, G^{\rceil}, G^{\lceil}$, and G^{M}) considered in Section 13.2 now reduce to *four independent physical Green's functions* (such as the functions ($G^{\mathrm{R}}, G^{<}, G^{\rceil}$, and G^{M}) as selected in Ref. [40] to the purpose).

13.4 Keldysh Formalism and Keldysh Rotation

As discussed in Sections 8.2 and 12.2, the Keldysh formalism restricts the integration contour to the two sectors C_1 and C_2 shown in Fig. 13.1, which are then both prolonged to $\rightarrow -\infty$, as shown in Fig. 13.13,

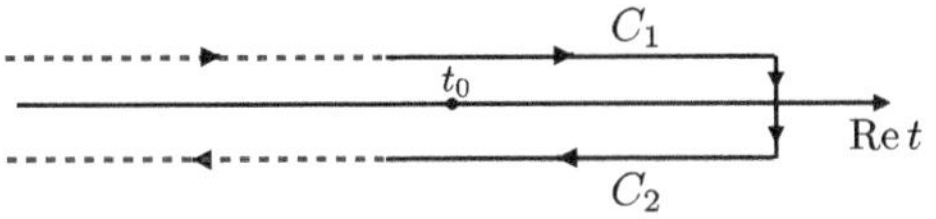

Figure 13.13 The Keldysh contour, again.

with C_1 running forward from $-\infty$ to $+\infty$ and C_2 running backward from $+\infty$ to $-\infty$,

where the dynamics over C_1 and C_2 is now governed by the adiabatic Hamiltonian (8.8).

The 3×3 matrix (13.2) representing the contour single-particle Green's function is then replaced by the following 2×2 matrix (where again the space and spin coordinates $\mathbf{x}$ are implicit):

$$\hat{G} = \begin{bmatrix} G_{11} & G_{12} \\ G_{21} & G_{22} \end{bmatrix} . \tag{13.29}$$

In addition, the thermal averages $\langle \cdots \rangle$ that define the components of the matrix (13.2) are now replaced by noninteracting thermal averages $\langle \cdots \rangle_0$ (cf. Eq. (8.9)).

Under these circumstances, when considering the Dyson equation (11.40), the time variables z_3 and z_4 therein, which are integrated over the contour $C_1 + C_2$, can be rearranged in the following way. When one of these variables (say, z_3) runs along C_2, we write

$$\int_{C_2} dz_3 \ \cdots = \int_{+\infty}^{-\infty} dt_3 \ \cdots = - \int_{-\infty}^{+\infty} dt_3 \ \cdots . \tag{13.30}$$

As a consequence, the second term on the right-hand side of the Dyson equation (11.40) can be rewritten in the explicit form (where $i = 1, 2$ is a contour index):

$$\begin{aligned} \int d34\, G_0(1,3)\, \Sigma(3,4)\, G(4,2) \ &\longrightarrow \sum_{i_3 i_4} \int d\mathbf{x}_3 d\mathbf{x}_4 \int_{C_1+C_2} dz_3 dz_4 \\ &\times G_0(\mathbf{x}_1 z_1, \mathbf{x}_3 z_3)_{i_1 i_3}\, \Sigma(\mathbf{x}_3 z_3, \mathbf{x}_4 z_4)_{i_3 i_4}\, G(\mathbf{x}_4 z_4, \mathbf{x}_2 z_2)_{i_4 i_2} \\ &= \sum_{i_3 i_4} \int d\mathbf{x}_3 d\mathbf{x}_4 \int_{-\infty}^{+\infty} dt_3 \int_{-\infty}^{+\infty} dt_4\, G_0(\mathbf{x}_1 z_1, \mathbf{x}_3 z_3)_{i_1 i_3} \\ &\times \left[\tau^3\, \Sigma(\mathbf{x}_3 t_3, \mathbf{x}_4 t_4)\right]_{i_3 i_4} \left[\tau^3\, G(\mathbf{x}_4 t_4, \mathbf{x}_2 t_2)\right]_{i_4 i_2}, \end{aligned} \tag{13.31}$$

where the presence of the third Pauli matrix $\tau^3 = \begin{bmatrix} 1 & 0 \\ 0 & -1 \end{bmatrix}$ absorbs the minus sign associated with the return contour C_2 in Eq. (13.30). In addition, both sides of the Dyson equation (11.40) can be multiplied by τ^3, yielding:

$$\left[\tau^3\, G(\mathbf{x}_1t_1,\mathbf{x}_2t_2)\right]_{i_1i_2} = \left[\tau^3\, G_0(\mathbf{x}_1t_1,\mathbf{x}_2t_2)\right]_{i_1i_2} + \sum_{i_3i_4}\int d\mathbf{x}_3 d\mathbf{x}_4 \int_{-\infty}^{+\infty} dt_3 \int_{-\infty}^{+\infty} dt_4$$
$$\times \left[\tau^3\, G_0(\mathbf{x}_1t_1,\mathbf{x}_3t_3)\right]_{i_1i_3} \left[\tau^3\, \Sigma(\mathbf{x}_3t_3,\mathbf{x}_4t_4)\right]_{i_3i_4} \left[\tau^3\, G(\mathbf{x}_4t_4,\mathbf{x}_2t_2)\right]_{i_4i_2}. \quad (13.32)$$

At this point, we introduce the matrix

$$L = \frac{1}{\sqrt{2}}\begin{bmatrix}1 & -1\\ 1 & 1\end{bmatrix} \quad \text{such that} \quad L^\dagger = \frac{1}{\sqrt{2}}\begin{bmatrix}1 & 1\\ -1 & 1\end{bmatrix} \quad \text{and} \quad L^\dagger L = \begin{bmatrix}1 & 0\\ 0 & 1\end{bmatrix}, \quad (13.33)$$

as we did in Eq. (13.14), and multiply both sides of Eq. (13.32) by L from the left and by $L^\dagger$ from the right, to obtain:

$$\left[L\tau^3\, G(\mathbf{x}_1t_1,\mathbf{x}_2t_2)L^\dagger\right]_{i_1i_2} = \left[L\tau^3\, G_0(\mathbf{x}_1t_1,\mathbf{x}_2t_2)L^\dagger\right]_{i_1i_2} + \sum_{i_3i_4}\int d\mathbf{x}_3 d\mathbf{x}_4 \int_{-\infty}^{+\infty} dt_3 \int_{-\infty}^{+\infty} dt_4$$
$$\times \left[L\tau^3\, G_0(\mathbf{x}_1t_1,\mathbf{x}_3t_3)L^\dagger\right]_{i_1i_3} \left[L\tau^3\, \Sigma(\mathbf{x}_3t_3,\mathbf{x}_4t_4)L^\dagger\right]_{i_3i_4} \left[L\tau^3\, G(\mathbf{x}_4t_4,\mathbf{x}_2t_2)L^\dagger\right]_{i_4i_2} \quad (13.34)$$

where according to Eqs. (13.14)–(13.17)

$$L\tau^3\, G(\mathbf{x}t,\mathbf{x}t')L^\dagger = \begin{bmatrix} G^{\mathrm{R}}(\mathbf{x}t,\mathbf{x}t') & G^{\mathrm{K}}(\mathbf{x}t,\mathbf{x}t')\\ 0 & G^{\mathrm{A}}(\mathbf{x}t,\mathbf{x}t')\end{bmatrix}, \quad (13.35)$$

which is called the *Keldysh rotation*. A similar transformation applied to the self-energy yields

$$L\tau^3\, \Sigma(\mathbf{x}t,\mathbf{x}'t')L^\dagger = \begin{bmatrix} \Sigma^{\mathrm{R}}(\mathbf{x}t,\mathbf{x}'t') & \Sigma^{\mathrm{K}}(\mathbf{x}t,\mathbf{x}'t')\\ 0 & \Sigma^{\mathrm{A}}(\mathbf{x}t,\mathbf{x}'t')\end{bmatrix}, \quad (13.36)$$

where a property analogous to Eq. (13.13) has been assumed. Note that matrix multiplication preserves the upper-triangular form of matrices since

$$\begin{bmatrix}a_{11} & a_{12}\\ 0 & a_{22}\end{bmatrix}\begin{bmatrix}b_{11} & b_{12}\\ 0 & b_{22}\end{bmatrix} = \begin{bmatrix}a_{11}b_{11} & a_{11}b_{12}+a_{12}b_{22}\\ 0 & a_{22}b_{22}\end{bmatrix}. \quad (13.37)$$

Note also that, owing to the symmetry property (13.25), only two elements ("K" and either "R" or "A") are independent in Eqs. (13.35) and (13.36).

Finally, for the noninteracting Green's function G_0, we obtain from Eq. (11.33) and the properties of the contour Dirac delta function of Section 7.3

$$G_0^{-1}(\mathbf{x}_1t_1,\mathbf{x}_2t_2)_{i_1i_2} = \left[i\frac{d}{dt_1} - h(\mathbf{x}_1,t_1)\right] \delta(\mathbf{x}_1,\mathbf{x}_2)\, \delta(t_1-t_2)\, \tau^3_{i_1i_2}, \quad (13.38)$$

where t_1 and t_2 are now meant to be real times. This implies that

$$\int d3\, G_0^{-1}(1,3)\, G_0(3,2)$$
$$\longrightarrow \sum_{i_3} \int d\mathbf{x}_3 \int_{-\infty}^{+\infty} dt_3\, G_0^{-1}(\mathbf{x}_1 t_1, \mathbf{x}_3 t_3)_{i_1 i_3} \left[\tau^3\, G_0(\mathbf{x}_3 t_3, \mathbf{x}_2 t_2)\right]_{i_3 i_2}$$
$$= \left[i\frac{d}{dt_1} - h(\mathbf{x}_1, t_1)\right] G_0(\mathbf{x}_1 t_1, \mathbf{x}_2 t_2)_{i_1 i_2} \underset{\substack{\uparrow \\ \text{[Eq.(11.34)]}}}{=} \delta(1,2) = \delta(\mathbf{x}_1, \mathbf{x}_2)\, \delta(t_1 - t_2)\, \tau^3_{i_1 i_2}, \tag{13.39}$$

and thus that $\left[\tau^3 G_0\right]_{i_1 i_2}$ satisfies the equation

$$\left[i\frac{d}{dt_1} - h(\mathbf{x}_1, t_1)\right] \left[\tau^3\, G_0(\mathbf{x}_1 t_1, \mathbf{x}_2 t_2)\right]_{i_1 i_2} = \delta(\mathbf{x}_1, \mathbf{x}_2)\, \delta(t_1 - t_2) \begin{bmatrix} 1 & 0 \\ 0 & 1 \end{bmatrix}, \tag{13.40}$$

or else (cf. Eq. (13.35) when $G \to G_0$)

$$\left[i\frac{d}{dt_1} - h(\mathbf{x}_1, t_1)\right] \left[L\tau^3\, G_0(\mathbf{x}_1 t_1, \mathbf{x}_2 t_2) L^\dagger\right]_{i_1 i_2}$$
$$= \left[i\frac{d}{dt_1} - h(\mathbf{x}_1, t_1)\right] \begin{bmatrix} G_0^{\mathrm{R}}(\mathbf{x}_1 t_1, \mathbf{x}_2 t_2) & G_0^{\mathrm{K}}(\mathbf{x}_1 t_1, \mathbf{x}_2 t_2) \\ 0 & G_0^{\mathrm{A}}(\mathbf{x}_1 t_1, \mathbf{x}_2 t_2) \end{bmatrix}$$
$$= \delta(\mathbf{x}_1, \mathbf{x}_2)\, \delta(t_1 - t_2) \begin{bmatrix} 1 & 0 \\ 0 & 1 \end{bmatrix}. \tag{13.41}$$

This equation will again be considered in Chapter 20 under homogeneity conditions in space and time.

14

Langreth Rules: Convolutions and Products

In the theory of the contour-ordered Green's functions, one encounters *convolutions* (like those occurring in the Dyson equation, either in the form (11.26) and (11.28) or in the form (11.40) and (11.41)) and *products* (like in the Fock expression (11.31) of the self-energy, which evolves into the so-called GW expression of the self-energy – to be discussed in Chapter 17 – upon replacing the bare potential v by the screened potential W). The task here is to obtain the corresponding expressions in terms of the real-time functions considered in detail in Chapter 13, which can be calculated analytically or implemented numerically. This task will be accomplished in terms of the so-called *Langreth* (or Langreth–Wilkins) *rules* [47].

To simplify the notation, in the present chapter, we shall omit to indicate explicitly the space and spin arguments $\mathbf{x}$ and concentrate only on the time variable z. In addition, we shall take the contour of the general form of the total contour γ_{T} of Fig. 8.1, which we split again into three branches, as shown in Fig. 13.1, as reproduced for convenience in Fig. 14.1.

14.1 A Few Preliminaries: Components in the Keldysh Space

We begin by establishing the relevant notation first in terms of the single-particle Green's function $G(z, z')$ on the contour γ_{T} of Fig.14.1. With the use of the contour Heaviside unit step function $\theta(z, z')$ (cf. Eq. (7.4), where $\tau \to z$), this function can be written in the compact form

$$\begin{aligned} G(z, z') &= -i \left\langle \mathcal{T}_{\gamma_{\mathrm{T}}} \left\{ \psi_{\mathrm{H}}(z)\, \psi_{\mathrm{H}}^{\dagger}(z') \right\} \right\rangle \\ &= -i\, \theta(z, z')\, \langle \psi_{\mathrm{H}}(z)\, \psi_{\mathrm{H}}^{\dagger}(z') \rangle \qquad (z \overset{\gamma_{\mathrm{T}}}{>} z') \\ &\mp i\, \theta(z', z)\, \langle \psi_{\mathrm{H}}^{\dagger}(z')\, \psi_{\mathrm{H}}(z) \rangle \qquad (z' \overset{\gamma_{\mathrm{T}}}{>} z) \\ &\equiv \theta(z, z')\, G^{>}(z, z') + \theta(z', z)\, G^{<}(z, z'), \end{aligned} \tag{14.1}$$

where the upper (lower) sign holds for bosons (fermions).

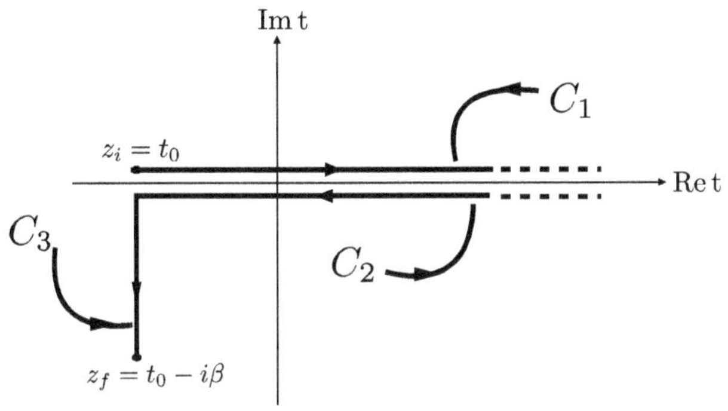

Figure 14.1 The total contour γ_T of Fig. 8.1 is split into three branches as shown in Fig. 13.1, such that $\gamma_T = \gamma_- + \gamma_+ + \gamma^M$ with $C_1 = \gamma_-$, $C_2 = \gamma_+$, and $C_3 = \gamma^M$.

Note how the present definition for $G^<(z,z')$ and $G^>(z,z')$ complies with the definitions (13.5) for the lesser Green's function $G^<(z,z')$ and (13.6) for the greater Green's function $G^>(z,z')$ but now with the variables z and z' running on the whole contour γ_T. Accordingly, from the functions $G^<(z,z')$ and $G^>(z,z')$, we can recover all components of the contour single-particle Green's function considered in Section 13.1 as follows:

$$\begin{cases} G^>(z \to t_+ \in \gamma_+, z' \to t_- \in \gamma_-) = G^>(t,t') & \text{[cf. Eq. (13.6)]} \\ G^<(z \to t_- \in \gamma_-, z' \to t_+ \in \gamma_+) = G^<(t,t') & \text{[cf. Eq. (13.5)]} \\ G^>(z \to t_0 - i\tau \in \gamma^M, z' \to t_\pm \in \gamma_\pm) = G^\lceil(\tau,t) & \text{[cf. Eqs. (13.9) and (13.10)]} \\ G^<(z \to t_\pm \in \gamma_\pm, z' \to t_0 - i\tau \in \gamma^M) = G^\rceil(t,\tau) & \text{[cf. Eqs. (13.7) and (13.8)]} \\ G(z \to t_0 - i\tau \in \gamma^M, z' \to t_0 - i\tau \in \gamma^M) & \text{[cf. Eq. (13.11)]} \\ = \theta(\tau,\tau')\, G^>(z \to t_0 - i\tau \in \gamma^M, z' \to t_0 - i\tau \in \gamma^M) & \\ + \theta(\tau',\tau)\, G^<(z \to t_0 - i\tau \in \gamma^M, z' \to t_0 - i\tau \in \gamma^M) & \\ \equiv \theta(\tau,\tau')\, G^>(\tau,\tau') + \theta(\tau',\tau)\, G^<(\tau,\tau') \equiv G^M(\tau,\tau')\,. & \end{cases} \tag{14.2}$$

In addition, in line with what we have done in Section 13.2, we have the freedom of moving the variables z and z' along the horizontal part $\gamma_- + \gamma_+$ of the contour

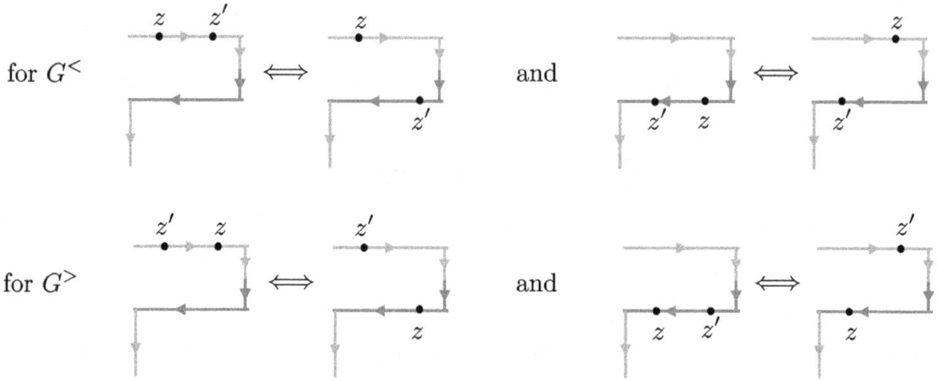

Figure 14.2 Possible ways of moving the variables z and z' along the horizontal part $\gamma_- + \gamma_+$ of the contour γ_T. These properties will be utilized also in Section 14.2.

γ_T in the way shown in Fig. 14.2, so as to generate from $G^<(z, z')$ and $G^>(z, z')$ also the standard time-ordered (13.3) and anti-time-ordered (13.4) components for real times.

A decomposition similar to Eq. (14.1) can be made also for the self-energy. In this case, however, an additional "singular" term Σ^δ has to be introduced (due to the Hartree (11.30) and Fock (11.31) contributions) [39], such that

$$\Sigma(z, z') = \delta(z, z')\, \Sigma^\delta(z) + \theta(z, z')\, \Sigma^>(z, z') + \theta(z', z)\, \Sigma^<(z, z') \tag{14.3}$$

with the contour Dirac delta function $\delta(z, z')$ defined in Section 7.3.

Generally, any function $k(z, z')$ of the variables z and z' running on the contour γ_T is said to belong to the *Keldysh space* if it can be written in the form

$$k(z, z') = \delta(z, z')\, k^\delta(z) + \theta(z, z')\, k^>(z, z') + \theta(z', z)\, k^<(z, z'), \tag{14.4}$$

where:

(i) The first term on the right-hand side is referred to as the *singular contribution* to $k(z, z')$, while the remaining two terms are together referred to as the *regular contribution* to $k(z, z')$ and indicated by $k_r(z, z')$;

(ii) The singular contribution proportional to $\delta(z, z')$ is active only when both z and z' belong to either $C_1 = \gamma_-$, or $C_2 = \gamma_+$, or $C_3 = \gamma^M$. When transforming to real times, this singular contribution will be attributed to the retarded and advanced components.

In this case, we write:

$$\begin{cases} k^{>}(z \to t_{+} \in \gamma_{+}, z' \to t_{-} \in \gamma_{-}) = k^{>}(t, t') \\ k^{<}(z \to t_{-} \in \gamma_{-}, z' \to t_{+} \in \gamma_{+}) = k^{<}(t, t') \\ k^{>}(z \to t_0 - i\tau \in \gamma^{M}, z' \to t_{\pm} \in \gamma_{\pm}) = k^{\lceil}(\tau, t) \\ k^{<}(z \to t_{\pm} \in \gamma_{\pm}, z' \to t_0 - i\tau \in \gamma^{M}) = k^{\rceil}(t, \tau) \end{cases} \tag{14.5}$$

in analogy to Eq. (14.2).

In addition, we can also define the retarded (R) and advanced (A) components for *real-time arguments* of a generic function (14.4) belonging to the Keldysh space, as follows:

$$\begin{cases} k^{R}(t, t') = \delta(t - t')\, k^{\delta}(t) \; + \; \theta(t - t')\, [k^{>}(t, t') - k^{<}(t, t')] \; , \\ k^{A}(t, t') = \delta(t - t')\, k^{\delta}(t) \; - \; \theta(t' - t)\, [k^{>}(t, t') - k^{<}(t, t')] \; . \end{cases} \tag{14.6}$$

As anticipated, the singular part k^{δ} of k has been attributed to its retarded and advanced components [39], while the regular parts of k^{R} and k^{A} are in line with the definitions (13.15) of G^{R} and (13.17) of G^{A}. In general, a singular part is attributed also to the *Matsubara component*, in the form

$$k^{M}(\tau, \tau') = i\, \delta(\tau - \tau')\, k^{\delta}(\tau) + \theta(\tau - \tau')\, k^{>}(\tau, \tau') \; + \; \theta(\tau' - \tau)\, k^{<}(\tau, \tau'), \tag{14.7}$$

where both variables τ and τ' run along the vertical track γ^{M}.

14.2 Langreth Rules for Convolutions

We are in a position to consider the *convolution* $c(z, z')$ of two functions $a(z, z')$ and $b(z, z')$ belonging to the Keldysh space, in the form

$$\boxed{c(z, z') = \int_{\gamma_T} d\bar{z}\; a(z, \bar{z})\, b(\bar{z}, z')} \tag{14.8}$$

where

$$\begin{cases} a(z, z') = \delta(z, z')\, a^{\delta}(z) + \; \theta(z, z')\, a^{>}(z, z') + \theta(z', z)\, a^{<}(z, z'), \\ b(z, z') = \delta(z, z')\, b^{\delta}(z) + \; \theta(z, z')\, b^{>}(z, z') + \theta(z', z)\, b^{<}(z, z'), \end{cases} \tag{14.9}$$

according to Eq. (14.4). Out of the convolution (14.8), we are going to extract the various components of $c(z, z')$ for real or imaginary times in line with Eq. (14.5), namely,

$$\begin{cases} c^>(t,t') = c^>(z \to t_+ \in \gamma_+, z' \to t_- \in \gamma_-) \\ c^<(t,t') = c^<(z \to t_- \in \gamma_-, z' \to t_+ \in \gamma_+) \\ c^\lceil(\tau,t) = c^>(z \to t_0 - i\tau \in \gamma^M, z' \to t_\pm \in \gamma_\pm) \\ c^\rceil(t,\tau) = c^<(z \to t_\pm \in \gamma_\pm, z' \to t_0 - i\tau \in \gamma^M) \end{cases} \tag{14.10}$$

besides $c^M(\tau,\tau')$ given as in Eq. (14.7), in terms of the corresponding components of a and b. In addition, it will be useful to extract the retarded (c^R) and advanced (c^A) components of c defined similar to Eq. (14.6). It will be instructive to carry out these calculations separately in some details since this will also help us to understand the relationships among the various components (Fig. 14.3–14.7).

Component $c^>(t,t')$

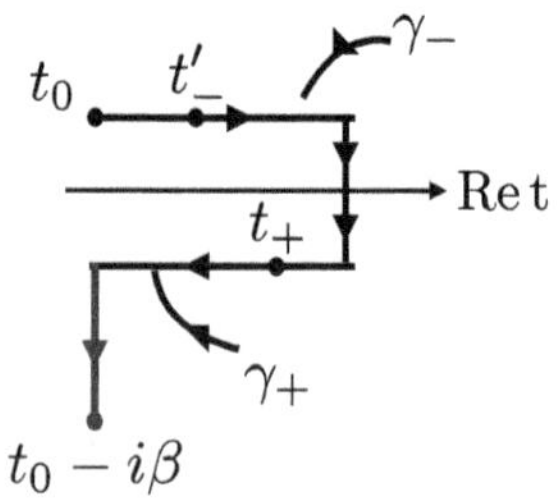

$$c^>(t,t') = c(t_+, t'_-) \tag{14.11}$$

Figure 14.3 Graphical view of Eq. (14.11).

We first consider the singular part of $c^>$.[1] From Eqs. (14.8) and (14.9), we obtain two contributions for this singular part. This is because in Eq. (14.8) $z \in \gamma_+$ and $z' \in \gamma_-$, such that in the integral over the contour γ_T

$$\begin{cases} \bar{z} \in \gamma_- & \Longrightarrow \text{ the singular part of b is active} \\ \bar{z} \in \gamma_+ & \Longrightarrow \text{ the singular part of a is active} \end{cases} \tag{14.12}$$

and the singular part of $c^>$ is given by

$$b^\delta(t'_-)\, a^>(t_+, t'_-) \;+\; a^\delta(t_+)\, b^>(t_+, t'_-)\,. \tag{14.13}$$

[1] By "singular part" and "singular contribution" for $c^>$, $c^<$, $c^\lceil$, and $c^\rceil$, we mean the contribution originating from the singular parts of a and b. On the other hand, c^R, c^A, and c^M have a truly singular contribution proportional to a Dirac delta function in their respective variables.

[Note that the singular parts of a and b are never active simultaneously because, in the present case, z and z' belong to different branches.]

For the regular part of $c^>$, we obtain:

$$\int_{\gamma_T} d\bar{z}\; a(t_+,\bar{z})\, b(\bar{z},t'_-) \longrightarrow \int_{t_0}^{t'_-} d\bar{z}\, a^>(t_+,\bar{z})\, b^<(\bar{z},t'_-)$$
$$+ \int_{t'_-}^{t_+} d\bar{z}\, a^>(t_+,\bar{z})\, b^>(\bar{z},t'_-) + \int_{t_+}^{t_0-i\beta} d\bar{z}\, a^<(t_+,\bar{z})\, b^>(\bar{z},t'_-), \qquad (14.14)$$

where the options given in Fig. 14.2 have been utilized. Note that in Eq. (14.14), it does not matter whether the numerical value of t_+ is larger or smaller than the numerical value of t'_-. Accordingly, we may directly identify $t_+ \leftrightarrow t$ and $t'_- \leftrightarrow t'$ and manipulate the second and third terms on the right-hand side therein as follows:

$$\int_{t'_-}^{t_+} d\bar{z}\, a^>(t_+,\bar{z})\, b^>(\bar{z},t'_-) \longrightarrow \int_{t'}^{t_0} d\bar{t}\, a^>(t,\bar{t})\, b^>(\bar{t},t') + \int_{t_0}^{t} d\bar{t}\, a^>(t,\bar{t})\, b^>(\bar{t},t') \qquad (14.15)$$

and

$$\int_{t_+}^{t_0-i\beta} d\bar{z}\, a^<(t_+,\bar{z})\, b^>(\bar{z},t'_-) \longrightarrow \int_{t}^{t_0} d\bar{t}\, a^<(t,\bar{t})\, b^>(\bar{t},t') - i\int_0^{\beta} d\bar{\tau}\, a^{\rceil}(t,\bar{\tau})\, b^{\lceil}(\bar{\tau},t'), \qquad (14.16)$$

where the relations (14.5) applied to $a^<$ and $b^>$ have been utilized in the last integral over γ^{M}.

Entering the results (14.15) and (14.16) into Eq. (14.14) for the regular part of $c^>$ and adding the contribution (14.13) from the singular part of $c^>$, we obtain (with $t > t_0$ and $t' > t_0$):

$$\begin{aligned} c^>(t,t') &= b^{\delta}(t')\, a^>(t,t') + a^{\delta}(t)\, b^>(t,t') \\ &\quad - \int_{t_0}^{+\infty} d\bar{t}\, a^>(t,\bar{t})\, \theta(t'-\bar{t})\, \left[b^>(\bar{t},t') - b^<(\bar{t},t')\right] \\ &\quad + \int_{t_0}^{+\infty} d\bar{t}\, \theta(t-\bar{t})\, \left[a^>(t,\bar{t}) - a^<(t,\bar{t})\right]\, b^>(\bar{t},t') \\ &\quad - i\int_0^{\beta} d\bar{\tau}\, a^{\rceil}(t,\bar{\tau})\, b^{\lceil}(\bar{\tau},t'), \end{aligned} \qquad (14.17)$$

where the integrations over $\bar{t}$ have been extended from t_0 to $+\infty$ by introducing the Heaviside unit step function for real time. Finally, the first two terms on the right-hand side of Eq. (14.17) can be brought inside the integral signs by recalling the definitions (14.6) of the retarded and advanced functions, here applied to a and b.

We obtain eventually:

$$\boxed{c^{>}(t,t')=\int_{t_0}^{+\infty} d\bar{t}\left[a^{>}(t,\bar{t})b^{\mathrm{A}}(\bar{t},t')+a^{\mathrm{R}}(t,\bar{t})b^{>}(\bar{t},t')\right]-i\int_0^{\beta} d\bar{\tau}a^{\rceil}(t,\bar{\tau})b^{\lceil}(\bar{\tau},t').} \tag{14.18}$$

Remark: A compact notation

The result (14.18) can be rewritten in a compact way by introducing the notation [15]

$$f \bullet g \equiv \int_{t_0}^{+\infty} d\bar{t} f(\bar{t})\, g(\bar{t}) \quad \text{and} \quad f \star g \equiv -i \int_0^{\beta} d\bar{\tau} f(\bar{\tau})\, g(\bar{\tau})\,. \tag{14.19}$$

In this way, Eq. (14.18) becomes

$$\boxed{c^{>} = a^{>} \bullet b^{\mathrm{A}} + a^{\mathrm{R}} \bullet b^{>} + a^{\rceil} \star b^{\lceil}} \tag{14.20}$$

where the “external” time variables (t at left and t' at right) are implicit.

Component $c^{<}(t,t')$

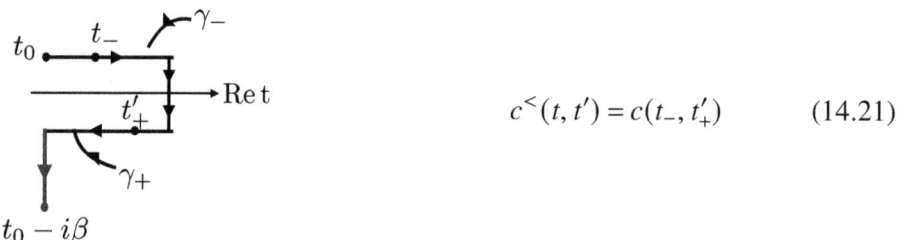

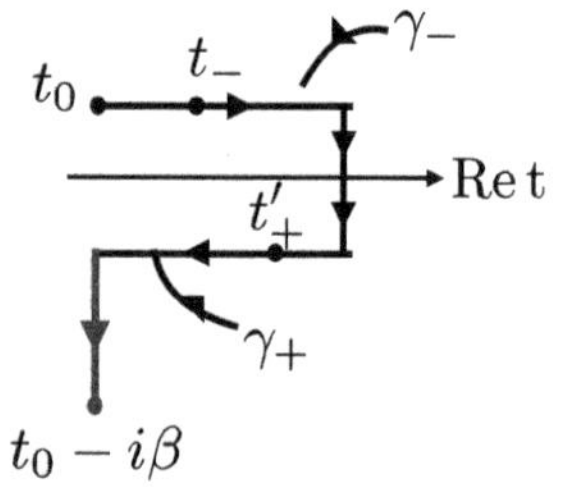

$$c^{<}(t,t') = c(t_-, t'_+) \tag{14.21}$$

Figure 14.4 Graphical view of Eq. (14.21).

The component $c^{<}$ of the convolution (14.8) can be obtained along similar lines, which we sketch here for completeness. In this case, $z \in \gamma_-$ and $z' \in \gamma_+$, such that in the integral over the contour γ_{T}

$$\begin{cases} \bar{z} \in \gamma_- & \Longrightarrow \text{ the singular part of a is active} \\ \bar{z} \in \gamma_+ & \Longrightarrow \text{ the singular part of b is active} \end{cases} \tag{14.22}$$

and the singular part of $c^{<}$ is given by

$$a^{\delta}(t_-)\, b^{<}(t_-, t'_+) \;+\; b^{\delta}(t'_+)\, a^{<}(t_-, t'_+)\,. \tag{14.23}$$

For the regular part of $c^{<}$, we obtain:

$$\int_{\gamma_T} d\bar{z}\; a(t_-, \bar{z})\, b(\bar{z}, t'_+) \longrightarrow \int_{t_0}^{t_-} d\bar{z}\, a^>(t_-, \bar{z})\, b^<(\bar{z}, t'_+)$$
$$+ \int_{t_-}^{t'_+} d\bar{z}\, a^<(t_-, \bar{z})\, b^<(\bar{z}, t'_+) + \int_{t'_+}^{t_0 - i\beta} d\bar{z}\, a^<(t_-, \bar{z})\, b^>(\bar{z}, t'_+)$$
$$= - \int_{t_0}^{+\infty} d\bar{t}\, a^<(t, \bar{t})\, \theta(t' - \bar{t})\, \left[b^>(\bar{t}, t') - b^<(\bar{t}, t')\right]$$
$$+ \int_{t_0}^{+\infty} d\bar{t}\, \theta(t - \bar{t})\, \left[a^>(t, \bar{t}) - a^<(t, \bar{t})\right]\, b^<(\bar{t}, t') \; - i \int_0^{\beta} d\bar{\tau}\, a^{\rceil}(t, \bar{\tau})\, b^{\lceil}(\bar{\tau}, t'), \tag{14.24}$$

where we have identified $t_- \leftrightarrow t$ and $t'_+ \leftrightarrow t'$, extended the integrations over $\bar{t}$ from t_0 to $+\infty$ through the Heaviside unit step function, and exploited again the results (14.5) for the functions a and b.

Adding together the singular part (14.23) and the regular part (14.24) of $c^<$ and introducing the retarded and advanced functions as in Eqs. (14.6), we obtain eventually

$$\boxed{c^<(t, t') = \int_{t_0}^{+\infty} d\bar{t}\, \left[a^<(t, \bar{t}) b^{A}(\bar{t}, t') + a^{R}(t, \bar{t}) b^<(\bar{t}, t')\right] - i \int_0^{\beta} d\bar{\tau} a^{\rceil}(t, \bar{\tau}) b^{\lceil}(\bar{\tau}, t')}, \tag{14.25}$$

which in terms of the compact notation (14.19) reads

$$\boxed{c^< = a^< \bullet b^{A} + a^{R} \bullet b^< + a^{\rceil} \star b^{\lceil}}. \tag{14.26}$$

Component $c^{\rceil}(t, \tau)$

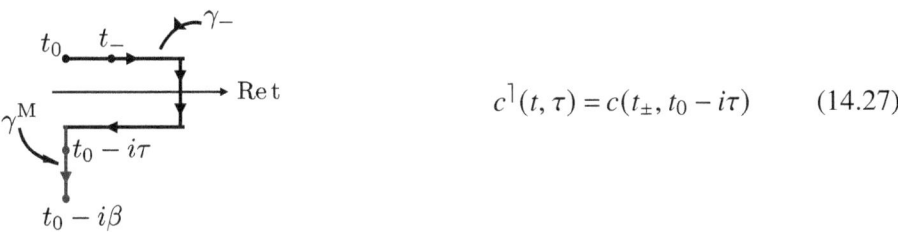

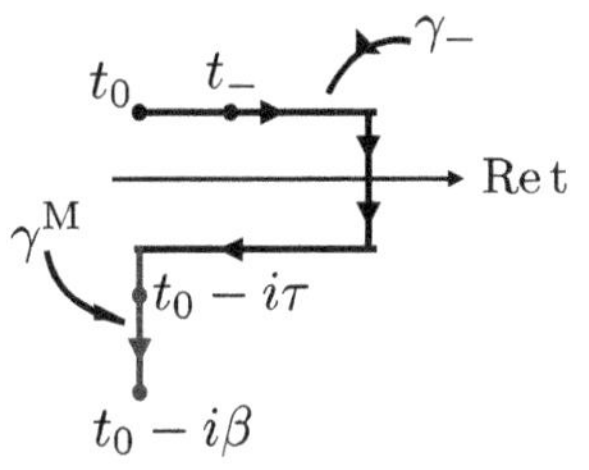

Figure 14.5 Graphical view of Eq. (14.27).

$$c^{\rceil}(t, \tau) = c(t_\pm, t_0 - i\tau) \tag{14.27}$$

In this case, $z \in \gamma_\pm$ and $z' \in \gamma^{\mathrm{M}}$, such that in the integral over the contour γ_{T}

$$\begin{cases} \bar{z} \in \gamma_\pm & \Longrightarrow \text{ the singular part of a is active} \\ \bar{z} \in \gamma^{\mathrm{M}} & \Longrightarrow \text{ the singular part of b is active} \end{cases} \tag{14.28}$$

and the singular part of $c^{\rceil}$ is given by

$$a^{\delta}(t_\pm)\, b^{<}(t_\pm, t_0 - i\tau) + b^{\delta}(t_0 - i\tau)\, a^{<}(t_\pm, t_0 - i\tau) \longrightarrow a^{\delta}(t)\, b^{\rceil}(t,\tau) + b^{\delta}(\tau)\, a^{\rceil}(t,\tau)\,. \tag{14.29}$$

For the regular part of $c^{\rceil}$, we obtain:

$$\begin{aligned} \int_{\gamma_{\mathrm{T}}} d\bar{z}\; a(t_\pm, \bar{z})\, b(\bar{z}, t_0 - i\tau) &\longrightarrow \int_{t_0}^{t} d\bar{z}\, a^{>}(t,\bar{z})\, b^{<}(\bar{z}, t_0 - i\tau) \\ &+ \int_{t}^{t_0} d\bar{z}\, a^{<}(t,\bar{z})\, b^{<}(\bar{z}, t_0 - i\tau) + \int_{t_0}^{t_0 - i\tau} d\bar{z}\, a^{<}(t,\bar{z})\, b^{<}(\bar{z}, t_0 - i\tau) \\ &+ \int_{t_0 - i\tau}^{t_0 - i\beta} d\bar{z}\, a^{<}(t,\bar{z})\, b^{>}(\bar{z}, t_0 - i\tau) = \int_{t_0}^{t} d\bar{t}\, \left[a^{>}(t,\bar{t}) - a^{<}(t,\bar{t})\right] b^{\rceil}(\bar{t},\tau) \\ &- i \int_{0}^{\beta} d\bar{\tau}\, a^{\rceil}(t,\bar{\tau}) \left[\theta(\tau - \bar{\tau})\, b^{<}(\bar{\tau},\tau) + \theta(\bar{\tau} - \tau)\, b^{>}(\bar{\tau},\tau)\right]. \end{aligned} \tag{14.30}$$

Adding together the singular part (14.29) and the regular part (14.30) of $c^{\rceil}$, extending the integrations over $\bar{t}$ from t_0 to $+\infty$ through the Heaviside unit step function, and introducing the retarded function as in Eq. (14.6) and the Matsubara function as in Eq. (14.7), we obtain eventually

$$\boxed{c^{\rceil}(t,\tau) = \int_{t_0}^{+\infty} d\bar{t}\, a^{\mathrm{R}}(t,\bar{t})\, b^{\rceil}(\bar{t},\tau) - i \int_{0}^{\beta} d\bar{\tau}\, a^{\rceil}(t,\bar{\tau})\, b^{\mathrm{M}}(\bar{\tau},\tau)}\,, \tag{14.31}$$

which in terms of the compact notation (14.19) reads

$$\boxed{c^{\rceil} = a^{\mathrm{R}} \bullet b^{\rceil} + a^{\rceil} \star b^{\mathrm{M}}}\,. \tag{14.32}$$

Component $c^{\lceil}(\tau, t)$

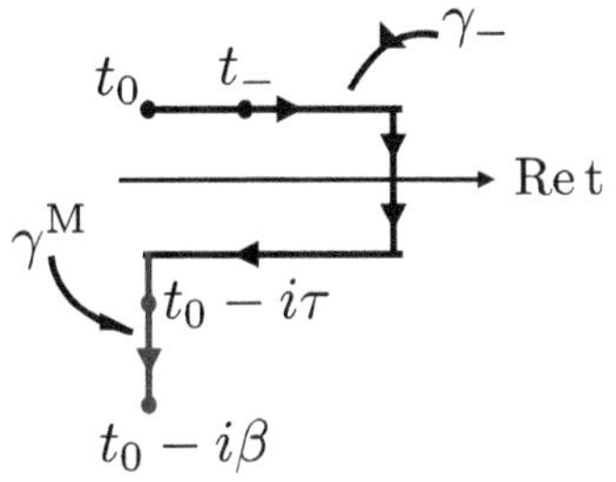

$$c^{\lceil}(\tau, t) = c(t_0 - i\tau, t_\pm) \tag{14.33}$$

Figure 14.6 Graphical view of Eq. (14.33).

In this case, $z \in \gamma^{\mathrm{M}}$ and $z' \in \gamma_\pm$, such that in the integral over the contour γ_{T}

$$\begin{cases} \bar{z} \in \gamma^{\mathrm{M}} & \Longrightarrow \text{ the singular part of a is active} \\ \bar{z} \in \gamma_\pm & \Longrightarrow \text{ the singular part of b is active} \end{cases} \tag{14.34}$$

and the singular part of $c^{\lceil}$ is given by

$$a^\delta(\tau)\, b^>(t_0 - i\tau, t_\pm) + b^\delta(t_\pm)\, a^>(t_0 - i\tau, t_\pm) \longrightarrow a^\delta(\tau)\, b^{\lceil}(\tau, t) + b^\delta(t)\, a^{\lceil}(\tau, t)\,. \tag{14.35}$$

For the regular part of $c^{\lceil}$, we obtain:

$$\begin{aligned} \int_{\gamma_{\mathrm{T}}} d\bar{z}\; a(t_0 - i\tau, \bar{z})\, b(\bar{z}, t_\pm) &\longrightarrow \int_{t_0}^{t} d\bar{z}\, a^>(t_0 - i\tau, \bar{z})\, b^<(\bar{z}, t) \\ &+ \int_{t}^{t_0} d\bar{z}\, a^>(t_0 - i\tau, \bar{z})\, b^>(\bar{z}, t) + \int_{t_0}^{t_0 - i\tau} d\bar{z}\, a^>(t_0 - i\tau, \bar{z})\, b^>(\bar{z}, t) \\ &+ \int_{t_0 - i\tau}^{t_0 - i\beta} d\bar{z}\, a^<(t_0 - i\tau, \bar{z})\, b^>(\bar{z}, t) = \int_{t_0}^{t} d\bar{t}\, a^{\lceil}(\tau, \bar{t})\left[b^<(\bar{t}, t) - b^>(\bar{t}, t)\right] \\ &- i\int_0^\beta d\bar{\tau}\, \left[\theta(\tau - \bar{\tau})\, a^>(\tau, \bar{\tau}) + \theta(\bar{\tau} - \tau)\, a^<(\tau, \bar{\tau})\right] b^{\lceil}(\bar{\tau}, t)\,. \end{aligned} \tag{14.36}$$

Adding together the singular part (14.35) and the regular part (14.36) of $c^{\lceil}$, extending the integrations over $\bar{t}$ from t_0 to $+\infty$ through the Heaviside unit step function, and introducing the advanced function as in Eq. (14.6) and the Matsubara function as in Eq. (14.7), we obtain eventually

$$\boxed{c^{\lceil}(\tau, t) = \int_{t_0}^{+\infty} d\bar{t}\, a^{\lceil}(\tau, \bar{t})\, b^{\mathrm{A}}(\bar{t}, t) - i\int_0^\beta d\bar{\tau}\, a^{\mathrm{M}}(\tau, \bar{\tau})\, b^{\lceil}(\bar{\tau}, t)}\,, \tag{14.37}$$

which in terms of the compact notation (14.19) reads

$$\boxed{c^{\rceil} = a^{\rceil} \bullet b^{\mathrm{A}} + a^{\mathrm{M}} \star b^{\rceil}}\,. \tag{14.38}$$

Component $c^{\mathrm{M}}(\tau, \tau')$

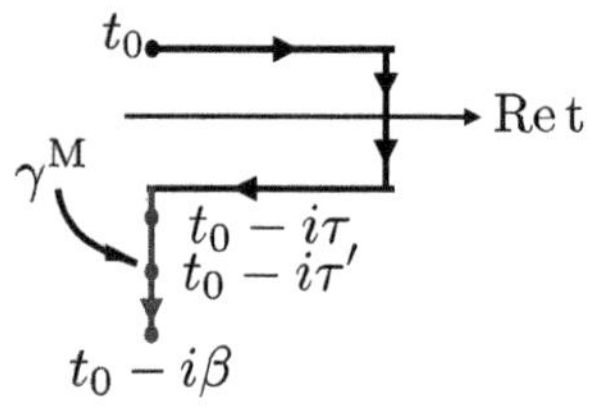

Figure 14.7 Graphical view of Eq. (14.39).

$$c^{\mathrm{M}}(\tau, \tau') = c(t_0 - i\tau, t_0 - i\tau') \tag{14.39}$$

In this case, one obtains a truly "singular contribution" in accordance with the definition (14.7) since both a and b have singular contributions, which are active in the same interval from t_0 to $t_0 - i\beta$. This singular contribution is given by:

$$\int_{\gamma_{\mathrm{T}}} d\bar{z}\; a^{\delta}(z)\, \delta(z, \bar{z})\, b^{\delta}(\bar{z})\, \delta(\bar{z}, z') = a^{\delta}(z)\, b^{\delta}(z')\, \delta(z, z') \longrightarrow$$
$$a^{\delta}(t_0 - i\tau)\, b^{\delta}(t_0 - i\tau')\, \delta(t_0 - i\tau, t_0 - i\tau') = i\, a^{\delta}(\tau)\, b^{\delta}(\tau')\, \delta(\tau - \tau')\,. \tag{14.40}$$

To obtain the regular part of c^{M}, we consider initially the case $\tau' > \tau$ (but one can readily verify that the final expression we will obtain for $c^{\mathrm{M}}(\tau, \tau')$ holds also for $\tau > \tau'$):

$$\int_{\gamma_{\mathrm{T}}} d\bar{z}\; a(t_0 - i\tau, \bar{z})\, b(\bar{z}, t_0 - i\tau') \longrightarrow \int_{t_0}^{t_0 - i\tau} d\bar{z}\, a^{>}(t_0 - i\tau, \bar{z})\, b^{<}(\bar{z}, t_0 - i\tau')$$
$$+ \int_{t_0 - i\tau}^{t_0 - i\tau'} d\bar{z}\, a^{<}(t_0 - i\tau, \bar{z})\, b^{<}(\bar{z}, t_0 - i\tau') + \int_{t_0 - i\tau'}^{t_0 - i\beta} d\bar{z}\, a^{<}(t_0 - i\tau, \bar{z}) b^{>}(\bar{z}, t_0 - i\tau')$$
$$+ a^{\delta}(t_0 - i\tau) b^{<}(t_0 - i\tau, t_0 - i\tau') + a^{<}(t_0 - i\tau, t_0 - i\tau') b^{\delta}(t_0 - i\tau'), \tag{14.41}$$

where the two terms in the last line originate from the crossed contributions of the regular and singular terms in the expressions (14.4) for a and b.

Adding together the results (14.40) and (14.41) and extending each integral on the right-hand side of Eq. (14.41) over the whole interval from t_0 to $t_0 - i\beta$ by means of Heaviside unit step function in the variables τ and τ', we obtain eventually

$$c^{\mathrm{M}}(\tau,\tau') = -i\int_0^{\beta} d\bar{\tau}\left[ia^{\delta}(\tau)\,\delta(\tau-\bar{\tau}) + \theta(\tau-\bar{\tau})\,a^{>}(\tau,\bar{\tau}) + \theta(\bar{\tau}-\tau)\,a^{<}(\tau,\bar{\tau})\right]$$
$$\times\left[ib^{\delta}(\tau')\,\delta(\bar{\tau}-\tau') + \theta(\bar{\tau}-\tau')\,b^{>}(\bar{\tau},\tau') + \theta(\tau'-\bar{\tau})\,b^{<}(\bar{\tau},\tau')\right], \tag{14.42}$$

where, in the expressions within brackets, we recognize the Matsubara components (14.7) for a and b, in the order. We thus write

$$\boxed{c^{\mathrm{M}}(\tau,\tau') = -i\int_0^{\beta} d\bar{\tau}\,a^{\mathrm{M}}(\tau,\bar{\tau})\,b^{\mathrm{M}}(\bar{\tau},\tau'),} \tag{14.43}$$

as it should have been expected, which, in terms of the compact notation (14.19), reads

$$\boxed{c^{\mathrm{M}} = a^{\mathrm{M}} \star b^{\mathrm{M}}}\,. \tag{14.44}$$

Component $c^{\mathrm{R}}(t,t')$

From the general expression (14.6), we expect $c^{\mathrm{R}}(t,t')$ to have the form

$$\begin{aligned} c^{\mathrm{R}}(t,t') &= \delta(t-t')\,c^{\delta}(t) + \theta(t-t')\left[c^{>}(t,t') - c^{<}(t,t')\right] \\ &\equiv \delta(t-t')\,c^{\delta}(t) + c^{\mathrm{R}}_{\mathrm{r}}(t,t'), \end{aligned} \tag{14.45}$$

where the suffix r stands for the "regular part", and both t and t' can be taken in the forward branch γ_- without loss of generality.

In analogy to Eq. (14.40), the singular contribution is here given by:

$$\begin{aligned} \int_{\gamma_{\mathrm{T}}} d\bar{z}\; a^{\delta}(z)\,\delta(z,\bar{z})\,b^{\delta}(\bar{z})\,\delta(\bar{z},z') &= a^{\delta}(z)\,b^{\delta}(z')\,\delta(z,z') \\ &\longrightarrow a^{\delta}(t)\,b^{\delta}(t')\,\delta(t-t'). \end{aligned} \tag{14.46}$$

The regular part of c^{R} can be obtained directly from the results (14.18) for $c^{>}$ and (14.25) for $c^{<}$, yielding

$$\begin{aligned} c^{\mathrm{R}}_{\mathrm{r}}(t,t') &= \theta(t-t')\left[c^{>}(t,t') - c^{<}(t,t')\right] \\ &= \theta(t-t')\int_{t_0}^{+\infty} d\bar{t}\,\left\{a^{\mathrm{R}}(t,\bar{t})\left[b^{>}(\bar{t},t') - b^{<}(\bar{t},t')\right]\right. \\ &\quad \left. + \left[a^{>}(t,\bar{t}) - a^{<}(t,\bar{t})\right] b^{\mathrm{A}}(\bar{t},t')\right\}, \end{aligned} \tag{14.47}$$

where the contributions from the left and right Keldysh components in Eqs. (14.18) and (14.25) have canceled out. The expression within braces in Eq. (14.47) can be conveniently manipulated, by writing

$$\begin{cases} a^{\mathrm{R}}(t,t') = \delta(t-t')\, a^{\delta}(t) + \theta(t-t')\, [a^{>}(t,t') - a^{<}(t,t')] \\ \qquad \equiv \delta(t-t')\, a^{\delta}(t) + a_{\mathrm{r}}^{\mathrm{R}}(t,t') \\ \\ b^{\mathrm{R}}(t,t') = \delta(t-t')\, b^{\delta}(t) + \theta(t-t')\, [b^{>}(t,t') - b^{<}(t,t')] \\ \qquad \equiv \delta(t-t')\, b^{\delta}(t) + b_{\mathrm{r}}^{\mathrm{R}}(t,t') \\ \\ a^{\mathrm{A}}(t,t') = \delta(t-t')\, a^{\delta}(t) - \theta(t'-t)\, [a^{>}(t,t') - a^{<}(t,t')] \\ \qquad \equiv \delta(t-t')\, a^{\delta}(t) + a_{\mathrm{r}}^{\mathrm{A}}(t,t') \\ \\ b^{\mathrm{A}}(t,t') = \delta(t-t')\, b^{\delta}(t) - \theta(t'-t)\, [b^{>}(t,t') - b^{<}(t,t')] \\ \qquad \equiv \delta(t-t')\, b^{\delta}(t) + b_{\mathrm{r}}^{\mathrm{A}}(t,t') \end{cases} \tag{14.48}$$

like in Eq. (14.45), such that

$$\begin{aligned} b^{>}(t,t') - b^{<}(t,t') &= \theta(t-t')\, \left[b^{>}(t,t') - b^{<}(t,t')\right] + \theta(t'-t)\, \left[b^{>}(t,t') - b^{<}(t,t')\right] \\ &= b_{\mathrm{r}}^{\mathrm{R}}(t,t') - b_{\mathrm{r}}^{\mathrm{A}}(t,t') = b^{\mathrm{R}}(t,t') - b^{\mathrm{A}}(t,t') \end{aligned} \tag{14.49}$$

$$\begin{aligned} a^{>}(t,t') - a^{<}(t,t') &= \theta(t-t')\, \left[a^{>}(t,t') - a^{<}(t,t')\right] + \theta(t'-t)\, \left[a^{>}(t,t') - a^{<}(t,t')\right] \\ &= a_{\mathrm{r}}^{\mathrm{R}}(t,t') - a_{\mathrm{r}}^{\mathrm{A}}(t,t') = a^{\mathrm{R}}(t,t') - a^{\mathrm{A}}(t,t') \end{aligned} \tag{14.50}$$

in analogy to Eq. (13.19). In this way, the expression (14.47) becomes

$$\begin{aligned} c_{\mathrm{r}}^{\mathrm{R}}(t,t') =&\ a^{\delta}(t)\, b_{\mathrm{r}}^{\mathrm{R}}(t,t') + a_{\mathrm{r}}^{\mathrm{R}}(t,t')\, b^{\delta}(t') \\ &+ \theta(t-t') \int_{t_0}^{+\infty} d\bar{t}\, \left\{a_{\mathrm{r}}^{\mathrm{R}}(t,\bar{t})\, \left[b_{\mathrm{r}}^{\mathrm{R}}(\bar{t},t') - b_{\mathrm{r}}^{\mathrm{A}}(\bar{t},t')\right]\right. \\ &+ \left.\left[a_{\mathrm{r}}^{\mathrm{R}}(t,\bar{t}) - a_{\mathrm{r}}^{\mathrm{A}}(t,\bar{t})\right] b^{\mathrm{A}}(\bar{t},t')\right\} \\ =&\ a^{\delta}(t)\, b_{\mathrm{r}}^{\mathrm{R}}(t,t') + a_{\mathrm{r}}^{\mathrm{R}}(t,t')\, b^{\delta}(t') \\ &+ \theta(t-t') \int_{t_0}^{+\infty} d\bar{t}\, a_{\mathrm{r}}^{\mathrm{R}}(t,\bar{t})\, b_{\mathrm{r}}^{\mathrm{R}}(\bar{t},t')\, . \end{aligned} \tag{14.51}$$

Adding together the singular part (14.46) to the regular part (14.51) of c^{R}, we obtain eventually

$$\boxed{\begin{aligned} c^{\mathrm{R}}(t,t') &= \int_{t_0}^{+\infty} d\bar{t}\, \left[a^{\delta}(t)\, \delta(t-\bar{t}) + a_{\mathrm{r}}^{\mathrm{R}}(t,\bar{t})\right] \left[b^{\delta}(\bar{t})\, \delta(\bar{t}-t') + b_{\mathrm{r}}^{\mathrm{R}}(\bar{t},t')\right] \\ &= \int_{t_0}^{+\infty} d\bar{t}\, a^{\mathrm{R}}(t,\bar{t})\, b^{\mathrm{R}}(\bar{t},t') \end{aligned}}\ , \tag{14.52}$$

which in terms of the compact notation (14.19) reads

$$\boxed{c^{\mathrm{R}} = a^{\mathrm{R}} \bullet b^{\mathrm{R}}}\,. \tag{14.53}$$

Component $c^{\mathrm{A}}(t, t')$

From the general expression (14.6), we expect $c^{\mathrm{A}}(t, t')$ to have the form

$$\begin{aligned} c^{\mathrm{A}}(t, t') &= \delta(t - t')\, c^{\delta}(t) \;-\; \theta(t' - t)\left[c^{>}(t, t') - c^{<}(t, t')\right] \\ &\equiv \delta(t - t')\, c^{\delta}(t) \;+\; c^{\mathrm{A}}_{\mathrm{r}}(t, t'), \end{aligned} \tag{14.54}$$

where t and t' can again be taken in the forward branch γ_{-}, and the singular contribution is the same of Eq. (14.46).

The regular part of c^{A} can be obtained following steps similar to those from Eq. (14.47) to Eq. (14.51). Accordingly, we write:

$$\begin{aligned} c^{\mathrm{A}}_{\mathrm{r}}(t, t') = &\; a^{\delta}(t)\, b^{\mathrm{A}}_{\mathrm{r}}(t, t') + a^{\mathrm{A}}_{\mathrm{r}}(t, t')\, b^{\delta}(t') \\ &- \theta(t' - t) \int_{t_0}^{+\infty} d\bar{t}\, \left\{a^{\mathrm{R}}_{\mathrm{r}}(t, \bar{t})\left[b^{\mathrm{R}}_{\mathrm{r}}(\bar{t}, t') - b^{\mathrm{A}}_{\mathrm{r}}(\bar{t}, t')\right]\right. \\ &\left. + \left[a^{\mathrm{R}}_{\mathrm{r}}(t, \bar{t}) - a^{\mathrm{A}}_{\mathrm{r}}(t, \bar{t})\right] b^{\mathrm{A}}(\bar{t}, t')\right\} \\ = &\; a^{\delta}(t)\, b^{\mathrm{A}}_{\mathrm{r}}(t, t') + a^{\mathrm{A}}_{\mathrm{r}}(t, t')\, b^{\delta}(t') \\ &+ \theta(t' - t) \int_{t_0}^{+\infty} d\bar{t}\, a^{\mathrm{A}}_{\mathrm{r}}(t, \bar{t})\, b^{\mathrm{A}}_{\mathrm{r}}(\bar{t}, t')\,. \end{aligned} \tag{14.55}$$

Adding together the singular part (14.46) to the regular part (14.55) of c^{A}, we obtain eventually

$$\boxed{\begin{aligned} c^{\mathrm{A}}(t, t') &= \int_{t_0}^{+\infty} d\bar{t}\, \left[a^{\delta}(t)\, \delta(t - \bar{t}) + a^{\mathrm{A}}_{\mathrm{r}}(t, \bar{t})\right]\left[b^{\delta}(\bar{t})\, \delta(\bar{t} - t') + b^{\mathrm{A}}_{\mathrm{r}}(\bar{t}, t')\right] \\ &= \int_{t_0}^{+\infty} d\bar{t}\, a^{\mathrm{A}}(t, \bar{t})\, b^{\mathrm{A}}(\bar{t}, t') \end{aligned}}\,, \tag{14.56}$$

which in terms of the compact notation (14.19) reads

$$\boxed{c^{\mathrm{A}} = a^{\mathrm{A}} \bullet b^{\mathrm{A}}}\,. \tag{14.57}$$

Component $c^{\mathrm{K}}(t,t')$

We extend the definition (13.18) to a generic function $c(z,z')$ and write for real-time arguments

$$
\begin{aligned}
c^{\mathrm{K}}(t,t') = c^{>}(t,t') + c^{<}(t,t') &\underset{\substack{\uparrow \\ \text{[Eqs.(14.18) and (14.25)]}}}{=} \int_{t_0}^{+\infty} d\bar{t}\,\left[a^{>}(t,\bar{t})b^{\mathrm{A}}(\bar{t},t') + a^{\mathrm{R}}(t,\bar{t})b^{>}(\bar{t},t')\right] \\
&+ \int_{t_0}^{+\infty} d\bar{t}\,\left[a^{\mathrm{R}}(t,\bar{t})b^{<}(\bar{t},t') + a^{<}(t,\bar{t})b^{\mathrm{A}}(\bar{t},t')\right] - 2i\int_0^{\beta} d\bar{\tau}\,a^{\rceil}(t,\bar{\tau})b^{\lceil}(\bar{\tau},t') \\
&= \int_{t_0}^{+\infty} d\bar{t}\left[a^{\mathrm{R}}(t,\bar{t})b^{\mathrm{K}}(\bar{t},t') + a^{\mathrm{K}}(t,\bar{t})b^{\mathrm{A}}(\bar{t},t')\right] - 2i\int_0^{\beta} d\bar{\tau}\,a^{\rceil}(t,\bar{\tau})b^{\lceil}(\bar{\tau},t'),
\end{aligned}
\tag{14.58}
$$

which in terms of the compact notation (14.19) reads

$$
\boxed{c^{\mathrm{K}} = a^{\mathrm{R}} \bullet b^{\mathrm{K}} + a^{\mathrm{K}} \bullet b^{\mathrm{A}} + 2\,a^{\rceil} \star b^{\lceil}}\,. \tag{14.59}
$$

This result is especially meaningful in the context of the Keldysh formalism, whereby $t_0 \longrightarrow -\infty$ and the vertical track γ^{M} of the contour γ_{K} in Fig. 14.1 is neglected (cf. Section 13.4). In this case, the multiplication of triangular matrices (13.37) corresponds to

$$
\begin{bmatrix} a^{\mathrm{R}} & a^{\mathrm{K}} \\ 0 & a^{\mathrm{A}} \end{bmatrix}
\begin{bmatrix} b^{\mathrm{R}} & b^{\mathrm{K}} \\ 0 & b^{\mathrm{A}} \end{bmatrix}
= \begin{bmatrix} a^{\mathrm{R}}b^{\mathrm{R}} & a^{\mathrm{R}}b^{\mathrm{K}} + a^{\mathrm{K}}b^{\mathrm{A}} \\ 0 & a^{\mathrm{A}}b^{\mathrm{A}} \end{bmatrix}
= \begin{bmatrix} c^{\mathrm{R}} & c^{\mathrm{K}} \\ 0 & c^{\mathrm{A}} \end{bmatrix}, \tag{14.60}
$$

which is equivalent to Eq. (14.53) for the element $(1,1)$, to Eq. (14.57) for the element $(2,2)$, and to Eq. (14.59) with the neglect of the last term therein for the element $(1,2)$.

Remark: Extension of the Langreth rules to multiple convolutions

The above results can be readily extended to multiple convolutions, due to the associative property satisfied by convolutions. Suppose we have to calculate the convolution of three functions

$$
f(z,z') = \int d\bar{z}\,d\bar{z}'\,a(z,\bar{z})\,b(\bar{z},\bar{z}')\,d(\bar{z}',z') \underset{\substack{\uparrow \\ \text{[Eq.(14.8)]}}}{=} \int d\bar{z}'\,c(z,\bar{z}')\,d(\bar{z}',z')\,, \tag{14.61}
$$

such that, for instance, the component $f^{>}$ is given by:

$$
\begin{aligned}
f^{>} &\underset{\substack{\uparrow \\ \text{[Eq.(14.20)]}}}{=} c^{>} \bullet d^{\mathrm{A}} + c^{\mathrm{R}} \bullet d^{>} + c^{\rceil} \star d^{\lceil} \underset{\substack{\uparrow \\ \text{[Eqs.(14.20), (14.53), and (14.32)]}}}{=} \left(a^{>} \bullet b^{\mathrm{A}} + a^{\mathrm{R}} \bullet b^{>} + a^{\rceil} \star b^{\lceil}\right) \bullet d^{\mathrm{A}} \\
&+ a^{\mathrm{R}} \bullet b^{\mathrm{R}} \bullet d^{>} + \left(a^{\mathrm{R}} \bullet b^{\rceil} + a^{\rceil} \star b^{\mathrm{M}}\right) \star d^{\lceil}\,.
\end{aligned}
\tag{14.62}
$$

14.3 Langreth Rules for Particle–Hole–Type Products

We next consider "particle–hole" (p-h) products of two functions, of the form (Fig. 14.8)

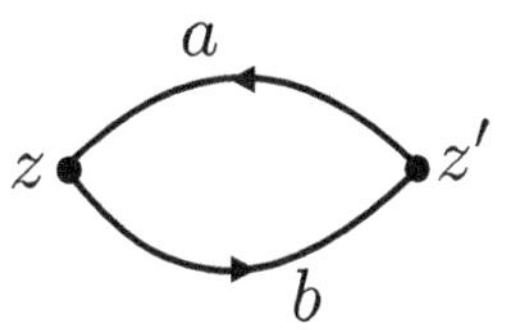

Figure 14.8 Graphical view of Eq. (14.63).

$$c_{\text{p-h}}(z, z') = a(z, z')\, b(z', z), \quad (14.63)$$

with the standard convention that the arrows run from the second to the first argument of a given function, where the terminology originates from the particle–hole rungs of the diagrammatic theory.

In the following, we shall consider the singular parts a^{δ} and b^{δ} to be identically zero (cf. the general expression (14.4)). From a physical point of view, this is the case that one encounters in practice. From a mathematical point of view, we would not be able to deal with the product (14.63) otherwise.

We consider the various components of the particle–hole product (14.63) separately, as reported in the expressions (14.64)–(14.68) and also shown pictorially in the accompanying Figs. 14.9–14.13:

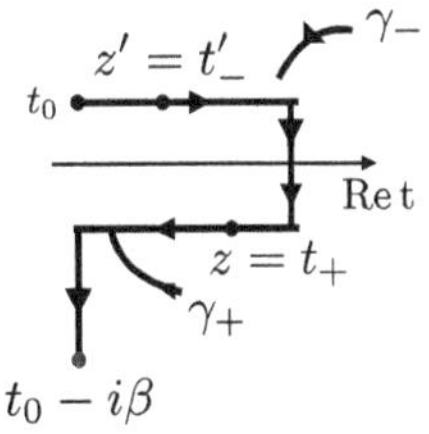

Figure 14.9 Graphical view of Eq. (14.64).

$$c^{>}_{\text{p-h}}(z = t_{+}, z' = t'_{-})$$
$$\longrightarrow \quad c^{>}(t, t') = a^{>}(t, t')\, b^{<}(t', t) \quad (14.64)$$

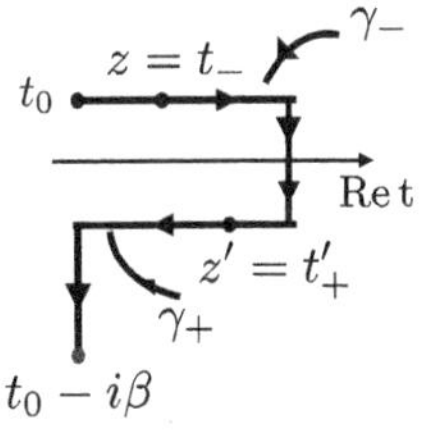

Figure 14.10 Graphical view of Eq. (14.65).

$$c^{<}_{\text{p-h}}(z = t_{-}, z' = t'_{+})$$
$$\longrightarrow \quad c^{<}(t, t') = a^{<}(t, t')\, b^{>}(t', t) \quad (14.65)$$

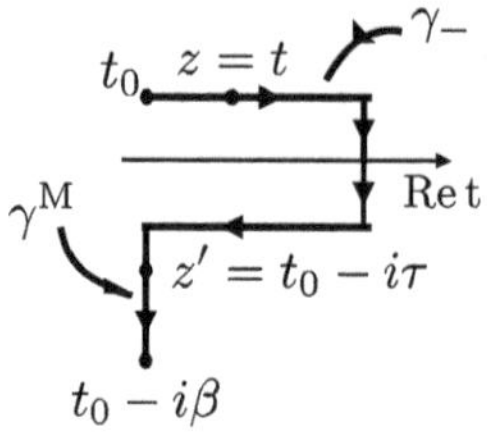

Figure 14.11 Graphical view of Eq. (14.66).

$$c^{\rceil}_{\text{p-h}}(z = t, z' = t_0 - i\tau) \longrightarrow c^{\rceil}(t,\tau) = a^{\rceil}(t,\tau)\, b^{\lceil}(\tau, t) \tag{14.66}$$

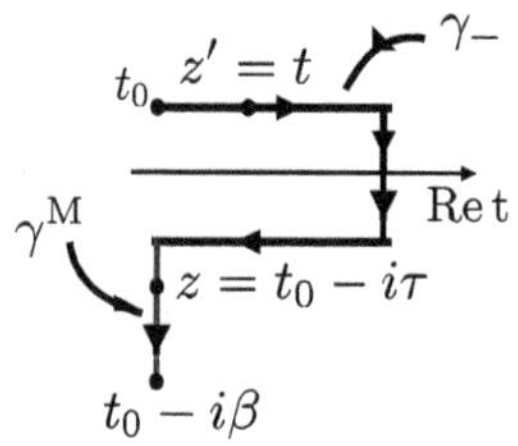

Figure 14.12 Graphical view of Eq. (14.67).

$$c^{\lceil}_{\text{p-h}}(z = t_0 - i\tau, z' = t) \longrightarrow c^{\lceil}(\tau, t) = a^{\lceil}(\tau, t)\, b^{\rceil}(t,\tau) \tag{14.67}$$

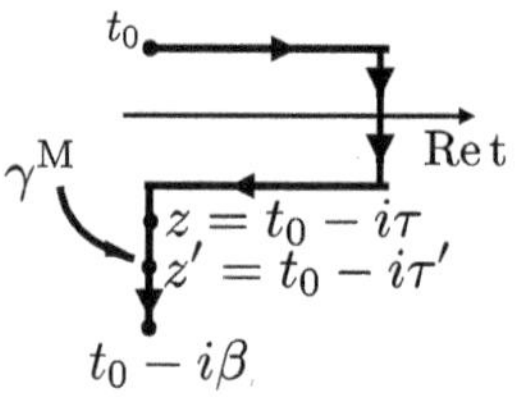

Figure 14.13 Graphical view of Eq. (14.68).

$$c^{\text{M}}_{\text{p-h}}(z = t_0 - i\tau, z' = t_0 - i\tau') \longrightarrow c^{\text{M}}(\tau,\tau') = a^{\text{M}}(\tau,\tau')\, b^{\text{M}}(\tau',\tau). \tag{14.68}$$

With the aforementioned results, we can readily obtain the expressions for the retarded $c^{\text{R}}_{\text{p-h}}(t,t')$, advanced $c^{\text{A}}_{\text{p-h}}(t,t')$, and Keldysh $c^{\text{K}}_{\text{p-h}}(t,t')$ components of the particle–hole product (14.63). From the definitions (14.6), we obtain for the retarded and advanced functions (where we drop the singular terms):

$$\begin{aligned} c^{\text{R}}_{\text{p-h}}(t,t') &= \theta(t-t')\left(c^{>}_{\text{p-h}}(t,t') - c^{<}_{\text{p-h}}(t,t')\right) \\ &\underset{\text{[Eqs.(14.64) and (14.65)]}}{\underset{\uparrow}{=}} \theta(t-t')\left(a^{>}(t,t')\, b^{<}(t',t) - a^{<}(t,t')\, b^{>}(t',t)\right) \\ &\underset{\text{[either]}}{\underset{\uparrow}{=}} a^{\text{R}}(t,t')\, b^{<}(t',t) + a^{<}(t,t')\, b^{\text{A}}(t',t) \\ &\underset{\text{[or]}}{\underset{\uparrow}{=}} a^{>}(t,t')\, b^{\text{A}}(t',t) + a^{\text{R}}(t,t')\, b^{>}(t',t) \end{aligned} \tag{14.69}$$

and

$$\begin{aligned}
c^{\mathrm{A}}_{\mathrm{p-h}}(t,t') &= -\theta(t'-t)\left(c^{>}_{\mathrm{p-h}}(t,t') - c^{<}_{\mathrm{p-h}}(t,t')\right)\\
&\underset{\substack{\uparrow\\ \text{[Eqs.(14.64) and (14.65)]}}}{=} -\theta(t'-t)\left(a^{>}(t,t')\,b^{<}(t',t) - a^{<}(t,t')\,b^{>}(t',t)\right)\\
&\underset{\substack{\uparrow\\ \text{[either]}}}{=} a^{\mathrm{A}}(t,t')\,b^{<}(t',t) + a^{<}(t,t')\,b^{\mathrm{R}}(t',t)\\
&\underset{\substack{\uparrow\\ \text{[or]}}}{=} a^{>}(t,t')\,b^{\mathrm{R}}(t',t) + a^{\mathrm{A}}(t,t')\,b^{>}(t',t)\,.
\end{aligned} \tag{14.70}$$

For the Keldysh component, we instead obtain:

$$\begin{aligned}
c^{\mathrm{K}}_{\mathrm{p-h}}(t,t') &= c^{>}_{\mathrm{p-h}}(t,t') + c^{<}_{\mathrm{p-h}}(t,t') \underset{\substack{\uparrow\\ \text{[Eqs.(14.64) and (14.65)]}}}{=} a^{>}(t,t')\,b^{<}(t',t) + a^{<}(t,t')\,b^{>}(t',t)\\
&= \frac{1}{2}\left\{a^{\mathrm{K}}(t,t')\,b^{\mathrm{K}}(t',t) + a^{\mathrm{R}}(t,t')\,b^{\mathrm{A}}(t',t) + a^{\mathrm{A}}(t,t')\,b^{\mathrm{R}}(t',t)\right\}.
\end{aligned} \tag{14.71}$$

Alternatively, we can make use of the equivalent to the expressions (13.20) and (13.21) to rewrite Eqs. (14.69) and (14.70) in the form:

$$\begin{aligned}
c^{\mathrm{R}}_{\mathrm{p-h}}(t,t') &= a^{>}(t,t')\,b^{\mathrm{A}}(t',t) + a^{\mathrm{R}}(t,t')\,b^{>}(t',t)\\
&= \frac{1}{2}\left\{a^{\mathrm{K}}(t,t')\,b^{\mathrm{A}}(t',t) + a^{\mathrm{R}}(t,t')\,b^{\mathrm{K}}(t',t)\right\}
\end{aligned} \tag{14.72}$$

and

$$\begin{aligned}
c^{\mathrm{A}}_{\mathrm{p-h}}(t,t') &= a^{\mathrm{A}}(t,t')\,b^{>}(t',t) + a^{>}(t,t')\,b^{\mathrm{R}}(t',t)\\
&= \frac{1}{2}\left\{a^{\mathrm{A}}(t,t')\,b^{\mathrm{K}}(t',t) + a^{\mathrm{K}}(t,t')\,b^{\mathrm{R}}(t',t)\right\}.
\end{aligned} \tag{14.73}$$

The above alternative expressions turn out to be useful in different contexts.

14.4 Langreth Rules for Particle–Particle-Type Products

We finally consider "particle–particle" (p–p) products of two functions of the form

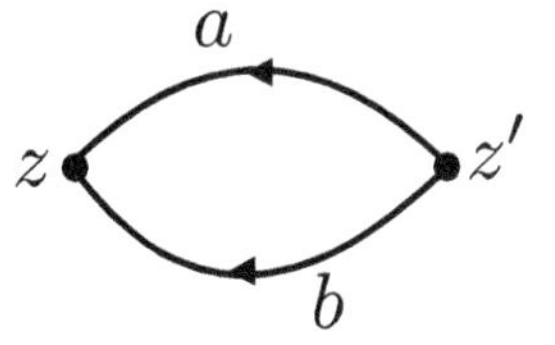

$$c_{\mathrm{p-p}}(z,z') = a(z,z')\,b(z,z')\,. \tag{14.74}$$

Figure 14.14 Graphical view of Eq. (14.74).

The various components of $c_{\text{p-p}}(z, z')$ can be obtained along similar lines to Eqs. (14.64)–(14.68). The results are as follows:

$$\begin{cases} c^{>}_{\text{p-p}}(t, t') = a^{>}(t, t')\, b^{>}(t, t') & \text{[to be compared with Eq. (14.64)]} \\ c^{<}_{\text{p-p}}(t, t') = a^{<}(t, t')\, b^{<}(t, t') & \text{[to be compared with Eq. (14.65)]} \\ c^{\rceil}_{\text{p-p}}(t, \tau) = a^{\rceil}(t, \tau)\, b^{\rceil}(t, \tau) & \text{[to be compared with Eq. (14.66)]} \\ c^{\lceil}_{\text{p-p}}(\tau, t) = a^{\lceil}(\tau, t)\, b^{\lceil}(\tau, t) & \text{[to be compared with Eq. (14.67)]} \\ c^{\text{M}}_{\text{p-p}}(\tau, \tau') = a^{\text{M}}(\tau, \tau')\, b^{\text{M}}(\tau, \tau') & \text{[to be compared with Eq. (14.68)]}\ . \end{cases} \tag{14.75}$$

In addition, we also obtain in the place of Eqs. (14.69)–(14.73):

$$\begin{aligned} c^{\text{R}}_{\text{p-p}}(t, t') &= \theta(t - t') \left(c^{>}_{\text{p-p}}(t, t') - c^{<}_{\text{p-p}}(t, t') \right) \\ &= \theta(t - t') \left(a^{>}(t, t')\, b^{>}(t, t') - a^{<}(t, t')\, b^{<}(t, t') \right) \\ &\underset{\substack{\uparrow \\ \text{[either]}}}{=} a^{\text{R}}(t, t')\, b^{>}(t, t') + a^{<}(t, t')\, b^{\text{R}}(t, t') \\ &\underset{\substack{\uparrow \\ \text{[or]}}}{=} a^{>}(t, t')\, b^{\text{R}}(t, t') + a^{\text{R}}(t, t')\, b^{<}(t, t'), \end{aligned} \tag{14.76}$$

$$\begin{aligned} c^{\text{A}}_{\text{p-p}}(t, t') &= -\theta(t' - t) \left(c^{>}_{\text{p-p}}(t, t') - c^{<}_{\text{p-p}}(t, t') \right) \\ &= -\,\theta(t' - t)\ \left(a^{>}(t, t')\, b^{>}(t, t') - a^{<}(t, t')\, b^{<}(t, t') \right), \\ &\underset{\substack{\uparrow \\ \text{[either]}}}{=} a^{\text{A}}(t, t')\, b^{>}(t, t') + a^{<}(t, t')\, b^{\text{A}}(t, t') \\ &\underset{\substack{\uparrow \\ \text{[or]}}}{=} a^{>}(t, t')\, b^{\text{A}}(t, t') + a^{\text{A}}(t, t')\, b^{<}(t, t'), \end{aligned} \tag{14.77}$$

$$\begin{aligned} c^{\text{K}}_{\text{p-p}}(t, t') &= c^{>}_{\text{p-p}}(t, t') + c^{<}_{\text{p-p}}(t, t')\ =\ a^{>}(t, t')\, b^{>}(t, t') + a^{<}(t, t')\, b^{<}(t, t') \\ &= \frac{1}{2} \left\{ a^{\text{K}}(t, t')\, b^{\text{K}}(t, t') + a^{\text{R}}(t, t')\, b^{\text{R}}(t, t') + a^{\text{A}}(t, t')\, b^{\text{A}}(t, t') \right\}, \end{aligned} \tag{14.78}$$

$$\begin{aligned} c^{\text{R}}_{\text{p-p}}(t, t') &= a^{>}(t, t')\, b^{\text{R}}(t, t') + a^{\text{R}}(t, t')\, b^{<}(t, t') \\ &= \frac{1}{2} \left\{ a^{\text{K}}(t, t')\, b^{\text{R}}(t, t') + a^{\text{R}}(t, t')\, b^{\text{K}}(t, t') \right\}, \end{aligned} \tag{14.79}$$

$$c^{\mathrm{A}}_{\mathrm{p-p}}(t,t') = a^{\mathrm{A}}(t,t')\, b^{<}(t,t') + a^{>}(t,t')\, b^{\mathrm{A}}(t,t')$$
$$= \frac{1}{2}\{a^{\mathrm{K}}(t,t')\, b^{\mathrm{A}}(t,t') + a^{\mathrm{A}}(t,t')\, b^{\mathrm{K}}(t,t')\}\,. \tag{14.80}$$

Remark: A possible symmetry

There are cases when the symmetry property $a(z,z') = a(z',z)$ holds, such that $c_{\mathrm{p-h}}(z,z') = a(z',z)\, b(z',z) = c_{\mathrm{p-p}}(z',z)$. This is what occurs, for instance, for the GW approximation (cf. Section 17.4), where $a(z',z) \to W(z,z')$ is the dynamically screened two-particle interaction, which takes the place of the bare interparticle potential $v(z,z')$ in the Fock self-energy (11.31). In this case, however, it appears appropriate on physical grounds to interpret the GW self-energy as a particle–hole product [15] since the screening properties it depends on are due to particle–hole excitations across the Fermi surface [32].

In conclusion, with these tools at our disposal, we can convert any expression in terms of contour variables, or any diagram for contour-ordered Green's functions, into product of real-time functions.

15

The Kadanoff–Baym Equations

In Chapter 11, we have derived the nonequilibrium Dyson equations (11.26) and (11.28) in integro-differential form, which we recall for convenience:

$$\begin{cases} \left[i\frac{d}{dz_1} - h(1)\right] G(1,2) - \int d3\, \Sigma(1,3)\, G(3,2) = \delta(1,2), \\ \\ \left[-i\frac{d}{dz_2} - h(2)\right] G(1,2) - \int d3\, G(1,3)\, \Sigma(3,2) = \delta(1,2)\,. \end{cases} \tag{15.1}$$

Here, the time variables run over a generic contour C, the single-particle Hamiltonian $h(1) = h(\mathbf{r}_1, z_1) = h_0(\mathbf{r_1}) + V_{\text{ext}}(\mathbf{r}_1, z_1)$ contains the time-dependent external potential $V_{\text{ext}}(\mathbf{r}_1, z_1)$ (cf. Eq. (11.1)), and we have used the property $\bar{\Sigma} = \Sigma$ that was justified in Section 11.5. In addition, the boundary conditions (12.2) and (12.3) have to be imposed on the solutions of Eqs. (15.1), in the order.

When dealing with nonequilibrium situations, the two equations (15.1) are not equivalent to each other since they convey different information. On physical grounds, this is because, between the time z_1 of the first argument and the time z_2 of the second argument of the contour single-particle Green's function $G(1,2)$, the external time-dependent potential V_{ext} (and possibly also the interparticle interaction v) has, in general, evolved. Nonetheless, one can determine the full time dependence of $G(1,2)$ from either one of these equations, provided the two times z_1 and z_2 run over all possible values along the contour C. Consideration of both equations (15.1) will, however, be important, either *in practice* when implementing their numerical solution (cf. Section 36.1 in Part III) or *in principle*, for instance, when deriving from them the Boltzmann equation, as it was done in Ref. [5]. For definiteness, we limit considering only the first of Eqs. (15.1) here.

15.1 Converting the Dyson Equation for the Contour Single-Particle Green's Function to Real-Time Variables

The Dyson equations (15.1) hold *formally* with all time variables running over the contour C. To make their solution useful *in practice*, these time variables should be converted into standard real-time arguments, following what was done in Chapter 13 with the contour C taken like in Fig. 13.1. There, we have concluded that it is sufficient to consider only four independent physical Green's functions (out of initial nine functions). As anticipated at the end of Section 13.3, we may take these four functions to be (G^{R}, $G^{<}$, $G^{\rceil}$, and G^{M}). In the following, we shall then consider how the first of Eqs. (15.1) reads when specified to these four Green's functions. Nonetheless, for completeness and later purposes, we will consider the first of Eqs. (15.1) also for the remaining Green's functions (G^{A}, $G^{>}$, $G^{\lceil}$, and G^{K}).

Dyson equation for G^{M}

In this case, $z_1 = t_0 - i\tau_1$ and $z_2 = t_0 - i\tau_2$ (in line with Eq. (13.11)), such that in the first of Eq. (15.1) $i\frac{d}{dz_1} \rightarrow i\frac{d}{(-id\tau_1)} = -\frac{d}{d\tau_1}$ and $\delta(z_1 - z_2) = i\delta(\tau_1 - \tau_2)$ as in Eq. (14.40). In addition, owing to the property (7.2), all time variables in the Dyson equation for G^{M} can be restricted to the vertical track γ^{M} (cf. Fig. 14.1). With the result (14.43) for the Matsubara component of the convolution entering the first of Eq. (15.1), we obtain eventually:

$$\left[-\frac{d}{d\tau_1} - h^{\mathrm{M}}\right] G^{\mathrm{M}}(\tau_1, \tau_2) + i\int_0^{\beta} d\tau_3\, \Sigma^{\mathrm{M}}(\tau_1, \tau_3)\, G^{\mathrm{M}}(\tau_3, \tau_2) = i\,\delta(\tau_1 - \tau_2), \tag{15.2}$$

where

(i) for simplicity, here and in the following spatial and spin variables $\mathbf{x}$ are suppressed (implying an implicit matrix multiplication in these variables);
(ii) the single-particle Hamiltonian h^{M} does not contain the time-dependent external potential (cf. Section 8.1); and
(iii) we have identified G^{M} with the component G_{33} of Section 13.1 and Σ^{M} with the corresponding self-energy.

Remark: Connection with a standard notation utilized in thermodynamic equilibrium

Standard textbooks on quantum field theory in statistical physics [6, 7] adopt a different convention for the temperature single-particle Green's function, namely,

$$G^{\mathrm{M}}_{\mathrm{FW}}(\tau, \tau') = -\theta(\tau - \tau')\,\langle \psi_{\mathrm{H}}(\tau)\, \psi^{\dagger}_{\mathrm{H}}(\tau')\rangle \pm \theta(\tau' - \tau)\,\langle \psi^{\dagger}_{\mathrm{H}}(\tau')\, \psi_{\mathrm{H}}(\tau)\rangle, \tag{15.3}$$

where the field operators in the Heisenberg representation along the vertical track γ^{M} have the form (12.6) and the plus (minus) sign refers to fermions (bosons). On the other hand, from the relations (14.7) and (13.5)–(13.6), we write in our notation

$$\begin{aligned} G_{33}(\tau,\tau') \;\longleftrightarrow\; & G^{\mathrm{M}}(\tau,\tau') = \theta(\tau-\tau')\,G^{>}(\tau,\tau') + \theta(\tau'-\tau)\,G^{<}(\tau,\tau') \\ = \;& \theta(\tau-\tau')\,(-i)\,\langle \psi_{\mathrm{H}}(\tau)\,\psi^{\dagger}_{\mathrm{H}}(\tau')\rangle \mp \theta(\tau'-\tau)\,i\,\langle \psi^{\dagger}_{\mathrm{H}}(\tau')\,\psi_{\mathrm{H}}(\tau)\rangle \\ \equiv \;& i\,G^{\mathrm{M}}_{\mathrm{FW}}(\tau,\tau'), \end{aligned} \tag{15.4}$$

as it was already anticipated at the end of Section 13.1. With this notation, Eq. (15.2) becomes

$$\left[-\frac{d}{d\tau_1} - h^{\mathrm{M}}\right] G^{\mathrm{M}}_{\mathrm{FW}}(\tau_1,\tau_2) - \int_0^{\beta} d\tau_3\, \Sigma^{\mathrm{M}}_{\mathrm{FW}}(\tau_1,\tau_3)\, G^{\mathrm{M}}_{\mathrm{FW}}(\tau_3,\tau_2) = \delta(\tau_1-\tau_2), \tag{15.5}$$

provided that we also identify $\Sigma^{\mathrm{M}} = i\,\Sigma^{\mathrm{M}}_{\mathrm{FW}}$. That this is the correct identification can be verified already at the level of the Fock self-energy (11.31), for which (upon restoring the space and spin variables $\mathbf{x}$), we write $\Sigma^{\mathrm{M}}(\mathbf{x}_1\tau_1,\mathbf{x}_2\tau_2) = iv(\mathbf{x}_1 t_0 - i\tau_1, \mathbf{x}_2 t_0 - i\tau_2)\, G^{\mathrm{M}}(\mathbf{x}_1\tau_1,\mathbf{x}_2\tau_2) = -v(\mathbf{r}_1-\mathbf{r}_2)\,\delta(\tau_1-\tau_2)\,G^{\mathrm{M}}(\mathbf{x}_1\tau_1,\mathbf{x}_2\tau_2)$, to be compared with $\Sigma^{\mathrm{M}}_{\mathrm{FW}}(\mathbf{x}_1\tau_1,\mathbf{x}_2\tau_2) = -v(\mathbf{r}_1-\mathbf{r}_2)\,\delta(\tau_1-\tau_2)\,G^{\mathrm{M}}_{\mathrm{FW}}(\tau_1,\tau_2)$ [6, 7].

It is important to emphasize that, in Eq. (15.2), there is *no mixing* between G^{M} and the other components of the Green's function (while the equations for the other components couple with G^{M} and with each other). When represented diagrammatically, the self-energy Σ^{M} also depends only on G^{M}. In addition, since the single-particle Hamiltonian h^{M} is constant along the vertical track, $G^{\mathrm{M}}(\tau,\tau')$ (and thus $\Sigma^{\mathrm{M}}(\tau,\tau')$) depends on the time difference $\tau-\tau'$ and not on τ and τ' separately. Taking further into account the boundary condition (12.2), which in the present context reads

$$G^{\mathrm{M}}(\tau=0,\tau') = \pm\, G^{\mathrm{M}}(\tau=\beta,\tau'), \tag{15.6}$$

where the plus (minus) sign refers to bosons (fermions), we then write (cf. Eqs. (12.7) and (12.8))

$$G^{\mathrm{M}}(\tau,\tau') = G^{\mathrm{M}}(\tau-\tau') = \frac{i}{\beta}\sum_{n=-\infty}^{+\infty} e^{-i\omega_n(\tau-\tau')}\, G^{\mathrm{M}}(\omega_n) \tag{15.7}$$

with Matsubara frequencies $\omega_n = 2n\frac{\pi}{\beta}$ for bosons and $\omega_n = (2n+1)\frac{\pi}{\beta}$ for fermions (n integer). This is because entering the expression (15.7) in the boundary condition (15.6) yields

$$\begin{aligned} G^{\mathrm{M}}(\tau=\beta,\tau') &= \frac{i}{\beta}\sum_{n=-\infty}^{+\infty} e^{-i\omega_n(\beta-\tau')}\, G^{\mathrm{M}}(\omega_n) \\ &= \frac{i}{\beta}\sum_{n=-\infty}^{+\infty} e^{i\omega_n\tau'}\, G^{\mathrm{M}}(\omega_n) \begin{cases} e^{-i2\pi n} & \text{(bosons)} \\ e^{-i\pi(2n+1)} & \text{(fermions)} \end{cases} \\ &= \pm\frac{i}{\beta}\sum_{n=-\infty}^{+\infty} e^{-i\omega_n(0-\tau')}\, G^{\mathrm{M}}(\omega_n) \;=\; \pm\, G^{\mathrm{M}}(\tau=0,\tau'), \end{aligned} \tag{15.8}$$

as it should be. In this way, the convolution in Eq. (15.2) becomes

$$\int_0^\beta d\tau_3\, \Sigma^{\mathrm{M}}(\tau_1-\tau_3)\, G^{\mathrm{M}}(\tau_3-\tau_2) \;=\; -\frac{1}{\beta}\sum_{n=-\infty}^{+\infty} e^{-i\omega_n(\tau_1-\tau_2)}\, \Sigma^{\mathrm{M}}(\omega_n)\, G^{\mathrm{M}}(\omega_n)\,, \tag{15.9}$$

such that the Dyson equation (15.2) acquires the simple form

$$\left[i\omega_n - h^{\mathrm{M}}\right] G^{\mathrm{M}}(\omega_n) - \Sigma^{\mathrm{M}}(\omega_n)\, G^{\mathrm{M}}(\omega_n) \;=\; 1, \tag{15.10}$$

where we recall that the dependence on spatial and spin variables has been left implicit. The solution to Eq. (15.10) determines the initial preparation of the system, before it is acted upon by an external time-dependent potential $V_{\mathrm{ext}}(t)$ starting at time t_0.

Remark: Several shades of self-consistency

Quite generally, the self-energy $\Sigma^{\mathrm{M}}[G^{\mathrm{M}}]$ entering Eq. (15.10) is a *functional* of the single-particle Green's function G^{M}, which is also the solution to that equation. This entails that, for a given choice of Σ^{M}, the solution to Eq. (15.10) should in principle be determined in *a self-consistent way*, to the extent that G^{M} obtained as the "output" of the solution should coincide (or, at least, should be as close as possible) with the G^{M} inserted as the "input" in $\Sigma^{\mathrm{M}}[G^{\mathrm{M}}]$. Leaving aside for the moment the specific choice of the self-energy $\Sigma^{\mathrm{M}}[G^{\mathrm{M}}]$ (as well as the kind of spatial environment in which the many-particle system is embedded), it is clear that the process of obtaining a fully (or even a partially) self-consistent solution of Eq. (15.10) is highly nontrivial.

A detailed description of how the path toward a full (or, at least, a partial) self-consistent solution of the Dyson equation (15.10) in Matsubara space can be implemented in practice is reported in Ref. [48], where a comparative study of various degrees of self-consistency was performed within the t-matrix approximation, to describe a homogeneous Fermi gas in the normal phase that evolves throughout the BCS–BEC crossover. While the t-matrix approximation in the normal phase will be described in detail in Chapter 16, consideration of the BCS–BEC crossover is postponed to Chapter 26 in Part II when dealing with the superfluid phase.

Dyson equation for $G^{\rceil}$

Here, $G^{\rceil}$ stands for the right Keldysh component given by Eq. (13.7), whereby $z_1 = t$ and $z_2 = t_0 - i\tau$. With the result (14.31), we then obtain from the first of Eqs. (15.1)

$$\boxed{\begin{aligned} &\left[i\frac{d}{dt} - h(t) \right] G^{\rceil}(t,\tau) - \int_{t_0}^{+\infty} d\bar{t}\; \Sigma^{R}(t,\bar{t})\, G^{\rceil}(\bar{t},\tau) \\ &+ i \int_0^{\beta} d\bar{\tau}\, \Sigma^{\rceil}(t,\bar{\tau})\, G^{M}(\bar{\tau},\tau) = 0, \end{aligned}} \tag{15.11}$$

where:

(i) the single-particle Hamiltonian $h(t)$ now contains the time-dependent external potential $V_{ext}(t)$ (cf. Eq. (11.1));
(ii) the Dirac delta function is missing from the right-hand side because the variables t and τ belong to different branches of the contour (cf. Section 7.3 for the definition of the contour Delta function);
(iii) the integral in the second term on the left-hand side actually runs up to t since the function $\Sigma^{R}(t,\bar{t})$ therein is subject to the constraint that $t > \bar{t}$;
(iv) the function $G^{M}(\bar{\tau},\tau)$ in the third term on the left-hand side is the solution to Eq. (15.2);
(v) to solve the equation, one also needs to know the self-energy $\Sigma^{\rceil}$, which transfers the information about the initial state at equilibrium for $t = t_0^-$ to the real-time propagation for $t > t_0$, as well as the self-energy Σ^{R}, which affects the real-time propagation starting from t_0; and
(vi) owing to the definitions (13.7) and (15.3)–(15.4), the *initial condition* to be utilized at time t_0 reads $G^{\rceil}(t_0,\tau) = G^{M}(0,\tau)$ for a given value of τ.

Dyson equation for $G^{\lceil}$

Here, $G^{\lceil}$ stands for the left Keldysh component given by Eq. (13.9), whereby $z_1 = t_0 - i\tau$ and $z_2 = t$. With the result (14.37), we then obtain from the first of Eqs. (15.1)

$$\boxed{\begin{aligned} &\left[-\frac{d}{d\tau} - h^{M} \right] G^{\lceil}(\tau,t) - \int_{t_0}^{+\infty} d\bar{t}\; \Sigma^{\lceil}(\tau,\bar{t})\, G^{A}(\bar{t},t) \\ &+ i \int_0^{\beta} d\bar{\tau}\, \Sigma^{M}(\tau,\bar{\tau})\, G^{\lceil}(\bar{\tau},t) = 0, \end{aligned}} \tag{15.12}$$

where:

(i) the single-particle Hamiltonian h^{M} does not contain the time-dependent external potential like in Eq. (15.2);
(ii) the Dirac delta function is again missing from the right-hand side;
(iii) the integral in the second term on the left-hand side runs up to t owing to the presence of the function $G^{\mathrm{A}}(\bar{t}, t)$;
(iv) the function $G^{\mathrm{A}}(\bar{t}, t)$ in the second term on the left-hand side has to be determined simultaneously (see below);
(v) the *initial condition* to be utilized at time t_0 reads $G^{\ulcorner}(\tau, t_0) = G^{\mathrm{M}}(\tau, 0)$ for any value of τ; and
(vi) alternatively, one may obtain $G^{\ulcorner}(\tau, t)$ directly from the solution $G^{\urcorner}(t, \tau)$ to Eq. (15.11) using the symmetry property (13.28).

Dyson equation for $G^<$

Here, $G^<$ stands for the lesser Green's function given by Eq. (13.5), whereby $z_1 = t_-$ and $z_2 = t'_+$. With the result (14.25), we then obtain from the first of Eqs. (15.1)

$$\left[i\frac{d}{dt} - h(t)\right] G^<(t, t') - \int_{t_0}^{+\infty} d\bar{t}\; \Sigma^{\mathrm{R}}(t, \bar{t})\, G^<(\bar{t}, t') - \int_{t_0}^{+\infty} d\bar{t}\; \Sigma^<(t, \bar{t})\, G^{\mathrm{A}}(\bar{t}, t') + i \int_0^{\beta} d\bar{\tau}\, \Sigma^{\urcorner}(t, \bar{\tau})\, G^{\ulcorner}(\bar{\tau}, t') = 0, \tag{15.13}$$

where:

(i) the single-particle Hamiltonian $h(t)$ contains the time-dependent external potential $V_{\mathrm{ext}}(t)$ (cf. Eq. (11.1));
(ii) the Dirac delta function is missing from the right-hand side because the variables t and t', although both spanning all real-time values from t_0 up to $+\infty$, belong to different branches of the contour;
(iii) the integral in the second term on the left-hand side runs up to t since the function $\Sigma^{\mathrm{R}}(t, \bar{t})$ implies that $t > \bar{t}$, while the integral in the third term on the left-hand side runs up to t' since the function $G^{\mathrm{A}}(\bar{t}, t')$ implies that $t' > \bar{t}$; and
(iv) the *initial condition* to be utilized at time t_0 reads $G^<(t_0, t_0^+) = G^{\mathrm{M}}(0, 0^+)$. [We defer to Section 36.1 in Part III for the way to solve for $G^<$ in an organic way simultaneously with $G^>$ in the context of the "time-stepping technique."]

Dyson equation for $G^>$

Here, $G^>$ stands for the greater Green's function given by Eq. (13.6), whereby $z_1 = t_+$ and $z_2 = t'_-$ belong again to different branches of the contour. With the result (14.18), we then obtain from the first of Eq. (15.1)

$$\left[i\frac{d}{dt} - h(t)\right] G^>(t,t') - \int_{t_0}^{+\infty} d\bar{t}\, \Sigma^>(t,\bar{t})\, G^{\mathrm{A}}(\bar{t},t') - \int_{t_0}^{+\infty} d\bar{t}\, \Sigma^{\mathrm{R}}(t,\bar{t})\, G^>(\bar{t},t') + i\int_0^{\beta} d\bar{\tau}\, \Sigma^{\rceil}(t,\bar{\tau})\, G^{\lceil}(\bar{\tau},t') = 0, \tag{15.14}$$

where:

(i) the Dirac delta function is missing from the right-hand side because the variables t and t' belong to different branches of the contour;
(ii) the integral in the second term on the left-hand side runs up to t' owing to $G^{\mathrm{A}}(\bar{t},t')$, while the integral in the third term on the left-hand side runs up to t owing to $\Sigma^{\mathrm{R}}(t,\bar{t})$;
(iii) the last term on the left-hand side coincides with the corresponding term of Eq. (15.13); and
(iv) the *initial condition* to be utilized at time t_0 reads $G^>(t_0^+, t_0) = G^{\mathrm{M}}(0^+, 0)$. [We defer again to Section 36.1 in Part III for the way to solve for $G^>$ in an organic way simultaneously with $G^<$ in the context of the "time-stepping technique."]

Dyson equation for G^{R}

The retarded Green's function G^{R} can be obtained directly from the definition (13.15)

$$G^{\mathrm{R}}(t,t') = \theta(t-t')\ \left(G^>(t,t') - G^<(t,t')\right), \tag{15.15}$$

once $G^<$ and $G^>$ have been determined beforehand by solving for Eqs. (15.13) and (15.14) (spatial and spin variables have again been suppressed as before). It may, however, be useful to determine the equation of motion obeyed by G^{R}, starting from the result

$$i\frac{d}{dt}G^{\mathrm{R}}(t,t') = i\,\delta(t-t')\left(G^>(t,t) - G^<(t,t)\right) + \theta(t-t')\, i\frac{d}{dt}\left(G^>(t,t') - G^<(t,t')\right). \tag{15.16}$$

From the definitions (13.5) and (13.6) (and momentarily restoring the space and spin variables), the factor in the first term on the right-hand side is given by:

$$
\begin{aligned}
G^{>}(t,t) - G^{<}(t,t) &\longleftrightarrow -i\,\langle \psi_{\mathrm{H}}(\mathbf{x}t)\,\psi_{\mathrm{H}}^{\dagger}(\mathbf{x}'t)\rangle \pm i\,\langle \psi_{\mathrm{H}}^{\dagger}(\mathbf{x}'t)\,\psi_{\mathrm{H}}(\mathbf{x}t)\rangle \\
&\underset{\substack{\uparrow \\ \text{[Eq.(2.25)]}}}{=} -i\,\Big\langle U(t_0,t)\,\Big(\psi(\mathbf{x})\,\psi^{\dagger}(\mathbf{x}') \mp \psi^{\dagger}(\mathbf{x}')\,\psi(\mathbf{x})\Big)\,U(t,t_0)\Big\rangle = -i\,\delta(\mathbf{x}-\mathbf{x}'),
\end{aligned}
\tag{15.17}
$$

where the upper (lower) sign holds for bosons (fermions), and the commutation/anti-commutation relations (2.2) have been utilized. Note that the result (15.17) determines also the *initial condition* $G^{\mathrm{R}}(\mathbf{x}t_0^{+}, \mathbf{x}'t_0) = -i\,\delta(\mathbf{x}, \mathbf{x}')$. On the other hand, from Eqs. (15.13) and (15.14), the second term on the right-hand side reads

$$
\begin{aligned}
\theta(t-t')\, i\frac{d}{dt}\left(G^{>}(t,t') - G^{<}(t,t')\right) &= \theta(t-t')\, h(t)\left(G^{>}(t,t') - G^{<}(t,t')\right) \\
&+ \theta(t-t') \int_{t_0}^{+\infty} d\bar{t}\, \left(\Sigma^{>}(t,\bar{t}) - \Sigma^{<}(t,\bar{t})\right) G^{\mathrm{A}}(\bar{t},t') \\
&+ \theta(t-t') \int_{t_0}^{+\infty} d\bar{t}\, \Sigma^{\mathrm{R}}(t,\bar{t}) \left(G^{>}(\bar{t},t') - G^{<}(\bar{t},t')\right),
\end{aligned}
\tag{15.18}
$$

where:

(i) in the first term, $\theta(t-t')\,(G^{>}(t,t') - G^{<}(t,t')) = G^{\mathrm{R}}(t,t')$ according to the definition (15.15);
(ii) in the second term $\bar{t} < t' < t$, such that we interpret $\theta(t-\bar{t})\,(\Sigma^{>}(t,\bar{t}) - \Sigma^{<}(t,\bar{t})) = \Sigma^{\mathrm{R}}_{\mathrm{r}}(t,\bar{t})$ as being the "regular part" of Σ^{R}, in line with Eq. (14.6);
(iii) in the third term, there can either be $t' < \bar{t} < t$ or $\bar{t} < t' < t$. When $t' < \bar{t} < t$, we interpret $\theta(\bar{t}-t')\,(G^{>}(\bar{t},t') - G^{<}(\bar{t},t')) = G^{\mathrm{R}}(\bar{t},t')$, while, when $\bar{t} < t' < t$, only the "regular part" of Σ^{R} contributes, and we interpret $\theta(t'-\bar{t})\,(G^{>}(\bar{t},t') - G^{<}(\bar{t},t')) = -G^{\mathrm{A}}(\bar{t},t')$. As a consequence, this last term cancels with the contribution (ii).

Collecting all these results, the equation of motion of the retarded Green's function G^{R} reads:

$$
\boxed{\left[i\frac{d}{dt} - h(t)\right] G^{\mathrm{R}}(t,t') - \int_{t'}^{t} d\bar{t}\, \Sigma^{\mathrm{R}}(t,\bar{t})\, G^{\mathrm{R}}(\bar{t},t') = \delta(t-t')\,.}
\tag{15.19}
$$

Dyson equation for G^{A}

The equation of motion of the advanced Green's function G^{A} given by Eq. (13.17), namely,

$$G^{\mathrm{A}}(t,t') = -\,\theta(t'-t)\,\left(G^{>}(t,t') - G^{<}(t,t')\right) \tag{15.20}$$

can be obtained along similar lines. In this case,

$$\begin{aligned} i\frac{d}{dt}\,G^{\mathrm{A}}(t,t') &= \delta(t-t') \;-\; \theta(t'-t)\,h(t)\,\left(G^{>}(t,t') - G^{<}(t,t')\right) \\ &-\; \theta(t'-t)\int_{t_0}^{+\infty} d\bar{t}\;\left(\Sigma^{>}(t,\bar{t}) - \Sigma^{<}(t,\bar{t})\right)\,G^{\mathrm{A}}(\bar{t},t') \\ &-\; \theta(t'-t)\int_{t_0}^{+\infty} d\bar{t}\;\Sigma^{\mathrm{R}}(t,\bar{t})\,\left(G^{>}(\bar{t},t') - G^{<}(\bar{t},t')\right), \end{aligned} \tag{15.21}$$

where:

(i) in the second term, $-\theta(t'-t)\,(G^{>}(t,t') - G^{<}(t,t')) = G^{\mathrm{A}}(t,t')$ according to the definition (15.20);

(ii) in the third term, there can either be $t<\bar{t}<t'$ or $\bar{t}<t<t'$. When $t<\bar{t}<t'$ we interpret $-\theta(\bar{t}-t)\,(\Sigma^{>}(t,\bar{t}) - \Sigma^{<}(t,\bar{t})) = \Sigma^{\mathrm{A}}_{\mathrm{r}}(t,\bar{t})$ as being the "regular part" of Σ^{A}, while when $\bar{t}<t<t'$ we interpret $\theta(t-\bar{t})\,(\Sigma^{>}(t,\bar{t}) - \Sigma^{<}(t,\bar{t})) = \Sigma^{\mathrm{R}}_{\mathrm{r}}(t,\bar{t})$ as being the "regular part" of Σ^{R};

(iii) in the fourth term $\bar{t}<t<t'$, such that we interpret $-\,\theta(t'-\bar{t})\,(G^{>}(\bar{t},t') - G^{<}(\bar{t},t')) = G^{\mathrm{A}}(\bar{t},t')$;

(iv) the singular part of $\Sigma^{\mathrm{R}}(t,\bar{t})$ present in the fourth term can be transferred to $\Sigma^{\mathrm{A}}_{\mathrm{r}}(t,\bar{t})$ in the third term to form the full $\Sigma^{\mathrm{A}}(t,\bar{t})$, while the part of the fourth term containing the remaining regular part of $\Sigma^{\mathrm{R}}(t,\bar{t})$ cancels with the contribution of the third term containing $\Sigma^{\mathrm{R}}_{\mathrm{r}}(t,\bar{t})$; and

(v) the *initial condition* reads $G^{\mathrm{A}}(\mathbf{x}t_0,\mathbf{x}'t_0^{+}) = i\,\delta(\mathbf{x},\mathbf{x}')$.

In this way, we end up with the following equation of motion for the advanced Green's function:

$$\boxed{\left[i\frac{d}{dt} - h(t)\right]G^{\mathrm{A}}(t,t') - \int_{t}^{t'} d\bar{t}\;\Sigma^{\mathrm{A}}(t,\bar{t})\,G^{\mathrm{A}}(\bar{t},t') = \delta(t-t')\,.} \tag{15.22}$$

Dyson equation for G^{K}

The equation of motion of the Keldysh Green's function G^{K} given by Eq. (13.16), namely,

$$G^{\mathrm{K}}(t,t') = G^{>}(t,t') + G^{<}(t,t')\,, \tag{15.23}$$

can readily be obtained from Eqs. (15.13) and (15.14) without the need for keeping track of the singular parts of the self-energies like before. The result is:

$$\boxed{\begin{aligned}&\left[i\frac{d}{dt}-h(t)\right]G^{\mathrm{K}}(t,t')-\int_{t_0}^{+\infty}d\bar{t}\left(\Sigma^{\mathrm{R}}(t,\bar{t})\,G^{\mathrm{K}}(\bar{t},t')+\Sigma^{\mathrm{K}}(t,\bar{t})\,G^{\mathrm{A}}(\bar{t},t')\right)\\&+2i\int_0^{\beta}d\bar{\tau}\,\Sigma^{\rceil}(t,\bar{\tau})\,G^{\lceil}(\bar{\tau},t')=0,\end{aligned}} \tag{15.24}$$

where we have identified $\Sigma^{\mathrm{K}}=\Sigma^{>}+\Sigma^{<}$ as the Keldysh self-energy.

Remark: The Dyson equation within the Keldysh formalism

To the extent that G^{K} and Σ^{K} are relevant in the context of the Keldysh formalism, as discussed in Section 13.4, the last term on the left-hand side of Eq. (15.24) can be neglected, and the three Dyson equations (15.19), (15.22), and (15.24) can be organized together in the following form [14]

$$\begin{aligned}&\left[i\frac{d}{dt}-h(t)\right]\begin{bmatrix}G^{\mathrm{R}}(t,t') & G^{\mathrm{K}}(t,t')\\ 0 & G^{\mathrm{A}}(t,t')\end{bmatrix}-\int_{-\infty}^{+\infty}d\bar{t}\begin{bmatrix}\Sigma^{\mathrm{R}}(t,\bar{t}) & \Sigma^{\mathrm{K}}(t,\bar{t})\\ 0 & \Sigma^{\mathrm{A}}(t,\bar{t})\end{bmatrix}\begin{bmatrix}G^{\mathrm{R}}(\bar{t},t') & G^{\mathrm{K}}(\bar{t},t')\\ 0 & G^{\mathrm{A}}(\bar{t},t')\end{bmatrix}\\&\quad=\begin{bmatrix}\delta(t-t') & 0\\ 0 & \delta(t-t')\end{bmatrix},\end{aligned} \tag{15.25}$$

consistently with the matrix multiplication (14.60) of triangular matrices. In this case, the dynamics from $t=-\infty$ to $t=t_0$ is governed by the adiabatic Hamiltonian (8.8), whose specific task is to build up correlations in the many-particle system starting from a noninteracting initial state at $t=-\infty$ and ending up at $t=t_0$ with a fully interacting state, on which a time-dependent perturbation then acts for $t\geq t_0$. [In practice, it may be convenient to utilize this adiabatic procedure to solve instead the two Dyson equations (15.13) for $G^{<}$ and (15.14) for $G^{>}$ from which (G^{R}, G^{A}, and G^{K}) can then be readily obtained, after having removed the last term on the left-hand side of those equations and let the lower limit of the time integrals recede to $-\infty$, by adapting to this case the "time- stepping technique" to be discussed in Part III.]

Although it may require long evolution times, the adiabatic procedure is meant to be an alternative (and, possibly, convenient) tool for achieving quantum control to drive the system to a definite target state, which in the present case should be the *fully* correlated state corresponding to the temperature single-particle Green's function G^{M} obtained by solving Eqs. (15.7) and (15.10) in a (possibly) *fully* self-consistent way. Which one of these alternative procedures (which refer, respectively, to the Konstantinov–Perel and Keldysh contours discussed in Section 12.2) is, in practice, the most convenient one, to end up with the required fully correlated state at $t=t_0$, can only be determined by dedicated numerical calculations.

15.2 Connection with the Original Kadanoff–Baym Equations

It can readily be verified that the Dyson equations (15.13) for $G^{<}$ and (15.14) for $G^{>}$ coincide, respectively, with the original Kadanoff–Baym equations given by Eqs. (8.27a) and (8.27b) of Ref. [5], apart from the following differences and restrictions. Specifically, in the original Kadanoff–Baym equations of Ref. [5]:

(i) terms such as the last ones on the left-hand side of Eqs. (15.13) and (15.14) are missing, thus implying that transient phenomena are not taken into account;
(ii) t_0 is let to recede to $-\infty$ in the terms analog to the second and third terms on the left-hand side of Eqs. (15.13) and (15.14), without, however, considering any adiabatic process to build up the interacting initial state;
(iii) the initial condition at $t = -\infty$ is that of an interacting state with the full many-particle Hamiltonian; and
(iv) the Fock term is missing in the analog of the retarded self-energy entering Eqs. (15.13) and (15.14), possibly owing to an incomplete handling of the singular terms that should be present in the retarded and advanced self-energies.

Recall that the original Kadanoff–Baym equations were formally obtained in Ref. [5] through a process of analytic continuation from the imaginary to the real axis, as it was briefly mentioned in Section 12.2. No process of analytic continuation was instead required to obtain the Dyson equations from (15.11) to (15.24). Nevertheless, in line with the Kadanoff–Baym approach, in Part II, we shall find it useful to resort to the process of analytic continuation from the imaginary to the real time axes, when dealing with formal manipulations involving the Dyson equations (15.1).

Remark: A mixed time representation

As mentioned at the beginning of this chapter, Kadanoff and Baym made *simultaneous* use of both Dyson equations (15.1) in the first (t) and second (t') time variables, while proceeding toward the derivation of the Boltzmann transport equation in chapter 9 of Ref. [5]. To this end, what they did is to subtract the second from the first of the equations (15.1) and change the time variables from (t, t') to (T, τ) defined by $T = \frac{(t+t')}{2}$ and $\tau = t - t'$ (with a similar transformation applied also to the spatial variables). In this way, $\frac{\partial}{\partial t} + \frac{\partial}{\partial t'} = \frac{\partial}{\partial T}$, while the dependence on the relative time τ was Fourier transformed, by exploiting the presence of a slowly varying external disturbance.

Finally, we report a list containing a few references where the numerical implementation of the solution of the Kadanoff–Baym equations was explicitly addressed [40]. The list includes:

(a) H. S. Köhler, N. H. Kwong, and H. A. Yousif, *A Fortran code for solving the Kadanoff–Baym equations for a homogeneous fermion system*, Comp. Phys. Commun. **123**, 123 (1999).

(b) M. Eckstein, M. Kollar, and P. Werner, *Interaction quench in the Hubbard model: Relaxation of the spectral function and the optical conductivity*, Phys. Rev. B **81**, 115131 (2010).

(c) K. Balzer and M. Bonitz, *Nonequilibrium Green's Functions Approach to Inhomogeneous Systems* (Springer, Heidelberg, 2013).

Additional references along these lines will be mentioned in Section 36.1 of Part III in the context of the "time-stepping technique."

16

The *t*-Matrix Approximation in the Normal Phase

When dealing with fermionic many-body systems at equilibrium in condensed matter, there are two main approximations for the single-particle self-energy that have often been utilized in two different contexts, namely, in the high- and low-density limits.

In the high-density limit of a degenerate electron gas, for which the interparticle interaction $v(\mathbf{r} - \mathbf{r}')$ in the interaction part (2.4) of the system Hamiltonian is of the Coulomb type, it is expected on physical grounds that the screening properties of the electron gas should play an important role. Accordingly, a meaningful approximation to the single-particle self-energy is obtained by replacing the "bare" Coulomb potential v in the Fock (exchange) term (11.31) by a "screened" potential W, thus resulting in the so-called GW *approximation* for the self-energy (which is sometimes referred to as the screened-exchange interaction – cf. section 5.8 of Ref. [10]). A systematic approach along these lines was provided by Hedin [49], where the homogeneous electron gas problem was formally re-investigated, aiming at estimating the convergence of the expansion from the high-density to the more realistic metallic-density regimes. To this end, a functional-derivative technique was used to generate an expansion of the diagrammatic structure in terms of a screened potential W rather than the bare Coulomb potential v. But it was for inhomogeneous systems (like insulators and semiconductors [32, 50, 51] and complex molecular systems in general [52, 53]) that the GW approximation has found its most successful applications. In Chapter 17, we shall consider the nonequilibrium (time-dependent) version of the GW approximation, in the process of reformulating the Schwinger–Keldysh approach using a time-dependent version of the functional-derivative technique.

For a low-density (or dilute) Fermi gas with a short-range interparticle interaction, on the other hand, the initial focus is on the scattering properties of two fermions *in vacuum*, which are then extended to the presence of the medium through which the fermions are self-consistently propagating. The resulting *t-matrix approximation* was originally considered by Galitskii for the case of a repulsive interparticle interaction [54] and later extended by Gorkov and Melik–Barkhudarov to the case of an attractive interparticle interaction to deal with the phenomenon of superfluidity [55]. Nowadays, routine use is made of the t-matrix approximation (and in part also of its more recent improved versions [56, 57]) in the context of the BCS–BEC crossover [20], which has been experimentally achieved in ultracold Fermi gases with an attractive interparticle interaction.[1] Below, we consider the nonequilibrium (time-dependent) version of the t-matrix approximation for fermions in the normal phase, leaving the form of interparticle interaction arbitrary. In Part II, we shall further consider its extension to the superfluid case in the context of the BCS–BEC crossover.

16.1 Integral Equation for the Effective Two-Particle Interaction and the Single-Particle Self-Energy

The t-matrix approximation is based on the *effective two-particle interaction* Γ depicted in Fig. 16.1, which represents an infinite sequence of repeated scattering processes between two fermions in the medium. It generalizes to the present nonequilibrium context the original Galitskii approach at equilibrium [54]. The corresponding integral equation reads

$$\Gamma(1,2;1',2') = \delta(1,1')\delta(2,2')iv(1,2) + iv(1,2)\int d1''d2''G(1,1'')G(2,2'')\Gamma(1'',2'';1',2'), \quad (16.1)$$

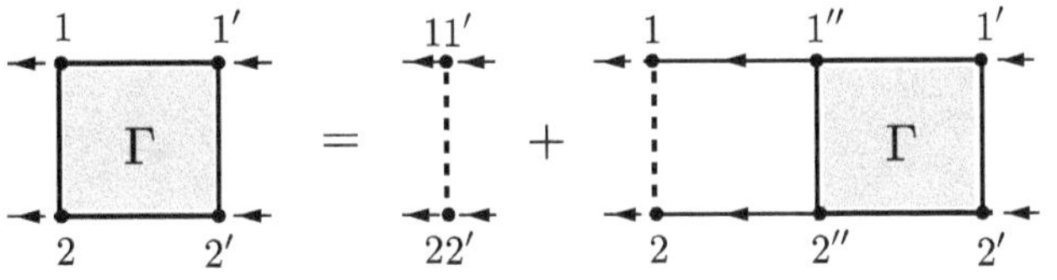

Figure 16.1 Graphical representation of the integral equation (16.1) that defines the effective two-particle interaction Γ. Full lines represent the single-particle Green's functions, with the arrow pointing from the second argument to the first, and dashed lines representing the interparticle interaction whose form is left unspecified for the moment. As before, the short-hand notation $1 \leftrightarrow (\mathbf{r}_1, \sigma_1, z_1)$ stands for the collection of spatial ($\mathbf{r}_1$), spin (σ_1), and contour time (z_1) variables.

[1] In this context, it is interesting to mention that an explicit numerical comparison between the GW and t-matrix approximations, performed for the neutral excited states of molecular systems [58], has suggested that the t-matrix approximation indeed performs best in few-electron systems for which the electron density remains low.

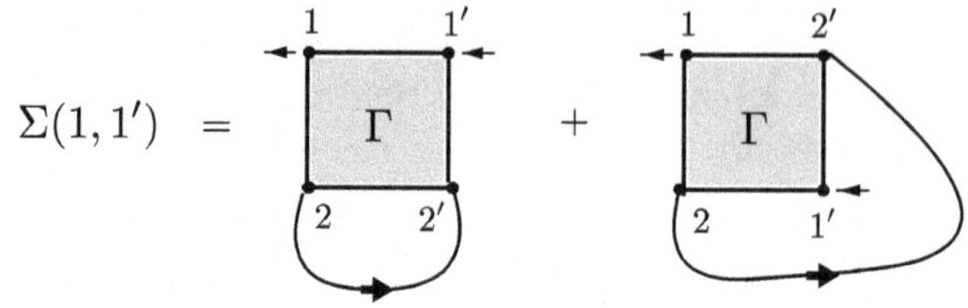

Figure 16.2 Single-particle self-energy within the *t*-matrix approximation. The notation is like in Fig. 16.1.

where the Dirac delta function $\delta(1,2)$ is defined in the last line of Eq. (11.16), the "bare" two-particle interaction $v(1,2)$ is given by Eq. (11.20) (or, more generally, by Eq. (11.21), with an explicit time dependence, which is here omitted for simplicity), and the presence of the imaginary unit is required to comply with the diagrammatic rules of Chapter 10. The single-particle self-energy within the *t*-matrix approximation, as depicted in Fig. 16.2, is then given by

$$\Sigma(1,1') = -\int d2d2' \left(\Gamma(1,2;1',2') - \Gamma(1,2;2',1')\right) G(2',2^+) , \qquad (16.2)$$

where the signs of the two terms take into account the presence of a fermionic loop in the first but not in the second term, again according to the diagrammatic rules of Chapter 10. In addition, the presence of an (infinitesimally) augmented time in the single-particle Green's function is required to deal with the first term of the integral equation (16.1), where the bare two-particle interaction v appears only once, to which there correspond the following contributions to the self-energy

$$\begin{aligned} \Sigma(1,1') \;\longrightarrow\; & -\int d2d2' \left(\delta(1,1')\delta(2,2') - \delta(1,2')\delta(2,1')\right) iv(1,2)\, G(2',2^+) \\ = \;& -i\,\delta(1,1') \int d2\, v(1,2)\, G(2,2^+) + iv(1,1')\, G(1,1'^+) \qquad (16.3) \end{aligned}$$

thereby recovering the Hartree (11.30) and Fock (11.31) approximations to the self-energy.

To proceed further,[2] we note that due to the two-particle interaction (11.20) we are considering throughout, $\Gamma(1,2;1',2')$ contains the product $\delta(z_1,z_2)\delta(z_{1'},z_{2'})$ of two contour delta functions (cf. Fig. 16.1). Accordingly, we set

$$\Gamma(1,2;1',2') = \delta(z_1,z_2)\delta(z_{1'},z_{2'}) \, \langle \mathbf{x}_1,\mathbf{x}_2 | \Gamma(z_1,z_{1'}) | \mathbf{x}_{1'},\mathbf{x}_{2'} \rangle , \qquad (16.4)$$

thereby reducing to two the number of independent contour time variables (like before, the notation $\mathbf{x} = (\mathbf{r},\sigma)$ refers to spatial and spin variables). With this position, the integral equation (16.1) becomes:

[2] In the following formal manipulations, we partly follow Appendix F of Ref. [39].

$$\begin{aligned}
&\delta(z_1, z_2)\delta(z_{1'}, z_{2'}) \langle \mathbf{x}_1, \mathbf{x}_2|\Gamma(z_1, z_{1'})|\mathbf{x}_{1'}, \mathbf{x}_{2'}\rangle \\
&= \delta(z_1, z_{1'})\delta(\mathbf{x}_1, \mathbf{x}_{1'})\delta(z_2, z_{2'})\delta(\mathbf{x}_2, \mathbf{x}_{2'})i\delta(z_1, z_2)v(\mathbf{r}_1 - \mathbf{r}_2) \\
&+ i\delta(z_1, z_2)v(\mathbf{r}_1 - \mathbf{r}_2)\int d1''d2''G(1, 1'')G(2, 2'')\delta(z_{1''}, z_{2''})\delta(z_{1'}, z_{2'}) \\
&\times \langle \mathbf{x}_{1''}, \mathbf{x}_{2''}|\Gamma(z_{1''}, z_{1'})|\mathbf{x}_{1'}, \mathbf{x}_{2'}\rangle, \qquad (16.5)
\end{aligned}$$

where in the first term on the right-hand side, we can rewrite

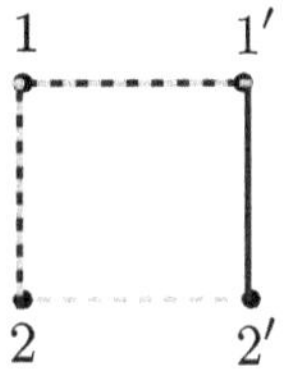

$$\begin{aligned}
&\delta(z_1, z_{1'})\,\delta(z_2, z_{2'})\,\delta(z_1, z_2) \\
&= \delta(z_1, z_{1'})\,\delta(z_{1'}, z_{2'})\,\delta(z_1, z_2)\,, \qquad (16.6)
\end{aligned}$$

Figure 16.3 Graphical view of Eq. (16.6).

as shown graphically in Fig. 16.3. We can then drop the product $\delta(z_1, z_2)\delta(z_{1'}, z_{2'})$ from both sides of Eq. (16.5), ending up with the following somewhat simplified integral equation:

$$\begin{aligned}
\langle \mathbf{x}_1, \mathbf{x}_2|\Gamma(z_1, z_{1'})|\mathbf{x}_{1'}, \mathbf{x}_{2'}\rangle &= v(\mathbf{r}_1 - \mathbf{r}_2)\delta(z_1, z_{1'})\delta(\mathbf{x}_1, \mathbf{x}_{1'})\delta(\mathbf{x}_2, \mathbf{x}_{2'}) \\
&+ iv(\mathbf{r}_1 - \mathbf{r}_2)\int dz_{1''}d\mathbf{x}_{1''}d\mathbf{x}_{2''}\, G(\mathbf{x}_1 z_1, \mathbf{x}_{1''}z_{1''})G(\mathbf{x}_2 z_1, \mathbf{x}_{2''}z_{1''}) \\
&\times \langle \mathbf{x}_{1''}, \mathbf{x}_{2''}|\Gamma(z_{1''}, z_{1'})|\mathbf{x}_{1'}, \mathbf{x}_{2'}\rangle\,. \qquad (16.7)
\end{aligned}$$

At this point, we can introduce the *particle–particle bubble* in real space

$$\langle \mathbf{x}_1, \mathbf{x}_2|\tilde{\chi}_{\mathrm{pp}}(z_1, z_{1''})|\mathbf{x}_{1''}, \mathbf{x}_{2''}\rangle \equiv G(\mathbf{x}_1 z_1, \mathbf{x}_{1''}z_{1''})\, G(\mathbf{x}_2 z_1, \mathbf{x}_{2''}z_{1''})\,, \qquad (16.8)$$

such that Eq. (16.7) can be cast in the "matrix" form

$$\begin{aligned}
\langle \mathbf{x}_1, \mathbf{x}_2|\Gamma(z_1, z_{1'})|\mathbf{x}_{1'}, \mathbf{x}_{2'}\rangle &= iv(\mathbf{r}_1 - \mathbf{r}_2)\,\delta(z_1, z_{1'})\delta(\mathbf{x}_1, \mathbf{x}_{1'})\delta(\mathbf{x}_2, \mathbf{x}_{2'}) \\
&+ iv(\mathbf{r}_1 - \mathbf{r}_2)\int dz_{1''}d\mathbf{x}_{1''}d\mathbf{x}_{2''}\, \langle \mathbf{x}_1, \mathbf{x}_2|\tilde{\chi}_{\mathrm{pp}}(z_1, z_{1''})|\mathbf{x}_{1''}, \mathbf{x}_{2''}\rangle \\
&\times \langle \mathbf{x}_{1''}, \mathbf{x}_{2''}|\Gamma(z_{1''}, z_{1'})|\mathbf{x}_{1'}, \mathbf{x}_{2'}\rangle. \qquad (16.9)
\end{aligned}$$

In addition, for later formal manipulations, it is also convenient to introduce the diagonal matrix

$$\langle \mathbf{x}_1, \mathbf{x}_2|\mathcal{V}(z_1, z_{1'})|\mathbf{x}_{1'}, \mathbf{x}_{2'}\rangle \equiv iv(\mathbf{r}_1 - \mathbf{r}_2)\,\delta(z_1, z_{1'})\delta(\mathbf{x}_1, \mathbf{x}_{1'})\delta(\mathbf{x}_2, \mathbf{x}_{2'})\,, \qquad (16.10)$$

such that Eq. (16.9) becomes eventually

$$\langle \mathbf{x}_1, \mathbf{x}_2 | \Gamma(z_1, z_{1'}) | \mathbf{x}_{1'}, \mathbf{x}_{2'} \rangle = \langle \mathbf{x}_1, \mathbf{x}_2 | \mathcal{V}(z_1, z_{1'}) | \mathbf{x}_{1'}, \mathbf{x}_{2'} \rangle$$
$$+ \int dz_{1''} d\mathbf{x}_{1''} d\mathbf{x}_{2''} \int dz_{1'''} d\mathbf{x}_{1'''} d\mathbf{x}_{2'''} \, \langle \mathbf{x}_1, \mathbf{x}_2 | \mathcal{V}(z_1, z_{1''}) | \mathbf{x}_{1''}, \mathbf{x}_{2''} \rangle$$
$$\times \langle \mathbf{x}_{1''}, \mathbf{x}_{2''} | \tilde{\chi}_{\mathrm{pp}}(z_{1''}, z_{1'''}) | \mathbf{x}_{1'''}, \mathbf{x}_{2'''} \rangle \, \langle \mathbf{x}_{1'''}, \mathbf{x}_{2'''} | \Gamma(z_{1'''}, z_{1'}) | \mathbf{x}_{1'}, \mathbf{x}_{2'} \rangle \qquad (16.11)$$

or in compact matrix form $\Gamma = \mathcal{V} + \mathcal{V} \, \tilde{\chi}_{\mathrm{pp}} \, \Gamma$, where each component depends on two contour times.

Finally, the position (16.4) simplifies also the expression of the self-energy (16.2), which then reads:

$$\Sigma(\mathbf{x}_1 z_1, \mathbf{x}_{1'} z_{1'}) = - \int d\mathbf{x}_2 d\mathbf{x}_{2'} \, \big(\langle \mathbf{x}_1, \mathbf{x}_2 | \Gamma(z_1, z_{1'}) | \mathbf{x}_{1'}, \mathbf{x}_{2'} \rangle$$
$$- \langle \mathbf{x}_1, \mathbf{x}_2 | \Gamma(z_1, z_{1'}) | \mathbf{x}_{2'}, \mathbf{x}_{1'} \rangle \big) \, G(\mathbf{x}_{2'} z_{1'}, \mathbf{x}_2 z_1^+) \, . \qquad (16.12)$$

16.2 Converting to Real-Time Variables via the Langreth Rules

The second term on the right-hand side of Eq. (16.9) contains a convolution of the type (14.8), such that the Langreth rules of Section 14.2 can be applied to the t-matrix solution to the integral equation (16.9) (or (16.11)). Here, the singular contribution given by the first term on the right-hand side of Eq. (16.9) is attributed to the retarded or advanced components [39], in line with Eq. (14.6). With the compact notation (16.10), we thus write $\Gamma^{\mathrm{R/A}} = \mathcal{V} + \mathcal{V} \, \tilde{\chi}_{\mathrm{pp}}^{\mathrm{R/A}} \, \Gamma^{\mathrm{R/A}}$. Similarly, for the greater and lesser components, we write schematically $\Gamma^{>} = \mathcal{V} \, \tilde{\chi}_{\mathrm{pp}}^{>} \, \Gamma^{\mathrm{A}} + \mathcal{V} \, \tilde{\chi}_{\mathrm{pp}}^{\mathrm{R}} \, \Gamma^{>} + \mathcal{V} \, \tilde{\chi}_{\mathrm{pp}}^{\rceil} \, \Gamma^{\lceil}$ and $\Gamma^{<} = \mathcal{V} \, \tilde{\chi}_{\mathrm{pp}}^{<} \, \Gamma^{\mathrm{A}} + \mathcal{V} \, \tilde{\chi}_{\mathrm{pp}}^{\mathrm{R}} \, \Gamma^{<} + \mathcal{V} \, \tilde{\chi}_{\mathrm{pp}}^{\rceil} \, \Gamma^{\lceil}$.

In particular, in the Keldysh formalism, whereby the term $\mathcal{V} \, \tilde{\chi}_{\mathrm{pp}}^{\rceil} \, \Gamma^{\lceil}$ is consistently disregarded, we can write for the Keldysh component $\Gamma^{\mathrm{K}} = \Gamma^{>} + \Gamma^{<} = \mathcal{V} \, (\tilde{\chi}_{\mathrm{pp}}^{<} + \tilde{\chi}_{\mathrm{pp}}^{<}) \, \Gamma^{\mathrm{A}} + \mathcal{V} \, \tilde{\chi}_{\mathrm{pp}}^{\mathrm{R}} \, (\Gamma^{<} + \Gamma^{<}) = \mathcal{V} \, \tilde{\chi}_{\mathrm{pp}}^{\mathrm{K}} \, \Gamma^{\mathrm{A}} + \mathcal{V} \, \tilde{\chi}_{\mathrm{pp}}^{\mathrm{R}} \, \Gamma^{\mathrm{K}}$. In this case, the relevant expressions for the t-matrix can be collected in the form of a 2×2 matrix:

$$\begin{bmatrix} \Gamma^{\mathrm{R}} & \Gamma^{\mathrm{K}} \\ 0 & \Gamma^{\mathrm{A}} \end{bmatrix} = \mathcal{V} \begin{bmatrix} 1 & 0 \\ 0 & 1 \end{bmatrix} + \mathcal{V} \begin{bmatrix} \tilde{\chi}_{\mathrm{pp}}^{\mathrm{R}} & \tilde{\chi}_{\mathrm{pp}}^{\mathrm{K}} \\ 0 & \tilde{\chi}_{\mathrm{pp}}^{\mathrm{A}} \end{bmatrix} \begin{bmatrix} \Gamma^{\mathrm{R}} & \Gamma^{\mathrm{K}} \\ 0 & \Gamma^{\mathrm{A}} \end{bmatrix} . \qquad (16.13)$$

Remark: Alternative forms of the integral equation for Γ

When solving the integral equation (16.11) for Γ, to avoid the need to deal with the singular term $\mathcal{V}$ on its right-hand side, we may subtract $\mathcal{V}$ from Γ and define $\bar{\Gamma} \equiv \Gamma - \mathcal{V}$ whose integral equation reads

$$\bar{\Gamma} = \Gamma - \mathcal{V} = \mathcal{V} \, \tilde{\chi}_{\mathrm{pp}} \, \Gamma = \mathcal{V} \, \tilde{\chi}_{\mathrm{pp}} \, (\Gamma - \mathcal{V} + \mathcal{V}) = \mathcal{V} \, \tilde{\chi}_{\mathrm{pp}} \, \mathcal{V} \, + \, \mathcal{V} \, \tilde{\chi}_{\mathrm{pp}} \, \bar{\Gamma} \, . \qquad (16.14)$$

Or else, we may cast Γ in a form analogous to the screened potential W entering the GW approximation (to be discussed in Chapter 17), by writing formally

$$\Gamma = \mathcal{V} + \mathcal{V}\,\tilde{\chi}_{\rm pp}\,\Gamma = \mathcal{V} + \mathcal{V}\,\tilde{\chi}_{\rm pp}\,\mathcal{V} + \mathcal{V}\,\tilde{\chi}_{\rm pp}\,\mathcal{V}\,\tilde{\chi}_{\rm pp}\,\mathcal{V} + \cdots \equiv \mathcal{V} + \mathcal{V}\,\chi_{\rm pp}\,\mathcal{V}, \tag{16.15}$$

where

$$\chi_{\rm pp} = \tilde{\chi}_{\rm pp} + \tilde{\chi}_{\rm pp}\mathcal{V}\,\chi_{\rm pp} \tag{16.16}$$

is the "reducible" particle–particle bubble.

Finally, we note that the expression (16.8) has the form of a particle–particle product (14.74), the last term in the integral equation (16.9) contains a convolution of the form (14.8), and the expression (16.12) for the self-energy contains a particle–hole product of the form (14.63). Accordingly, the various components of these quantities can be obtained in terms of the rules of Sections 14.2–14.4.

Remark: The simpler case of a spatially homogeneous system

The integral equation (16.9) for Γ simplifies considerably for a spatially homogeneous system, which admits translational invariance in real space. Leaving aside time and spin variables for the time being, in this case, we write

$$\Gamma(x_1, x_2; x_{1'}, x_{2'}) = \int dp \int dp' \int dq\; e^{i(p+q)x_1}\, e^{-ipx_2}\, e^{-i(p'+q)x_{1'}}\, e^{ip'x_{2'}}\; \Gamma(p, p'; q) \tag{16.17}$$

in terms of the wave vectors (p, p', q) (where, for simplicity, we limit to one spatial dimension). In particular, this yields the following representation for the first term on the right-hand side of Eq. (16.1):

$$\begin{aligned}
&\int dp \int dp' \int dq\; e^{i(p+q)x_1}\, e^{-ipx_2}\, e^{-i(p'+q)x_{1'}}\, e^{ip'x_{2'}}\; v(p-p') \\
&= \int dp \int dp'\, e^{ip(x_1-x_2)}\, e^{-ip'(x_{1'}-x_{2'})}\; v(p-p') \int dq\; e^{iq(x_1-x_{1'})} \\
&= \delta(x_1 - x_{1'}) \int d\mathrm{P} \int d\mathcal{P}\; e^{i(\mathrm{P}+\frac{\mathcal{P}}{2})(x_1-x_2)}\; e^{-i(\mathrm{P}-\frac{\mathcal{P}}{2})(x_{1'}-x_{2'})}\; v(\mathcal{P}) \\
&= \delta(x_1 - x_{1'})\,\delta(x_2 - x_{2'}) \int d\mathcal{P}\; e^{i\mathcal{P}(x_1-x_2)}\; v(\mathcal{P}) \\
&= \delta(x_1 - x_{1'})\,\delta(x_2 - x_{2'})\, v(x_1 - x_2)\,.
\end{aligned} \tag{16.18}$$

In the second term on the right-hand side of Eq. (16.1), we instead obtain:

$$\begin{aligned}
&\int dx_{1''} \int dx_{2''}\; G(x_1, x_{1''})\, G(x_2, x_{2''})\, \Gamma(x_{1''}, x_{2''}; x_{1'}, x_{2'}) \\
&= \int dp \int dp' \int dq\; e^{i(p+q)x_1}\, e^{-ipx_2}\, e^{-i(p'+q)x_{1'}}\, e^{ip'x_{2'}}\; G(p+q)\, G(-p)\, \Gamma(p, p'; q)\,.
\end{aligned} \tag{16.19}$$

The two results (16.18) and (16.19) can be combined together, ending up with the graphical representation for the integral equation for Γ shown in Fig. 16.4.

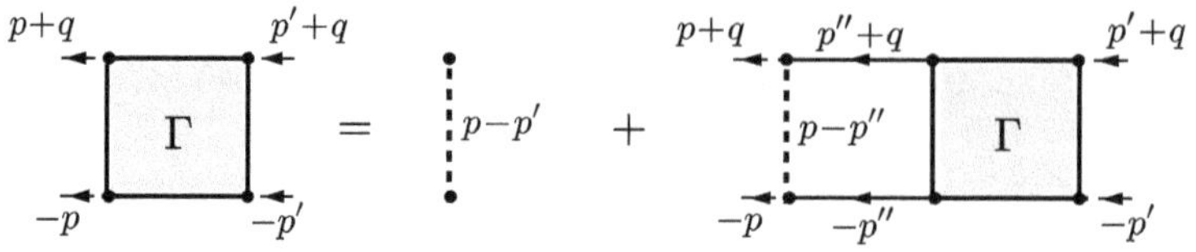

Figure 16.4 Graphical representation of the integral equation for the effective two-particle interaction Γ in wave-vector space for a spatially homogeneous system (where time and spin variables are not indicated for simplicity).

A further simplification occurs in Fig. 16.4 when the two-particle interaction υ is of the contact type. In this case, $\upsilon(p-p') \rightarrow \upsilon_0$ is a constant and $\Gamma(p, p'; q) \rightarrow \Gamma(q)$ depends only on q (after an appropriate handling of the ultraviolet cutoff, as it will be done in Part II when dealing with superfluid ultracold Fermi gases).

In Part II, we shall discuss the t-matrix approximation for the superfluid phase of a Fermi system, while, in Part III, we shall consider a specific implementation in the context of the Hubbard model [38]. In this context, possible diagrammatic extensions of the t-matrix approximation will also be highlighted, along the lines of what has recently been achieved for the equilibrium case.

17

Contour Diagrammatic Structure in Terms of Functional Derivatives

At the beginning of Chapter 9, we mentioned that, besides the more standard perturbative approach due to Feynman to calculate the single- and two-particle Green's functions of quantum many-body systems, there is an alternative procedure, originally due to Schwinger, which sets up a number of exact coupled integral equations that these Green's functions satisfy, avoiding in this way to found the formal theory on expansions in powers of the coupling constant [59]. In practice, an approximate solution of those equations entails a careful choice of the single-particle self-energy to best describe the physical system of interest. Accordingly, the single-particle self-energy represents the "trait d'union" between the Feynman and Schwinger approaches to the Green's functions method for quantum many-body systems, as it was originally pointed out by Dyson in the context of Quantum Electrodynamics [60].

For the time-independent (equilibrium) case, a compact presentation of the Schwinger approach within the zero-temperature formalism for nonrelativistic many-body systems can be found in section 3 of Ref. [32] (where the version of the Schwinger approach set up by Hedin [49] for the electron-gas problem was partly followed). Here, we extend that approach to the time-dependent (nonequilibrium) case, similar to what was done in Appendix B of Ref. [39] and Appendix L of Ref. [15]. In both (equilibrium and nonequilibrium) cases, the method that is adopted is the *source field method*, which rests on a procedure of functional differentiation with respect to a *source field* U.

Contrary to the physical external potential $V_{\text{ext}}(\mathbf{r}, t)$ of Eq. (11.1), which takes the same values on the forward (γ_-) and backward (γ_+) branches of the contour (cf. Fig. 14.1), the source field $\mathrm{U}(\mathbf{x}, z)$ is assumed to take different values on the forward and backward branches (and can further vary along the vertical track γ^{M} of the contour). As a consequence, while the result (7.2) holds for the physical external potential $V_{\text{ext}}(\mathbf{r}, t)$, an analogous result cannot hold for the source field $\mathrm{U}(\mathbf{x}, z)$. Accordingly, while for the single-particle Green's function defined on $V_{\text{ext}}(\mathbf{r}, t)$, one can let the contour extend up to $-\infty$ like in Fig. 14.1, an analogous procedure cannot be applied when considering the source field $\mathrm{U}(\mathbf{x}, z)$. We anticipate that the requirement for $U(\mathbf{x}, z)$ to take different values on the forward and backward branches (as well as on the vertical track) is dictated by the need to take *functional derivatives* with respect to $U(\mathbf{x}, z)$.

17.1 Generalized Single- and Two-Particle Contour Green's Functions: Their Connection via a Functional-Derivative Identity

We begin by supplementing the system Hamiltonian $\mathcal{H}(z)$, given by Eqs. (2.1) and (2.3)–(2.5), with an additional Hamiltonian that accounts for the interaction between the system and the source field $U(\mathbf{x}, z)$, such that now the total Hamiltonian becomes

$$\mathcal{H}_{\mathrm{U}}(z) = \mathcal{H}(z) + \int d\mathbf{x}\, U(\mathbf{x}, z)\, n(\mathbf{x}), \tag{17.1}$$

where $n(\mathbf{x}) = \psi^{\dagger}(\mathbf{x})\psi(\mathbf{x})$ is the density operator (for spin component, if any). We then define the *generalized single-particle contour Green's function*

$$G_{\mathrm{U}}(1,2) = -i\, \frac{\mathrm{Tr}\left\{\mathcal{T}_{\gamma_{\mathrm{U}}}\left\{e^{-i\int_{\gamma_{\mathrm{U}}} dz\, \mathcal{H}_{\mathrm{U}}(z)}\, \psi(1)\, \psi^{\dagger}(2)\right\}\right\}}{\mathrm{Tr}\left\{\mathcal{T}_{\gamma_{\mathrm{U}}}\left\{e^{-i\int_{\gamma_{\mathrm{U}}} dz\, \mathcal{H}_{\mathrm{U}}(z)}\right\}\right\}}, \tag{17.2}$$

where $\psi(1) = \psi(\mathbf{x}_1 z_1)$ is a field operator, and the representation (8.5) has been used. For the reasons mentioned earlier, the contour γ_{U} in Eq. (17.2) does not extend to $+\infty$ similar to the contour γ_{T} of Eq. (8.5), but it rather terminates at the largest real-time value along the contour (i.e., either t_1 or t_2). In other words, the definition (17.2) describes the four possible situations reported in Fig. 17.1, where in both cases $t_1 > t_2$ and $t_2 > t_1$, we can write the numerator of Eq. (17.2) as follows:

$$\begin{aligned}
&\mathcal{T}_{\gamma_{\mathrm{U}}}\left\{e^{-i\int_{\gamma_{\mathrm{U}}} dz\, \mathcal{H}_{\mathrm{U}}((z)}\, \psi(1)\, \psi^{\dagger}(2)\right\} \\
&= \theta(z_1, z_2)\, \mathcal{T}_{\gamma_{\mathrm{U}}}\left\{e^{-i\int_{z_1}^{t_0 - i\beta} dz\, \mathcal{H}_{\mathrm{U}}(z)}\right\} \psi(1)\, \mathcal{T}_{\gamma_{\mathrm{U}}}\left\{e^{-i\int_{t_0^-}^{z_1} dz\, \mathcal{H}_{\mathrm{U}}(z)}\, \psi^{\dagger}(2)\right\} \\
&\pm \theta(z_2, z_1)\, \mathcal{T}_{\gamma_{\mathrm{U}}}\left\{e^{-i\int_{z_1}^{t_0 - i\beta} dz\, \mathcal{H}_{\mathrm{U}}(z)}\, \psi^{\dagger}(2)\right\} \psi(1)\, \mathcal{T}_{\gamma_{\mathrm{U}}}\left\{e^{-i\int_{t_0^-}^{z_1} dz\, \mathcal{H}_{\mathrm{U}}(z)}\right\},
\end{aligned} \tag{17.3}$$

where the upper (lower) sign holds for bosons (fermions). This convention for the contour γ_{U} can be extended to the *generalized two-particle contour Green's function* defined in analogy to Eq. (7.13), namely,

$$G_{2\mathrm{U}}(1,2;1',2') = \frac{1}{i^2}\, \frac{\mathrm{Tr}\left\{\mathcal{T}_{\gamma_{\mathrm{U}}}\left\{e^{-i\int_{\gamma_{\mathrm{U}}} dz\, \mathcal{H}_{\mathrm{U}}(z)}\, \psi(1)\, \psi(2)\, \psi^{\dagger}(2')\, \psi^{\dagger}(1')\right\}\right\}}{\mathrm{Tr}\left\{\mathcal{T}_{\gamma_{\mathrm{U}}}\left\{e^{-i\int_{\gamma_{\mathrm{U}}} dz\, \mathcal{H}_{\mathrm{U}}(z)}\right\}\right\}}, \tag{17.4}$$

where four different times appear.

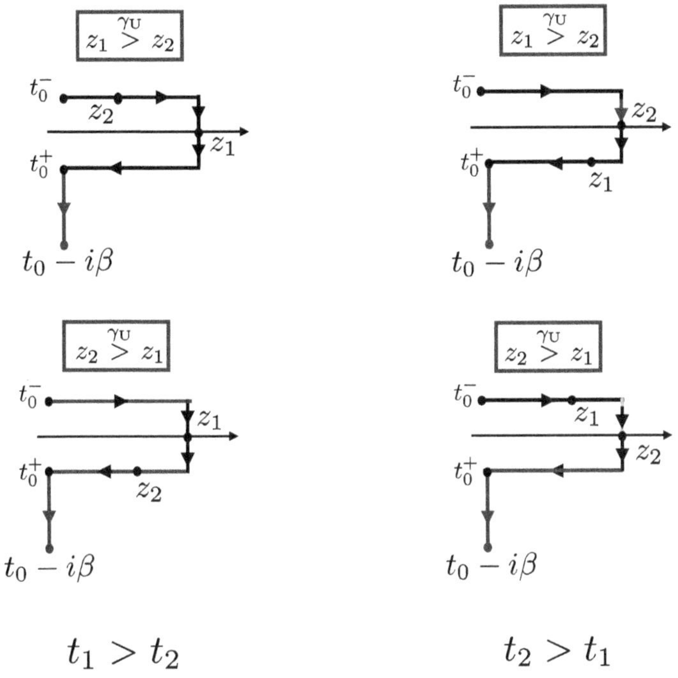

Figure 17.1 The four possible cases occurring in Eq. (17.2) from the correspondence between the variables (z_1, z_2) running along the contour γ_U and the real-time variables (t_1, t_2). Note the difference with what occurs in Fig. 14.2, where the result (7.2) can instead be utilized.

The form (17.3) evidences the position of the field operator $\psi(1)$ along the contour γ_U. Taking then the derivative of this expression with respect to z_1 yields the equation of motion of the generalized single-particle contour Green's function (17.2) similar to that in Section 11.3, in the form

$$\left[i\frac{d}{dz_1} - h(1) - U(1)\right] G_U(1,2) \pm i \int d3\, v(1,3)\, G_{2U}(1,3^+;2,3^{++}) = \delta(1,2), \tag{17.5}$$

where now the upper (lower) sign refers to fermions (bosons) and 3^+ implies that the time variable z_3 is augmented by a positive infinitesimal along the contour γ_U.

17.2 Generalized Self-Energy and Dyson Equation

The equation of motion (17.5) of the generalized single-particle contour Green's function G_U involves the generalized two-particle contour Green's function G_{2U}. As already mentioned in Section 11.4, if one were to continue along this way and look for the equation of motion of G_{2U}, this equation would contain the generalized three-particle contour Green's function, and so on. In Section 11.4, this hierarchy of equations was formally truncated by introducing the single-particle self-energy Σ via the definition (11.25), leaving in this way the selection of an approximate expression for Σ to the development of a diagrammatic expansion of the single-particle Green's function obtained by Wick's theorem in powers of the coupling constant.

In the present context of generalized Green's functions subject to the source field U, the generalized two-particle contour Green's function $G_{2\mathrm{U}}$ can be formally eliminated from the equation of motion (17.5) via the following *functional derivative identity*:

$$
\begin{aligned}
\frac{\delta G_{\mathrm{U}}(1,2)}{\delta U(3)} = \frac{1}{i}\, & \frac{\mathrm{Tr}\left\{\mathcal{T}_{\gamma_{\mathrm{U}}}\left\{e^{-i\int_{\gamma_{\mathrm{U}}} dz\, \mathcal{H}_{\mathrm{U}}(z)}\,(-i)\, n(3)\, \psi(1)\, \psi^{\dagger}(2)\right\}\right\}}{\mathrm{Tr}\left\{\mathcal{T}_{\gamma_{\mathrm{U}}}\left\{e^{-i\int_{\gamma_{\mathrm{U}}} dz\, \mathcal{H}_{\mathrm{U}}(z)}\right\}\right\}} \\
- \frac{1}{i}\, & \frac{\mathrm{Tr}\left\{\mathcal{T}_{\gamma_{\mathrm{U}}}\left\{e^{-i\int_{\gamma_{\mathrm{U}}} dz\, \mathcal{H}_{\mathrm{U}}(z)} \psi(1)\, \psi^{\dagger}(2)\right\}\right\}\, \mathrm{Tr}\left\{\mathcal{T}_{\gamma_{\mathrm{U}}}\left\{e^{-i\int_{\gamma_{\mathrm{U}}} dz\, \mathcal{H}_{\mathrm{U}}(z)}\,(-i)\, n(3)\right\}\right\}}{\left(\mathrm{Tr}\left\{\mathcal{T}_{\gamma_{\mathrm{U}}}\left\{e^{-i\int_{\gamma_{\mathrm{U}}} dz\, \mathcal{H}_{\mathrm{U}}(z)}\right\}\right\}\right)^{2}} \\
= \mp\, & \left(G_{2\mathrm{U}}(1,3;2,3^{+}) - G_{\mathrm{U}}(1,2)\, G_{\mathrm{U}}(3,3^{+})\right), \qquad (17.6)
\end{aligned}
$$

where again the upper (lower) sign holds for fermions (bosons). In this way, the term containing $G_{2\mathrm{U}}$ on the right-hand side of Eq. (17.5) can be cast in the form

$$
\begin{aligned}
& \pm\, i \int d3\, v(1,3)\, G_{2\mathrm{U}}(1,3^{+};2,3^{++}) \\
& = \pm\, i \int d3\, v(1^{+},3) \left(G_{\mathrm{U}}(1,2)\, G_{\mathrm{U}}(3,3^{+}) \mp \frac{\delta G_{\mathrm{U}}(1,2)}{\delta U(3)}\right) \\
& = \int d4 \Big[(\pm i)\, \delta(1,4) \int d3\, v(1,3)\, G_{\mathrm{U}}(3,3^{+})\, G_{\mathrm{U}}(4,2) \\
& \qquad - i \int d3\, v(1^{+},3)\, \frac{\delta G_{\mathrm{U}}(1,4)}{\delta U(3)}\, \delta(4,2)\Big] \\
& = \int d4 \Big[(\pm i)\, \delta(1,4) \int d3\, v(1,3)\, G_{\mathrm{U}}(3,3^{+}) \\
& \qquad - i \int d3 d5\, v(1^{+},3)\, \frac{\delta G_{\mathrm{U}}(1,5)}{\delta U(3)}\, G_{\mathrm{U}}^{-1}(5,4)\Big]\, G_{\mathrm{U}}(4,2) \\
& \equiv - \int d4\, \Sigma_{\mathrm{U}}(1,4)\, G_{\mathrm{U}}(4,2)\,, \qquad (17.7)
\end{aligned}
$$

where we have introduced the inverse G_{U}^{-1} of G_{U} such that

$$\int d3\, G_{\mathrm{U}}^{-1}(1,3)\, G_{\mathrm{U}}(3,2) = \int d3\, G_{\mathrm{U}}(1,3)\, G_{\mathrm{U}}^{-1}(3,2) = \delta(1,2) \tag{17.8}$$

and defined the *generalized single-particle self-energy* Σ_{U} by the expression

$$\begin{aligned}\Sigma_{\mathrm{U}}(1,2) = &\mp\, \delta(1,2)\, i \int d3\, v(1,3)\, G_{\mathrm{U}}(3,3^{+}) \\ &+ i \int d3d4\, v(1^{+},3)\, \frac{\delta G_{\mathrm{U}}(1,4)}{\delta U(3)}\, G_{\mathrm{U}}^{-1}(4,2),\end{aligned} \tag{17.9}$$

with the upper (lower) sign again holding for fermions (bosons). Note that the first term on the right-hand side recovers in the fermionic case the form (11.30) of the Hartree contribution to the self-energy, while the subtlety of keeping 1^{+} in the argument of $v(1^{+},3)$ in the second term allows us to accommodate the Fock contribution (11.31) to the self-energy. Quite generally, the two terms on the right-hand side of Eq. (17.9) are referred to as the generalized Hartree $\Sigma_{\mathrm{U}}^{\mathrm{H}}$ and "mass" M_{U} terms, in the order. Entering the result (17.9) into Eq. (17.5), we obtain eventually the *generalized Dyson equation* in the presence of the source field U:

$$\left[i\frac{d}{dz_1} - h(1) - U(1)\right] G_{\mathrm{U}}(1,2) - \int d3\, \Sigma_{\mathrm{U}}(1,3)\, G_{\mathrm{U}}(3,2) = \delta(1,2)\,. \tag{17.10}$$

17.3 Irreducible Vertex Function, Irreducible Polarizability, and Dynamically Screened Interaction

To complete our program, there still remains to eliminate any explicit reference to the source field U, in such a way that the limit $U \rightarrow 0$ can eventually be taken. This goal cannot be achieved with the form (17.9) of the generalized single-particle self-energy because it would entail an explicit knowledge of the functional dependence of G_{U} on U. To this end, we can again follow the treatment of Ref. [32], thereby limiting ourselves to consider the fermionic case of most physical interest. This treatment proceeds along the following steps.

Step # 1 In Eq. (17.10), combine the Hartree contribution to Σ_{U} with the source field U, thereby defining the *total potential*

$$V(1) \equiv U(1) - i \int d3\, v(1,3)\, G_{\mathrm{U}}(3,3^{+})\,. \tag{17.11}$$

The mass term in Eq. (17.9) can then be rewritten in the form

$$M_{\mathrm{U}}(1,2) = -\, i \int d3d4d5\, v(1^{+},3)\, G_{\mathrm{U}}(1,4)\, \frac{\delta G_{\mathrm{U}}^{-1}(4,2)}{\delta V(5)}\, \frac{\delta V(5)}{\delta U(3)}, \tag{17.12}$$

where the "chain rule" of functional differentiation has been utilized when regarding G_{U} as a functional of V instead of U.

Step # 2 With G_0^{-1} defined like in Section 11.5, write $G_{\mathrm{U}}^{-1}(1,2) = G_0^{-1}(1,2) - U(1)\delta(1,2) - \Sigma_{\mathrm{U}}(1,2) = G_0^{-1}(1,2) - V(1)\delta(1,2) - M_{\mathrm{U}}(1,2)$ and obtain for the functional derivative in Eq. (17.12)

$$-\frac{\delta G_{\mathrm{U}}^{-1}(1,2)}{\delta V(3)} = \delta(1,3)\,\delta(2,3) + \frac{\delta M_{\mathrm{U}}(1,2)}{\delta V(3)} \equiv \tilde{\Gamma}(1,2;3)\,, \tag{17.13}$$

which defines the *irreducible vertex function* $\tilde{\Gamma}$. Since M_{U} depends on V only through G_{U}, the chain rule can again be used to obtain the following integral equation for $\tilde{\Gamma}$:

$$\tilde{\Gamma}(1,2;3) = \delta(1,3)\,\delta(2,3) + \int d3d4d5d6d7\,\frac{\delta M_{\mathrm{U}}(1,2)}{\delta G_{\mathrm{U}}(4,5)} G_{\mathrm{U}}(4,6)\,G_{\mathrm{U}}(7,5)\,\tilde{\Gamma}(6,7;3)\,. \tag{17.14}$$

Step # 3 There remains to handle the last factor in the integrand of Eq. (17.12). To this end, we introduce the *polarizability*

$$\chi(1,2) = -i\,\frac{\delta G_{\mathrm{U}}(1,1^+)}{\delta U(2)} \tag{17.15}$$

and the *irreducible polarizability*

$$\tilde{\chi}(1,2) = -i\,\frac{\delta G_{\mathrm{U}}(1,1^+)}{\delta V(2)}\,, \tag{17.16}$$

which are mutually related through the integral equation

$$\chi(1,2) = \int d3\,\tilde{\chi}(1,3)\,\frac{\delta V(3)}{\delta U(2)} \underset{[\text{Eq.}(17.11)]}{\overset{\uparrow}{=}} \int d3\,\tilde{\chi}(1,3)\left[\delta(3,2) + \int d4\,v(3,4)\,\chi(4,2)\right]$$

$$= \tilde{\chi}(1,2) + \int d3d4\,\tilde{\chi}(1,3)\,v(3,4)\,\chi(4,2) \tag{17.17}$$

obtained by applying the chain rule once more. In addition, $\tilde{\chi}(1,2)$ can be rewritten in the form:

$$\tilde{\chi}(1,2) = i\int d3d4\,G_{\mathrm{U}}(1,3)\,\frac{\delta G_{\mathrm{U}}^{-1}(3,4)}{\delta V(2)}\,G_{\mathrm{U}}(4,1^+)$$

$$\underset{[\text{Eq.}(17.13)]}{\overset{\uparrow}{=}} -i\int d3d4\,G_{\mathrm{U}}(1,3)\,G_{\mathrm{U}}(4,1^+)\,\tilde{\Gamma}(3,4;2)\,. \tag{17.18}$$

The functional derivative identity (17.6) can be utilized to rewrite the polarizability (17.15) in the physically meaningful form of a density–density correlation function (which will be further considered in Chapter 18), namely,

$$\chi(1,2) = -i\,\frac{\delta G_{\mathrm{U}}(1,1^{+})}{\delta U(2)} = i\left(G_{2\mathrm{U}}(1,2;1^{+},2^{+}) - G_{\mathrm{U}}(1,1^{+})\,G_{\mathrm{U}}(2,2^{+})\right)$$
$$= -i\,\frac{\mathrm{Tr}\left\{\mathcal{T}_{\gamma_{\mathrm{U}}}\left\{e^{-i\int_{\gamma_{\mathrm{U}}} dz\,\mathcal{H}_{\mathrm{U}}(z)}\,n'(1)\,n'(2)\right\}\right\}}{\mathrm{Tr}\left\{\mathcal{T}_{\gamma_{\mathrm{U}}}\left\{e^{-i\int_{\gamma_{\mathrm{U}}} dz\,\mathcal{H}_{\mathrm{U}}(z)}\right\}\right\}}\,, \tag{17.19}$$

where

$$n'(1) = n(1) - \frac{\mathrm{Tr}\left\{\mathcal{T}_{\gamma_{\mathrm{U}}}\left\{e^{-i\int_{\gamma_{\mathrm{U}}} dz\,\mathcal{H}_{\mathrm{U}}(z)}\,n(1)\right\}\right\}}{\mathrm{Tr}\left\{\mathcal{T}_{\gamma_{\mathrm{U}}}\left\{e^{-i\int_{\gamma_{\mathrm{U}}} dz\,\mathcal{H}_{\mathrm{U}}(z)}\right\}\right\}} \tag{17.20}$$

is the density deviation operator. Note from Eq. (17.19) that $\chi(1,2)$ is symmetric under the interchange $1 \leftrightarrow 2$ and from Eq. (17.17) that a similar property holds for $\tilde{\chi}(1,2)$.

Step # 4 Introduce the *dynamically screened interaction*

$$W(1,2) \equiv \int d3\, v(1,3)\,\frac{\delta V(2)}{\delta U(3)} \underset{[\mathrm{Eq.}(17.11)]}{=} \int d3\, v(1,3)\left[\delta(2,3) + \int d4\, v(2,4)\,\chi(4,3)\right]$$
$$= v(1,2) + \int d3d4\, v(1,3)\,\chi(3,4)\,v(4,2) \tag{17.21}$$

since $v(1,2) = v(2,1)$ by definition.

Step # 5 With the help of Eqs. (17.13) and (17.21), the expression (17.12) for the mass term acquires its final form:

$$M_{\mathrm{U}}(1,2) = i\int d3d4d5\, v(1^{+},3)\,G_{\mathrm{U}}(1,4)\,\tilde{\Gamma}(4,2;5)\,\frac{\delta V(5)}{\delta U(3)}$$
$$= i\int d3d4\, W(1^{+},3)\,G_{\mathrm{U}}(1,4)\,\tilde{\Gamma}(4,2;3)\,. \tag{17.22}$$

Step # 6 We are eventually in a position to let $\mathrm{U} \to 0$ in (a) the integral equation (17.14) for $\tilde{\Gamma}$, (b) the ensuing definition (17.18) for $\tilde{\chi}$, (c) the integral equation (17.17) for χ, (d) the ensuing definition (17.21) for W, and (e) the final form (17.22) for M. All these quantities are represented graphically in Fig. 17.2. Note that, once the functional derivative $\delta M/\delta G$ has been specified in Eq. (17.14) for $\tilde{\Gamma}$, all other quantities directly follow.

The aforementioned formal achievements fulfill the Schwinger original plan for obtaining the (in the present case, contour) single-particle Green's function, by avoiding at a formal level any expansion in terms of the (interparticle) coupling constant. However, useful results can be obtained in practice only once a specific functional form for the dependence of mass term M on G is specified. To this end,

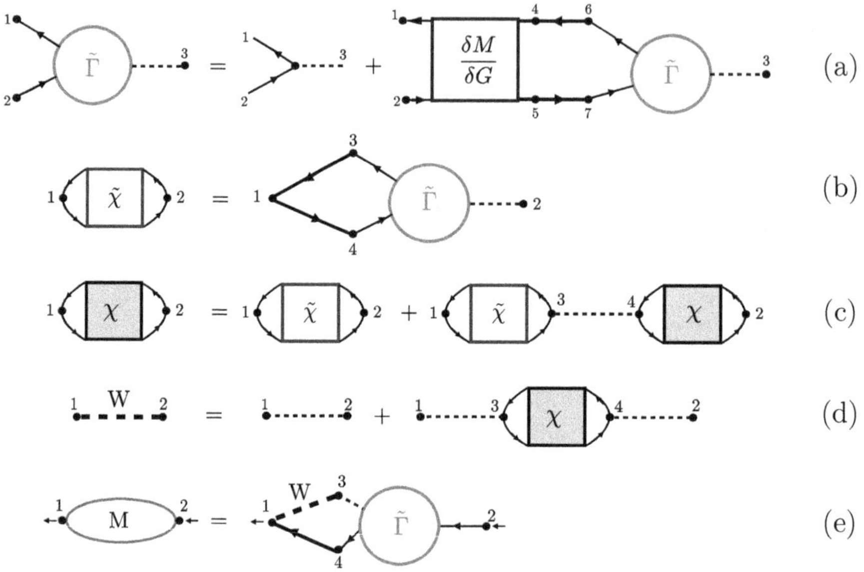

Figure 17.2 Graphical representation of (a) Eq. (17.14) for $\tilde{\Gamma}$, (b) Eq. (17.18) for $\tilde{\chi}$, (c) Eq. (17.17) for χ, (d) Eq. (17.21) for W, and (e) Eq. (17.22) for M.

knowledge of the topology of the Feynman diagrams obtained beforehand from perturbation theory turns out to be an essential hint. In particular, when the need arises to satisfy conservation laws, the choice of the diagrams for the self-energy has to be carefully gauged, as it was mentioned at the end of Chapter 11 in the context of conserving approximation.

17.4 The GW Approximation

In terms of the quantities depicted in Fig. 17.2, the simplest approximation to the mass term M is obtained by considering for $\tilde{\Gamma}$ only the first term on the right-hand side of Eq. (17.14). This choice affects the expression (17.22) for M_U also through the irreducible polarizability $\tilde{\chi}$ given by Eq. (17.18) since this in turn determines the polarizability χ from Eq. (17.17) and, eventually, the dynamically screened interaction W from Eq. (17.21). One thus ends up with the following expressions

$$M(1,2) \to i\, W(1^+,2)\, G(1,2) \quad \text{with} \quad \tilde{\chi}(1,2) \to -i\, G(1,2)\, G(2,1) \quad (17.23)$$

as represented graphically in Fig. 17.3. These expressions are referred to as the GW approximation for the self-energy, with the screening properties treated within the random-phase approximation (RPA) (cf., e.g., section 5.5 of Ref. [10] for a discussion of the screening properties of an electron gas within RPA). Note further

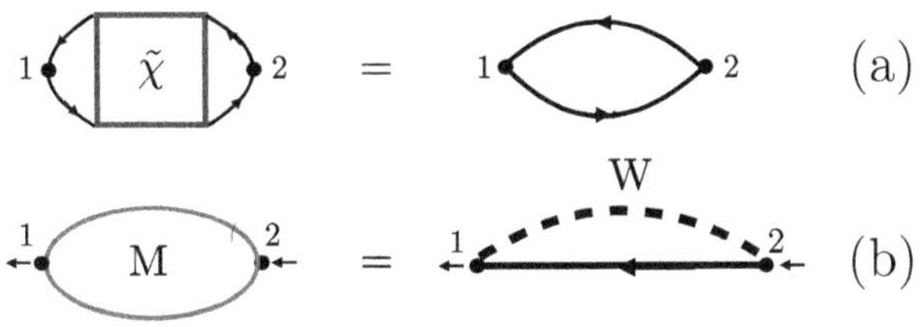

Figure 17.3 Graphical representation of (a) the irreducible polarizability $\tilde{\chi}$ and (b) the mass term M within the GW approximation.

that the approximation (17.23), as it stands, is "conserving" in the Baym–Kadanoff sense [3, 4, 32]. The GW approximation has extensively and successfully been utilized for the equilibrium case, as mentioned at the beginning of Chapter 16. Although to a much more limited extent, the GW approximation has also been implemented for the nonequilibrium case, so as to address physical problems related to time-dependent screening effects [61, 62].

17.5 The Bethe–Salpeter Equation for the Two-Particle Green's Function

In Section17.3, we have already pointed out that the two-particle Green's function is directly related to the correlation functions, such as the density–density correlation function of Eq. (17.19). Through this property, it is possible to perform systematic calculations of the correlation functions within linear-response theory in terms of the diagrammatic structure associated with the two-particle Green's function at equilibrium. However, when the time dependence, say, of the density under the action of a time-dependent external agent is accounted for in terms of the contour single-particle Green's function (as discussed in Chapter 18), there is apparently no need to resort to the two-particle Green's function for nonequilibrium situations. Nevertheless, there may be cases when the time evolution of two-particle correlations under the action of a time-dependent external agent is of specific interest, for example, when considering either the pair correlation function for opposite-spin fermions [63] or the two-particle reduced density matrix associated with the spatial emergence of off-diagonal long-range order [64]. In these cases, one needs to specify the diagrammatic structure of the out-of-equilibrium two-particle Green's function.

To this end, we go back to the functional derivative identity (17.6) in the fermionic case and identify it with the *two-particle correlation function*, in the form:

$$L_{\mathrm{U}}(1,3;2,3^{+}) \equiv -\,G_{2\mathrm{U}}(1,3;2,3^{+}) + G_{\mathrm{U}}(1,2)\,G_{\mathrm{U}}(3,3^{+}) = \frac{\delta G_{\mathrm{U}}(1,2)}{\delta U(3)}\,. \quad (17.24)$$

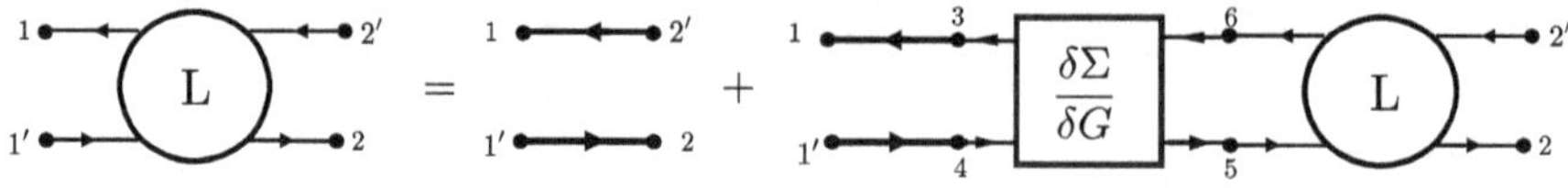

Figure 17.4 Graphical representation of the Bethe–Salpeter equation (17.26).

From the properties of the functional derivatives, we then write:

$$\begin{aligned} L_{\mathrm{U}}(1,3;2,3^{+}) &= -\int d4d5\, G_{\mathrm{U}}(1,4)\, \frac{\delta G_{\mathrm{U}}^{-1}(4,5)}{\delta U(3)}\, G_{\mathrm{U}}(5,2) \\ &= \int d4d5\, G_{\mathrm{U}}(1,4) \left(\delta(4,5)\delta(4,3) + \frac{\delta \Sigma_{\mathrm{U}}(4,5)}{\delta U(3)} \right) G_{\mathrm{U}}(5,2) \\ &= G_{\mathrm{U}}(1,3)\, G_{\mathrm{U}}(3,2) + \int d4d5d6d7\, G_{\mathrm{U}}(1,4)\, \frac{\delta \Sigma_{\mathrm{U}}(4,5)}{\delta G_{\mathrm{U}}(6,7)} \frac{\delta G_{\mathrm{U}}(6,7)}{\delta U(3)}\, G_{\mathrm{U}}(5,2) \\ &= G_{\mathrm{U}}(1,3)\, G_{\mathrm{U}}(3,2) + \int d4d5d6d7\, G_{\mathrm{U}}(1,4)\, G_{\mathrm{U}}(5,2)\, \frac{\delta \Sigma_{\mathrm{U}}(4,5)}{\delta G_{\mathrm{U}}(6,7)}\, L_{\mathrm{U}}(6,3;7,3^{+}), \end{aligned} \tag{17.25}$$

where the limit $\mathrm{U} \to 0$ can eventually be taken. In addition, the topology of the associated diagrammatic representation implies that Eq. (17.25) holds also when the variables 3 and 3^{+} therein span arbitrary and independent values. This consideration leads to the most general equation

$$\begin{aligned} L(1,2;1',2') &= G(1,2')\, G(2,1') \\ &\quad + \int d3d4d5d6\, G(1,3)\, G(4,1')\, \Xi(3,5;4,6)\, L(6,2;5,2') \end{aligned} \tag{17.26}$$

known as the *Bethe–Salpeter equation* [65] for L, with kernel

$$\Xi(3,5;4,6) = \frac{\delta \Sigma(3,4)}{\delta G(6,5)} . \tag{17.27}$$

This equation is depicted in Fig. 17.4, where we recall that, in line with the present context, all Gs are meant to be contour single-particle Green's functions.

18

Beyond Linear-Response Theory

In Chapter 15, the Kadanoff–Baym equations for the various real-time components of the contour single-particle Green's function were obtained under the action of an external time-dependent perturbing potential $V_{\text{ext}}(\mathbf{r}, t)$ of *arbitrary strength.* In particular, knowledge of the lesser Green's function $G^<(\mathbf{x}t, \mathbf{x}'t') = \mp i \langle \psi_{\text{H}}^\dagger(\mathbf{x}'t') \, \psi_{\text{H}}(\mathbf{x}t) \rangle$ (cf. Eq. (13.5) where $\mathbf{x} = (\mathbf{r}, \sigma)$), obtained by solving the Kadanoff–Baym equations of Chapter 15, provides the time development of the number and current densities, as given by the time-dependent averages of the respective operators (per spin component)

$$
\begin{aligned}
n(\mathbf{x}) &= \psi^\dagger(\mathbf{x}) \, \psi(\mathbf{x}) && (18.1)\\
\mathbf{j}(\mathbf{x}) &= \frac{1}{2im} \left[\psi^\dagger(\mathbf{x}) \nabla \psi(\mathbf{x}) - \left(\nabla \psi^\dagger(\mathbf{x}) \right) \psi(\mathbf{x}) \right], && (18.2)
\end{aligned}
$$

in the order. Similar to Eq. (8.1), let $\langle \cdots \rangle = \text{Tr} \{ \rho \cdots \}$ be the thermal average with the density matrix ρ corresponding to the thermodynamic equilibrium established before the external perturbation $V_{\text{ext}}(\mathbf{r}, t)$ begins to act on the system at time t_0, average that is thus taken over the eigenstates of the Hamiltonian H. Similar to Eq. (4.14), the real-time evolution of this thermal average is then given by

$$
n(\mathbf{x}, t) = \left\langle U^\dagger(t, t_0) \, \psi^\dagger(\mathbf{x}) \, \psi(\mathbf{x}) \, U(t, t_0) \right\rangle = \left\langle \psi_{\text{H}}^\dagger(\mathbf{x}, t) \, \psi_{\text{H}}(\mathbf{x}, t) \right\rangle = \pm i \, G^<(\mathbf{x}t, \mathbf{x}t) \tag{18.3}
$$

for the number density, and similarly

$$
\begin{aligned}
\mathbf{j}(\mathbf{x}, t) &= \frac{1}{2im} \left\langle U^\dagger(t, t_0) \left[\psi^\dagger(\mathbf{x}) \nabla \psi(\mathbf{x}) - \left(\nabla \psi^\dagger(\mathbf{x}) \right) \psi(\mathbf{x}) \right] U(t, t_0) \right\rangle \\
&= \frac{1}{2im} \left\langle \left[\psi_{\text{H}}^\dagger(\mathbf{x}, t) \nabla \psi_{\text{H}}(\mathbf{x}, t) - \left(\nabla \psi_{\text{H}}^\dagger(\mathbf{x}, t) \right) \psi_{\text{H}}(\mathbf{x}, t) \right] \right\rangle \\
&= \pm \frac{1}{2m} \left(\nabla_{\mathbf{r}} - \nabla_{\mathbf{r}'} \right) G^<(\mathbf{r}\sigma t, \mathbf{r}'\sigma t) \Big|_{\mathbf{r}'=\mathbf{r}}
\end{aligned} \tag{18.4}
$$

for the current density, where the time-evolution operator $U(t, t_0)$ is given by Eq. (2.8), the suffix H specifies the Heisenberg representation (2.25), and the upper (lower) sign refers to bosons (fermions).

The expressions (18.3) and (18.4) hold for an arbitrary strength of the external potential V_{ext}, which enters the time-dependent Hamiltonian $H^{\text{ext}}(t)$ of Eq. (2.5). This situation is appropriate when V_{ext} is so strong (like in the "pump and probe" experiments considered in Part III) that the system is brought much far from equilibrium. On the other hand, when the strength of V_{ext} is not so strong that the system is made to slightly deviate from the initial equilibrium, one may envisage expanding the expressions (18.3) and (18.4) in powers of V_{ext}, up to linear, quadratic, and so on, orders depending on the physical circumstances. For definiteness, here, we consider the expansion of the number density (18.3) up to linear order in V_{ext}.

To this end, it is convenient to rewrite the expression (18.3) by utilizing the Heisenberg picture of Section 3.1, whereby $U(t, t_0) = e^{-iHt} V(t, t_0)\, e^{iHt_0}$ according to Eq. (3.7) with the time-evolution operator $V(t, t_0)$ in the Heisenberg picture given by the expression (3.5). This yields

$$n(\mathbf{x}, t) = \left\langle V^{\dagger}(t, t_0)\, e^{iHt}\, \psi^{\dagger}(\mathbf{x})\, \psi(\mathbf{x})\, e^{-iHt} V(t, t_0) \right\rangle = \left\langle V^{\dagger}(t, t_0)\, n_{\text{h}}(\mathbf{x}, t)\, V(t, t_0) \right\rangle , \tag{18.5}$$

where $n_{\text{h}}(\mathbf{x}, t) = e^{iHt}\, \psi^{\dagger}(\mathbf{x})\, \psi(\mathbf{x})\, e^{-iHt}$ in line with Eq. (3.11). In addition, in terms of the (non-extended) oriented contour $\gamma = \gamma_- + \gamma_+$ depicted in Fig. 6.1, which does not include the vertical track, we can rewrite the expression (18.5) in the form

$$n(\mathbf{x}, t) = \text{Tr} \left\{ \rho\, \mathcal{T}_{\gamma} \left\{ e^{-i \int_{\gamma} d\bar{z}\, H_{\text{h}}^{\text{ext}}(\bar{z})}\, n_{\text{h}}(\mathbf{x}, t) \right\} \right\} , \tag{18.6}$$

which extends the result (6.8) for a generic operator to the case of a thermal average. Finally, due to the properties of the time-ordering operator $\mathcal{T}_{\gamma}$ along the contour γ, the expression (18.6) reads

$$n(\mathbf{x}, t) = \text{Tr} \left\{ \rho\, \mathcal{T}_{\gamma} \left\{ e^{-i \int_{t}^{t_0} dt'\, H_{\text{h}}^{\text{ext}}(t')} \right\} n_{\text{h}}(\mathbf{x}, t)\, \mathcal{T}_{\gamma} \left\{ e^{-i \int_{t_0}^{t} dt'\, H_{\text{h}}^{\text{ext}}(t')} \right\} \right\} , \tag{18.7}$$

where $H_{\text{h}}^{\text{ext}}(t) = e^{iHt}\, H^{\text{ext}}(t)\, e^{-iHt}$ takes the same values on the forward (γ_-) and backward (γ_+) branches of the contour γ for given t.

The expression (18.7), which still holds for arbitrary strength of the external disturbance V_{ext}, is in a suitable form to obtain the value of $\delta n(\mathbf{x}, t)$ (over and above the unperturbed value $n_0(\mathbf{x})$ before the action of the external disturbance) to *linear order* in the external disturbance. We thus obtain

$$\delta n(\mathbf{x},t) = -i\int_{t}^{t_0} dt'\,\mathrm{Tr}\left\{\rho\, H_{\mathrm{h}}^{\mathrm{ext}}(t')\, n_{\mathrm{h}}(\mathbf{x},t)\right\} - i\int_{t_0}^{t} dt'\,\mathrm{Tr}\left\{\rho\, n_{\mathrm{h}}(\mathbf{x},t)\, H_{\mathrm{h}}^{\mathrm{ext}}(t')\right\}$$
$$= i\int_{t_0}^{t} dt'\,\mathrm{Tr}\left\{\rho\,\left[H_{\mathrm{h}}^{\mathrm{ext}}(t'), n_{\mathrm{h}}(\mathbf{x},t)\right]\right\}, \tag{18.8}$$

which coincides with the standard expression obtained within linear-response theory (where it is known as the Kubo formula) [7]. Expressions beyond the linear order can also be obtained in this way, by proceeding further in the expansions in powers of $H_{\mathrm{h}}^{\mathrm{ext}}$ of the exponentials in Eq. (18.7) [66].

Considering specifically the linear-response regime, it should be emphasized that, when the external perturbation is sufficiently weak, knowledge of $G^{<}(\mathbf{x}t,\mathbf{x}t)$ enables us to obtain the (retarded) *density–density correlation function* for real times (and, accordingly, for real frequencies, when the Fourier analysis is utilized). This is an important remark because this correlation function can directly be measured, like in crystals by neutron diffraction [67]. This correlation function comes up by recalling from Eq. (2.5) that $H^{\mathrm{ext}}(t) = \int d\mathbf{x}\, n(\mathbf{x})\, V_{\mathrm{ext}}(\mathbf{r},t)$, such that the expression (18.8) becomes

$$\delta n(\mathbf{r},t) = -i\int d\mathbf{r}'\int_{t_0}^{t} dt'\,\mathrm{Tr}\left\{\rho\,[n_{\mathrm{h}}(\mathbf{r},t), n_{\mathrm{h}}(\mathbf{r}',t')]\right\} V_{\mathrm{ext}}(\mathbf{r}',t')$$
$$= \int d\mathbf{r}'\int_{t_0}^{+\infty} dt'\, D^{\mathrm{R}}(\mathbf{r}t,\mathbf{r}'t')\, V_{\mathrm{ext}}(\mathbf{r}',t'), \tag{18.9}$$

where $n(\mathbf{r}) = \sum_{\sigma} n(\mathbf{x})$ is the total density operator (including all spin components) and

$$D^{\mathrm{R}}(\mathbf{r}t,\mathbf{r}'t') \equiv -i\,\theta(t-t')\,\mathrm{Tr}\left\{\rho\,[n_{\mathrm{h}}(\mathbf{r},t), n_{\mathrm{h}}(\mathbf{r}',t')]\right\} \tag{18.10}$$

is the aforementioned (retarded) correlation function.

Apart from obtaining the limiting behavior of $G^{<}(\mathbf{x}t,\mathbf{x}t)$ for small V_{ext}, an alternative way to obtain the correlation function (18.10) by perturbation theory is to connect it with the corresponding *temperature correlation function*

$$\mathcal{D}(\mathbf{r}\tau,\mathbf{r}'\tau') \equiv -\,\mathrm{Tr}\left\{\rho\,\mathcal{T}_{\tau}\left\{n_{\mathrm{K}}(\mathbf{r},\tau)\, n_{\mathrm{K}}(\mathbf{r}',\tau')\right\}\right\}. \tag{18.11}$$

Here, $\mathcal{T}_{\tau}$ is the time-ordering operator for imaginary time τ as in Eq. (13.11) and $n_{\mathrm{K}}(\mathbf{r},\tau) = e^{K\tau} n(\mathbf{r})\, e^{-K\tau}$ as in Eq. (12.6), where $K = H - \mu N$ with thermodynamic chemical potential μ corresponds to the operator H^{M} used in Section 8.1 on the vertical track. To this end, we first note that the Hamiltonian H can be replaced by the grand-canonical Hamiltonian K in the definition of the operator $n_{\mathrm{h}}(\mathbf{r},t)$ entering Eq. (18.10), since we can write in line with Eq. (9.4)

$$e^{iHt}\,\psi(\mathbf{x})\,e^{-iHt} = e^{-i\mu t}\,e^{iKt}\,\psi(\mathbf{x})\,e^{-iKt} \tag{18.12}$$

$$e^{iHt}\,\psi^\dagger(\mathbf{x})\,e^{-iHt} = e^{i\mu t}\,e^{iKt}\,\psi^\dagger(\mathbf{x})\,e^{-iKt}\,, \tag{18.13}$$

in such a way that $n_{\rm h}(\mathbf{x},t) = e^{iHt}\,\psi^\dagger(\mathbf{x})\,\psi(\mathbf{x})\,e^{-iHt} = e^{iKt}\,\psi^\dagger(\mathbf{x})\,\psi(\mathbf{x})\,e^{-iKt} \equiv n_{\rm K}(\mathbf{x},t)$. We then note that the two functions (18.10) and (18.11) can be related to each other via the Lehmann representation, written in terms of the eigenstates $|N,n\rangle$ and eigenvalues $K_{N,n}$ of the grand-canonical Hamiltonian K. We obtain in the two cases:

$$D^{\rm R}(\mathbf{r},\mathbf{r}';\omega) = \frac{1}{Z}\sum_{N,n}\sum_{N',n'}\left(e^{-\beta K_{N,n}} - e^{-\beta K_{N',n'}}\right) \times \frac{\langle N,n|n(\mathbf{r})|N',n'\rangle\,\langle N',n'|n(\mathbf{r}')|N,n\rangle}{K_{N,n}-K_{N',n'}+\omega+i\eta}, \tag{18.14}$$

$$\mathcal{D}(\mathbf{r},\mathbf{r}';\Omega_\nu) = \frac{1}{Z}\sum_{N,n}\sum_{N',n'}\left(e^{-\beta K_{N,n}} - e^{-\beta K_{N',n'}}\right) \times \frac{\langle N,n|n(\mathbf{r})|N',n'\rangle\,\langle N',n'|n(\mathbf{r}')|N,n\rangle}{K_{N,n}-K_{N',n'}+i\Omega_\nu}, \tag{18.15}$$

where Z is the grand-partition function of Eqs. (4.12)–(4.13) and (8.4), $\eta = 0^+$, $\Omega_\nu = 2\pi\nu/\beta$ (ν integer) are bosonic Matsubara frequencies, and $N = N'$ because the density operator commutes with the total particle number (such that $K_{N,n} - K_{N',n'} \to E_{N,n} - E_{N,n'}$ in the expressions (18.14) and (18.15)). Note also that $\mathcal{D}(\mathbf{r},\mathbf{r}';-\Omega_\nu) = \mathcal{D}(\mathbf{r}',\mathbf{r};\Omega_\nu)$, so that one can limit to consider $\Omega_\nu \geq 0$ only. Accordingly, the expression (18.14) can be *formally* obtained from the expression (18.15) via the *analytic continuation* $i\Omega_\nu \to \omega + i\eta$ from imaginary to real frequencies, as sketched in Fig. 18.1. In practice, this prescription can directly be applied

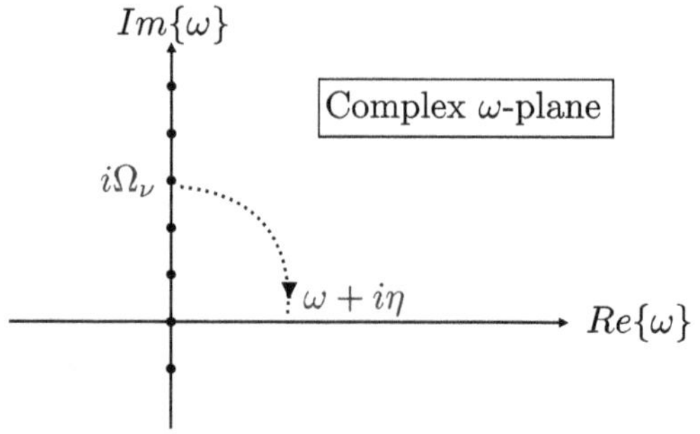

Figure 18.1 Process of analytic continuation from the Matsubara frequency Ω_ν on the imaginary axis to the frequency $\omega + i\eta$ just above the real axis.

when analytic results are obtained from perturbation theory for the temperature correlation function (18.11) [10]. On the other hand, when perturbation theory can only provide numerical results, one has to resort also to numerical analytic continuation on imaginary frequencies, which is a mathematically ill-defined procedure [68] and, most of the time, is dealt with through Padé approximants [48].

19

Time-Dependent Hartree–Fock Approximation and Mean-Field Decoupling

Mean-field theories are often regarded as a first-hand approach to deal with a variety of physical problems arising at equilibrium in condensed matter [69], to the extent that they can provide preliminary qualitative answers to those problems and are at the same time mathematically simple and relatively easy to implement numerically. In particular, equilibrium quantum many-body problems are usually first considered at zero temperature when a suitable form of the ground-state wave function is envisaged and then determined variationally by minimizing the corresponding average value of the Hamiltonian. Notable examples along these lines are the Bardeen–Cooper–Schrieffer (BCS) ground state of a superconductor [70] and the Laughlin state of the fractional quantum Hall effect [71]. These kinds of approaches date back to the pioneering Hartree–Fock method, which assumes that the exact N-body wave function of a Fermi system can be approximated by a single Slater determinant and then invokes the variational method [72], leading in this way to the *self-consistent field method* (as Hartree first named it [73]).

Still at equilibrium, in terms of quantum field theory, mean-field approaches amounts to a *mean-field decoupling* of the interacting part of the system Hamiltonian. To go beyond this initial level of approximation, alternative approaches are utilized in different contexts, like, for instance, the configuration–interaction method of quantum chemistry aimed at including correlation effects over and above the Hartree–Fock approximation [72]. Diagrammatic Green's functions methods prove especially suitable to the purpose, as they enable one to go readily beyond the Hartree–Fock approximation even at finite temperature, for instance, in terms of the t-matrix approximation for a dilute Fermi system (cf. Section 16.1) and of the GW approximation for a dense Fermi system (cf. Section 17.4). By a similar token, Gor'kov formulation of the BCS theory in terms of the temperature Green's functions [46] allows for a systematic inclusion of pairing fluctuations beyond BCS mean field [20].

Mean-field decouplings can also be utilized in time-dependent (nonequilibrium) situations. Here, we shall consider the time-dependent Hartree–Fock (TDHF) approximation for the normal phase (and postpone its Gor'kov generalization to the superfluid phase to Part II). Although in this context we could directly consider the Kadanoff–Baym equation (15.14) for the greater Green's function (13.6), for didactic purposes, we prefer to rederive this equation from scratch, starting from a time-dependent mean-field decoupling. Afterward, we shall connect with a more standard formulation of the TDHF approximation in terms of a set of time-dependent single-particle wave functions. In the following, only the fermion case will be considered.

Note also how the present treatment highlights the fact that, at the mean-field level, there would be essentially no need to set up the machinery of the Schwinger–Keldysh formalism (here in the normal phase and in Part II in the superfluid phase).

19.1 Time-Dependent Hartree–Fock Decoupling

Let's consider the interparticle interaction part (2.4) of the system Hamiltonian and perform on it the following mean-field decoupling:

$$H^{\text{int}} = \frac{1}{2} \sum_{\sigma,\sigma'} \int d\mathbf{r} d\mathbf{r}' \, \psi^{\dagger}_{\sigma}(\mathbf{r}) \psi^{\dagger}_{\sigma'}(\mathbf{r}') \, v(\mathbf{r}-\mathbf{r}') \, \psi_{\sigma'}(\mathbf{r}') \psi_{\sigma}(\mathbf{r})$$
$$\longrightarrow \int d\mathbf{r} d\mathbf{r}' \, v(\mathbf{r}-\mathbf{r}') \left\{ \sum_{\sigma} \langle \psi^{\dagger}_{\sigma}(\mathbf{r}) \psi_{\sigma}(\mathbf{r}) \rangle \sum_{\sigma'} \psi^{\dagger}_{\sigma'}(\mathbf{r}') \psi_{\sigma'}(\mathbf{r}') \right.$$
$$\left. - \sum_{\sigma} \langle \psi^{\dagger}_{\sigma}(\mathbf{r}) \psi_{\sigma}(\mathbf{r}') \rangle \, \psi^{\dagger}_{\sigma}(\mathbf{r}') \psi_{\sigma}(\mathbf{r}) \right\}, \tag{19.1}$$

where we have assumed that spin-flip processes are not allowed. In addition, in the present context, the averages $\langle \cdots \rangle$ depend themselves on time, such that

$$\langle \psi^{\dagger}_{\sigma}(\mathbf{r}) \, \psi_{\sigma}(\mathbf{r}') \rangle \rightarrow \left\langle \psi^{\dagger}_{\text{H}\sigma}(\mathbf{r}, t) \, \psi_{\text{H}\sigma}(\mathbf{r}', t) \right\rangle \tag{19.2}$$

with the definition (2.25) for the operators in the Heisenberg representation and where $\langle \cdots \rangle$ refers to the initial preparation of the system (before the time-dependent perturbation begins to act). With the above expressions and the commutation relation $\left[\psi_{\sigma''}(\mathbf{r}''), \psi^{\dagger}_{\sigma}(\mathbf{r}') \, \psi_{\sigma}(\mathbf{r}) \right] = \delta(\mathbf{r}' - \mathbf{r}'') \delta_{\sigma\sigma''} \psi_{\sigma}(\mathbf{r})$, the equation of motion of the field operator becomes

$$i \frac{\partial}{\partial t} \psi_{\text{H}\sigma}(\mathbf{r}, t) = h(\mathbf{r}, t) \, \psi_{\text{H}\sigma}(\mathbf{r}, t) + \int d\mathbf{r}' \, v(\mathbf{r}-\mathbf{r}')$$
$$\times \left\{ \sum_{\sigma'} \left\langle \psi^{\dagger}_{\text{H}\sigma'}(\mathbf{r}', t) \, \psi_{\text{H}\sigma'}(\mathbf{r}', t) \right\rangle \psi_{\text{H}\sigma}(\mathbf{r}, t) \right.$$
$$\left. - \left\langle \psi^{\dagger}_{\text{H}\sigma}(\mathbf{r}', t) \, \psi_{\text{H}\sigma}(\mathbf{r}, t) \right\rangle \psi_{\text{H}\sigma}(\mathbf{r}', t) \right\}, \tag{19.3}$$

where we note that all field operators share *the same time*. In this way, the equation of motion of the time-ordered single-particle Green's function (cf. Eq. (13.3))

$$G_\sigma(\mathbf{r}t, \mathbf{r}'t') = -\,i\,\theta(t-t')\,\left\langle \psi_{\mathrm{H}\sigma}(\mathbf{r},t)\,\psi^\dagger_{\mathrm{H}\sigma}(\mathbf{r}',t')\right\rangle + i\,\theta(t'-t)\,\left\langle \psi^\dagger_{\mathrm{H}\sigma}(\mathbf{r}',t')\,\psi_{\mathrm{H}\sigma}(\mathbf{r},t)\right\rangle \tag{19.4}$$

reads

$$i\frac{\partial}{\partial t}G_\sigma(\mathbf{r}t, \mathbf{r}'t') = \delta(t-t')\,\delta(\mathbf{r}-\mathbf{r}') + h(\mathbf{r},t)\,G_\sigma(\mathbf{r}t, \mathbf{r}'t') + \int d\mathbf{r}''\, v(\mathbf{r}-\mathbf{r}'') \times\left\{\sum_{\sigma''}\left\langle \psi^\dagger_{\mathrm{H}\sigma''}(\mathbf{r}'',t)\,\psi_{\mathrm{H}\sigma''}(\mathbf{r}'',t)\right\rangle G_\sigma(\mathbf{r}t, \mathbf{r}'t') - \left\langle \psi^\dagger_{\mathrm{H}\sigma}(\mathbf{r}'',t)\,\psi_{\mathrm{H}\sigma}(\mathbf{r},t)\right\rangle G_\sigma(\mathbf{r}''t, \mathbf{r}'t')\right\}. \tag{19.5}$$

This expression can be further manipulated by formally introducing an integral over the real-time t ranging from $-\infty$ to $+\infty$ and defining the *Hartree–Fock single-particle self-energy*

$$\Sigma^\sigma_{\mathrm{HF}}(\mathbf{r}t, \mathbf{r}'t') = -\,i\,\delta(t-t')\left\{\delta(\mathbf{r}-\mathbf{r}')\int d\mathbf{r}''\, v(\mathbf{r}-\mathbf{r}'')\sum_{\sigma''} G_{\sigma''}(\mathbf{r}''t, \mathbf{r}''t^+) - v(\mathbf{r}-\mathbf{r}')\,G_\sigma(\mathbf{r}t, \mathbf{r}'t^+)\right\} \tag{19.6}$$

where

$$\begin{aligned} G_{\sigma''}(\mathbf{r}''t, \mathbf{r}''t^+) &= i\left\langle \psi^\dagger_{\mathrm{H}\sigma''}(\mathbf{r}'',t)\,\psi_{\mathrm{H}\sigma''}(\mathbf{r}'',t)\right\rangle = G^<_{\sigma''}(\mathbf{r}''t, \mathbf{r}''t) \\ G_\sigma(\mathbf{r}t, \mathbf{r}''t^+) &= i\left\langle \psi^\dagger_{\mathrm{H}\sigma}(\mathbf{r}'',t)\,\psi_{\mathrm{H}\sigma}(\mathbf{r},t)\right\rangle = G^<_\sigma(\mathbf{r}t, \mathbf{r}''t)\,, \end{aligned} \tag{19.7}$$

in such a way that the expression $\int d\mathbf{r}'' \int_{-\infty}^{+\infty} dt''\, \Sigma_{\mathrm{HF}}(\mathbf{r}t, \mathbf{r}''t'')\, G^<_\sigma(\mathbf{r}''t'', \mathbf{r}'t')$ coincides with the last term on the right-hand side of Eq. (19.5).

Note that the expression (19.6) of the self-energy, which results from a mean-field decoupling, recovers the sum of the Hartree $\Sigma_{\mathrm{H}}(1,2) = \delta(1,2)(-i)\int d3\, v(1,3)\, G(3,3^+)$ (cf. Eq. (11.30)) and Fock $\Sigma_{\mathrm{F}}(1,2) = i\, v(1,2) G(1,2^+)$ (cf. Eq. (11.31)) approximations to the self-energy as obtained from the diagrammatic structure (or even directly from the zero-temperature formalism [7]). Note also that the presence of the Dirac delta function $\delta(t-t')$ renders the Hartree–Foch self-energy (19.6)) *instantaneous*, that is, without "memory effects" of what has happened at previous times.

19.2 Time-Dependent Hartree–Fock Single-Particle Wave Functions

The time-dependent Hartree–Fock approximation for a Fermi system is usually expressed in terms of time-dependent single-particle wave functions, when it is conventionally derived from a single Slater determinant constructed in terms of these functions [74, 75]. Here, we provide an alternative route to this derivation, which has the advantage of considering also an initial thermal preparation of the system and of being readily extended to the superfluid phase (as it will be done in Part II). This derivation goes through the following steps.

Step # 1 Consider the equation of motion for the greater Green's function

$$G^{>}_{\sigma}(\mathbf{r}t, \mathbf{r}'t') = -i \left\langle \psi_{\mathrm{H}\sigma}(\mathbf{r}, t)\, \psi^{\dagger}_{\mathrm{H}\sigma}(\mathbf{r}', t') \right\rangle \tag{19.8}$$

as obtained from the mean-field decoupling (19.3), namely,

$$\begin{aligned} i\frac{\partial}{\partial t} G^{>}_{\sigma}(\mathbf{r}t, \mathbf{r}'t') &= h(\mathbf{r}, t)\, G^{>}_{\sigma}(\mathbf{r}t, \mathbf{r}'t') + \int d\mathbf{r}''\, v(\mathbf{r} - \mathbf{r}'') \\ &\times \Bigg\{ \sum_{\sigma''} \left\langle \psi^{\dagger}_{\mathrm{H}\sigma''}(\mathbf{r}'', t)\, \psi_{\mathrm{H}\sigma''}(\mathbf{r}'', t) \right\rangle G^{>}_{\sigma}(\mathbf{r}t, \mathbf{r}'t') \\ &- \left\langle \psi^{\dagger}_{\mathrm{H}\sigma}(\mathbf{r}'', t)\, \psi_{\mathrm{H}\sigma}(\mathbf{r}, t) \right\rangle G^{>}_{\sigma}(\mathbf{r}''t, \mathbf{r}'t') \Bigg\} \end{aligned} \tag{19.9}$$

to the extent that this has a simpler structure than the equation of motion (19.5) for the time-ordered single-particle Green's function (19.4) (the equation of motion for the lesser Green's function $G^{<}_{\sigma}$ might alternatively be considered).

Step # 2 Consistently with the mean-field decoupling, the field operators in the Heisenberg representation, which occur in Eqs. (19.8) and (19.9), can be represented in the form

$$\psi_{\mathrm{H}\sigma}(\mathbf{r}, t) = \sum_{\nu} w_{\nu}(\mathbf{r}, t)\, \gamma_{\nu\sigma}\,. \tag{19.10}$$

In this way, the whole time evolution is assumed to be carried over by the single-particle orbitals $\{w_{\nu}(\mathbf{r}, t); \nu = 1, 2, \cdots\}$, which have to be self-consistently determined and evolve in time out of the orbitals $\{w_{\nu}(\mathbf{r}); \nu = 1, 2, \cdots\}$ solutions of the time-independent Hartree–Fock equations before the time-dependent perturbation begins to act at time t_0. The corresponding fermionic operators $\{\gamma_{\nu\sigma}; \nu = 1, 2, \cdots\}$ are such that

$$\left\langle \gamma^{\dagger}_{\nu\sigma}\, \gamma_{\nu'\sigma'} \right\rangle = \delta_{\nu\nu'}\, \delta_{\sigma\sigma'}\, f_F(\epsilon_{\nu})\,, \tag{19.11}$$

where $f_F(\epsilon_{\nu})$ is the Fermi–Dirac distribution function for the initial (thermal) configuration and ϵ_{ν} is the Hartree–Fock eigenvalue associated with the initial orbital $w_{\nu}(\mathbf{r})$. The average in Eq. (19.11) thus specifies the meaning of the averages

$\langle \cdots \rangle$ occurring in Eq. (19.9). We emphasize that Eq. (19.10) represents the *key assumption* to derive the time-dependent Hartree–Fock equations.

Step # 3 Prove that the orbitals $\{w_\nu(\mathbf{r},t); \nu = 1, 2, \cdots\}$ form a *complete set* at any time t. This property follows by representing the field operators at equal time in the expression

$$\left\langle \psi_{H\sigma}(\mathbf{r},t)\, \psi^\dagger_{H\sigma}(\mathbf{r}',t) \right\rangle + \left\langle \psi^\dagger_{H\sigma}(\mathbf{r}',t)\, \psi_{H\sigma}(\mathbf{r},t) \right\rangle = \delta(\mathbf{r}-\mathbf{r}') \tag{19.12}$$

in terms of Eq. (19.10), thereby yielding:

$$\begin{aligned} &\left\langle \psi_{H\sigma}(\mathbf{r},t)\, \psi^\dagger_{H\sigma}(\mathbf{r}',t) \right\rangle + \left\langle \psi^\dagger_{H\sigma}(\mathbf{r}',t)\, \psi_{H\sigma}(\mathbf{r},t) \right\rangle \\ &\quad = \sum_{\nu\nu'} w_\nu(\mathbf{r},t)\, w_{\nu'}(\mathbf{r}',t)^* \left(\left\langle \gamma_{\nu\sigma}\, \gamma^\dagger_{\nu'\sigma} \right\rangle + \left\langle \gamma^\dagger_{\nu'\sigma}\, \gamma_{\nu\sigma} \right\rangle \right) \\ &\quad = \sum_{\nu} w_\nu(\mathbf{r},t)\, w_\nu(\mathbf{r}',t)^* \underset{\substack{\uparrow \\ [\text{Eq.}(19.12)]}}{=} \delta(\mathbf{r}-\mathbf{r}')\,. \end{aligned} \tag{19.13}$$

Step # 4 Introduce the function

$$\rho(\mathbf{r},\mathbf{r}';t) \equiv \sum_\nu w_\nu(\mathbf{r},t)\, w_\nu(\mathbf{r}',t)^*\, f_F(\epsilon_\nu) \tag{19.14}$$

such that

$$\left\langle \psi^\dagger_{H\sigma}(\mathbf{r}',t)\, \psi_{H\sigma}(\mathbf{r},t) \right\rangle = \sum_{\nu\nu'} w_\nu(\mathbf{r}',t)^*\, w_{\nu'}(\mathbf{r},t) \left\langle \gamma^\dagger_{\nu\sigma}\, \gamma_{\nu'\sigma} \right\rangle \underset{\substack{\uparrow \\ [\text{Eqs.}(19.11)\text{ and }(19.14)]}}{=} \rho(\mathbf{r},\mathbf{r}';t)\,. \tag{19.15}$$

In this way, the equation of motion (19.9) simplifies to the form

$$\begin{aligned} &\left[i\frac{\partial}{\partial t} - h(\mathbf{r},t) \right] G^>_\sigma(\mathbf{r}t,\mathbf{r}'t') - \int d\mathbf{r}''\, v(\mathbf{r}-\mathbf{r}'') \left\{ 2\,\rho(\mathbf{r}'',\mathbf{r}'';t)\, G^>_\sigma(\mathbf{r}t,\mathbf{r}'t') \right. \\ &\quad \left. - \rho(\mathbf{r},\mathbf{r}'';t)\, G^>_\sigma(\mathbf{r}''t,\mathbf{r}'t') \right\} = 0 \end{aligned} \tag{19.16}$$

with the factor of 2 arising from the spin degeneracy.

Step # 5 In Eq. (19.16) express $G^>_\sigma$ given by Eq. (19.8) in terms of the representation (19.10) and use the property (19.11), to obtain:

$$\begin{aligned} &\sum_\nu w_\nu(\mathbf{r}',t')^*\, (1 - f_F(\epsilon_\nu)) \left\{ \left[i\frac{\partial}{\partial t} - h(\mathbf{r},t) \right] w_\nu(\mathbf{r},t) \right. \\ &\left. - \int d\mathbf{r}''\, v(\mathbf{r}-\mathbf{r}'') \left[2\,\rho(\mathbf{r}'',\mathbf{r}'';t)\, w_\nu(\mathbf{r},t) - \rho(\mathbf{r},\mathbf{r}'';t)\, w_\nu(\mathbf{r}'',t) \right] \right\} = 0\,. \end{aligned} \tag{19.17}$$

Owing to the completeness of the set $\{w_\nu(\mathbf{r}',t')^*; \nu = 1, 2, \cdots\}$ at any time t', the expression within braces on the left-hand side of Eq. (19.17) vanishes separately for each ν, yielding eventually the desired *TDHF equations* for the orbitals $w_\nu(\mathbf{r},t)$

$$i\frac{\partial}{\partial t}w_\nu(\mathbf{r},t) - \left[h(\mathbf{r},t) + 2\int d\mathbf{r}''v(\mathbf{r}-\mathbf{r}'')\,\rho(\mathbf{r}'',\mathbf{r}'';t)\right]w_\nu(\mathbf{r},t)$$
$$+\int d\mathbf{r}''v(\mathbf{r}-\mathbf{r}'')\,\rho(\mathbf{r},\mathbf{r}'';t)\,w_\nu(\mathbf{r}'',t) = 0 \qquad (19.18)$$

with $\rho(\mathbf{r},\mathbf{r}';t)$ to be *self-consistently* determined from Eq. (19.14). This concludes our derivation.

Remark: Connecting with the Dyson equation for $G^>$

The equation of motion (19.9) for the greater Green's function $G^>_\sigma$, which was obtained from the simple mean-field decoupling (19.1), could have been obtained directly from the Dyson equation (15.14), where only the retarded self-energy Σ^R given by Eq. (19.6) and active at equal times is retained. This would correspond to the singular term of the expression (14.3), whereby no memory effect is taken into account.

In Part II, the procedure discussed in the present chapter will be generalized to the fermionic superfluid phase, to obtain the time-dependent Bogoliubov–de Gennes equations.

20

Miscellany and Addenda to Part I

Now that Part I is coming to an end, it is worth appending here the treatment of a few topics, which are somewhat relevant to the general purposes of the book.

20.1 A Generic Pre-summation in the Dyson Equation

Integral equations such as the Dyson equation (11.40) may be subject to a partial pre-summation according to the following argument. Quite generally, consider the following integral equation (in compact matrix form)

$$A = a + a\,B\,A \qquad \text{with} \qquad B = b + \beta\,. \tag{20.1}$$

This integral equation can be split into two coupled integral equations, by writing

$$\begin{cases} A = \alpha + \alpha\,b\,A \\ \alpha = a + a\,\beta\,\alpha\,. \end{cases} \tag{20.2}$$

Proof of Eqs. (20.2)

Solve formally first of Eq. (20.2) $[\Rightarrow A = (1 - \alpha\,b)^{-1}\,\alpha]$, as well as the second of Eq. (20.2) $[\Rightarrow \alpha = (1 - a\,\beta)^{-1}\,a]$. Then, enter the solution for α into the solution for A, to obtain

$$\begin{aligned} A = (1 - \alpha\,b)^{-1}\,\alpha &= (1 - \alpha\,b)^{-1}\,(1 - a\,\beta)^{-1}\,a = [(1 - a\,\beta)\;(1 - \alpha\,b)]^{-1}\,a \\ &= [1 - (a + a\,\beta\alpha)\,b - a\beta + a\beta\alpha b]^{-1}\,a \\ &= [1 - a\,(b + \beta)]^{-1}\,a = [1 - aB]^{-1}\,a\,. \qquad \text{[QED]} \end{aligned} \tag{20.3}$$

This kind of procedure was utilized, for example, in Ref. [37], where the mean-field "singular" parts of the self-energy were included in the free propagators, such that the remaining self-energies are "regular" two-time functions. It will also be adopted in some formal manipulations considered later on in the book, specifically to a large extent in Section 28.1 of Part II and to a lesser extent in Section 34.2 of Part III. It is, however, clear that this procedure does not apply when, for a given approximate choice of the self-energy, the resulting Dyson equations for the various components of the single-particle contour Green's function are solved numerically in a fully self-consistent way because the chosen functional form of the self-energy to be split into two components depends itself on the Green's function, which is to be solved for.

Finally, we may mention that a kind of pre-summation was also performed when passing from the third to the second line of Eq. (6.8) (even when considering the replacement (8.1) for the thermal average), to the extent that in the third line the "perturbation" is $H_{\mathrm{I}}^{\mathrm{int}}(z) + H_{\mathrm{I}}^{\mathrm{ext}}(z)$, while in the second line it is only $H_{\mathrm{h}}^{\mathrm{ext}}(z)$.

20.2 Green's Functions within the Keldysh Formalism for a Homogeneous and Time-Independent Non-interacting System

At the end of Section 13.4, we arrived to the conclusion that, within the Keldysh formalism, the non-interacting Green's function satisfies Eq. (13.41), namely,

$$\begin{aligned}
&\left[i\frac{d}{dt_1} - h(\mathbf{x}_1, t_1)\right] \left[L\tau^3\, G_0(\mathbf{x}_1 t_1, \mathbf{x}_2 t_2) L^{\dagger}\right]_{i_1 i_2} \\
&= \left[i\frac{d}{dt_1} - h(\mathbf{x}_1, t_1)\right] \begin{bmatrix} G_0^{\mathrm{R}}(\mathbf{x}_1 t_1, \mathbf{x}_2 t_2) & G_0^{\mathrm{K}}(\mathbf{x}_1 t_1, \mathbf{x}_2 t_2) \\ 0 & G_0^{\mathrm{A}}(\mathbf{x}_1 t_1, \mathbf{x}_2 t_2) \end{bmatrix} \\
&= \delta(\mathbf{x}_1, \mathbf{x}_2)\, \delta(t_1 - t_2) \begin{bmatrix} 1 & 0 \\ 0 & 1 \end{bmatrix},
\end{aligned} \tag{20.4}$$

where the indices (i_1, i_2) refer to the (contour) components of the 2×2 matrix (13.29). Here, we specify these equations for a homogeneous and time-independent system, for which they can be solved in terms of Fourier transforms, whereby for a generic function, we write

$$g(\mathbf{x}t, \mathbf{x}'t') \to g(\mathbf{x} - \mathbf{x}', t - t') = \int \frac{d\mathbf{k}}{(2\pi)^3} \int_{-\infty}^{+\infty} \frac{d\omega}{2\pi}\, e^{i\mathbf{k}\cdot(\mathbf{x}-\mathbf{x}')} e^{-i\omega(t-t')}\, g(\mathbf{k}, \omega), \tag{20.5}$$

leaving aside the spin indices.

For the retarded (R) and advanced (A) components, we readily obtain

$$\begin{cases} G^{\mathrm{R}}(\mathbf{k}, \omega) = \dfrac{1}{\omega - \epsilon(\mathbf{k}) + i\eta} \\[2ex] G^{\mathrm{A}}(\mathbf{k}, \omega) = \dfrac{1}{\omega - \epsilon(\mathbf{k}) - i\eta} \end{cases} \tag{20.6}$$

where $\epsilon(\mathbf{k}) = \frac{\mathbf{k}^2}{2m}$, η is a positive infinitesimal, and the suffix 0 standing for "non-interacting" has been dropped for simplicity. Regarding the Fourier transform of the Keldysh (K) component, it is evident from Eq. (20.4) that it is proportional to $\delta(\omega - \epsilon(\mathbf{k}))$ (owing to the property $x\delta(x) = 0$ of the Dirac delta function). To determine the coefficient of proportionality, it is convenient to go back to the definition (13.18), thus writing $G^{\mathrm{K}}(\mathbf{x}t, \mathbf{x}'t') = G^{>}(\mathbf{x}t, \mathbf{x}'t') + G^{<}(\mathbf{x}t, \mathbf{x}'t')$. In particular, for a time-independent, homogeneous, and noninteracting system of interest, here we obtain for $G^{>}$:

$$
\begin{aligned}
G^{>}(\mathbf{x}t, \mathbf{x}'t') &= -\,i\,\langle \psi_{\mathrm{H}}(\mathbf{x}t)\,\psi_{\mathrm{H}}^{\dagger}(\mathbf{x}'t')\rangle \\
&= -\,i\,\langle e^{iH_0 t}\psi(\mathbf{x})\,e^{-iH_0 t}\,e^{iH_0 t'}\psi^{\dagger}(\mathbf{x}')\,e^{-iH_0 t'}\rangle \\
&= -\,i\int \frac{d\mathbf{k}}{(2\pi)^3}\,e^{i\mathbf{k}\cdot(\mathbf{x}-\mathbf{x}')}\,e^{-i\epsilon(\mathbf{k})(t-t')}\,\langle c_{\mathbf{k}} c_{\mathbf{k}}^{\dagger}\rangle \\
&= -\,i\int \frac{d\mathbf{k}}{(2\pi)^3}\,e^{i\mathbf{k}\cdot(\mathbf{x}-\mathbf{x}')}\,e^{-i\epsilon(\mathbf{k})(t-t')}\,(1 \pm f_{\mp}(\epsilon(\mathbf{k}) - \mu))\,,
\end{aligned}
\tag{20.7}
$$

where the upper (lower) sign holds for bosons (fermions), and $\langle c_{\mathbf{k}}^{\dagger} c_{\mathbf{k}}\rangle = f_{\mp}(\epsilon(\mathbf{k}) - \mu)$ contains the chemical potential μ of thermodynamic equilibrium with $f_{\mp}(E) = \left(e^{E/k_B T} \mp 1\right)^{-1}$. This yields in Fourier space

$$
G^{>}(\mathbf{k}, \omega) = -\,i\,2\pi\,\delta(\omega - \epsilon(\mathbf{k}))\,\left(1 \pm f_{\mp}(\epsilon(\mathbf{k}) - \mu)\right)\,. \tag{20.8}
$$

Correspondingly, we obtain for $G^{<}$:

$$
\begin{aligned}
G^{<}(\mathbf{x}t, \mathbf{x}'t') &= \mp\,i\,\langle \psi_{\mathrm{H}}^{\dagger}(\mathbf{x}'t')\,\psi_{\mathrm{H}}(\mathbf{x}t)\rangle = \mp\,i\int \frac{d\mathbf{k}}{(2\pi)^3}\,e^{i\mathbf{k}\cdot(\mathbf{x}-\mathbf{x}')}\,e^{-i\epsilon(\mathbf{k})(t-t')}\,\langle c_{\mathbf{k}}^{\dagger} c_{\mathbf{k}}\rangle \\
&= \mp\,i\int \frac{d\mathbf{k}}{(2\pi)^3}\,e^{i\mathbf{k}\cdot(\mathbf{x}-\mathbf{x}')}\,e^{-i\epsilon(\mathbf{k})(t-t')}\,f_{\mp}(\epsilon(\mathbf{k}) - \mu)\,,
\end{aligned}
\tag{20.9}
$$

yielding in Fourier space

$$
G^{<}(\mathbf{k}, \omega) = \mp\,i\,2\pi\,\delta(\omega - \epsilon(\mathbf{k}))\,f_{\mp}(\epsilon(\mathbf{k}) - \mu)\,. \tag{20.10}
$$

In this way, we obtain for the Keldysh component in Fourier space:

$$
\begin{aligned}
G^{\mathrm{K}}(\mathbf{k}, \omega) &= G^{>}(\mathbf{k}, \omega) + G^{<}(\mathbf{k}, \omega) = -\,i\,2\pi\,\delta(\omega - \epsilon(\mathbf{k}))\,\left(1 \pm 2 f_{\mp}(\epsilon(\mathbf{k}) - \mu)\right) \\
&= -\,i\,2\pi\,\delta(\omega - \epsilon(\mathbf{k}))\left\{
\begin{array}{ll}
\coth\left(\frac{(\epsilon(\mathbf{k})-\mu)}{2k_B T}\right) & \text{(bosons)} \\
\tanh\left(\frac{(\epsilon(\mathbf{k})-\mu)}{2k_B T}\right) & \text{(fermions)}\,.
\end{array}
\right.
\end{aligned}
\tag{20.11}
$$

Note that, contrary to the retarded and advanced components (20.6), the Keldysh component (20.11) contains reference to the thermodynamic equilibrium in terms of the temperature T and the chemical potential μ.

Collecting the results (20.6) and (20.11), we write eventually

$$\begin{bmatrix} G^{\mathrm{R}}(\mathbf{k},\omega) & G^{\mathrm{K}}(\mathbf{k},\omega) \\ 0 & G^{\mathrm{A}}(\mathbf{k},\omega) \end{bmatrix} = \begin{bmatrix} \frac{1}{\omega-\epsilon(\mathbf{k})+i\eta} & -i2\pi\delta(\omega-\epsilon(\mathbf{k}))\begin{cases} \coth\left(\frac{(\epsilon(\mathbf{k})-\mu)}{2k_BT}\right) \\ \tanh\left(\frac{(\epsilon(\mathbf{k})-\mu)}{2k_BT}\right) \end{cases} \\ 0 & \frac{1}{\omega-\epsilon(\mathbf{k})-i\eta} \end{bmatrix}, \tag{20.12}$$

such that

$$\begin{bmatrix} \omega-\epsilon(\mathbf{k}) & 0 \\ 0 & \omega-\epsilon(\mathbf{k}) \end{bmatrix}\begin{bmatrix} G^{\mathrm{R}}(\mathbf{k},\omega) & G^{\mathrm{K}}(\mathbf{k},\omega) \\ 0 & G^{\mathrm{A}}(\mathbf{k},\omega) \end{bmatrix} = \begin{bmatrix} (\omega-\epsilon(\mathbf{k}))G^{\mathrm{R}}(\mathbf{k},\omega) & (\omega-\epsilon(\mathbf{k}))G^{\mathrm{K}}(\mathbf{k},\omega) \\ 0 & (\omega-\epsilon(\mathbf{k}))G^{\mathrm{A}}(\mathbf{k},\omega) \end{bmatrix} = \begin{bmatrix} 1 & 0 \\ 0 & 1 \end{bmatrix}, \tag{20.13}$$

in accordance with the Fourier transform of Eq. (20.4).

Part II

Fermionic Superfluid Phase

21

Time-Dependent Version of the BCS Hamiltonian: Gor'kov Equations for the Normal and Anomalous Single-Particle Green's Functions

Although all our knowledge arises with experience, not all of it arises from experience

I. Kant Critique of Pure Reason

In this chapter, we consider the analog of the time-dependent Hartree–Fock (mean-field) decoupling described in Section 19.1 of Part I and extend it to the broken-symmetry phase for superfluid fermions, while postponing to Chapter 25 the consideration of the associated single-particle wave functions of the Bogoliubov–de Gennes equations. [Here and in the rest of Part II, only systems of interacting fermions will be considered.] In this way, we will end up with two coupled equations for the "normal" and "anomalous" time-dependent single-particle Green's functions, which extend to nonequilibrium situations the equations originally obtained at equilibrium by Gor'kov [46], soon after the BCS original article on the theory of superconductivity [70]. Similar to the normal phase considered in Section 19.1 of Part I, by limiting here to the mean-field decoupling also for the superfluid phase, we can avoid dwelling at the outset on the full machinery of the more general Schwinger–Keldysh approach once properly extended to the superfluid phase. This more general approach will be considered in detail in Chapter 23, where it will confirm the present results within the simplest (BCS-like) approximation. Accordingly, the material contained in this chapter should be regarded as a necessary prolegomenon to later developments, as it is almost invariably the case with mean-field decouplings.

21.1 Time-Dependent Mean-Field Decoupling in the Superfluid Phase

We assume that the fermionic system we are considering is still described by the Hamiltonian given by Eqs. (2.1)–(2.5) of Part I, where in the interacting part (2.4), we:

(i) take the interparticle interaction to be of the *contact type*, namely,

$$v(\mathbf{r}-\mathbf{r}') = v_0\,\delta(\mathbf{r}-\mathbf{r}'), \tag{21.1}$$

with $v_0 < 0$ corresponding to an attractive interaction;

(ii) factorize the string of four field operators (with the appropriate spin structure) in terms of the mean-field decoupling

$$\begin{aligned} H^{\text{int}} &= \frac{1}{2} v_0 \sum_{\sigma,\sigma'} \int d\mathbf{r}\, \psi^\dagger_\sigma(\mathbf{r}) \psi^\dagger_{\sigma'}(\mathbf{r}) \psi_{\sigma'}(\mathbf{r}) \psi_\sigma(\mathbf{r}) \\ &= v_0 \int d\mathbf{r}\, \psi^\dagger_\uparrow(\mathbf{r}) \psi^\dagger_\downarrow(\mathbf{r}) \psi_\downarrow(\mathbf{r}) \psi_\uparrow(\mathbf{r}) \longrightarrow v_0 \int d\mathbf{r}\, \Big(\big\langle \psi^\dagger_\uparrow(\mathbf{r}) \psi^\dagger_\downarrow(\mathbf{r}) \big\rangle \psi_\downarrow(\mathbf{r}) \psi_\uparrow(\mathbf{r}) \\ &\quad + \psi^\dagger_\uparrow(\mathbf{r})\, \psi^\dagger_\downarrow(\mathbf{r}) \, \big\langle \psi_\downarrow(\mathbf{r}) \psi_\uparrow(\mathbf{r}) \big\rangle \Big) . \end{aligned} \tag{21.2}$$

To arrive at this expression, we have considered the averages $\langle \cdots \rangle$ over a distribution that depends on H^{int} itself (as we did in Section 19.1 of Part I) and retained only *anomalous averages* as Gor'kov first did in Ref. [46] (with the argument that these averages make the superfluid phase to differ from the normal phase through the occurrence of "off-diagonal long-range order" [76]). In addition, as in Section 19.1 of Part I, in the time-dependent case that we are considering these averages depend themselves on time, such that

$$\big\langle \psi_\downarrow(\mathbf{r})\, \psi_\uparrow(\mathbf{r}) \big\rangle \longrightarrow \big\langle \psi_{H\downarrow}(\mathbf{r},t)\, \psi_{H\uparrow}(\mathbf{r},t) \big\rangle \tag{21.3}$$

with the definition (2.25) of the operators in the Heisenberg representation, and where $\langle \cdots \rangle$ now refers to the initial preparation of the system (namely, before the time-dependent perturbation begins to act). With these identifications, the equation of motion of the field operator $\psi_{H\uparrow}(\mathbf{r},t)$ becomes:

$$i\frac{\partial}{\partial t}\psi_{H\uparrow}(\mathbf{r},t) = h(\mathbf{r},t)\,\psi_{H\uparrow}(\mathbf{r},t) + v_0\, \psi^\dagger_{H\downarrow}(\mathbf{r},t)\psi_{H\downarrow}(\mathbf{r},t)\psi_{H\uparrow}(\mathbf{r},t) \tag{21.4}$$

$$\longrightarrow h(\mathbf{r},t)\,\psi_{H\uparrow}(\mathbf{r},t) + v_0 \big\langle \psi_{H\downarrow}(\mathbf{r},t)\,\psi_{H\uparrow}(\mathbf{r},t) \big\rangle\, \psi^\dagger_{H\downarrow}(\mathbf{r},t) . \tag{21.5}$$

One is thus led to introduce the *time-dependent gap (order) parameter* defined by

$$\Delta(\mathbf{r},t) = v_0 \big\langle \psi_{H\downarrow}(\mathbf{r},t)\,\psi_{H\uparrow}(\mathbf{r},t) \big\rangle , \tag{21.6}$$

which generalizes to the present time-dependent context the standard definition of the BCS theory at equilibrium [6, 7].

Recall that, in Section 9.3 of Part I, "anomalous" averages were considered on the basis of the η-ensemble, whereby an external field "breaks the symmetry" of the physical system, made of either bosons or fermions. As a consequence, we have shown there that for bosons it is directly the average of the field operator that acquires a nonvanishing value, as originally envisaged by Bogoliubov [43], while for fermions one needs pairs of field operators to do the job, as in Eqs. (21.2)–(21.6). In these equations, however, we have not explicitly mentioned the presence of an η-ensemble, but rather assumed at the outset the occurrence of nonvanishing anomalous averages, thereby deferring their sustainability under appropriate physical circumstances to a later explicit validation following the Gor'kov original approach [46]. While, in general, the concept of *spontaneous broken symmetry* dwells on the nontrivial complexity of large enough many-body systems, the phenomenon of superconductivity is considered, in particular, the most spectacular example of broken symmetry that ordinary macroscopic bodies can undergo [77] and as such has inspired also the extension to the physics of elementary particles [78].

21.2 Equations of Motion for the Normal and Anomalous Single-Particle Green's Functions

We can now introduce the "normal" single-particle Green's function for spin-up fermions

$$\begin{aligned} G(\mathbf{r}t, \mathbf{r}'t') &= -i \left\langle \mathcal{T} \left\{ \psi_{\mathrm{H}\uparrow}(\mathbf{r}, t)\, \psi^{\dagger}_{\mathrm{H}\uparrow}(\mathbf{r}', t') \right\} \right\rangle \\ &= -i\,\theta(t-t') \left\langle \psi_{\mathrm{H}\uparrow}(\mathbf{r}, t)\, \psi^{\dagger}_{\mathrm{H}\uparrow}(\mathbf{r}', t') \right\rangle \\ &\quad + i\,\theta(t'-t) \left\langle \psi^{\dagger}_{\mathrm{H}\uparrow}(\mathbf{r}', t')\, \psi_{\mathrm{H}\uparrow}(\mathbf{r}, t) \right\rangle , \end{aligned} \tag{21.7}$$

where $\mathcal{T}$ is the standard time-ordering operator for real time in Section 2.2 of Part I. [We could as well consider spin-down fermions since no magnetic effects are included here.] With the equation of motion (21.5) of the field operator, the equation of motion of the Green's function (21.7) then reads

$$\begin{aligned} i\frac{\partial}{\partial t} G(\mathbf{r}t, \mathbf{r}'t') &= \delta(t-t')\,\delta(\mathbf{r}-\mathbf{r}') \\ &\quad + \theta(t-t') \left\langle \left(\frac{\partial}{\partial t} \psi_{\mathrm{H}\uparrow}(\mathbf{r}, t) \right) \psi^{\dagger}_{\mathrm{H}\uparrow}(\mathbf{r}', t') \right\rangle \\ &\quad - \theta(t'-t) \left\langle \psi^{\dagger}_{\mathrm{H}\uparrow}(\mathbf{r}', t') \left(\frac{\partial}{\partial t} \psi_{\mathrm{H}\uparrow}(\mathbf{r}, t) \right) \right\rangle \\ &\underset{[\mathrm{Eq.}(21.5)]}{\overset{\uparrow}{=}} \delta(t-t')\,\delta(\mathbf{r}-\mathbf{r}') + h(\mathbf{r}, t)\, G(\mathbf{r}t, \mathbf{r}'t') + \Delta(\mathbf{r}, t)\, F(\mathbf{r}t, \mathbf{r}'t') , \end{aligned} \tag{21.8}$$

where the anti-commutation relations (2.2) have been used, $\Delta(\mathbf{r}, t)$ is given by Eq. (21.6), and

$$F(\mathbf{r}t, \mathbf{r}'t') = -i \left\langle \mathcal{T} \left\{ \psi^{\dagger}_{\mathrm{H}\downarrow}(\mathbf{r}, t)\, \psi^{\dagger}_{\mathrm{H}\uparrow}(\mathbf{r}', t') \right\} \right\rangle \tag{21.9}$$

is the "anomalous" single-particle Green's function associated with the broken symmetry, such that

$$\Delta(\mathbf{r}, t)^* = -i\, v_0\, F(\mathbf{r}t^+, \mathbf{r}t)\,. \tag{21.10}$$

To obtain the time evolution of $G(\mathbf{r}t, \mathbf{r}'t')$, it is thus necessary to know also the time evolution of $F(\mathbf{r}t, \mathbf{r}'t')$. This is obtained similarly to Eq. (21.8), in the form:

$$\begin{aligned} i\frac{\partial}{\partial t} F(\mathbf{r}t, \mathbf{r}'t') &= \delta(t-t') \left\langle \left(\psi^{\dagger}_{\mathrm{H}\downarrow}(\mathbf{r}, t)\, \psi^{\dagger}_{\mathrm{H}\uparrow}(\mathbf{r}', t) + \psi^{\dagger}_{\mathrm{H}\uparrow}(\mathbf{r}', t)\, \psi^{\dagger}_{\mathrm{H}\downarrow}(\mathbf{r}, t) \right) \right\rangle \\ &\quad + \theta(t-t') \left\langle \left(\frac{\partial}{\partial t} \psi^{\dagger}_{\mathrm{H}\downarrow}(\mathbf{r}, t) \right) \psi^{\dagger}_{\mathrm{H}\uparrow}(\mathbf{r}', t') \right\rangle \\ &\quad - \theta(t'-t) \left\langle \psi^{\dagger}_{\mathrm{H}\uparrow}(\mathbf{r}', t') \left(\frac{\partial}{\partial t} \psi^{\dagger}_{\mathrm{H}\downarrow}(\mathbf{r}, t) \right) \right\rangle \\ &\underset{[\text{Eq.}(21.5)]}{\underset{\uparrow}{=}} -h(\mathbf{r}, t)\, F(\mathbf{r}t, \mathbf{r}'t') + \Delta(\mathbf{r}, t)^*\, G(\mathbf{r}t, \mathbf{r}'t'), \end{aligned} \tag{21.11}$$

where the anti-commutation relations (2.2) have again been used, together with the definition (21.6) of the time-dependent gap parameter. One is thus led to solve the *two coupled equations* (21.8) and (21.11) subject to the constraint (21.10), with appropriate initial conditions to be specified at the time t_0 at which the external perturbation begins to act on the system. Note how these equations extend to the time-dependent (nonequilibrium) case the Gor'kov equations originally obtained for imaginary time in the equilibrium case [6, 7, 46], which will be considered in detail in Section 25.1.

As already mentioned, at the mean-field level for the superfluid phase considered in this chapter, there is apparently no need to introduce the closed-path Green's functions that were exploited at length for the normal phase in Part I. Similar to the time-dependent Hartree–Fock decoupling of Chapter 19 of Part I, on physical grounds, the simplifying feature here is the lack of memory effects in Eqs. (21.8) and (21.11), whereby the many-body dynamics is absorbed in the single function $\Delta(\mathbf{r}, t)$ of one-time variable. Recourse to the closed-path Green's functions will, however, be required when considering the inclusion of beyond-mean-field effects (like pairing fluctuations). Of course, the mean-field approximation considered in this chapter can also be reframed in terms of closed-path Green's functions. However, this will not be accomplished before having introduced the Nambu representation of the field operators, as it will be done in Chapter 22.

22

The Hamiltonian in the Nambu Representation and the Role of the Hartree–Fock–BCS Self-Energy

In Chapter 21, we have pointed out that a crucial pillar of the BCS theory of superconductivity is the introduction of *anomalous averages* in the superfluid phase, as defined in Eq. (21.6). This kind of averages vanishes in the normal phase. However, the corresponding anomalous single-particle Green's function (21.9) does not match the form of the single-particle Green's functions treated extensively in Part I, where an even number of creation and destruction operators appear.

The solution to this problem stems from considering that anomalous averages such as (21.6) can be self-consistently sustained only provided opposite-spin fermions are coupled in *pairs*, such that adding a fermion with spin up leads to the same state as removing a fermion with spin down. Besides being the physical basis of the BCS theory of superconductivity [70], this is also the basis of Nambu's idea for introducing a mixed spinor field in the place of the original field operators [79].

22.1 The Hamiltonian in the Nambu Representation

The Nambu representation for the *pseudo-spinor fields* reads

$$\mathbf{\Psi}(\mathbf{r}) = \begin{pmatrix} \psi_\uparrow(\mathbf{r}) \\ \psi_\downarrow^\dagger(\mathbf{r}) \end{pmatrix} , \quad \mathbf{\Psi}^\dagger(\mathbf{r}) = \left(\psi_\uparrow^\dagger(\mathbf{r}) \, , \, \psi_\downarrow(\mathbf{r}) \right) , \tag{22.1}$$

whose components are distinguished by the index $i = (1, 2)$ (instead of the original spin index $\sigma = (\uparrow, \downarrow)$). These pseudo-spinor fields preserve the anti-commutation relations (2.2) of Part I for the original field operators:

$$\begin{aligned} \left\{ \Psi_i(\mathbf{r}), \Psi_j^\dagger(\mathbf{r}') \right\} &= \delta_{ij}\, \delta(\mathbf{r} - \mathbf{r}') \\ \left\{ \Psi_i(\mathbf{r}), \Psi_j(\mathbf{r}') \right\} &= \left\{ \Psi_i^\dagger(\mathbf{r}), \Psi_j^\dagger(\mathbf{r}') \right\} = 0 \, . \end{aligned} \tag{22.2}$$

Accordingly, all physical operators will have to be expressed in terms of these new fields.

We start by considering the system Hamiltonian, which we take in the form (2.1)–(2.5) of Part I and reproduce here for convenience:

$$H(t) = \sum_{\sigma} \int d\mathbf{r}\, \psi^{\dagger}_{\sigma}(\mathbf{r})\, h(\mathbf{r},t)\, \psi_{\sigma}(\mathbf{r}) + \frac{1}{2} \sum_{\sigma,\sigma'} \int d\mathbf{r} d\mathbf{r}'\, \psi^{\dagger}_{\sigma}(\mathbf{r})\, \psi^{\dagger}_{\sigma'}(\mathbf{r}')\, v(\mathbf{r}-\mathbf{r}')\, \psi_{\sigma'}(\mathbf{r}')\, \psi_{\sigma}(\mathbf{r}), \tag{22.3}$$

where $h(\mathbf{r},t) = h_0(\mathbf{r}) + V_{\text{ext}}(\mathbf{r},t)$ contains the external perturbation. Note that we are considering here a generic interparticle (two-body) interaction $v(\mathbf{r}-\mathbf{r}')$ (also with a possible dependence on time – cf. Section 11.3), thus postponing to further developments the restriction to the contact interaction (21.1).

We now examine in detail separately how the one- and two-body terms of the Hamiltonian (22.3) transform when expressed in terms of the Nambu pseudo-spinor fields (22.1).

One-body term of Eq. (22.3)

$$\begin{aligned} H_0(t) &= \sum_{\sigma} \int d\mathbf{r}\, \psi^{\dagger}_{\sigma}(\mathbf{r})\, h(\mathbf{r},t)\, \psi_{\sigma}(\mathbf{r}) \\ &= \int d\mathbf{r} \left(\psi^{\dagger}_{\uparrow}(\mathbf{r})\, h(\mathbf{r},t)\, \psi_{\uparrow}(\mathbf{r}) + \psi^{\dagger}_{\downarrow}(\mathbf{r})\, h(\mathbf{r},t)\, \psi_{\downarrow}(\mathbf{r}) \right) \\ &= \int d\mathbf{r} \left(\Psi^{\dagger}_{1}(\mathbf{r})\, h(\mathbf{r},t)\, \Psi_{1}(\mathbf{r}) + \Psi_{2}(\mathbf{r})\, h(\mathbf{r},t)\, \Psi^{\dagger}_{2}(\mathbf{r}) \right) \\ &\underset{\uparrow}{=} \int d\mathbf{r} \left(\Psi^{\dagger}_{1}(\mathbf{r})\, h(\mathbf{r},t)\, \Psi_{1}(\mathbf{r}) + (h(\mathbf{r},t)\, \Psi_{2}(\mathbf{r}))\; \Psi^{\dagger}_{2}(\mathbf{r}) \right), \end{aligned} \tag{22.4}$$

[integrate by parts the second term]

where we rewrite

$$\begin{aligned} (h(\mathbf{r},t)\, \Psi_{2}(\mathbf{r}))\; \Psi^{\dagger}_{2}(\mathbf{r}) &= h(\mathbf{r},t)\, \Psi_{2}(\mathbf{r})\, \Psi^{\dagger}_{2}(\mathbf{r}') \Big|_{\mathbf{r}'=\mathbf{r}} \\ &= h(\mathbf{r},t) \left(\delta(\mathbf{r}-\mathbf{r}') - \Psi^{\dagger}_{2}(\mathbf{r}')\, \Psi_{2}(\mathbf{r}) \right) \Big|_{\mathbf{r}'=\mathbf{r}} \end{aligned} \tag{22.5}$$

and represent the Dirac delta function

$$\delta(\mathbf{r}-\mathbf{r}') = \sum_{n} \phi_n(\mathbf{r},t)\, \phi_n(\mathbf{r}',t)^{*} \tag{22.6}$$

in terms of the instantaneous (normalized) wave functions $\{\phi(\mathbf{r},t)\}$ of the time-dependent single-particle Hamiltonian $h(\mathbf{r},t)$, such that

$$h(\mathbf{r},t)\,\phi_n(\mathbf{r},t) = \varepsilon_n(t)\,\phi_n(\mathbf{r},t)\,. \tag{22.7}$$

In this way, the second term in the last line of Eq. (22.4) becomes:

$$\begin{aligned}\int d\mathbf{r}\,(h(\mathbf{r},t)\,\Psi_2(\mathbf{r}))\,\Psi_2^\dagger(\mathbf{r}) &= -\int d\mathbf{r}\,\Psi_2^\dagger(\mathbf{r})\,h(\mathbf{r},t)\,\Psi_2(\mathbf{r})\\ &\quad + \int d\mathbf{r}\,h(\mathbf{r},t)\sum_n \phi_n(\mathbf{r},t)\phi_n(\mathbf{r}',t)^*|_{\mathbf{r}'=\mathbf{r}}\\ &= -\int d\mathbf{r}\,\Psi_2^\dagger(\mathbf{r})\,h(\mathbf{r},t)\,\Psi_2(\mathbf{r}) + \sum_n \varepsilon_n(t)\,.\end{aligned} \tag{22.8}$$

In conclusion, we obtain for the one-body term (22.4):

$$\begin{aligned}H_0(t) - \sum_n \varepsilon_n(t) &= \int d\mathbf{r}\left(\Psi_1^\dagger(\mathbf{r})\,h(\mathbf{r},t)\,\Psi_1(\mathbf{r}) - \Psi_2^\dagger(\mathbf{r})\,h(\mathbf{r},t)\,\Psi_2(\mathbf{r})\right)\\ &= \int d\mathbf{r}\,\Psi^\dagger(\mathbf{r})\,\tau^3\,h(\mathbf{r},t)\,\Psi(\mathbf{r}),\end{aligned} \tag{22.9}$$

where the spinor notation (22.1) has been utilized, and $\tau^3 = \begin{pmatrix}1 & 0\\ 0 & -1\end{pmatrix}$ is the third Pauli matrix. Note that the term $\sum_n \varepsilon_n(t)$ on the left-hand side of Eq. (22.9) is a diverging quantity, which will be disposed of (together with other diverging terms originating from the two-body term of the system Hamiltonian (22.3)) when introducing the Green's functions.

Two-body term of Eq. (22.3)

We begin by rewriting the kernel of the interacting Hamiltonian in the Nambu representation (22.1):

$$\begin{aligned}&\sum_{\sigma,\sigma'} \psi_\sigma^\dagger(\mathbf{r})\,\psi_{\sigma'}^\dagger(\mathbf{r}')\,\psi_{\sigma'}(\mathbf{r}')\,\psi_\sigma(\mathbf{r})\\ &= \Psi_1^\dagger(\mathbf{r})\,\Psi_1^\dagger(\mathbf{r}')\,\Psi_1(\mathbf{r}')\,\Psi_1(\mathbf{r}) + \Psi_1^\dagger(\mathbf{r})\,\Psi_2(\mathbf{r}')\,\Psi_2^\dagger(\mathbf{r}')\,\Psi_1(\mathbf{r})\\ &\quad + \Psi_2(\mathbf{r})\,\Psi_1^\dagger(\mathbf{r}')\,\Psi_1(\mathbf{r}')\,\Psi_2^\dagger(\mathbf{r}) + \Psi_2(\mathbf{r})\,\Psi_2(\mathbf{r}')\,\Psi_2^\dagger(\mathbf{r}')\,\Psi_2^\dagger(\mathbf{r})\\ &= \sum_{ij} \Psi_i^\dagger(\mathbf{r})\,\tau_{ij}^3\,\Psi_j(\mathbf{r}) \sum_{i'j'} \Psi_{i'}^\dagger(\mathbf{r}')\,\tau_{i'j'}^3\,\Psi_{j'}(\mathbf{r}') - \delta(\mathbf{r}-\mathbf{r}') \sum_{ij} \Psi_i^\dagger(\mathbf{r})\,\tau_{ij}^3\,\Psi_j(\mathbf{r})\end{aligned}$$

$$+\,\delta(0)\left(\sum_{ij}\Psi_i^\dagger(\mathbf{r})\,\tau_{ij}^3\,\Psi_j(\mathbf{r})+\sum_{ij}\Psi^\dagger(\mathbf{r}')\,\tau_{i'j'}^3\,\Psi_{j'}(\mathbf{r}')\right)$$
$$+\,\delta(0)\,\delta(0)-\delta(\mathbf{r}-\mathbf{r}')\,\delta(\mathbf{r}-\mathbf{r}'), \tag{22.10}$$

where suitable regularizations of the Dirac delta functions are implied.

We now enter the result (22.10) into the interacting part of the Hamiltonian (22.3) and combine the resulting expression with Eq. (22.9) for the noninteracting part of the Hamiltonian, to obtain eventually:

$$H(t)=\int d\mathbf{r}\,d\mathbf{r}'\sum_{ij}\Psi_i^\dagger(\mathbf{r})\,\tau_{ij}^3\,\left(h(\mathbf{r},t)\,\delta(\mathbf{r}-\mathbf{r}')+\mho(\mathbf{r},\mathbf{r}')\right)\Psi_j(\mathbf{r}')\;+\;C_\infty(t)$$
$$+\frac{1}{2}\sum_{\sigma,\sigma'}\int d\mathbf{r}d\mathbf{r}'\sum_{ij}\Psi_i^\dagger(\mathbf{r})\,\tau_{ij}^3\,\Psi_j(\mathbf{r})\,v(\mathbf{r}-\mathbf{r}')\sum_{i'j'}\Psi_{i'}^\dagger(\mathbf{r}')\,\tau_{i'j'}^3\,\Psi_{j'}(\mathbf{r}')\,. \tag{22.11}$$

Here,

$$\mho(\mathbf{r},\mathbf{r}')\equiv\delta(\mathbf{r}-\mathbf{r}')\left(\delta(0)\int d\mathbf{r}''v(\mathbf{r}-\mathbf{r}'')-\frac{1}{2}v(0)\right) \tag{22.12}$$

is a single-particle (local) operator and

$$C_\infty(t)\equiv\frac{1}{2}\int d\mathbf{r}d\mathbf{r}'v(\mathbf{r}-\mathbf{r}')\;\left(\delta(0)\,\delta(0)-\delta(\mathbf{r}-\mathbf{r}')\,\delta(\mathbf{r}-\mathbf{r}')\right)+\sum_n\varepsilon_n(t) \tag{22.13}$$

is a diverging c-number. Both the quantity $C_\infty(t)$ and the operator $\mho(\mathbf{r},\mathbf{r}')$ can be disposed of according to the following considerations:

(i) The quantity $C_\infty(t)$ drops off in the definition of the Green's functions of interest (cf. Eqs. (7.12), (7.13), and (8.1) of Part I). Specifically, its contributions to the forward and backward branches of the oriented contour cancel out.
(ii) The "potential" $\mho(\mathbf{r},\mathbf{r}')$ is also a diverging quantity. It will turn out that this quantity cancels with an analogous infinite term appearing in the Hartree–Fock self-energy at the lowest order in the interaction, once this is suitably handled by the "regularization" procedure considered in detail in Section 22.2.

Note, in addition, that the last term on the right-hand side of Eq. (22.11), which corresponds to the interaction part of the Hamiltonian, contains a τ^3 matrix at each elementary interaction vertex, as depicted in Fig. 22.1. This represents one *additional diagrammatic rule* for superfluid fermions, which has to be added to those considered in Chapter 10 of Part I. Note finally that the first term on the right-hand side of Eq. (22.11) also contains a τ^3 matrix.

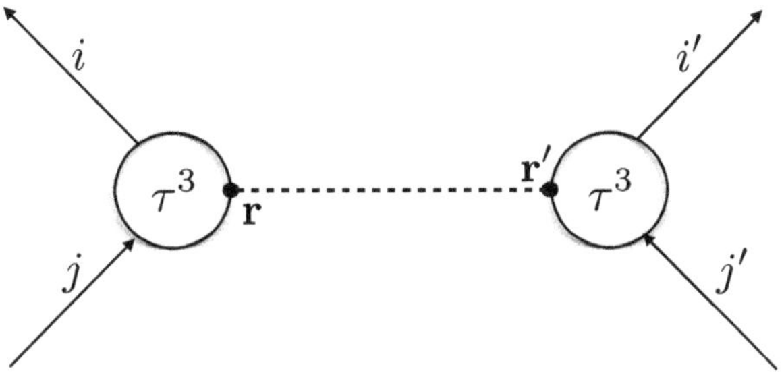

Figure 22.1 Diagrammatic representation of the interacting part of the system Hamiltonian (22.11).

22.2 A Proper Handling of the Hartree–Fock Self-Energy

By a proper handling of the Hartree–Fock self-energy, we mean that it should be defined as follows:

$$\Sigma_{\text{HF}}(1,2) \longrightarrow -i\,\delta(x_1-x_2)\,\tau^3_{i_1 i_2}\int dx_3\,v(x_1-x_3)\sum_{ij}\tau^3_{ij}\,\frac{1}{2}\left(G_{ij}(x_3,x_3^+)+G_{ij}(x_3,x_3^-)\right)$$
$$+\,i\,v(x_1-x_2)\sum_{ij}\tau^3_{i_1 i}\,\frac{1}{2}\left(G_{ij}(x_1,x_2^+)+G_{ij}(x_1,x_2^-)\right)\,\tau^3_{j i_2}, \qquad (22.14)$$

where $x=(\mathbf{r},t)$ and $v(x_1-x_2)=\delta(t_1-t_2)\,v(\mathbf{r}_1-\mathbf{r}_2)$ (and with the understanding that the time $t \leftrightarrow z$ runs over the total contour γ_{T} – cf. Fig. 8.1 of Part I). Note that, in the expression (22.14), we have modified the equal-time convention, in the form,

$$\begin{cases} G_{ij}(x_3,x_3^+) \rightarrow \frac{1}{2}\left(G_{ij}(x_3,x_3^+)+G_{ij}(x_3,x_3^-)\right) & \text{for the Hartree term} \\ G_{ij}(x_1,x_2^+) \rightarrow \frac{1}{2}\left(G_{ij}(x_1,x_2^+)+G_{ij}(x_1,x_2^-)\right) & \text{for the Fock term} \end{cases}$$

with respect to the convention previously adopted for the Hartree and Fock self-energies in the normal phase (cf. Eqs. (11.30) and (11.31) of Part I), thereby upgrading to the present needs the rule #4 of Chapter 10 in Part I. In the above expressions,

$$G_{ij}(x_1,x_2^+)=i\left\langle \Psi_j^\dagger(x_2)\,\Psi_i(x_1)\right\rangle \quad , \quad G_{ij}(x_1,x_2^-)=-\,i\left\langle \Psi_i(x_1)\,\Psi_j^\dagger(x_2)\right\rangle \quad , \qquad (22.15)$$

where the meaning of the average $\langle \cdots \rangle$ will be specified in Chapter 23. With the definition (22.14), we obtain for the various Nambu-spin elements of the Hartree–Fock self-energy:

Element 11

$$
\begin{aligned}
\Sigma_{\mathrm{HF}}(x_1,x_2)_{11} &= \frac{1}{2}\,\delta(x_1-x_2)\int d\mathbf{r}_3\, v(\mathbf{r}_1-\mathbf{r}_3)\,\Big(\big\langle \Psi_1^\dagger(x_3)\,\Psi_1(x_3)\big\rangle \\
&\quad - \big\langle \Psi_1(x_3)\Psi_1^\dagger(x_3)\big\rangle - \big\langle \Psi_2^\dagger(x_3)\Psi_2(x_3)\big\rangle + \big\langle \Psi_2(x_3)\Psi_2^\dagger(x_3)\big\rangle\Big)_{x_3=(\mathbf{r}_3,t_1)} \\
&\quad - \frac{1}{2}\,v(x_1-x_2)\,\Big(\big\langle \Psi_1^\dagger(x_2)\,\Psi_1(x_1)\big\rangle - \big\langle \Psi_1(x_1)\,\Psi_1^\dagger(x_2)\big\rangle\Big) \\
&\underset{\substack{\uparrow \\ [\text{Eq.}(22.1)]}}{=} \frac{1}{2}\,\delta(x_1-x_2)\int d\mathbf{r}_3\, v(\mathbf{r}_1-\mathbf{r}_3)\,\Big(\big\langle \psi_\uparrow^\dagger(x_3)\,\psi_\uparrow(x_3)\big\rangle \\
&\quad - \big\langle \psi_\uparrow(x_3)\,\psi_\uparrow^\dagger(x_3)\big\rangle - \big\langle \psi_\downarrow(x_3)\,\psi_\downarrow^\dagger(x_3)\big\rangle + \big\langle \psi_\downarrow^\dagger(x_3)\,\psi_\downarrow(x_3)\big\rangle\Big)_{x_3=(\mathbf{r}_3,t_1)} \\
&\quad - \frac{1}{2}\,v(x_1-x_2)\,\Big(\big\langle \psi_\uparrow^\dagger(x_2)\,\psi_\uparrow(x_1)\big\rangle - \big\langle \psi_\uparrow(x_1)\,\psi_\uparrow^\dagger(x_2)\big\rangle\Big)_{x_2=(\mathbf{r}_2,t_1)} \\
&= \delta(t_1-t_2)\,\Big\{\delta(\mathbf{r}_1-\mathbf{r}_2)\int d\mathbf{r}_3\, v(\mathbf{r}_1-\mathbf{r}_3)\,\Big(\big\langle \psi_\uparrow^\dagger(\mathbf{r}_3,t_1)\,\psi_\uparrow(\mathbf{r}_3,t_1)\big\rangle \\
&\quad + \big\langle \psi_\downarrow^\dagger(\mathbf{r}_3,t_1)\,\psi_\downarrow(\mathbf{r}_3,t_1)\big\rangle\Big) - v(\mathbf{r}_1-\mathbf{r}_2)\,\big\langle \psi_\uparrow^\dagger(\mathbf{r}_2,t_1)\,\psi_\uparrow(\mathbf{r}_1,t_1)\big\rangle \\
&\quad - \mho(\mathbf{r}_1,\mathbf{r}_2)\Big\} = \Sigma_{\mathrm{HF}}(x_1,x_2)_{\uparrow\uparrow} - \delta(t_1-t_2)\,\mho(\mathbf{r}_1,\mathbf{r}_2),
\end{aligned}
\tag{22.16}
$$

where $\Sigma_{\mathrm{HF}}(x_1,x_2)_{\uparrow\uparrow}$ is the Hartree–Fock self-energy in the original spin representation $\sigma=(\uparrow,\downarrow)$ (cf. Eq. (19.6)), and $\mho(\mathbf{r}_1,\mathbf{r}_2)$ is the operator given by Eq. (22.12).

Element 12

$$
\begin{aligned}
\Sigma_{\mathrm{HF}}(x_1,x_2)_{12} &= -i\,v(x_1-x_2)\,\frac{1}{2}\,\big(G_{12}(x_1,x_2^+) + G_{12}(x_1,x_2^-)\big) \\
&= v(x_1-x_2)\,\frac{1}{2}\,\Big(\big\langle \Psi_2^\dagger(\mathbf{r}_2,t_1)\,\Psi_1(\mathbf{r}_1,t_1)\big\rangle - \big\langle \Psi_1(\mathbf{r}_1,t_1)\,\Psi_2^\dagger(\mathbf{r}_2,t_1)\big\rangle\Big) \\
&= v(x_1-x_2)\,\frac{1}{2}\,\big(\langle \psi_\downarrow(\mathbf{r}_2,t_1)\,\psi_\uparrow(\mathbf{r}_1,t_1)\rangle - \langle \psi_\uparrow(\mathbf{r}_1,t_1)\,\psi_\downarrow(\mathbf{r}_2,t_1)\rangle\big) \\
&= v(x_1-x_2)\,\langle \psi_\downarrow(\mathbf{r}_2,t_1)\,\psi_\uparrow(\mathbf{r}_1,t_1)\rangle = \Sigma_{\mathrm{HF}}(x_1,x_2)_{\downarrow\uparrow},
\end{aligned}
\tag{22.17}
$$

which does not contain the operator $\mho(\mathbf{r}_1,\mathbf{r}_2)$, consistently with Eq. (22.11), where $\mho$ enters only the diagonal elements. As before, $\Sigma_{\mathrm{HF}}(x_1,x_2)_{\downarrow\uparrow}$ is expressed in the original spin representation.

Element 21

$$
\begin{aligned}
\Sigma_{\mathrm{HF}}(x_1, x_2)_{21} &= -\,i\, v(x_1 - x_2)\, \frac{1}{2}\, \left(G_{21}(x_1, x_2^+) + G_{21}(x_1, x_2^-)\right) \\
&= v(x_1 - x_2)\, \frac{1}{2}\, \left(\left\langle \Psi_1^\dagger(\mathbf{r}_2, t_1)\, \Psi_2(\mathbf{r}_1, t_1) \right\rangle - \left\langle \Psi_2(\mathbf{r}_1, t_1)\, \Psi_1^\dagger(\mathbf{r}_2, t_1) \right\rangle \right) \\
&= v(x_1 - x_2)\, \frac{1}{2}\, \left(\left\langle \psi_\uparrow^\dagger(\mathbf{r}_2, t_1)\, \psi_\downarrow^\dagger(\mathbf{r}_1, t_1) \right\rangle - \left\langle \psi_\downarrow^\dagger(\mathbf{r}_1, t_1)\, \psi_\uparrow^\dagger(\mathbf{r}_2, t_1) \right\rangle \right) \\
&= v(x_1 - x_2)\, \left\langle \psi_\uparrow^\dagger(\mathbf{r}_2, t_1)\, \psi_\downarrow^\dagger(\mathbf{r}_1, t_1) \right\rangle \\
&= v(x_1 - x_2)\, \left\langle \psi_\downarrow(\mathbf{r}_1, t_1) \psi_\uparrow(\mathbf{r}_2, t_1) \right\rangle^* \\
&= \Sigma_{\mathrm{HF}}(x_2, x_1)_{12}^* = \Sigma_{\mathrm{HF}}(x_2, x_1)_{\downarrow\uparrow}^* \,.
\end{aligned}
\tag{22.18}
$$

Element 22

$$
\begin{aligned}
\Sigma_{\mathrm{HF}}(x_1, x_2)_{22} &= -\,\frac{1}{2}\, \delta(x_1 - x_2) \int d\mathbf{r}_3\, v(\mathbf{r}_1 - \mathbf{r}_3)\, \Big(\left\langle \Psi_1^\dagger(x_3)\, \Psi_1(x_3) \right\rangle \\
&\quad - \left\langle \Psi_1(x_3) \Psi_1^\dagger(x_3) \right\rangle - \left\langle \Psi_2^\dagger(x_3)\, \Psi_2(x_3) \right\rangle + \left\langle \Psi_2(x_3) \Psi_2^\dagger(x_3) \right\rangle \Big)_{x_3 = (\mathbf{r}_3, t_1)} \\
&\quad - \frac{1}{2}\, v(x_1 - x_2)\, \left(\left\langle \Psi_2^\dagger(x_2)\, \Psi_2(x_1) \right\rangle - \left\langle \Psi_2(x_1)\, \Psi_2^\dagger(x_2) \right\rangle \right) \\
&\underset{[\mathrm{Eq.(22.1)}]}{\overset{\uparrow}{=}} -\,\frac{1}{2}\, \delta(x_1 - x_2) \int d\mathbf{r}_3\, v(\mathbf{r}_1 - \mathbf{r}_3)\, \Big(\left\langle \psi_\uparrow^\dagger(x_3)\, \psi_\uparrow(x_3) \right\rangle \\
&\quad - \left\langle \psi_\uparrow(x_3)\, \psi_\uparrow^\dagger(x_3) \right\rangle - \left\langle \psi_\downarrow(x_3)\, \psi_\downarrow^\dagger(x_3) \right\rangle + \left\langle \psi_\downarrow^\dagger(x_3)\, \psi_\downarrow(x_3) \right\rangle \Big)_{x_3 = (\mathbf{r}_3, t_1)} \\
&\quad - \frac{1}{2}\, v(x_1 - x_2)\, \left(\left\langle \psi_\downarrow(x_2)\, \psi_\downarrow^\dagger(x_1) \right\rangle - \left\langle \psi_\downarrow^\dagger(x_1)\, \psi_\downarrow(x_2) \right\rangle \right)_{x_2 = (\mathbf{r}_2, t_1)} \\
&= -\Sigma_{\mathrm{HF}}(x_2, x_1)_{11}
\end{aligned}
\tag{22.19}
$$

provided that

$$
\left\langle \psi_\downarrow^\dagger(\mathbf{r}_1, t_1)\, \psi_\downarrow(\mathbf{r}_2, t_1) \right\rangle = \left\langle \psi_\uparrow^\dagger(\mathbf{r}_1, t_1)\, \psi_\uparrow(\mathbf{r}_2, t_1) \right\rangle \,, \tag{22.20}
$$

which is the case when no magnetic effects (or spin imbalance in ultracold Fermi gases) are present.

We have thus proved that the diagonal part of the Hartree–Fock self-energy (22.14) in the Nambu representation cancels the contribution of the operator $\mho$, given by Eq. (22.12), to the system Hamiltonian (22.11) and that what is left coincides with the diagonal part of the Hartree–Fock self-energy for the normal phase in the ordinary spin representation. Once this cancellation has occurred, there are no longer infinities afflicting the theory. Note also that the off-diagonal elements (22.17) and (22.18) correspond to the mean-field (BCS) approximation introduced in Chapter 21.

On physical grounds, the "regularization" procedure that we have adopted is consistent with the assumption that the standard (diagonal) Hartree–Fock self-energy is the same in both normal and superconducting phases, such that it is often omitted altogether when describing the properties of the superconducting phase [7].

The special role played by the Hartree–Fock self-energy for a superfluid Fermi system was also emphasized in Ref. [13], where a particular limiting procedure for the single-particle Green's function at equal time was also postulated. All this implies that, when dealing with superfluid Fermi systems, the choices one makes for the self-energy should always at least contain the Hartree–Fock contribution. This point was further emphasized in Ref. [57], where beyond-mean-field effects were explicitly considered in the equation for the gap parameter in the context of the BCS–BEC crossover (to be briefly overviewed in Chapter 26).

Remark: Normal and anomalous external agents

In Chapter 17 of Part I, a "source field" $U(\mathbf{x}, z)$ was introduced, in order to obtain the diagrammatic structure in terms of functional derivatives. In that context, we were concerned only with the normal phase. In the present context, on the other hand, where we are concerned with the fermionic superfluid phase as appropriately described in terms of the Nambu pseudo-spinor field operators (22.1), the introduction of the source term accordingly has to be generalized to accommodate anomalous terms as well. To this end, we introduce not only the *normal interaction*

$$H'_{\mathrm{n}}(z) = \sum_{\sigma} \int d\mathbf{r}\, d\mathbf{r}'\, \psi^{\dagger}_{\sigma}(\mathbf{r})\, U_{\mathrm{n}}(\mathbf{r}, \mathbf{r}'; z)\, \psi_{\sigma}(\mathbf{r}') \tag{22.21}$$

in analogy to what we did in Eq. (17.1) of Part I but also the *anomalous interaction*

$$H'_{\mathrm{a}}(z) = \int d\mathbf{r}\, d\mathbf{r}' \left(\psi_{\downarrow}(\mathbf{r})\, U_{\mathrm{a}}(\mathbf{r}, \mathbf{r}'; z)\, \psi_{\uparrow}(\mathbf{r}') + \psi^{\dagger}_{\uparrow}(\mathbf{r})\, U_{\mathrm{a}}(\mathbf{r}, \mathbf{r}'; z)^{*}\, \psi^{\dagger}_{\downarrow}(\mathbf{r}') \right) . \tag{22.22}$$

Here:

(i) In the normal term (22.21), we have generalized the form (17.1) of Part I by considering a *non-local* "normal" source field $U_{\mathrm{n}}(\mathbf{r}, \mathbf{r}'; z)$, such that $U_{\mathrm{n}}(\mathbf{r}, \mathbf{r}'; z)^{*} = U_{\mathrm{n}}(\mathbf{r}', \mathbf{r}; z)$ for the Hamiltonian (22.21) to be Hermitian. The normal interaction (17.1) of Part I is then recovered by taking $U_{\mathrm{n}}(\mathbf{r}, \mathbf{r}'; z) \to \delta(\mathbf{r} - \mathbf{r}')\, U(\mathbf{r}, z)$.

(ii) In the anomalous term (22.22), the "anomalous" source field $U_{\mathrm{a}}(\mathbf{r}, \mathbf{r}'; z)$ satisfies the property $U_{\mathrm{a}}(\mathbf{r}, \mathbf{r}'; z) = U_{\mathrm{a}}(\mathbf{r}', \mathbf{r}; z)$ for the Hamiltonian (22.22) to be Hermitian.

In addition, as we did for the system Hamiltonian (22.3), also in the interactions (22.21) and (22.22), the field operators can be rewritten in terms of the Nambu pseudo-spinor fields (22.1), yielding the expression

$$
\begin{aligned}
H'(z) &= H'_{\mathrm{n}}(z) + H'_{\mathrm{a}}(z) \\
&= \int d\mathbf{r}\, d\mathbf{r}' \left(\Psi_1^\dagger(\mathbf{r}), \Psi_2^\dagger(\mathbf{r}) \right) \begin{pmatrix} U_{\mathrm{n}}(\mathbf{r}, \mathbf{r}'; z) & U_{\mathrm{a}}(\mathbf{r}, \mathbf{r}'; z)^* \\ U_{\mathrm{a}}(\mathbf{r}, \mathbf{r}'; z) & -U_{\mathrm{n}}(\mathbf{r}, \mathbf{r}'; z)^* \end{pmatrix} \begin{pmatrix} \Psi_1(\mathbf{r}') \\ \Psi_2(\mathbf{r}') \end{pmatrix}, \qquad (22.23)
\end{aligned}
$$

where the irrelevant c-number term $\int d\mathbf{r}\, U_{\mathrm{n}}(\mathbf{r}, \mathbf{r}; z)$ has been dropped.

As a final manipulation, we can also adopt the notation $(\mathbf{r}, i) \to \mathbf{x}$ for Nambu spinors and rewrite the expression (22.23) in the compact form

$$
H'(z) = \int d\mathbf{x}\, d\mathbf{x}'\, \mathbf{\Psi}^\dagger(\mathbf{x})\, \overleftrightarrow{U}(\mathbf{x}, \mathbf{x}'; z)\, \mathbf{\Psi}(\mathbf{x}), \qquad (22.24)
$$

where $\int d\mathbf{x} = \int d\mathbf{r} \sum_i$, and $\overleftrightarrow{U}$ is the 2×2 matrix of Eq. (22.23). In this way, the manipulations we made in Chapter 17 of Part I can readily be generalized to the superfluid phase [80].

23

Contour-Ordered Green's Functions in the Nambu Representation

In this chapter, we adapt to the language of the Nambu pseudo-spinor field operators (22.1) for the superfluid phase (i) the closed time path Green's functions introduced in Chapter 7 of Part I, (ii) the ensuing nonequilibrium Dyson equations considered in Chapter 11 of Part I, (iii) the conversion of contour-time to real-time arguments made in Chapter 13 of Part I, and (iv) the Langreth rules of Chapter 14 of Part I. In this way, we will frame in a more general context the results obtained previously in Chapter 21 that were specific to a mean-field decoupling, which will make it possible the inclusion of beyond-mean-field effects for the superfluid phase, as considered later in Chapter 24.

23.1 The Schwinger–Keldysh Approach for Superfluid Fermions

We begin by recalling the definition of the contour single-particle Green's function (cf. Section 11.3 of Part I)

$$G(1,2) = -i \left\langle \mathcal{T}_{\mathrm{C}} \left\{ \psi(1)\, \psi^{\dagger}(2) \right\} \right\rangle , \tag{23.1}$$

where the suffix "H" of the Heisenberg representation has been omitted in the field operators for simplicity, the short-hand notation $1 \leftrightarrow (\mathbf{r}_1, \sigma_1, z_1)$ refers to the original spin label $\sigma = (\uparrow, \downarrow)$, and $\mathcal{T}_{\mathrm{C}}$ is the contour time-ordering operator along the generic contour C, which may (like $\mathrm{C} \to \gamma_{\mathrm{T}}$ of Fig. 8.1 of Part I) or may not (like $\mathrm{C} \to \gamma_{\mathrm{K}}$ of Fig. 8.2 of Part I) include transient phenomena. Recall further that the function (23.1) satisfies the Dyson equation in (integro-)differential form (cf. Eq. (11.26) of Part I)

$$\left[i\frac{d}{dz_1} - h(1) \right] G(1,2) - \int d3\, \Sigma(1,3)\, G(3,2) = \delta(1,2) , \tag{23.2}$$

where $\delta(1,2) = \delta(z_1, z_2)\, \delta(\mathbf{r}_1 - \mathbf{r}_2)\, \delta_{\sigma_1,\sigma_2}$ (cf. Eq. (11.16) of Part I), and the time component of the integral runs over the contour C.

Next, we follow the procedure of Chapter 22 and convert from the original fermion field operators $\psi_\sigma(\mathbf{r})$ to the Nambu pseudo-spinor fields $\Psi_i(\mathbf{r})$ with $i=(1,2)$, as defined in Eq. (22.1). Accordingly, we also transform the system Hamiltonian to the form (22.11) that, once depurated from the diverging terms $C_\infty(t)$ and $\mho(\mathbf{r},\mathbf{r}')$ in agreement with the prescriptions of Section 22.2, reads:

$$
\begin{aligned}
H(t) = &\int d\mathbf{r} \sum_{ij} \Psi_i^\dagger(\mathbf{r})\, \tau_{ij}^3\, h(\mathbf{r},t)\, \Psi_j(\mathbf{r}) \\
&+ \frac{1}{2} \int d\mathbf{r} d\mathbf{r}' \sum_{ij} \Psi_i^\dagger(\mathbf{r})\, \tau_{ij}^3\, \Psi_j(\mathbf{r})\, v(\mathbf{r}-\mathbf{r}') \sum_{i'j'} \Psi_{i'}^\dagger(\mathbf{r}')\, \tau_{i'j'}^3\, \Psi_{j'}(\mathbf{r}') \,. \qquad (23.3)
\end{aligned}
$$

In this way, the single-particle Green's function in the Nambu representation, namely,

$$
G(\mathbf{r}_1 z_1, \mathbf{r}_2 z_2)_{i_1 i_2} = -i \left\langle \mathcal{T}_C \left\{ \Psi_{i_1}(\mathbf{r}_1 z_1)\, \Psi_{i_2}^\dagger(\mathbf{r}_2 z_2) \right\} \right\rangle \qquad (23.4)
$$

satisfies the following equation of motion

$$
\begin{aligned}
&\sum_{i_3} \left[i\delta_{i_1 i_3} \frac{d}{dz_1} - \tau_{i_1 i_3}^3 h(\mathbf{r}_1, z_1) \right] G(\mathbf{r}_1 z_1, \mathbf{r}_2 z_2)_{i_3 i_2} \\
&\quad - \int_C dz_3 \int d\mathbf{r}_3 \sum_{i_3} \Sigma(\mathbf{r}_1 z_1, \mathbf{r}_3 z_3)_{i_1 i_3} G(\mathbf{r}_3 z_3, \mathbf{r}_2 z_2)_{i_3 i_2} \\
&= \delta(z_1, z_2)\, \delta(\mathbf{r}_1 - \mathbf{r}_2)\, \delta_{i_1 i_2}, \qquad (23.5)
\end{aligned}
$$

where no spin dependence of the single-particle Hamiltonian $h(\mathbf{r},t)$ has been assumed. Note the following differences between the Dyson equation (23.5) in terms of the Nambu pseudo-spin fields $\Psi_i(\mathbf{r})$ and the Dyson equation (23.2) in terms of the original field operators $\psi_\sigma(\mathbf{r})$:

(i) the presence of the Pauli matrix τ^3 in the term containing the single-particle Hamiltonian $h(\mathbf{r},t)$;
(ii) the presence of τ^3 also in the elementary interaction vertex (depicted in Fig. 22.1), which results in the additional diagrammatic rule mentioned at the end of Section 22.1.

Equation (23.5) in the variables $(\mathbf{r}_1, z_1)$ and its companion in the variables $(\mathbf{r}_2, z_2)$ for the fermionic superfluid phase will be briefly derived in Section 31.1, adapting the steps that led to the corresponding equations (11.26) and (11.28) of Part I for the normal phase, but in a way that will highlight the emergence of the self-energy through a different route.

23.2 Matrix Structure of the Dyson Equation for the Single-Particle Green's Functions in the Nambu–Keldysh Space

The Dyson equation (23.5) has a matrix structure in the Nambu index "i." This suggests introducing the following 2×2 matrices in Nambu space:

$$\overleftrightarrow{\mathbb{1}} = \begin{pmatrix} 1 & 0 \\ 0 & 1 \end{pmatrix}, \quad \overleftrightarrow{\tau_3} = \begin{pmatrix} 1 & 0 \\ 0 & -1 \end{pmatrix}, \quad \overleftrightarrow{h}(\mathbf{r}, z) = \overleftrightarrow{\tau_3}\, h(\mathbf{r}, z), \tag{23.6}$$

$$\overleftrightarrow{G}(\mathbf{r}_1 z_1, \mathbf{r}_2 z_2) = \begin{pmatrix} G(\mathbf{r}_1 z_1, \mathbf{r}_2 z_2)_{11} & G(\mathbf{r}_1 z_1, \mathbf{r}_2 z_2)_{12} \\ G(\mathbf{r}_1 z_1, \mathbf{r}_2 z_2)_{21} & G(\mathbf{r}_1 z_1, \mathbf{r}_2 z_2)_{22} \end{pmatrix}, \tag{23.7}$$

$$\overleftrightarrow{\Sigma}(\mathbf{r}_1 z_1, \mathbf{r}_2 z_2) = \begin{pmatrix} \Sigma(\mathbf{r}_1 z_1, \mathbf{r}_2 z_2)_{11} & \Sigma(\mathbf{r}_1 z_1, \mathbf{r}_2 z_2)_{12} \\ \Sigma(\mathbf{r}_1 z_1, \mathbf{r}_2 z_2)_{21} & \Sigma(\mathbf{r}_1 z_1, \mathbf{r}_2 z_2)_{22} \end{pmatrix}, \tag{23.8}$$

such that Eq. (23.5) can be rewritten in matrix notation

$$\left[\overleftrightarrow{\mathbb{1}}\, i\frac{d}{dz_1} - \overleftrightarrow{h}(\mathbf{r}_1, z_1)\right] \overleftrightarrow{G}(\mathbf{r}_1 z_1, \mathbf{r}_2 z_2)$$
$$- \int_C dz_3 \int d\mathbf{r}_3\, \overleftrightarrow{\Sigma}(\mathbf{r}_1 z_1, \mathbf{r}_3 z_3)\, \overleftrightarrow{G}(\mathbf{r}_3 z_3, \mathbf{r}_2 z_2) = \delta(z_1, z_2)\, \delta(\mathbf{r}_1 - \mathbf{r}_2)\, \overleftrightarrow{\mathbb{1}}, \tag{23.9}$$

where matrix multiplication is implied. For instance, in the mean-field (BCS) approximation, one has (cf. Eqs. (21.6) and (22.17)–(22.18))

$$\overleftrightarrow{\Sigma}_{\mathrm{BCS}}(\mathbf{r}_1 z_1, \mathbf{r}_2 z_2) = \begin{pmatrix} 0 & \Delta(\mathbf{r}_1 z_1) \\ \Delta(\mathbf{r}_1 z_1)^* & 0 \end{pmatrix} \delta(z_1, z_2)\, \delta(\mathbf{r}_1 - \mathbf{r}_2) \tag{23.10}$$

when the interparticle interaction is assumed to be of the contact type (21.1).

Remark: An alternative matrix convention

A different convention can be found in the literature for the contour single-particle Green's functions in the Nambu–Keldysh space [14], according to which

$$\underset{\sim}{\overleftrightarrow{G}}(\mathbf{r}_1 z_1, \mathbf{r}_2 z_2) = \overleftrightarrow{\tau_3}\, \overleftrightarrow{G}(\mathbf{r}_1 z_1, \mathbf{r}_2 z_2)\,. \tag{23.11}$$

Taking into account that $\overleftrightarrow{\tau_3}\,\overleftrightarrow{\tau_3} = \overleftrightarrow{\mathbb{1}}$, the Dyson equation (23.5) can then be cast in the form

$$\left[\overleftrightarrow{\tau_3}\, i\frac{d}{dz_1} - \overleftrightarrow{\mathbb{1}}\, h(\mathbf{r}, z)\right] \underset{\sim}{\overleftrightarrow{G}}(\mathbf{r}_1 z_1, \mathbf{r}_2 z_2)$$
$$- \int_C dz_3 \int d\mathbf{r}_3\, \underset{\sim}{\overleftrightarrow{\Sigma}}(\mathbf{r}_1 z_1, \mathbf{r}_3 z_3)\, \underset{\sim}{\overleftrightarrow{G}}(\mathbf{r}_3 z_3, \mathbf{r}_2 z_2) = \delta(z_1, z_2)\, \delta(\mathbf{r}_1 - \mathbf{r}_2)\, \overleftrightarrow{\mathbb{1}}, \tag{23.12}$$

where we have noted that $\overleftrightarrow{h}(\mathbf{r},z)\overleftrightarrow{\tau}_3 = \overleftrightarrow{\tau}_3\, h(\mathbf{r},z)\,\overleftrightarrow{\tau}_3 = \overleftrightarrow{\mathbb{1}}\, h(\mathbf{r},z)$ and introduced the notation

$$\underset{\sim}{\overleftrightarrow{\Sigma}}(\mathbf{r}_1 z_1, \mathbf{r}_2 z_2) = \overleftrightarrow{\Sigma}(\mathbf{r}_1 z_1, \mathbf{r}_2 z_2)\,\overleftrightarrow{\tau}_3 \,. \tag{23.13}$$

For instance, the BCS self-energy (23.10) becomes in this language:

$$\begin{aligned}\underset{\sim}{\overleftrightarrow{\Sigma}}(\mathbf{r}_1 z_1, \mathbf{r}_2 z_2) &= \begin{pmatrix} 0 & \Delta(\mathbf{r}_1 z_1) \\ \Delta(\mathbf{r}_1 z_1)^* & 0 \end{pmatrix} \begin{pmatrix} 1 & 0 \\ 0 & -1 \end{pmatrix} \delta(z_1, z_2)\,\delta(\mathbf{r}_1 - \mathbf{r}_2) \\ &= \begin{pmatrix} 0 & -\Delta(\mathbf{r}_1 z_1) \\ \Delta(\mathbf{r}_1 z_1)^* & 0 \end{pmatrix} \delta(z_1, z_2)\delta(\mathbf{r}_1 - \mathbf{r}_2)\,. \end{aligned} \tag{23.14}$$

Nonetheless, we will prefer to stick with the original definitions (23.7) and (23.8) and the Dyson equation (23.9), instead of adopting the alternative definitions (23.11) with (23.13) and the Dyson equation (23.12). This is because the structure of the first term within brackets on the left-hand side of Eq. (23.9) coincides with that of the temperature Green's function in Nambu space, as given in standard textbooks (cf. e.g., section 51 of Ref. [7]). Accordingly, when considering beyond-mean-field nonequilibrium approximations (like in Chapter 24), it will be easier to check the formal structure of the beyond-mean-field nonequilibrium self-energy with that obtained correspondingly at equilibrium within the Matsubara formalism (which, in addition, will be relevant for setting up the initial conditions at thermodynamic equilibrium for the nonequilibrium problem). On the other hand, when limiting to consider only the BCS approximation (like in Ref. [14]), the use of either one of the two formulations (23.9) and (23.12) makes in practice no difference.

One of the advantages for using the matrix structure (23.9) of the Dyson equation is that the Langreth rules discussed in Chapter 14 of Part I remain the same, provided that they are now applied to the 2×2 matrix structure related to the Nambu indices as in Eq. (23.9). This consideration allows us to convert the contour-time arguments in Eq. (23.9) into standard real-time arguments. In practice, this implies that each 2×2 matrix in Eq. (23.9) can be considered as an element in the 3×3 space introduced in Chapter 13 of Part I (which reduces to a 2×2 space in the Keldysh approach). Let's consider a few examples here:

Component 11 (cf. Eq. (13.3) of Part I)

$$\overleftrightarrow{G}_{11}(\mathbf{r}t, \mathbf{r}'t') = -\,i \begin{pmatrix} \left\langle \mathcal{T}\left\{\Psi_1(\mathbf{r}t)\,\Psi_1^\dagger(\mathbf{r}'t')\right\}\right\rangle & \left\langle \mathcal{T}\left\{\Psi_1(\mathbf{r}t)\,\Psi_2^\dagger(\mathbf{r}'t')\right\}\right\rangle \\ \left\langle \mathcal{T}\left\{\Psi_2(\mathbf{r}t)\,\Psi_1^\dagger(\mathbf{r}'t')\right\}\right\rangle & \left\langle \mathcal{T}\left\{\Psi_2(\mathbf{r}t)\,\Psi_2^\dagger(\mathbf{r}'t')\right\}\right\rangle \end{pmatrix}, \tag{23.15}$$

where $\Psi_i(\mathbf{r}t)$ are Nambu pseudo-spinor field operators, whose time dependence is given in the Heisenberg representation (cf. Eq. (2.25) of Part I), and $\mathcal{T}$ is the standard time-ordering operator.

Component 12 (cf. Eq. (13.5) of Part I)

$$\overleftrightarrow{G}_{12}(\mathbf{r}t, \mathbf{r}'t') = \overleftrightarrow{G}^{<}(\mathbf{r}t, \mathbf{r}'t') = i \begin{pmatrix} \left\langle \Psi_1^\dagger(\mathbf{r}'t')\, \Psi_1(\mathbf{r}t) \right\rangle & \left\langle \Psi_2^\dagger(\mathbf{r}'t')\, \Psi_1(\mathbf{r}t) \right\rangle \\ \left\langle \Psi_1^\dagger(\mathbf{r}'t')\, \Psi_2(\mathbf{r}t) \right\rangle & \left\langle \Psi_2^\dagger(\mathbf{r}'t')\, \Psi_2(\mathbf{r}t) \right\rangle \end{pmatrix}, \tag{23.16}$$

where only the fermion case has been considered. Note that confusion must be avoided between the indices $(1, 2)$ in Keldysh and Nambu spaces.

Component 21 (cf. Eq. (13.6) of Part I)

$$\overleftrightarrow{G}_{21}(\mathbf{r}t, \mathbf{r}'t') = \overleftrightarrow{G}^{>}(\mathbf{r}t, \mathbf{r}'t') = -i \begin{pmatrix} \left\langle \Psi_1(\mathbf{r}t)\, \Psi_1^\dagger(\mathbf{r}'t') \right\rangle & \left\langle \Psi_1(\mathbf{r}t)\, \Psi_2^\dagger(\mathbf{r}'t') \right\rangle \\ \left\langle \Psi_2(\mathbf{r}t)\, \Psi_1^\dagger(\mathbf{r}'t') \right\rangle & \left\langle \Psi_2(\mathbf{r}t)\, \Psi_2^\dagger(\mathbf{r}'t') \right\rangle \end{pmatrix}. \tag{23.17}$$

In all these cases, the inner indices refer to the matrix of Eq. (13.2) in Keldysh space and the outer indices to the matrix of Eq. (23.7) in Nambu space, as shown pictorially in Fig. 23.1.

As a consequence of the aforementioned considerations, the Kadanoff–Baym equations for the various components of the (contour-ordered) single-particle

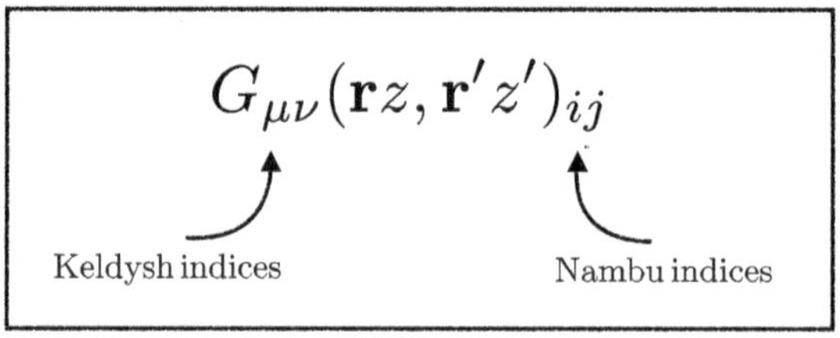

Figure 23.1 Double-matrix structure for the single-particle Green's function. Here, the inner (Keldysh) indices $(\mu, \nu) = (1, 2, 3)$ refer to the elements of the 3×3 matrix (13.2), while the outer (Nambu) indices $(i, j) = (1, 2)$ refer to the elements of the 2×2 matrix (23.7). [Similar considerations apply to the self-energy as well.]

Green's function maintain the structure of the equations discussed in Chapter 15 of Part I, provided that:

(i) We replace (whenever it occurs) the operator $\left[i\frac{d}{dt} - h(\mathbf{r}, t)\right]$ by the matrix structure in Nambu space, namely,

$$\left[\overleftrightarrow{\mathbb{1}}\, i\frac{d}{dt} - \overleftrightarrow{h}(\mathbf{r}, t)\right], \tag{23.18}$$

where (cf. Eq. (23.6))

$$\overleftrightarrow{\mathbb{1}} = \begin{pmatrix} 1 & 0 \\ 0 & 1 \end{pmatrix} \quad \text{and} \quad \overleftrightarrow{h}(\mathbf{r}, t) = \begin{pmatrix} h(\mathbf{r}, t) & 0 \\ 0 & -h(\mathbf{r}, t) \end{pmatrix}; \tag{23.19}$$

(ii) We consider all G and Σ components as matrices in Nambu space, that is,

$$G \longrightarrow \overleftrightarrow{G} \quad \text{and} \quad \Sigma \longrightarrow \overleftrightarrow{\Sigma} \tag{23.20}$$

with appropriate space ($\mathbf{r}$) and time (t or τ) variables.

23.3 The Mean-Field (BCS) Approximation, Again

As a first check on the internal consistency of the aforementioned formalism, it is convenient to consider again the example of the mean-field (BCS) approximation. In this case, $\overleftrightarrow{\Sigma}$ in Nambu space is given by Eq. (23.10) when the interparticle interaction is of the contact type, where

$$\Delta(\mathbf{r}, t) = v_0 \left\langle \Psi_2^\dagger(\mathbf{r}, t)\, \Psi_1(\mathbf{r}, t) \right\rangle = -\,i\, v_0\, G_{12}(\mathbf{r}t, \mathbf{r}t^+)_{12} = -\,i\, v_0\, G^<(\mathbf{r}t, \mathbf{r}t)_{12} \tag{23.21}$$

in the language of Eq. (23.16). According to Eq. (14.3) of Part I, the BCS self-energy (23.10) contains only the singular contribution $\Sigma^\delta(z)$, which, upon transforming to real time, is attributed to either the retarded Σ^{R} or the advanced Σ^{R} (cf. Eq. (14.6) of Part I).

For definiteness, we shall limit ourselves to considering the equations of motion for the (upper-left) 11-components of the matrices (23.15), (23.16), and (23.17) (the cases of the other components readily follow).

Equation of motion of $G^>(\mathbf{r}t, \mathbf{r}'t')_{11}$ From Eq. (15.14) of Part I and the rules (23.18)–(23.20) above, we obtain:

$$\left[i\frac{d}{dt} - h(\mathbf{r}, t)\right] G^>(\mathbf{r}t, \mathbf{r}'t')_{11} - \int_{t_0}^{+\infty} dt'' \int d\mathbf{r}''\, \Sigma^{\mathrm{R}}(\mathbf{r}t, \mathbf{r}''t'')_{12}\, G^>(\mathbf{r}''t'', \mathbf{r}'t')_{21} = 0, \tag{23.22}$$

that is

$$\boxed{\left[i\frac{d}{dt} - h(\mathbf{r},t)\right] G^{>}(\mathbf{r}t,\mathbf{r}'t')_{11} \;-\; \Delta(\mathbf{r},t)\, G^{>}(\mathbf{r}t,\mathbf{r}'t')_{21} = 0\,.} \tag{23.23}$$

A further check on this equation can be readily made in terms of the definitions (cf. Eqs. (21.6) and (23.17))

$$\begin{aligned}
\Delta(\mathbf{r},t) &= v_0 \left\langle \psi_{\downarrow}(\mathbf{r},t)\, \psi_{\uparrow}(\mathbf{r},t)\right\rangle \\
G^{>}(\mathbf{r}t,\mathbf{r}'t')_{11} &= -\,i \left\langle \Psi_1(\mathbf{r}t)\, \Psi_1^{\dagger}(\mathbf{r}'t')\right\rangle = -\,i\left\langle \psi_{\uparrow}(\mathbf{r}t)\, \psi_{\uparrow}^{\dagger}(\mathbf{r}'t')\right\rangle \\
G^{>}(\mathbf{r}t,\mathbf{r}'t')_{21} &= -\,i \left\langle \Psi_2(\mathbf{r}t)\, \Psi_1^{\dagger}(\mathbf{r}'t')\right\rangle = -\,i\left\langle \psi_{\downarrow}^{\dagger}(\mathbf{r}t)\, \psi_{\uparrow}^{\dagger}(\mathbf{r}'t')\right\rangle ,
\end{aligned} \tag{23.24}$$

such that Eq. (23.23) becomes

$$\left[i\frac{d}{dt} - h(\mathbf{r},t)\right] \left\langle \psi_{\uparrow}(\mathbf{r}t)\, \psi_{\uparrow}^{\dagger}(\mathbf{r}'t')\right\rangle - v_0 \left\langle \psi_{\downarrow}(\mathbf{r},t)\, \psi_{\uparrow}(\mathbf{r},t)\right\rangle \left\langle \psi_{\downarrow}^{\dagger}(\mathbf{r}t)\, \psi_{\uparrow}^{\dagger}(\mathbf{r}'t')\right\rangle = 0 \tag{23.25}$$

consistently with the second line of the equation of motion (21.5), where the mean-field decoupling is considered and the suffix "H" for the Heisenberg representation is again dropped for simplicity.

Equation of motion of $G^{<}(\mathbf{r}t,\mathbf{r}'t')_{11}$ In a similar way, from Eq. (15.13) of Part I, we obtain:

$$\left[i\frac{d}{dt} - h(\mathbf{r},t)\right] G^{<}(\mathbf{r}t,\mathbf{r}'t')_{11} - \int_{t_0}^{+\infty} dt'' \int d\mathbf{r}''\, \Sigma^{\mathrm{R}}(\mathbf{r}t,\mathbf{r}''t'')_{12}\, G^{<}(\mathbf{r}''t'',\mathbf{r}'t')_{21} = 0, \tag{23.26}$$

that is

$$\boxed{\left[i\frac{d}{dt} - h(\mathbf{r},t)\right] G^{<}(\mathbf{r}t,\mathbf{r}'t')_{11} \;-\; \Delta(\mathbf{r},t)\, G^{<}(\mathbf{r}t,\mathbf{r}'t')_{21} = 0\,.} \tag{23.27}$$

Equation of motion of $G_{11}(\mathbf{r}t,\mathbf{r}'t')_{11}$ With reference to Eq. (23.15), we have

$$\begin{aligned}
G_{11}(\mathbf{r}t,\mathbf{r}'t')_{11} &= -\,i\,\theta(t-t')\, \left\langle \Psi_1(\mathbf{r}t)\, \Psi_1^{\dagger}(\mathbf{r}'t')\right\rangle + i\,\theta(t'-t)\, \left\langle \Psi_1^{\dagger}(\mathbf{r}'t')\, \Psi_1(\mathbf{r}t)\right\rangle \\
&\underset{\substack{\uparrow \\ \text{[Eqs.(23.16) and (23.17)]}}}{=} \theta(t-t')\, G^{>}(\mathbf{r}t,\mathbf{r}'t')_{11} + \theta(t'-t)\, G^{<}(\mathbf{r}t,\mathbf{r}'t')_{11}\,,
\end{aligned} \tag{23.28}$$

such that

$$i\frac{\partial}{\partial t}G_{11}(\mathbf{r}t,\mathbf{r}'t')_{11} = i\,\delta(t-t')\,\left(G^{>}(\mathbf{r}t,\mathbf{r}'t)_{11} - G^{<}(\mathbf{r}t,\mathbf{r}'t)_{11}\right) + \theta(t-t')\,i\frac{\partial}{\partial t}G^{>}(\mathbf{r}t,\mathbf{r}'t')_{11} + \theta(t'-t)\,i\frac{\partial}{\partial t}G^{<}(\mathbf{r}t,\mathbf{r}'t')_{11} \tag{23.29}$$

where

$$G^{>}(\mathbf{r}t,\mathbf{r}'t)_{11} - G^{<}(\mathbf{r}t,\mathbf{r}'t)_{11} = -\,i\left\langle\left(\Psi_1(\mathbf{r}t)\,\Psi_1^{\dagger}(\mathbf{r}'t) + \Psi_1^{\dagger}(\mathbf{r}'t)\,\Psi_1(\mathbf{r}t)\right)\right\rangle = -\,i\,\delta(\mathbf{r}-\mathbf{r}') \tag{23.30}$$

owing to the anti-commutation relations (22.2). Taking into account the equations of motion (23.23) for $G^{>}_{11}$ and (23.27) for $G^{<}_{11}$, we obtain eventually:

$$\begin{aligned} i\frac{\partial}{\partial t}G_{11}(\mathbf{r}t,\mathbf{r}'t')_{11} &= \delta(t-t')\,\delta(\mathbf{r}-\mathbf{r}') \\ &\quad + \theta(t-t')\,\left(h(\mathbf{r},t)\,G^{>}(\mathbf{r}t,\mathbf{r}'t')_{11} + \Delta(\mathbf{r},t)\,G^{>}(\mathbf{r}t,\mathbf{r}'t')_{21}\right) \\ &\quad + \theta(t'-t)\,\left(h(\mathbf{r},t)\,G^{<}(\mathbf{r}t,\mathbf{r}'t')_{11} + \Delta(\mathbf{r},t)\,G^{<}(\mathbf{r}t,\mathbf{r}'t')_{21}\right) \\ &= \delta(t-t')\,\delta(\mathbf{r}-\mathbf{r}') + h(\mathbf{r},t)\,G_{11}(\mathbf{r}t,\mathbf{r}'t')_{11} \\ &\quad + \Delta(\mathbf{r},t)\,G_{11}(\mathbf{r}t,\mathbf{r}'t')_{21}, \end{aligned} \tag{23.31}$$

where the expression (23.28) has been used together with

$$\begin{aligned} G_{11}(\mathbf{r}t,\mathbf{r}'t')_{21} &= -\,i\,\theta(t-t')\,\left\langle\Psi_2(\mathbf{r}t)\,\Psi_1^{\dagger}(\mathbf{r}'t')\right\rangle + i\,\theta(t'-t)\,\left\langle\Psi_1^{\dagger}(\mathbf{r}'t')\,\Psi_2(\mathbf{r}t)\right\rangle \\ &\underset{\substack{\uparrow \\ \text{[Eqs. (23.16) and (23.17)]}}}{=} \theta(t-t')\,G^{>}(\mathbf{r}t,\mathbf{r}'t')_{21} + \theta(t'-t)\,G^{<}(\mathbf{r}t,\mathbf{r}'t')_{21}\,. \end{aligned} \tag{23.32}$$

If we now convert the Nambu pseudo-spinor fields $\Psi_i(\mathbf{r})$ back to the original field operators $\psi_\sigma(\mathbf{r})$ according to Eq. (22.1), such that $G_{11}(\mathbf{r}t,\mathbf{r}'t')_{11} \leftrightarrow -\,i\left\langle\mathcal{T}\left\{\psi_\uparrow(\mathbf{r}t)\,\psi_\uparrow^{\dagger}(\mathbf{r}'t')\right\}\right\rangle$ coincides with $G(\mathbf{r}t,\mathbf{r}'t')$ of Eq. (21.7) and $G_{11}(\mathbf{r}t,\mathbf{r}'t')_{21} \leftrightarrow -\,i\left\langle\mathcal{T}\left\{\psi_\downarrow^{\dagger}(\mathbf{r}t)\,\psi_\uparrow^{\dagger}(\mathbf{r}'t')\right\}\right\rangle$ coincides with $F(\mathbf{r}t,\mathbf{r}'t')$ of Eq. (21.9), we recognize that the equation of motion (23.31) coincides with Eq. (21.8), which was obtained in Chapter 21 in terms of a simple mean-field decoupling without invoking the machinery of the Schwinger–Keldysh approach. [The complementary equation of motion (21.11) for $F(\mathbf{r}t,\mathbf{r}'t')$ can be similarly obtained from Eqs. (23.9), (23.10), and (23.15).]

Recourse to the Schwinger–Keldysh approach is instead required when considering beyond-mean-field approximations that give rise to memory effects, as discussed in Chapter 24.

24

The T-matrix Approximation in the Superfluid Phase

The t-matrix approximation for the normal phase was considered in Chapter 16 of Part I with a generic interparticle interaction $v(\mathbf{r} - \mathbf{r}')$. Here, we consider the extension of this approximation to the superfluid phase, for which it is convenient to restrict from the outset to a contact-type inter-particle interaction (cf. Eq. (21.1)).

At thermodynamic equilibrium, Gorkov and Melik-Barkhudarov were the first ones to consider the phenomenon of superfluidity in a dilute (neutral) Fermi gas [55], which they handled by extending the original Galitskii t-matrix approach [54] to the case of an attractive interparticle interaction. Later, various versions of this approach were extensively utilized for the superfluid phase in the context of the BCS–BEC crossover [57, 81–85], a topic that will be briefly considered in Chapter 26. When addressing nonequilibrium (time-dependent) situations, the extension of the fermionic t-matrix approach from the normal to the superfluid phase requires us to carefully account for the Nambu indices in the two-particle channels, owing to the presence of the "anomalous" single-particle Green's functions (cf. Eqs. (23.15)–(23.17)). This task will be greatly simplified by the use of a contact interaction.

24.1 Integral Equation for the T-matrix and the Single-Particle Self-Energy

The use of the contact-type interparticle interaction (21.1) implies that only fermions with opposite spins can interact with each other due to the Pauli principle. This feature considerably simplifies the *elementary interaction vertex* represented in Fig. 22.1, for which only two possibilities now exist in terms of the Nambu pseudo-spinor fields (22.1), as indicated in Fig. 24.1.

Note here the presence of two τ^3 matrices with opposite values of the pseudo-spin, which gives an overall minus sign to each elementary vertex. Note also that, in the lower lines of Fig. 24.1, "ingoing" and "outgoing" arrows refer explicitly

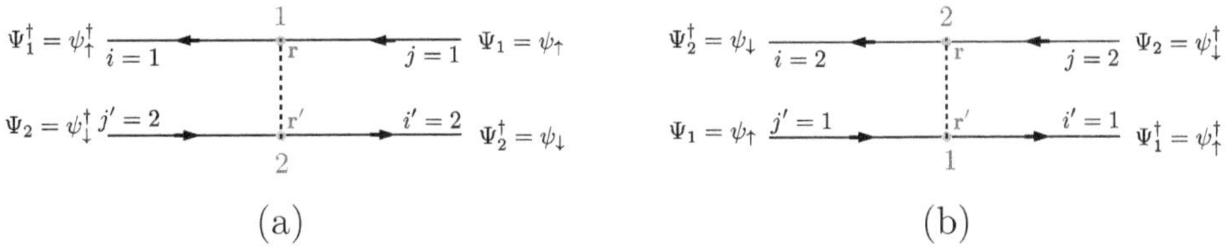

Figure 24.1 The two possible [(a) and (b)] cases of the elementary interaction vertex for a contact interparticle interaction, occurring between different spins or else different Nambu indices. Here, the dots (•) signal the presence of a τ^3 matrix.

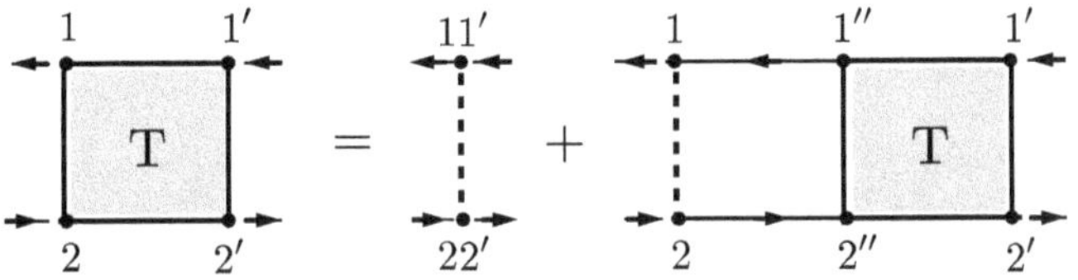

Figure 24.2 Graphical representation of the integral equation (24.1) for the T-matrix in the ladder approximation. Here, the short-hand notation $1 \leftrightarrow (\mathbf{r}_1, i_1, z_1)$ stands for the collection of spatial ($\mathbf{r}_1$), Nambu index (i_1), and contour time (z_1) variables.

to the Nambu pseudo-spinor fields Ψ_i, while for the ordinary spin fields ψ_σ, these arrows should run in the opposite direction. It is for this reason that in Chapter 16 of Part I, the t-matrix approach for the normal phase was formulated in terms of the effective two-particle interaction Γ, given by Eq. (16.1) and depicted in Fig.16.1 of Part I.

For the same reason, to describe a dilute superfluid Fermi system, it is instead preferable to utilize the many-particle T-matrix [32] in the *ladder approximation*, whereby the arrows run in opposite directions as in Fig. 24.1, similar to what is done in the equilibrium superfluid case [57, 83]. Specifically, in the ladder approximation, the T-matrix satisfies the following integral equation

$$\mathrm{T}(1,2;1',2') = \delta(1,1')\delta(2,2')V(1,2) \\ + V(1,2)\int d1''d2''G(1,1'')\,G(2'',2)\,\mathrm{T}(1'',2'';1',2'), \quad (24.1)$$

as represented graphically in Fig. 24.2. Note here the following features:

(i) $V(1,2) = -i\,v(1,2)$ has opposite sign with respect to Eq. (16.1) of Part I, owing to the presence of a τ^3 matrix at each vertex as in Fig. 24.1;
(ii) The Nambu indices associated with the variables 1 and 2, as well as with 1″ and 2″ in the single-particle Green's functions, are opposite to each other;

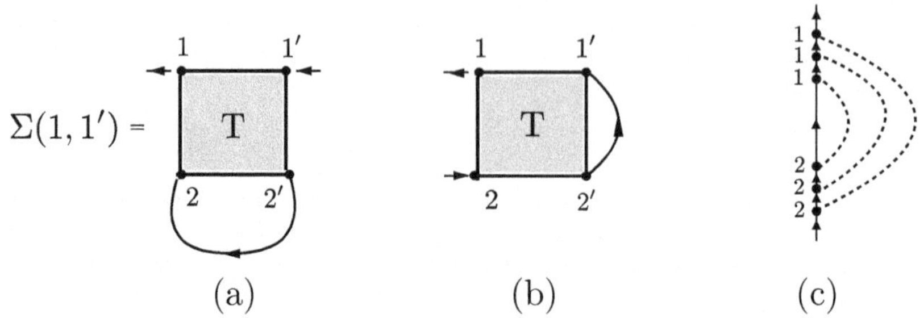

Figure 24.3 (a) Single-particle self-energy within the T-matrix in the ladder approximation. (b) Possible additional diagram that need not be considered in the present approximation because it is nothing but the mean-field (BCS) self-energy, as represented in (c).

(iii) Although we have assumed $v(1,2)$ to be of the contact type (21.1), we have not yet taken full advantage of the presence of a Dirac delta function in space (as well in contour time); and

(iv) Due to the particle–hole mixing intrinsic to the BCS pairing theory [70], the orientation of the arrows of the single-particle lines in the diagrams for the superfluid phase is only conventional to the extent that it is not directly related to either particle–particle or particle–hole lines in the two-particle channel, as it occurs for the normal phase.

Accordingly, in the superfluid case, the single-particle self-energy in the ladder approximation for the many-particle T-matrix is given by

$$\Sigma(1,1') = -\int d2d2' \ \mathrm{T}(1,2;1',2') \ G(2^{(-1)^{i_2}},2') \tag{24.2}$$

as depicted diagrammatically in Fig. 24.3(a). Note that, in Eq. (24.2), the arguments of the single-particle Green's function have been interchanged with respect to Eq. (16.1) of Part I, owing to the different orientations of the lines in Γ and T and that the "exchange diagram" (which appears in Fig. 16.2 of Part I) is not present in Eq. (24.2) (and thus in Fig. 24.3(a)), in accordance also with its absence in the normal phase owing to the contact interaction that we are now restricting to. Note further that the additional self-energy diagram shown in Fig. 24.3(b), which could in principle be built up from the T-matrix of Fig. 24.2, is nothing but a different representation of the mean-field (BCS) contribution to the off-diagonal self-energy (23.21) once self-consistency is taken into account, as depicted in Fig. 24.3(c). This BCS contribution is thus not taken into account by the T-matrix self-energy (24.2) but has to be considered separately (as emphasized near the end of Chapter 22). Note, finally, that the sign $(-1)^{i_2}$ in the first argument of the single-particle

Green's function on the right-hand side of the expression (24.2) is needed to formally recover the Hartree–Fock approximation at the lowest order in the contact potential (although this would not be strictly required when solving for the integral equation (24.1)).

Remark: The Hartree–Fock self-energy for a contact potential

For a contact potential, the expression (19.6) of Part I simplifies to the form

$$\Sigma_{\rm HF}^{\sigma}(\mathbf{r}_1 t_1, \mathbf{r}_{1'} t_{1'}) = v_0\, \delta(t_1 - t_{1'})\, \delta(\mathbf{r}_1 - \mathbf{r}_{1'}) \left\langle \psi_{\bar{\sigma}}^{\dagger}(\mathbf{r}_1, t_1)\, \psi_{\bar{\sigma}}(\mathbf{r}_1, t_1) \right\rangle, \tag{24.3}$$

where $\bar{\sigma} = -\sigma$ and the suffix H of the Heisenberg representation is again omitted. Note that the expression (24.3) does not depend on σ, provided that the condition (22.20) holds. On the other hand, from the expression (24.2) in the Nambu representation, we obtain at the lowest order in the interparticle interaction $V(1, 2)$:

$$\begin{aligned}
\Sigma_{\rm HF}(\mathbf{r}_1 t_1, \mathbf{r}_{1'} t_{1'})_{11} &= i\, v_0\, \delta(t_1 - t_{1'})\, \delta(\mathbf{r}_1 - \mathbf{r}_{1'})\, G(\mathbf{r}_1 t_1^{+}, \mathbf{r}_1 t_1)_{22} \\
G(\mathbf{r}_1 t_1^{+}, \mathbf{r}_1 t_1)_{22} &= -\, i \left\langle \mathcal{T} \left[\Psi_2(\mathbf{r}_1, t_1^{+})\, \Psi_2^{\dagger}(\mathbf{r}_1, t_1) \right] \right\rangle \\
&= -\, i \left\langle \psi_{\downarrow}^{\dagger}(\mathbf{r}_1, t_1)\, \psi_{\downarrow}(\mathbf{r}_1, t_1) \right\rangle
\end{aligned} \tag{24.4}$$

as well as

$$\begin{aligned}
\Sigma_{\rm HF}(\mathbf{r}_1 t_1, \mathbf{r}_{1'} t_{1'})_{22} &= i\, v_0\, \delta(t_1 - t_{1'})\, \delta(\mathbf{r}_1 - \mathbf{r}_{1'})\, G(\mathbf{r}_1 t_1^{-}, \mathbf{r}_1 t_1)_{11} \\
G(\mathbf{r}_1 t_1^{-}, \mathbf{r}_1 t_1)_{11} &= -\, i \left\langle \mathcal{T} \left[\Psi_1(\mathbf{r}_1, t_1^{-})\, \Psi_1^{\dagger}(\mathbf{r}_1, t_1) \right] \right\rangle \\
&= i \left\langle \psi_{\uparrow}^{\dagger}(\mathbf{r}_1, t_1)\, \psi_{\uparrow}(\mathbf{r}_1, t_1) \right\rangle ,
\end{aligned} \tag{24.5}$$

such that $\Sigma_{\rm HF}(\mathbf{r}_1 t_1, \mathbf{r}_{1'} t_{1'})_{11} = \Sigma_{\rm HF}^{\sigma}(\mathbf{r}_1 t_1, \mathbf{r}_{1'} t_{1'})$ and $\Sigma_{\rm HF}(\mathbf{r}_1 t_1, \mathbf{r}_{1'} t_{1'})_{22} = -\Sigma_{\rm HF}^{\sigma}(\mathbf{r}_{1'} t_{1'}, \mathbf{r}_1 t_1)_{11}$ as expected. Note, finally, that the different conditions we have adopted in Eq. (24.2) for G_{11} and G_{22} with equal time variables coincide with those considered in section 7-2 of Ref. [13].

24.2 Labeling the T-matrix for a Contact Interaction

Proceeding as in Chapter 16 of Part I, we arrive at the expression for $\langle \mathbf{x}_{1''}, \mathbf{x}_{2''} | \mathrm{T}(z_{1''}, z_{1'}) | \mathbf{x}_{1'}, \mathbf{x}_{2'} \rangle$ depicted graphically in Fig. 24.4(a), which is similar to that entering the right-hand side of Eq. (16.7) of Part I, apart from the following changes:

(i) The variable $\mathbf{x}$ now stands for $(\mathbf{r}, i)$ in terms of the Nambu spinor index;
(ii) The imaginary unit i on the right-hand side is replaced by $-\, i$; and
(iii) $\mathbf{x}_2 z_1 \rightleftarrows \mathbf{x}_{2''} z_{1''}$ in the argument of the second single-particle Green's function therein.

$$\langle \mathbf{x}_{1''}, \mathbf{x}_{2''} | \mathrm{T}(z_{1''}, z_{1'}) | \mathbf{x}_{1'}, \mathbf{x}_{2'} \rangle = \quad \text{(a)}$$

$$\Sigma(\mathbf{r}_1 z_1, \mathbf{r}_{1'} z_{1'})_{\mathrm{I,I}} = \quad \text{with } G(\mathbf{r}_2 z_1, \mathbf{r}_{2'} z_{1'})_{\mathrm{II,II}} \quad \text{(b)}$$

$$\Sigma(\mathbf{r}_1 z_1, \mathbf{r}_{1'} z_{1'})_{\mathrm{II,II}} = \quad \text{with } G(\mathbf{r}_2 z_1, \mathbf{r}_{2'} z_{1'})_{\mathrm{I,I}} \quad \text{(c)}$$

$$\Sigma(\mathbf{r}_1 z_1, \mathbf{r}_{1'} z_{1'})_{\mathrm{I,II}} = \quad \text{with } G(\mathbf{r}_2 z_1, \mathbf{r}_{2'} z_{1'})_{\mathrm{II,I}} \quad \text{(d)}$$

$$\Sigma(\mathbf{r}_1 z_1, \mathbf{r}_{1'} z_{1'})_{\mathrm{II,I}} = \quad \text{with } G(\mathbf{r}_2 z_1, \mathbf{r}_{2'} z_{1'})_{\mathrm{I,II}} \quad \text{(e)}$$

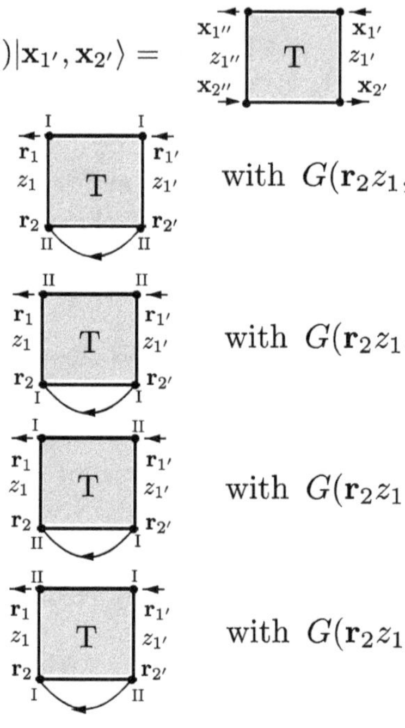

Figure 24.4 Graphical representation of (a) the T-matrix and (b)–(e) the matrix elements of the corresponding self-energy in Nambu space, in line with the form of the 2×2 matrix (23.8).

In addition, to comply with the structure of the Dyson equation (23.5) in Nambu space, we have to consider in detail the Nambu spin structure of the T-matrix and of the related self-energy. To avoid confusion, it is convenient to indicate the two Nambu spinor components by the symbols I and II (instead of 1 and 2). Accordingly,

$$1 \longleftrightarrow (\mathbf{x}_1, z_1) \quad \text{where} \quad \mathbf{x}_1 = (\mathbf{r}_1, i_1) \quad \text{with} \quad i_1 = (\mathrm{I}, \mathrm{II}) , \tag{24.6}$$

and so on. The corresponding matrix elements of the T-matrix self-energy (24.2) in the Nambu representation are depicted graphically in Fig. 24.4(b)–(e), together with the associated single-particle Green's function that closes the loop.

Figure 24.5 shows instead the four possible combinations for the product of the two single-particle Green's functions entering the second term on the right-hand side of the integral equation (24.1) for the T-matrix, which identify the four matrix elements of the *bubble* $\tilde{\chi}$ defined in analogy to Eq. (16.8) of Part I (although now there is no distinction between particle–particle and particle–hole bubbles owing to the particle–hole mixing of the pairing theory). Note further that the "pair" indices "a" and "b" utilized in Fig. 24.5 are identified in accordance with Fig. 24.6.

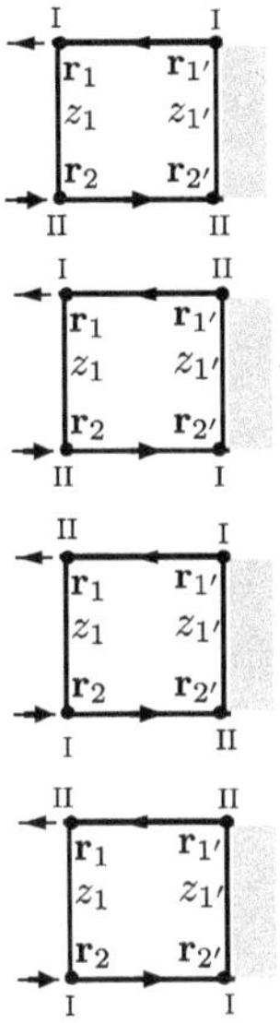

$$\leftrightarrow \; G(\mathbf{r}_1 z_1, \mathbf{r}_{1''} z_{1''})_{\mathrm{I,I}}\, G(\mathbf{r}_{2''} z_{1''}, \mathbf{r}_2 z_1)_{\mathrm{II,II}} \equiv \langle \mathbf{r}_1, \mathbf{r}_2 | \tilde{\chi}(z_1, z_2) | \mathbf{r}_{1''}, \mathbf{r}_{2''} \rangle_{\mathrm{aa}} \quad \text{(a)}$$

$$\leftrightarrow \; G(\mathbf{r}_1 z_1, \mathbf{r}_{1''} z_{1''})_{\mathrm{I,II}}\, G(\mathbf{r}_{2''} z_{1''}, \mathbf{r}_2 z_1)_{\mathrm{I,II}} \equiv \langle \mathbf{r}_1, \mathbf{r}_2 | \tilde{\chi}(z_1, z_2) | \mathbf{r}_{1''}, \mathbf{r}_{2''} \rangle_{\mathrm{ab}} \quad \text{(b)}$$

$$\leftrightarrow \; G(\mathbf{r}_1 z_1, \mathbf{r}_{1''} z_{1''})_{\mathrm{II,I}}\, G(\mathbf{r}_{2''} z_{1''}, \mathbf{r}_2 z_1)_{\mathrm{II,I}} \equiv \langle \mathbf{r}_1, \mathbf{r}_2 | \tilde{\chi}(z_1, z_2) | \mathbf{r}_{1''}, \mathbf{r}_{2''} \rangle_{\mathrm{ba}} \quad \text{(c)}$$

$$\leftrightarrow \; G(\mathbf{r}_1 z_1, \mathbf{r}_{1''} z_{1''})_{\mathrm{II,II}}\, G(\mathbf{r}_{2''} z_{1''}, \mathbf{r}_2 z_1)_{\mathrm{I,I}} \equiv \langle \mathbf{r}_1, \mathbf{r}_2 | \tilde{\chi}(z_1, z_2) | \mathbf{r}_{1''}, \mathbf{r}_{2''} \rangle_{\mathrm{bb}} \quad \text{(d)}$$

Figure 24.5 Four possible [from (a) to (d)] combinations for the product of two single-particle Green's functions entering the integral equation (24.1), which identify the matrix elements of the bubble $\tilde{\chi}$ with indices "a" and "b" specified as in Fig. 24.6. The shaded area signals the presence of the T-matrix to the right of the two single-particle Green's functions.

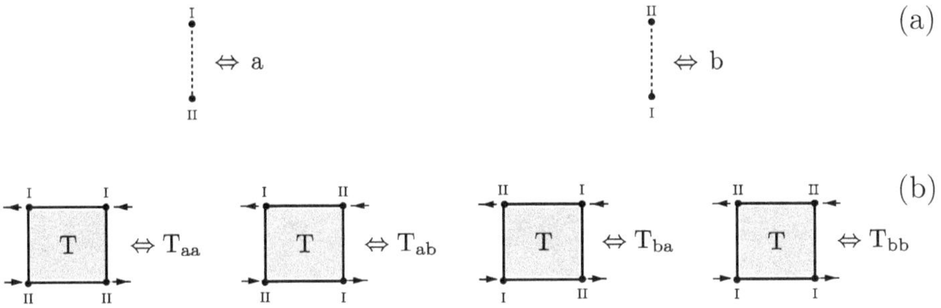

Figure 24.6 (a) Pair indices "a" and "b," which specify (b) the four possible combinations of pairs of Nambu indices occurring in the T-matrix approximation for a contact interparticle interaction.

Resting on the above analysis of the Nambu spin structure, the equation (24.1) for the T-matrix reduces to (in analogy to Eq. (16.9) of Part I):

$$\begin{aligned}
\langle \mathbf{r}_1, \mathbf{r}_2 | \mathrm{T}(z_1, z_{1'}) | \mathbf{r}_{1'}, \mathbf{r}_{2'} \rangle_{\alpha\beta} &= -i\, v(\mathbf{r}_1 - \mathbf{r}_2)\, \delta(z_1, z_{1'})\, \delta(\mathbf{r}_1 - \mathbf{r}_{1'})\, \delta(\mathbf{r}_2 - \mathbf{r}_{2'})\, \delta_{\alpha\beta} \\
&- iv(\mathbf{r}_1 - \mathbf{r}_2) \int dz_{1''} d\mathbf{r}_{1''} d\mathbf{r}_{2''} \sum_{\gamma} \langle \mathbf{r}_1, \mathbf{r}_2 | \tilde{\chi}(z_1, z_{1''}) | \mathbf{r}_{1''}, \mathbf{r}_{2''} \rangle_{\alpha\gamma} \\
&\times \langle \mathbf{r}_{1''}, \mathbf{r}_{2''} | \mathrm{T}(z_{1''}, z_{1'}) | \mathbf{r}_{1'}, \mathbf{r}_{2'} \rangle_{\gamma\beta},
\end{aligned} \tag{24.7}$$

where $\alpha, \beta, \gamma = (\mathrm{a}, \mathrm{b})$. Note that a diagonal pre-summation in these indices can be done in Eq. (24.7), in analogy to what was done in Ref. [83] (cf. Appendix A therein).

24.3 Simplifying the Integral Equation of the T-matrix for a Contact Interaction

Thus far, the contact interparticle interaction has been utilized only to specify the Nambu spin structure. However, the presence of a contact interaction can be further exploited to considerably simplify also the spatial structure of the original equation (24.1), which defines the T-matrix. To this end, by combining Eq. (21.1) for the contact interaction with the compact definition (11.20) of Part I, we write $v(1,2) = v_0\, \delta(z_1, z_2)\, \delta(\mathbf{r}_1 - \mathbf{r}_2)$ and set

$$\begin{aligned} &\mathrm{T}(1,2;1',2') \\ &\quad = \delta(z_1,z_2)\, \delta(\mathbf{r}_1 - \mathbf{r}_2)\, \delta(z_{1'},z_{2'})\, \delta(\mathbf{r}_{1'} - \mathbf{r}_{2'})\, \langle \mathbf{r}_1; i_1, i_2 | \mathrm{T}(z_1, z_{1'}) | \mathbf{r}_{1'}; i_{1'}, i_{2'} \rangle, \end{aligned} \tag{24.8}$$

where the convention for the labels $1, 2, 1', 2'$ is the same as in Fig. 24.2, going in this way one step further with respect to Eq. (16.4) of Part I. Utilizing at this point also for the spatial variables the manipulation made in Eq. (16.6) of Part I for the product of three delta functions, and dropping the product $\delta(z_1, z_2)\, \delta(\mathbf{r}_1 - \mathbf{r}_2)\, \delta(z_{1'}, z_{2'})\, \delta(\mathbf{r}_{1'} - \mathbf{r}_{2'})$ from both sides of the resulting equation, we end up with the following simplified form of the integral equation

$$\begin{aligned} \langle \mathbf{r}_1; i_1, i_2 | \mathrm{T}(z_1, z_{1'}) | \mathbf{r}_{1'}; i_{1'}, i_{2'} \rangle &= -i\, v_0\, \delta(z_1, z_{1'})\, \delta(\mathbf{r}_1 - \mathbf{r}_{1'})\, \delta_{i_1, i_{1'}}\, \delta_{i_2, i_{2'}} \\ - i v_0 \int dz_{1''} d\mathbf{r}_{1''} &\sum_{i_{1''} i_{2''}} G(\mathbf{r}_1 z_1 i_1, \mathbf{r}_{1''} z_{1''} i_{1''})\, G(\mathbf{r}_{1''} z_{1''} i_{2''}, \mathbf{r}_1 z_1 i_2) \\ \times \langle \mathbf{r}_{1''}; i_{1''}, i_{2''} | &\mathrm{T}(z_{1''}, z_{1'}) | \mathbf{r}_{1'}; i_{1'}, i_{2'} \rangle, \end{aligned} \tag{24.9}$$

whose graphical representation is given in Fig. 24.7.

There remains to rephrase the Nambu spin structure of Eq. (24.9), following the steps that have led to Eq. (24.7). Using again the notation of Fig. 24.6, we then end up with the compact equation

$$\begin{aligned} \langle \mathbf{r}_1 | \mathrm{T}(z_1, z_{1'}) | \mathbf{r}_{1'} \rangle_{\alpha\beta} &= -i\, v_0\, \delta(\mathbf{r}_1 - \mathbf{r}_{1'})\, \delta(z_1, z_{1'})\, \delta_{\alpha\beta} \\ - i v_0 \int dz_{1''} d\mathbf{r}_{1''} &\sum_{\gamma} \langle \mathbf{r}_1 | \tilde{\chi}(z_1, z_{1''}) | \mathbf{r}_{1''} \rangle_{\alpha\gamma}\, \langle \mathbf{r}_{1''} | \mathrm{T}(z_{1''}, z_{1'}) | \mathbf{r}_{1'} \rangle_{\gamma\beta} \end{aligned} \tag{24.10}$$

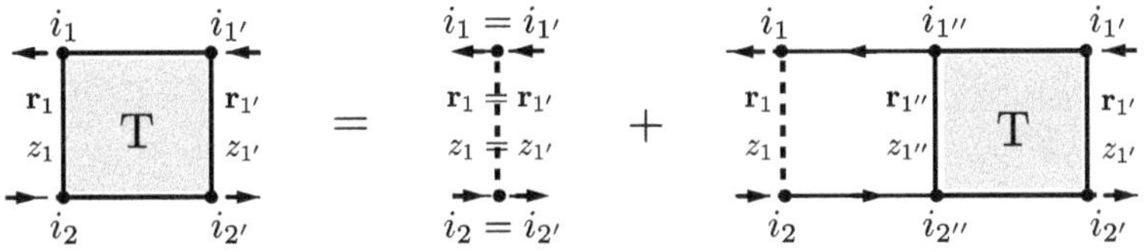

Figure 24.7 Graphical representation of the simplified form (24.9) of the integral equation for the T-matrix, where the Nambu-spin (i), space ($\mathbf{r}$), and contour-time (z) variables are indicated in detail.

where $\langle \mathbf{r}_1 | \tilde{\chi}(z_1, z_{1''}) | \mathbf{r}_{1''} \rangle_{\alpha\gamma}$ is given by the expressions reported in Fig. 24.5 with $\mathbf{r}_1 = \mathbf{r}_2$ and $\mathbf{r}_{1''} = \mathbf{r}_{2''}$. Note, finally, that by setting

$$\langle \mathbf{r}_1 | \mathcal{V}(z_1, z_{1'}) | \mathbf{r}_{1'} \rangle_{\alpha\beta} \equiv -i v_0 \, \delta(\mathbf{r}_1 - \mathbf{r}_{1'}) \, \delta(z_1, z_{1'}) \, \delta_{\alpha\beta} \tag{24.11}$$

in analogy with Eq. (16.10) of Part I, Eq. (24.10) becomes eventually

$$\begin{aligned} \langle \mathbf{r}_1 | \mathrm{T}(z_1, z_{1'}) | \mathbf{r}_{1'} \rangle_{\alpha\beta} &= \langle \mathbf{r}_1 | \mathcal{V}(z_1, z_{1'}) | \mathbf{r}_{1'} \rangle_{\alpha\beta} + \int dz_{1''} d\mathbf{r}_{1''} \sum_{\gamma''} \int dz_{1'''} d\mathbf{r}_{1'''} \sum_{\gamma'''} \\ &\times \langle \mathbf{r}_1 | \mathcal{V}(z_1, z_{1''}) | \mathbf{r}_{1''} \rangle_{\alpha\gamma''} \langle \mathbf{r}_{1''} | \tilde{\chi}(z_{1''}, z_{1'''}) | \mathbf{r}_{1'''} \rangle_{\gamma''\gamma'''} \langle \mathbf{r}_{1'''} | \mathrm{T}(z_{1'''}, z_{1'}) | \mathbf{r}_{1'} \rangle_{\gamma'''\beta} , \end{aligned} \tag{24.12}$$

or in compact matrix form $\mathrm{T} = \mathcal{V} + \mathcal{V} \tilde{\chi} \mathrm{T}$, where each component depends on two time variables. This equation admits the formal solution $\mathrm{T} = (\mathcal{V}^{-1} - \tilde{\chi})^{-1}$, where $\mathcal{V}^{-1} \leftrightarrow \frac{i}{v_0} \delta(\mathbf{r}_1 - \mathbf{r}_{1'}) \, \delta(z_1, z_{1'}) \, \delta_{\alpha\beta}$, owing to the properties of the Dirac delta function.

In this framework, the expression (24.2) of the self-energy simplifies as follows:

$$\begin{aligned} \Sigma(1, 1') &= \Sigma(\mathbf{r}_1 z_1, \mathbf{r}_{1'} z_{1'})_{i_1, i_{1'}} \\ &= - \sum_{i_2, i_{2'}} \langle \mathbf{r}_1; i_1, i_2 | \mathrm{T}(z_1, z_{1'}) | \mathbf{r}_{1'}; i_{1'}, i_{2'} \rangle \, G(\mathbf{r}_1 z_1, \mathbf{r}_{1'} z_{1'})_{i_2, i_{2'}} , \end{aligned} \tag{24.13}$$

where the Nambu spin structure can now be handled as in Fig. 24.4.

24.4 The Case of a Spatially Homogeneous System

The integral equation (24.10) for the T-matrix can be further simplified when the physical system is spatially homogeneous (while still keeping the full time-dependent structure). In this case, we set

$$\langle \mathbf{r}_1 | \mathrm{T}(z_1, z_{1'}) | \mathbf{r}_{1'} \rangle_{\alpha\beta} = \int \frac{d\mathbf{q}}{(2\pi)^3} \, e^{i\mathbf{q} \cdot (\mathbf{r}_1 - \mathbf{r}_{1'})} \, T(\mathbf{q}; z_1, z_{1'})_{\alpha\beta}, \tag{24.14}$$

as well as

$$\langle \mathbf{r}_1 | \tilde{\chi}(z_1, z_{1'}) | \mathbf{r}_{1'} \rangle_{\alpha\beta} = \int \frac{d\mathbf{q}}{(2\pi)^3} \, e^{i\mathbf{q}\cdot(\mathbf{r}_1 - \mathbf{r}_{1'})} \, \tilde{\chi}(\mathbf{q}; z_1, z_{1'})_{\alpha\beta} \tag{24.15}$$

and write

$$\delta(\mathbf{r}_1 - \mathbf{r}_{1'}) = \int \frac{d\mathbf{q}}{(2\pi)^3} \, e^{i\mathbf{q}\cdot(\mathbf{r}_1 - \mathbf{r}_{1'})} \, . \tag{24.16}$$

In this way, Eq. (24.10) becomes eventually:

$$\begin{aligned} T(\mathbf{q}; z_1, z_{1'})_{\alpha\beta} = & - i\, v_0 \, \delta(z_1, z_{1'}) \, \delta_{\alpha\beta} \\ & - i\, v_0 \sum_{\gamma} \int dz_{1''} \, \tilde{\chi}(\mathbf{q}; z_1, z_{1''})_{\alpha\gamma} \, T(\mathbf{q}; z_{1''}, z_{1'})_{\gamma\beta} \, . \end{aligned} \tag{24.17}$$

Once transformed to real times according to the Langreth rules of Chapter 14 of Part I, one may further Fourier transform the relative time $t_{\text{rel}} = t - t'$ to the real frequency ω, while keeping the total (average) time $t_{\text{tot}} = \left(\frac{t+t'}{2}\right)$ fixed. That is to say, for a generic function $f(t, t')$ of the real-time variables t and t', one may write

$$f(t, t') = f\left(t_{\text{tot}} + \frac{t_{\text{rel}}}{2}, t_{\text{tot}} - \frac{t_{\text{rel}}}{2}\right) \equiv \bar{f}(t_{\text{tot}}, t_{\text{rel}}) \longrightarrow \bar{f}(t_{\text{tot}}, \omega) \, . \tag{24.18}$$

Going back to Eq. (24.12) for the inhomogeneous case, a transformation as (24.18) could as well be applied to the space variables $(\mathbf{r}_1, \mathbf{r}_{1'})$.

Remark: The mean-field self-energy always comes in first

Although at the formal level, the T-matrix approximation for the self-energy has an independent relevance in the framework of conserving approximations [3, 4], on physical grounds, the fermionic pairing theory requires the presence of the mean-field (BCS) anomalous self-energy contribution (cf. Eq. (23.10) and Fig. 24.3(c)), which breaks the symmetry relevant to the fermionic superfluid phase to begin with. This implies that any choice for the self-energy beyond the mean-field level (like the T-matrix approximation considered here) must necessarily be accompanied by the mean-field contribution itself [57].

It is thus worth considering in some detail (as it will be done in Chapter 25) the time-dependent mean-field approximation for the general case of a spatially inhomogeneous system, in terms of which numerous practical applications have also been implemented in the context of fermionic superfluidity. Its outcomes could then be of help (and also of use as a reference level) for more sophisticated approximations like the T-matrix itself.

Finally, it is expected that a combination of BCS and T-matrix self-energies in the superfluid phase should be required to account for the experimental findings with ultracold Fermi gases (like those mentioned in Section 32.6 of Part III) when a sudden quench of the interparticle interaction leads to crossing the boundary between the normal and superfluid phases, in regions of parameters where the inclusion of pairing fluctuations beyond mean field is expected to be important [20]. A full implementation along these lines appears still lacking at the time of writing.

25

Derivation of the Time-Dependent Bogoliubov–deGennes Equations

In Chapter 19 of Part I, we have considered the time-dependent Hartree–Fock approximation associated with a mean-field decoupling in the normal phase. In this chapter, we consider its extension to the superfluid phase for a fermionic system whose symmetry breaking is specified by anomalous averages such as that given by Eq. (21.6). This will lead us to the so-called *time-dependent Bogoliubov–deGennes* (TDBdG) *equations* in real time, which extend to nonequilibrium situations the widely used stationary BdG equations that deal with equilibrium superfluid phenomena [86].

Accordingly, we will begin by briefly recalling the time-independent BdG equations for the equilibrium case and show that they are equivalent to the Gor'kov approach for inhomogeneous fermionic superfluidity at equilibrium (whose time-dependent version was already considered in Section 21.2). We will then show how the BdG approach can be extended to the nonequilibrium case in the framework of the general theory of Chapter 23, once implemented at the mean-field level. The extension of the Gor'kov approach to the nonequilibrium case will instead be reconsidered after the procedure of analytic continuation from the imaginary to the real-time axis will be discussed in Chapter 27.

25.1 The Bogoliubov–deGennes Equations at Equilibrium and Related Matters

In the present context, consideration of the time-independent BdG equations is relevant for specifying the initial equilibrium condition that precedes the time evolution in terms of the TDBdG equations. In the same context, for future reference, it is also convenient to consider the equivalence of the time-independent BdG equations with the original Gor'kov approach to inhomogeneous superfluidity [46] (as conveniently described in Ref. [7] – cf. section 51 therein). This section briefly summarizes the relevant features of the two approaches.

Summary of the Gor'kov approach at equilibrium

At equilibrium, the Gor'kov equations for the normal ($\mathcal{G}$) and anomalous ($\mathcal{F}$) single-particle Green's functions can be formally obtained from Eqs. (21.8) and (21.11) by replacing therein $t \longrightarrow -i\tau$ with imaginary time τ and neglecting the time dependence of the gap parameter (21.6). We obtain in this way:

$$-\frac{\partial}{\partial\tau}\mathcal{G}(\mathbf{r}\tau,\mathbf{r}'\tau') - (h_0(\mathbf{r}) - \mu)\,\mathcal{G}(\mathbf{r}\tau,\mathbf{r}'\tau') - \Delta(\mathbf{r})\mathcal{F}(\mathbf{r}\tau,\mathbf{r}'\tau') = \delta(\tau-\tau')\delta(\mathbf{r}-\mathbf{r}') \tag{25.1}$$

$$-\frac{\partial}{\partial\tau}\mathcal{F}(\mathbf{r}\tau,\mathbf{r}'\tau') + (h_0(\mathbf{r}) - \mu)\,\mathcal{F}(\mathbf{r}\tau,\mathbf{r}'\tau') - \Delta(\mathbf{r})^*\,\mathcal{G}(\mathbf{r}\tau,\mathbf{r}'\tau') = 0, \tag{25.2}$$

where $h_0(\mathbf{r}) = -\frac{1}{2m}\nabla^2 + V(\mathbf{r})$ (with $\hbar = 1$ for the reduced Planck's constant) contains a static potential $V(\mathbf{r})$, and μ is the equilibrium chemical potential (whose presence will be justified later on in this section). The gap parameter is, in turn, self-consistently determined by the condition $\Delta(\mathbf{r})^* = v_0\langle\psi_\uparrow^\dagger(\mathbf{r})\psi_\downarrow^\dagger(\mathbf{r})\rangle = v_0\mathcal{F}(\mathbf{r}\tau,\mathbf{r}\tau^+)$ (cf. Eqs. (21.6) and (21.10)). Note how the Gor'kov equations (25.1) and (25.2) extend the equation for the temperature single-particle Green's function along the vertical track (given in Section 15.1 of Part I) to the presence of anomalous averages.

All quantities in Eqs. (25.1) and (25.2) depend on the difference $\tau - \tau'$ on the vertical track. We can then make use of an expansion as in Eq. (15.7) of Part I in terms of the fermionic Matsubara frequencies $\omega_n = (2n+1)\pi/\beta$ (n is an integer and $\beta = \frac{1}{k_B T}$ the inverse temperature, with k_B the Boltzmann constant), such that $-\frac{\partial}{\partial\tau} \rightarrow i\omega_n$ in Eqs. (25.1) and (25.2), yielding

$$i\omega_n\,\mathcal{G}(\mathbf{r},\mathbf{r}';\omega_n) - (h_0(\mathbf{r}) - \mu)\,\mathcal{G}(\mathbf{r},\mathbf{r}';\omega_n) - \Delta(\mathbf{r})\,\mathcal{F}(\mathbf{r},\mathbf{r}';\omega_n) = \delta(\mathbf{r}-\mathbf{r}') \tag{25.3}$$

$$i\omega_n\,\mathcal{F}(\mathbf{r},\mathbf{r}';\omega_n) + (h_0(\mathbf{r}) - \mu)\,\mathcal{F}(\mathbf{r},\mathbf{r}';\omega_n) - \Delta(\mathbf{r})^*\,\mathcal{G}(\mathbf{r},\mathbf{r}';\omega_n) = 0, \tag{25.4}$$

where now $\Delta(\mathbf{r})^* = \frac{v_0}{\beta}\sum_n e^{i\omega_n\eta}\mathcal{F}(\mathbf{r},\mathbf{r};\omega_n)$, with η a positive infinitesimal.

In the present context, the BdG approach at equilibrium [86] amounts to expressing the two functions $\mathcal{G}(\mathbf{r},\mathbf{r}';\omega_n)$ and $\mathcal{F}(\mathbf{r},\mathbf{r}';\omega_n)$, solutions to the equations (25.3) and (25.4), in terms of (pairs of) single-particle wave functions, in an analogous way to what is done in the quantum theory of scattering when the corresponding Green's functions are obtained via an eigenfunction expansion technique [87].

Summary of the BdG approach at equilibrium

An alternative to the Gor'kov approach, to deal with a superfluid Fermi system at equilibrium, was introduced by deGennes in terms of the so-called self-consistent field method [86, 88]. This method amounts to solving the following differential equations

$$\begin{pmatrix} h_0(\mathbf{r}) - \mu & \Delta(\mathbf{r}) \\ \Delta(\mathbf{r})^* & -h_0(\mathbf{r}) + \mu \end{pmatrix} \begin{pmatrix} u_\nu(\mathbf{r}) \\ v_\nu(\mathbf{r}) \end{pmatrix} = \varepsilon_\nu \begin{pmatrix} u_\nu(\mathbf{r}) \\ v_\nu(\mathbf{r}) \end{pmatrix} \tag{25.5}$$

for the *pair* of wave functions $(u_\nu(\mathbf{r}), v_\nu(\mathbf{r}))$. Here, the gap parameter is obtained from the *local self-consistency condition*

$$\Delta(\mathbf{r}) = -\, v_0 \sum_\nu u_\nu(\mathbf{r}) v_\nu(\mathbf{r})^* \left(1 - 2 f_F(\varepsilon_\nu)\right), \tag{25.6}$$

where $f_F(x) = (e^{\beta x} + 1)^{-1}$ is the Fermi–Dirac distribution function. [Throughout, we are assuming fermionic spin-balanced populations, such that the functions $\{u_\nu(\mathbf{r}), v_\nu(\mathbf{r})\}$ do not depend on spin.] The set $\{u_\nu(\mathbf{r}), v_\nu(\mathbf{r}); \nu = 1, 2, \cdots\}$ satisfies the *orthonormality condition*

$$\int d\mathbf{r} \left(u_\nu(\mathbf{r})^* u_{\nu'}(\mathbf{r}) + v_\nu(\mathbf{r})^* v_{\nu'}(\mathbf{r})\right) = \delta_{\nu\nu'}, \tag{25.7}$$

as well as the *completeness condition*

$$\sum_\nu \left[\begin{pmatrix} u_\nu(\mathbf{r}) \\ v_\nu(\mathbf{r}) \end{pmatrix} (u_\nu(\mathbf{r}')^*, v_\nu(\mathbf{r}')^*) + \begin{pmatrix} -v_\nu(\mathbf{r})^* \\ u_\nu(\mathbf{r})^* \end{pmatrix} (-v_\nu(\mathbf{r}'), u_\nu(\mathbf{r}')) \right] = \delta(\mathbf{r} - \mathbf{r}') \begin{pmatrix} 1 & 0 \\ 0 & 1 \end{pmatrix}. \tag{25.8}$$

It should be pointed out that in Eqs. (25.6) and (25.8) (as well as in Eq. (25.9)), the sum over ν is limited to solutions of the BdG equations (25.5) with positive eigenvalues ε_ν.

The key step for arriving at Eqs. (25.5) is the diagonalization of the "effective" Hamiltonian (cf. Eq. (21.2)) obtained within the mean-field approach, in terms of the unitary transformation

$$\begin{pmatrix} \psi_\uparrow(\mathbf{r}) \\ \psi_\downarrow^\dagger(\mathbf{r}) \end{pmatrix} = \sum_\nu \begin{pmatrix} u_\nu(\mathbf{r}) & -v_\nu(\mathbf{r})^* \\ v_\nu(\mathbf{r}) & u_\nu(\mathbf{r})^* \end{pmatrix} \begin{pmatrix} \gamma_{\nu 1} \\ \gamma_{\nu 2}^\dagger \end{pmatrix}, \tag{25.9}$$

such that the effective grand-canonical Hamiltonian becomes

$$K_{\text{eff}} = \sum_\nu \sum_{\alpha=1}^{2} \varepsilon_\nu \, \gamma_{\nu\alpha}^\dagger \, \gamma_{\nu\alpha} + C, \tag{25.10}$$

where C is the associated ground-state energy (whose expression will be given in Section 31.5, where it will be relevant for obtaining the average total energy of the system). In addition, thermal averages $\langle \cdots \rangle$ are here taken in terms of $K_{\rm eff}$ itself, such that (cf. the results of Section 9.2 of Part I)

$$\langle \gamma^{\dagger}_{\nu\alpha}\, \gamma_{\nu'\alpha'} \rangle = f_F(\varepsilon_\nu)\, \delta_{\nu\nu'}\, \delta_{\alpha\alpha'} \tag{25.11}$$

$$\langle \gamma_{\nu\alpha}\, \gamma_{\nu'\alpha'} \rangle = 0\,. \tag{25.12}$$

In this way, we obtain for the gap parameter:

$$\begin{aligned}
\Delta(\mathbf{r}) &= v_0\, \langle \psi_\downarrow(\mathbf{r})\, \psi_\uparrow(\mathbf{r}) \rangle \\
&\underset{[\text{Eq.}(25.9)]}{\underset{\uparrow}{=}} v_0 \sum_{\nu\nu'} \left\langle \left(v_\nu(\mathbf{r})^*\, \gamma^{\dagger}_{\nu 1} + u_\nu(\mathbf{r})\, \gamma_{\nu 2} \right) \left(u_{\nu'}(\mathbf{r})\, \gamma_{\nu' 1} - v_{\nu'}(\mathbf{r})^*\, \gamma^{\dagger}_{\nu' 2} \right) \right\rangle \\
&\underset{[\text{Eqs.}(25.11)\text{-}(25.12)]}{\underset{\uparrow}{=}} v_0 \sum_{\nu} \left[v_\nu(\mathbf{r})^*\, u_\nu(\mathbf{r})\, f_F(\varepsilon_\nu) - u_\nu(\mathbf{r})\, v_\nu(\mathbf{r})^*\, (1 - f_F(\varepsilon_\nu)) \right] \\
&= -v_0 \sum_\nu u_\nu(\mathbf{r})\, v_\nu(\mathbf{r})^*\, (1 - 2 f_F(\varepsilon_\nu))\,,
\end{aligned} \tag{25.13}$$

which recovers the expression (25.6). By a similar token, we obtain for the τ-propagator of the fermionic operators $\gamma_{\nu\alpha}$:

$$\begin{aligned}
&\left\langle T_\tau \left\{ \gamma_{\nu\alpha}(\tau)\, \gamma^{\dagger}_{\nu'\alpha'}(\tau') \right\} \right\rangle = \left\langle T_\tau \left\{ e^{K_{\rm eff}\tau}\, \gamma_{\nu\alpha} e^{-K_{\rm eff}\tau}\; e^{K_{\rm eff}\tau'}\, \gamma^{\dagger}_{\nu'\alpha'} e^{-K_{\rm eff}\tau'} \right\} \right\rangle \\
&= e^{-\varepsilon_\nu \tau}\, e^{\varepsilon_{\nu'}\tau'} \left(\theta(\tau - \tau') \left\langle \gamma_{\nu\alpha}\, \gamma^{\dagger}_{\nu'\alpha'} \right\rangle - \theta(\tau' - \tau) \left\langle \gamma^{\dagger}_{\nu'\alpha'}\, \gamma_{\nu\alpha} \right\rangle \right) \\
&= \delta_{\nu\nu'}\, \delta_{\alpha\alpha'}\, e^{-\varepsilon_\nu(\tau - \tau')} \left[\theta(\tau - \tau')\, (1 - f_F(\varepsilon_\nu)) - \theta(\tau' - \tau)\, f_F(\varepsilon_\nu) \right],
\end{aligned} \tag{25.14}$$

where the results of Sections 9.1 and 9.2 of Part I have again been utilized. The corresponding Matsubara Fourier coefficients are then given by

$$\begin{aligned}
&\int_0^\beta d(\tau - \tau')\, e^{i\omega_n(\tau - \tau')}\, \delta_{\nu\nu'}\, \delta_{\alpha\alpha'}\, e^{-\varepsilon_\nu(\tau - \tau')}\, (1 - f_F(\varepsilon_\nu)) \\
&\quad = \delta_{\nu\nu'}\, \delta_{\alpha\alpha'}\, \frac{(1 - f_F(\varepsilon_\nu))}{i\omega_n - \varepsilon_\nu}\, e^{(i\omega_n - \varepsilon_\nu)(\tau - \tau')} \Big|_0^\beta \\
&\quad = -\, \delta_{\nu\nu'}\, \delta_{\alpha\alpha'}\, \frac{1}{i\omega_n - \varepsilon_\nu}\,.
\end{aligned} \tag{25.15}$$

Equivalence of the BdG and Gor'kov approaches

The Gor'kov and BdG approaches are equivalent to each other, in the sense that the Gor'kov normal ($\mathcal{G}$) and anomalous ($\mathcal{F}$) Green's functions, which are solutions to the equations (25.3) and (25.4), can be expressed in terms of the pair wave functions $\{u_\nu(\mathbf{r}), v_\nu(\mathbf{r}); \nu = 1, 2, \cdots\}$ solutions to the BdG equations (25.5).

To this end, we begin by defining quite generally

$$\overleftrightarrow{\mathcal{G}}(\mathbf{r}\tau,\mathbf{r}'\tau') = -\left\langle \mathcal{T}_\tau\left\{\begin{pmatrix}\psi_\uparrow(\mathbf{r},\tau)\\ \psi_\downarrow^\dagger(\mathbf{r},\tau)\end{pmatrix}\left(\psi_\uparrow^\dagger(\mathbf{r}',\tau'),\psi_\downarrow(\mathbf{r}',\tau')\right)\right\}\right\rangle$$

$$= -\begin{pmatrix}\left\langle\mathcal{T}_\tau\left\{\psi_\uparrow(\mathbf{r},\tau)\,\psi_\uparrow^\dagger(\mathbf{r}',\tau')\right\}\right\rangle & \left\langle\mathcal{T}_\tau\left\{\psi_\uparrow(\mathbf{r},\tau)\,\psi_\downarrow(\mathbf{r}',\tau')\right\}\right\rangle\\ \left\langle\mathcal{T}_\tau\left\{\psi_\downarrow^\dagger(\mathbf{r},\tau)\,\psi_\uparrow^\dagger(\mathbf{r}',\tau')\right\}\right\rangle & \left\langle\mathcal{T}_\tau\left\{\psi_\downarrow^\dagger(\mathbf{r},\tau)\,\psi_\downarrow(\mathbf{r}',\tau')\right\}\right\rangle\end{pmatrix}, \tag{25.16}$$

where we recognize the Gor'kov normal function $\mathcal{G}(\mathbf{r}\tau,\mathbf{r}'\tau')$ in the 11-element and the anomalous function $\mathcal{F}(\mathbf{r}\tau,\mathbf{r}'\tau')$ in the 21-element. With the BdG transformation (25.9) of the field operators, we can rewrite these functions in the following form:

$$\begin{aligned}\mathcal{G}(\mathbf{r}\tau,\mathbf{r}'\tau') &= -\sum_{\nu\nu'}\left\langle\mathcal{T}_\tau\left\{\left(u_\nu(\mathbf{r})\,\gamma_{\nu1}(\tau)-v_\nu(\mathbf{r})^*\,\gamma_{\nu2}^\dagger(\tau)\right)\right.\right.\\ &\quad\left.\left.\times\left(u_{\nu'}(\mathbf{r}')^*\,\gamma_{\nu'1}^\dagger(\tau')-v_{\nu'}(\mathbf{r}')\,\gamma_{\nu'2}(\tau')\right)\right\}\right\rangle\\ &= -\sum_\nu\left(u_\nu(\mathbf{r})\,u_\nu(\mathbf{r}')^*\left\langle\mathcal{T}_\tau\left\{\gamma_{\nu1}(\tau)\,\gamma_{\nu1}^\dagger(\tau')\right\}\right\rangle\right.\\ &\quad\left.+\,v_\nu(\mathbf{r})^*\,v_\nu(\mathbf{r}')\left\langle\mathcal{T}_\tau\left\{\gamma_{\nu2}^\dagger(\tau)\,\gamma_{\nu2}(\tau')\right\}\right\rangle\right)\\ &\xrightarrow{(\tau>\tau')} -\sum_\nu\left[u_\nu(\mathbf{r})\,u_\nu(\mathbf{r}')^*\,(1-f_F(\varepsilon_\nu))\,e^{-\varepsilon_\nu(\tau-\tau')}\right.\\ &\quad\left.+\,v_\nu(\mathbf{r})^*\,v_\nu(\mathbf{r}')f_F(\varepsilon_\nu)\,e^{\varepsilon_\nu(\tau-\tau')}\right],\end{aligned} \tag{25.17}$$

where only the contribution for $\tau > \tau'$ has been retained in the last line, as well as

$$\begin{aligned}\mathcal{F}(\mathbf{r}\tau,\mathbf{r}'\tau') &= -\sum_{\nu\nu'}\left\langle\mathcal{T}_\tau\left\{\left(u_\nu(\mathbf{r})^*\,\gamma_{\nu2}^\dagger(\tau)+v_\nu(\mathbf{r})\,\gamma_{\nu1}(\tau)\right)\right.\right.\\ &\quad\left.\left.\times\left(u_{\nu'}(\mathbf{r}')^*\,\gamma_{\nu'1}^\dagger(\tau')-v_{\nu'}(\mathbf{r}')\,\gamma_{\nu'2}(\tau')\right)\right\}\right\rangle\\ &= \sum_\nu\left(u_\nu(\mathbf{r})^*\,v_\nu(\mathbf{r}')\left\langle\mathcal{T}_\tau\left\{\gamma_{\nu2}^\dagger(\tau)\,\gamma_{\nu2}(\tau')\right\}\right\rangle\right.\\ &\quad\left.-v_\nu(\mathbf{r})\,u_\nu(\mathbf{r}')^*\left\langle\mathcal{T}_\tau\left\{\gamma_{\nu1}(\tau)\,\gamma_{\nu1}^\dagger(\tau')\right\}\right\rangle\right)\\ &\xrightarrow{(\tau>\tau')} \sum_\nu\left[u_\nu(\mathbf{r})^*\,v_\nu(\mathbf{r}')\,f_F(\varepsilon_\nu)\,e^{\varepsilon_\nu(\tau-\tau')}\right.\\ &\quad\left.-v_\nu(\mathbf{r})\,u_\nu(\mathbf{r}')^*\,(1-f_F(\varepsilon_\nu))\,e^{-\varepsilon_\nu(\tau-\tau')}\right],\end{aligned} \tag{25.18}$$

where only the contribution for $\tau > \tau'$ has again been retained in the last line. The corresponding Matsubara Fourier coefficients are then given by

$$\mathcal{G}(\mathbf{r}, \mathbf{r}'; \omega_n) = \sum_\nu \left(\frac{u_\nu(\mathbf{r})\, u_\nu(\mathbf{r}')^*}{i\omega_n - \varepsilon_\nu} + \frac{v_\nu(\mathbf{r})^*\, v_\nu(\mathbf{r}')}{i\omega_n + \varepsilon_\nu} \right) \tag{25.19}$$

$$\mathcal{F}(\mathbf{r}, \mathbf{r}'; \omega_n) = \sum_\nu \left(\frac{v_\nu(\mathbf{r})\, u_\nu(\mathbf{r}')^*}{i\omega_n - \varepsilon_\nu} - \frac{u_\nu(\mathbf{r})^*\, v_\nu(\mathbf{r}')}{i\omega_n + \varepsilon_\nu} \right) . \tag{25.20}$$

Similar expressions can be obtained for the remaining elements of the matrix (25.16). The full expression for the Matsubara Fourier coefficients of this matrix can eventually be cast in the form:

$$\begin{aligned} \overleftrightarrow{\mathcal{G}}(\mathbf{r}, \mathbf{r}'; \omega_n) = \sum_\nu \Bigg\{ & \begin{pmatrix} u_\nu(\mathbf{r}) \\ v_\nu(\mathbf{r}) \end{pmatrix} \frac{1}{i\omega_n - \varepsilon_\nu} \left(u_\nu(\mathbf{r}')^*, v_\nu(\mathbf{r}')^* \right) \\ & + \begin{pmatrix} -v_\nu(\mathbf{r})^* \\ u_\nu(\mathbf{r})^* \end{pmatrix} \frac{1}{i\omega_n + \varepsilon_\nu} \left(-v_\nu(\mathbf{r}'), u_\nu(\mathbf{r}') \right) \Bigg\} . \end{aligned} \tag{25.21}$$

At this point, it can be readily verified that the expressions (25.19) and (25.20), obtained within the BdG approach, satisfy the Gor'kov equations (25.3) and (25.4), provided the functions $(u_\nu(\mathbf{r}), v_\nu(\mathbf{r}))$ therein satisfy the BdG equations (25.5) and the completeness condition (25.8) is taken into account.

The equivalence between the BdG and Gor'kov approaches was explicitly exploited in Ref. [89] in the context of the BCS–BEC crossover, to show that, in the BEC limit of the crossover (where composite bosons form out of fermion pairs), the solutions of the BdG equations at zero temperature provide the same physical results obtained alternatively by an independent solution of the (time-independent) Gross–Pitaevskii equation for condensed bosons, which was originally derived in that reference directly from the Gor'kov equations (25.3) and (25.4). This equivalence can be carried on to the time-dependent case, as shown in Section 25.2.

About the role of the chemical potential

Note that, in both the Gor'kov equations (25.1) and (25.2) and the BdG equations (25.5), the chemical potential μ of thermodynamic equilibrium has been subtracted from the single-particle Hamiltonian $h_0(\mathbf{r})$. This procedure is inherited from the original BCS article [70], where the single-particle energies were measured from the Fermi energy E_F, close to which a gap opens in the single-particle spectrum at low temperature, and with which μ coincides in the weak-coupling limit relevant to that article. In practice, replacing $h_0(\mathbf{r}) \rightarrow h_0(\mathbf{r}) - \mu$ as in

Eqs. (25.1)–(25.2) and (25.5) amounts to measuring the single-particle energies from the chemical potential μ of the initial equilibrium state.

Keeping track of the value of μ is especially important when dealing with the BCS–BEC crossover (which will be briefly summarized in Chapter 26) since μ plays the key role of a "driving field" that makes the fermionic system evolve from the BCS to the BEC limit of the crossover [20]. Accordingly, in this chapter we shall assume that the replacement $h_0(\mathbf{r}) \rightarrow h_0(\mathbf{r}) - \mu$ has been made in the single-particle Hamiltonian, even when dealing with time-dependent effects, whereby $h(\mathbf{r}, t) \rightarrow h(\mathbf{r}, t) - \mu$ in the language of Eq. (11.1) of Part I. For internal consistency, we apply this replacement back to Chapter 21 at the level of the time-dependent mean-field decoupling (in accordance with Ref. [14]), as well as to Chapter 24 at the level of the T-matrix approximation.

Regularization of the BdG equations for a contact interaction

The self-consistent condition (25.6) for the gap parameter $\Delta(\mathbf{r})$ diverges in the ultraviolet for a contact interparticle potential with coupling constant v_0 and thus requires a suitable regularization. For a homogeneous system with uniform Δ, this regularization is simply achieved in three dimensions by relating v_0 to the scattering length a_F of the two-fermion problem (which is experimentally accessible in ultracold Fermi gases), via the expression [20]

$$\frac{m}{4\pi a_F} = \frac{1}{v_0} + \int_{|\mathbf{k}|\leq k_0} \frac{d\mathbf{k}}{(2\pi)^3} \frac{m}{\mathbf{k}^2} = \frac{1}{v_0} + \frac{m k_0}{2\pi^2}, \tag{25.22}$$

where k_0 is an ultraviolet cutoff. On the other hand, when the gap parameter $\Delta(\mathbf{r})$ has a spatial dependence (like in the presence of an external potential $V(\mathbf{r})$), a more general procedure is, in principle, required. A quite exhaustive procedure in this respect was implemented in Ref. [90] (cf. Appendix B therein), where the gap equation (25.6) was transformed into a nonlinear Gross–Pitaevskii–like differential equation with the addition of a suitable source term. A simpler version of this regularization introduces a cutoff energy E_c in the sum over ν on the right-hand side of Eq. (25.6), such that $\varepsilon_\nu < E_c - \mu$, and further identifies $E_c = \frac{k_0^2}{2m}$. In this way, Eq. (25.22) becomes

$$\frac{1}{k_F a_F} = \frac{8\pi E_F}{v_0 k_F^3} + \frac{2}{\pi}\sqrt{\frac{E_c}{E_F}}, \tag{25.23}$$

where $E_F = \frac{k_F^2}{2m}$ is the Fermi energy associated with the Fermi wave vector $k_F = (3\pi^2 n)^{1/3}$ of a noninteracting Fermi gas with (average) density n. For chosen E_c, the value of v_0 is readily extracted from this expression once the values of a_F and k_F are known.

Time evolution in the absence of a time-dependent perturbation

The effective grand-canonical Hamiltonian (25.10) has the form of the generic grand-canonical Hamiltonian (9.3) of Part I (apart from the irrelevant constant C in Eq. (25.10)). Accordingly, the time development of the operators $\gamma_{\nu\alpha}$ and $\gamma^{\dagger}_{\nu\alpha}$ is similar to that of Eq. (9.12) of Part I, namely,

$$e^{iK_{\text{eff}}t}\,\gamma_{\nu\alpha}\,e^{-iK_{\text{eff}}t} = e^{-i\varepsilon_\nu t}\,\gamma_{\nu\alpha}, \tag{25.24}$$

$$e^{iK_{\text{eff}}t}\,\gamma^{\dagger}_{\nu\alpha}\,e^{-iK_{\text{eff}}t} = e^{i\varepsilon_\nu t}\,\gamma^{\dagger}_{\nu\alpha}\,. \tag{25.25}$$

As a consequence, the time development of the field operators (25.9) becomes

$$\begin{aligned}\psi_\uparrow(\mathbf{r},t) &= e^{iK_{\text{eff}}t}\psi_\uparrow(\mathbf{r})e^{-iK_{\text{eff}}t} = \sum_\nu \left(u_\nu(\mathbf{r})e^{iK_{\text{eff}}t}\gamma_{\nu 1}e^{-iK_{\text{eff}}t} - v_\nu(\mathbf{r})^* e^{iK_{\text{eff}}t}\gamma^{\dagger}_{\nu 2}e^{-iK_{\text{eff}}t}\right)\\ &= \sum_\nu \left(u_\nu(\mathbf{r})e^{-i\varepsilon_\nu t}\gamma_{\nu 1} - v_\nu(\mathbf{r})^* e^{i\varepsilon_\nu t}\gamma^{\dagger}_{\nu 2}\right) = \sum_\nu \left(u_\nu(\mathbf{r},t)\gamma_{\nu 1} - v_\nu(\mathbf{r},t)^*\gamma^{\dagger}_{\nu 2}\right),\end{aligned} \tag{25.26}$$

$$\begin{aligned}\psi^{\dagger}_\downarrow(\mathbf{r},t) &= e^{iK_{\text{eff}}t}\psi^{\dagger}_\downarrow(\mathbf{r})e^{-iK_{\text{eff}}t} = \sum_\nu \left(v_\nu(\mathbf{r})e^{iK_{\text{eff}}t}\gamma_{\nu 1}e^{-iK_{\text{eff}}t} + u_\nu(\mathbf{r})^* e^{iK_{\text{eff}}t}\gamma^{\dagger}_{\nu 2}e^{-iK_{\text{eff}}t}\right)\\ &= \sum_\nu \left(v_\nu(\mathbf{r})e^{-i\varepsilon_\nu t}\gamma_{\nu 1} + u_\nu(\mathbf{r})^* e^{i\varepsilon_\nu t}\gamma^{\dagger}_{\nu 2}\right) = \sum_\nu \left(v_\nu(\mathbf{r},t)\gamma_{\nu 1} + u_\nu(\mathbf{r},t)^*\gamma^{\dagger}_{\nu 2}\right),\end{aligned} \tag{25.27}$$

where we have set

$$u_\nu(\mathbf{r},t) = e^{-i\varepsilon_\nu t}\,u_\nu(\mathbf{r}) \quad \text{and} \quad v_\nu(\mathbf{r},t) = e^{-i\varepsilon_\nu t}\,v_\nu(\mathbf{r})\,. \tag{25.28}$$

Consistently, at this level, no time dependence can be introduced in the expression (25.13) for the gap parameter. Nevertheless, the results (25.26) and (25.27) will serve as a guide for the TDBdG treatment in the presence of a time-dependent perturbation.

25.2 Time-Dependent Bogoliubov–deGennes Equations

When a time-dependent perturbation is switched on for $t \geq t_0$, either by activating a time-dependent single-particle potential $V_{\text{ext}}(\mathbf{r},t)$ or by letting the strength v_0 of the interparticle contact potential to depend on time (as it is the case for ultracold Fermi gases, where interaction quenches are routinely applied [24]), the real-time evolution operator $U(t, t_0)$ introduced in Chapter 2 of Part I has to be considered in the place of $e^{-iK_{\text{eff}}(t-t_0)}$ utilized earlier. [Recall that we now consistently subtract the chemical potential μ of the initial equilibrium state for $t \le t_0$ from the single-particle Hamiltonian $h(\mathbf{r}, t)$.] This will result in a nontrivial time evolution $\Delta(\mathbf{r}, t)$ of the gap parameter out of its expression (25.13), holding for $t \le t_0$. Within the mean-field decoupling (21.2), this evolution can be determined in terms of the TDBdG equations, as discussed later in this chapter.

Assumptions underlying the TDBdG equations

The *key assumption* to derive the TDBdG equations is that, in the presence of a time-dependent perturbation, the time dependence of the field operator is given by an expression similar to (25.26) and (25.27), namely,

$$\begin{pmatrix} \psi_\uparrow(\mathbf{r}, t) \\ \psi_\downarrow^\dagger(\mathbf{r}, t) \end{pmatrix} = \sum_\nu \begin{pmatrix} u_\nu(\mathbf{r}, t) & -v_\nu(\mathbf{r}, t)^* \\ v_\nu(\mathbf{r}, t) & u_\nu(\mathbf{r}, t)^* \end{pmatrix} \begin{pmatrix} \gamma_{\nu 1} \\ \gamma_{\nu 2}^\dagger \end{pmatrix}, \tag{25.29}$$

where, however, the functions $\{u_\nu(\mathbf{r}, t), v_\nu(\mathbf{r}, t); \nu = 1, 2, \cdots\}$ are *not* given by Eq. (25.28) but have to be determined from the equation of motion of the single-particle Green's function within a mean-field decoupling. A further assumption is that the operators $\{\gamma_{\nu 1}, \gamma_{\nu 2}; \nu = 1, 2, \cdots\}$ are the same as those entering the expression (25.9) for the time-independent case, such that when averaging over the initial equilibrium configuration for $t \le t_0$ expressions like (25.11) and (25.12) still hold. Note that Eq. (25.29) generalizes to the superfluid phase Eq. (19.10) of Part I that holds in the normal phase.

As a first example, we consider the time-dependent gap parameter given by Eq. (21.6), for which we obtain

$$\begin{aligned}
\Delta(\mathbf{r}, t) &= v_0 \, \langle \psi_\downarrow(\mathbf{r}, t) \, \psi_\uparrow(\mathbf{r}, t) \rangle \\
&\underset{[\text{Eq.}(25.29)]}{\underset{\uparrow}{=}} v_0 \sum_{\nu\nu'} \left\langle \left(v_\nu(\mathbf{r}, t)^* \, \gamma_{\nu 1}^\dagger + u_\nu(\mathbf{r}, t) \, \gamma_{\nu 2} \right) \left(u_{\nu'}(\mathbf{r}, t) \, \gamma_{\nu' 1} - v_{\nu'}(\mathbf{r}, t)^* \, \gamma_{\nu' 2}^\dagger \right) \right\rangle \\
&\underset{[\text{Eqs.}(25.11)\text{-}(25.12)]}{\underset{\uparrow}{=}} - v_0 \sum_\nu u_\nu(\mathbf{r}, t) \, v_\nu(\mathbf{r}, t)^* \, (1 - 2 f_F(\varepsilon_\nu)) \, ,
\end{aligned} \tag{25.30}$$

in analogy with the result (25.13). Note how in this quantity the imprint of the initial equilibrium configuration for $t \le t_0$ appears through the Fermi–Dirac distribution function $f_F(\varepsilon_\nu)$, which depends on the initial equilibrium temperature and chemical potential, in addition to the initial condition $\{u_\nu(\mathbf{r}, t_0), v_\nu(\mathbf{r}, t_0); \nu = 1, 2, \cdots\}$ on the wave functions.

Lesser, greater, and retarded BdG Green's functions

The real-time *lesser* (cf. Eq. (23.16)), *greater* (cf. Eq. (23.17)), and *retarded* (cf. Eq. (13.15) of Part I) Green's functions can all be expressed in terms of the time-dependent field operators in the form (25.29), with initial conditions specified by the thermal averages (25.11) and (25.12), as well as by $\{u_\nu(\mathbf{r}, t_0), v_\nu(\mathbf{r}, t_0);\ \nu = 1, 2, \cdots\}$. Here, we examine the ensuing expressions in detail.

A – Lesser BdG Green's functions in Nambu space

By definition (cf. Eq. (23.16)),

$$G^<(\mathbf{r}t, \mathbf{r}'t')_{ij} = i\left\langle \Psi_j^\dagger(\mathbf{r}', t')\, \Psi_i(\mathbf{r}, t)\right\rangle\ . \qquad (25.31)$$

We then obtain for the various components:

$$\begin{aligned} G^<(\mathbf{r}t, \mathbf{r}'t')_{11} &= i\left\langle \Psi_1^\dagger(\mathbf{r}', t')\, \Psi_1(\mathbf{r}, t)\right\rangle = i\left\langle \psi_\uparrow^\dagger(\mathbf{r}', t')\, \psi_\uparrow(\mathbf{r}, t)\right\rangle \\ &= i\sum_\nu \left[u_\nu(\mathbf{r}', t')^*\, u_\nu(\mathbf{r}, t)\, f_F(\varepsilon_\nu) + v_\nu(\mathbf{r}', t')\, v_\nu(\mathbf{r}, t)^*\, (1 - f_F(\varepsilon_\nu))\right], \end{aligned} \qquad (25.32)$$

$$\begin{aligned} G^<(\mathbf{r}t, \mathbf{r}'t')_{12} &= i\left\langle \Psi_2^\dagger(\mathbf{r}', t')\, \Psi_1(\mathbf{r}, t)\right\rangle = i\left\langle \psi_\downarrow(\mathbf{r}', t')\, \psi_\uparrow(\mathbf{r}, t)\right\rangle \\ &= i\sum_\nu \left[v_\nu(\mathbf{r}', t')^*\, u_\nu(\mathbf{r}, t)\, f_F(\varepsilon_\nu) - u_\nu(\mathbf{r}', t')\, v_\nu(\mathbf{r}, t)^*\, (1 - f_F(\varepsilon_\nu))\right], \end{aligned} \qquad (25.33)$$

$$\begin{aligned} G^<(\mathbf{r}t, \mathbf{r}'t')_{21} &= i\left\langle \Psi_1^\dagger(\mathbf{r}', t')\, \Psi_2(\mathbf{r}, t)\right\rangle = i\left\langle \psi_\uparrow^\dagger(\mathbf{r}', t')\, \psi_\downarrow^\dagger(\mathbf{r}, t)\right\rangle \\ &= i\sum_\nu \left[u_\nu(\mathbf{r}', t')^*\, v_\nu(\mathbf{r}, t)\, f_F(\varepsilon_\nu) - v_\nu(\mathbf{r}', t')\, u_\nu(\mathbf{r}, t)^*\, (1 - f_F(\varepsilon_\nu))\right], \end{aligned} \qquad (25.34)$$

$$\begin{aligned} G^<(\mathbf{r}t, \mathbf{r}'t')_{22} &= i\left\langle \Psi_2^\dagger(\mathbf{r}', t')\, \Psi_2(\mathbf{r}, t)\right\rangle = i\left\langle \psi_\downarrow(\mathbf{r}', t')\, \psi_\downarrow^\dagger(\mathbf{r}, t)\right\rangle \\ &= i\sum_\nu \left[v_\nu(\mathbf{r}', t')^*\, v_\nu(\mathbf{r}, t)\, f_F(\varepsilon_\nu) + u_\nu(\mathbf{r}', t')\, u_\nu(\mathbf{r}, t)^*\, (1 - f_F(\varepsilon_\nu))\right], \end{aligned} \qquad (25.35)$$

such that Eq. (25.30) is consistent with the form of Eq. (23.21), namely,

$$\Delta(\mathbf{r}, t) = -\, i\, v_0\, G^<(\mathbf{r}t, \mathbf{r}t^+)_{12}\ . \qquad (25.36)$$

B – Greater BdG Green's functions in Nambu space

By definition (cf. Eq. (23.17)),

$$G^>(\mathbf{r}t, \mathbf{r}'t')_{ij} = -i \left\langle \Psi_i(\mathbf{r}, t)\, \Psi_j^\dagger(\mathbf{r}', t') \right\rangle . \tag{25.37}$$

We then obtain for the various components:

$$\begin{aligned} G^>(\mathbf{r}t, \mathbf{r}'t')_{11} &= -i \left\langle \Psi_1(\mathbf{r}, t)\, \Psi_1^\dagger(\mathbf{r}', t') \right\rangle = -i \left\langle \psi_\uparrow(\mathbf{r}, t)\, \psi_\uparrow^\dagger(\mathbf{r}', t') \right\rangle \\ &= -i \sum_\nu \, [u_\nu(\mathbf{r}', t')^*\, u_\nu(\mathbf{r}, t)\, (1 - f_F(\varepsilon_\nu)) \\ &\quad + v_\nu(\mathbf{r}', t')\, v_\nu(\mathbf{r}, t)^*\, f_F(\varepsilon_\nu)] , \end{aligned} \tag{25.38}$$

$$\begin{aligned} G^>(\mathbf{r}t, \mathbf{r}'t')_{12} &= -i \left\langle \Psi_1(\mathbf{r}, t)\, \Psi_2^\dagger(\mathbf{r}', t') \right\rangle = -i \left\langle \psi_\uparrow(\mathbf{r}, t)\, \psi_\downarrow(\mathbf{r}', t') \right\rangle \\ &= -i \sum_\nu \, [v_\nu(\mathbf{r}', t')^*\, u_\nu(\mathbf{r}, t)\, (1 - f_F(\varepsilon_\nu)) \\ &\quad - u_\nu(\mathbf{r}', t')\, v_\nu(\mathbf{r}, t)^*\, f_F(\varepsilon_\nu)] , \end{aligned} \tag{25.39}$$

$$\begin{aligned} G^>(\mathbf{r}t, \mathbf{r}'t')_{21} &= -i \left\langle \Psi_2(\mathbf{r}, t)\, \Psi_1^\dagger(\mathbf{r}', t') \right\rangle = i \left\langle \psi_\downarrow^\dagger(\mathbf{r}, t)\, \psi_\uparrow^\dagger(\mathbf{r}', t') \right\rangle \\ &= -i \sum_\nu \, [u_\nu(\mathbf{r}', t')^*\, v_\nu(\mathbf{r}, t)\, (1 - f_F(\varepsilon_\nu)) \\ &\quad - v_\nu(\mathbf{r}', t')\, u_\nu(\mathbf{r}, t)^*\, f_F(\varepsilon_\nu)] , \end{aligned} \tag{25.40}$$

$$\begin{aligned} G^>(\mathbf{r}t, \mathbf{r}'t')_{22} &= -i \left\langle \Psi_2(\mathbf{r}, t)\, \Psi_2^\dagger(\mathbf{r}', t') \right\rangle = -i \left\langle \psi_\downarrow^\dagger(\mathbf{r}, t) \psi_\downarrow(\mathbf{r}', t') \right\rangle \\ &= -i \sum_\nu \, [v_\nu(\mathbf{r}', t')^*\, v_\nu(\mathbf{r}, t)\, (1 - f_F(\varepsilon_\nu)) \\ &\quad + u_\nu(\mathbf{r}', t')\, u_\nu(\mathbf{r}, t)^*\, f_F(\varepsilon_\nu)] . \end{aligned} \tag{25.41}$$

C – Retarded BdG Green's functions in Nambu space

From Eq. (13.15) of Part I (now adapted to the Nambu notation), we write

$$G^{\mathrm{R}}(\mathbf{r}t, \mathbf{r}'t')_{ij} = \theta(t - t') \, \left(G^>(\mathbf{r}t, \mathbf{r}'t')_{ij} - G^<(\mathbf{r}t, \mathbf{r}'t')_{ij} \right) . \tag{25.42}$$

Making use of the aforementioned results for $G^>$ and $G^<$, we obtain for the various components:

$$G^{\mathrm{R}}(\mathbf{r}t, \mathbf{r}'t')_{11} = -i\, \theta(t - t') \, \sum_\nu \, [u_\nu(\mathbf{r}', t')^*\, u_\nu(\mathbf{r}, t) + v_\nu(\mathbf{r}', t')\, v_\nu(\mathbf{r}, t)^*] , \tag{25.43}$$

$$G^{\mathrm{R}}(\mathbf{r}t, \mathbf{r}'t')_{12} = -i\, \theta(t - t') \, \sum_\nu \, [v_\nu(\mathbf{r}', t')^*\, u_\nu(\mathbf{r}, t) - u_\nu(\mathbf{r}', t')\, v_\nu(\mathbf{r}, t)^*] , \tag{25.44}$$

$$G^{\mathrm{R}}(\mathbf{r}t,\mathbf{r}'t')_{21} = -i\,\theta(t-t')\,\sum_{\nu}\left[u_{\nu}(\mathbf{r}',t')^{*}\,v_{\nu}(\mathbf{r},t) - v_{\nu}(\mathbf{r}',t')\,u_{\nu}(\mathbf{r},t)^{*}\right]\,, \tag{25.45}$$

$$G^{\mathrm{R}}(\mathbf{r}t,\mathbf{r}'t')_{22} = -i\,\theta(t-t')\,\sum_{\nu}\left[v_{\nu}(\mathbf{r}',t')^{*}\,v_{\nu}(\mathbf{r},t) + u_{\nu}(\mathbf{r}',t')\,u_{\nu}(\mathbf{r},t)^{*}\right]\,. \tag{25.46}$$

Remark: Number densities for spin ↑ and ↓

From the above expressions, one can calculate, for instance, the number densities for both spin ↑

$$\begin{aligned} n_{\uparrow}(\mathbf{r},t) &= \left\langle \psi_{\uparrow}^{\dagger}(\mathbf{r},t)\,\psi_{\uparrow}(\mathbf{r},t)\right\rangle = -i\,G^{<}(\mathbf{r}t,\mathbf{r})_{11} \\ &= \sum_{\nu}\left[|u_{\nu}(\mathbf{r},t)|^{2} f_{F}(\varepsilon_{\nu}) + |v_{\nu}(\mathbf{r},t)|^{2}\,(1-f_{F}(\varepsilon_{\nu}))\right] \end{aligned} \tag{25.47}$$

and spin ↓

$$\begin{aligned} n_{\downarrow}(\mathbf{r},t) &= \left\langle \psi_{\downarrow}^{\dagger}(\mathbf{r},t)\,\psi_{\downarrow}(\mathbf{r},t)\right\rangle = i\,G^{>}(\mathbf{r}t,\mathbf{r})_{22} \\ &= \sum_{\nu}\left[|u_{\nu}(\mathbf{r},t)|^{2} f_{F}(\varepsilon_{\nu}) + |v_{\nu}(\mathbf{r},t)|^{2}\,(1-f_{F}(\varepsilon_{\nu}))\right]\,. \end{aligned} \tag{25.48}$$

Note that $n_{\uparrow}(\mathbf{r},t) = n_{\downarrow}(\mathbf{r},t)$ at any t, consistently with the assumption (22.20).

Derivation of the TDBdG equations from the Kadanoff–Baym equations

To derive the desired TDBdG equations from the Kadanoff–Baym equations in the superfluid phase at the level of a mean-field decoupling, it is sufficient to consider Eq. (23.23), which involves $G^{>}(\mathbf{r}t,\mathbf{r}'t')_{11}$ and $G^{>}(\mathbf{r}t,\mathbf{r}'t')_{21}$, as well as Eq. (23.27), which involves $G^{<}(\mathbf{r}t,\mathbf{r}'t')_{11}$ and $G^{<}(\mathbf{r}t,\mathbf{r}'t')_{21}$, where now the relevant Green's functions are taken of the form (25.38)–(25.40) and (25.32)–(25.34), respectively. We then obtain from Eq. (23.23)

$$\begin{aligned} &\left(i\frac{\partial}{\partial t} - h(\mathbf{r},t) + \mu\right)(-i)\sum_{\nu}\left[u_{\nu}(\mathbf{r}',t')^{*}u_{\nu}(\mathbf{r},t)(1-f_{F}(\varepsilon_{\nu})) + v_{\nu}(\mathbf{r}',t')v_{\nu}(\mathbf{r},t)^{*}f_{F}(\varepsilon_{\nu})\right] \\ &\quad - \Delta(\mathbf{r},t)(-i)\sum_{\nu}\left[u_{\nu}(\mathbf{r}',t')^{*}\,v_{\nu}(\mathbf{r},t)\,(1-f_{F}(\varepsilon_{\nu})) - v_{\nu}(\mathbf{r}',t')\,u_{\nu}(\mathbf{r},t)^{*}\,f_{F}(\varepsilon_{\nu})\right] \\ &= -i\sum_{\nu}\left\{\left[\left(i\frac{\partial}{\partial t} - h(\mathbf{r},t) + \mu\right)u_{\nu}(\mathbf{r},t) - \Delta(\mathbf{r},t)v_{\nu}(\mathbf{r},t)\right]u_{\nu}(\mathbf{r}',t')^{*}\,(1-f_{F}(\varepsilon_{\nu}))\right. \\ &\quad \left. + \left[\left(i\frac{\partial}{\partial t} - h(\mathbf{r},t) + \mu\right)v_{\nu}(\mathbf{r},t)^{*} + \Delta(\mathbf{r},t)\,u_{\nu}(\mathbf{r},t)^{*}\right]v_{\nu}(\mathbf{r}',t')\,f_{F}(\varepsilon_{\nu})\right\} = 0\,, \end{aligned} \tag{25.49}$$

as well as from Eq. (23.27)

$$\left(i\frac{\partial}{\partial t} - h(\mathbf{r},t) + \mu\right) i \sum_\nu \left[u_\nu(\mathbf{r}',t')^* u_\nu(\mathbf{r},t) f_F(\varepsilon_\nu) + v_\nu(\mathbf{r}',t') v_\nu(\mathbf{r},t)^* (1 - f_F(\varepsilon_\nu))\right]$$
$$- \Delta(\mathbf{r},t)\, i \sum_\nu \left[u_\nu(\mathbf{r}',t')^*\, v_\nu(\mathbf{r},t) f_F(\varepsilon_\nu) - v_\nu(\mathbf{r}',t')\, u_\nu(\mathbf{r},t)^*\, (1 - f_F(\varepsilon_\nu))\right]$$
$$= i \sum_\nu \left\{ \left[\left(i\frac{\partial}{\partial t} - h(\mathbf{r},t) + \mu\right) u_\nu(\mathbf{r},t) - \Delta(\mathbf{r},t)\, v_\nu(\mathbf{r},t) \right] u_\nu(\mathbf{r}',t')^* f_F(\varepsilon_\nu) \right.$$
$$\left. + \left[\left(i\frac{\partial}{\partial t} - h(\mathbf{r},t) + \mu\right) v_\nu(\mathbf{r},t)^* + \Delta(\mathbf{r},t)\, u_\nu(\mathbf{r},t)^* \right] v_\nu(\mathbf{r}',t')\, (1 - f_F(\varepsilon_\nu)) \right\} = 0, \tag{25.50}$$

where μ has consistently been subtracted from $h(\mathbf{r},t)$. Next, we can get rid of the Fermi–Dirac distribution functions in Eqs. (25.49) and (25.50) by taking the difference between these two equations, obtaining in this way

$$\sum_\nu \left\{ \left[\left(i\frac{\partial}{\partial t} - h(\mathbf{r},t) + \mu\right) u_\nu(\mathbf{r},t) - \Delta(\mathbf{r},t)\, v_\nu(\mathbf{r},t) \right] u_\nu(\mathbf{r}',t')^* \right.$$
$$\left. + \left[\left(i\frac{\partial}{\partial t} - h(\mathbf{r},t) + \mu\right) v_\nu(\mathbf{r},t)^* + \Delta(\mathbf{r},t)\, u_\nu(\mathbf{r},t)^* \right] v_\nu(\mathbf{r}',t') \right\} = 0\,. \tag{25.51}$$

At this point, we anticipate the validity of the completeness condition (which will consistently be proved in Eq. (25.55)) for the set of functions $\{u_\nu(\mathbf{r}',t'), v_\nu(\mathbf{r}',t'); \nu = 1, 2, \cdots\}$ at any time t', in terms of which we can *separately* set equal to zero each expression within brackets in Eq. (25.51) for given ν.[1] This yields eventually the desired *time-dependent Bogoliubov–deGennes equations*:

$$\begin{cases} \left(i\frac{\partial}{\partial t} - h(\mathbf{r},t) + \mu\right) u_\nu(\mathbf{r},t) - \Delta(\mathbf{r},t)\, v_\nu(\mathbf{r},t) = 0, \\ \left(i\frac{\partial}{\partial t} - h(\mathbf{r},t) + \mu\right) v_\nu(\mathbf{r},t)^* + \Delta(\mathbf{r},t)\, u_\nu(\mathbf{r},t)^* = 0\,. \end{cases} \tag{25.52}$$

Orthonormality of the solutions of the TDBdG equations

The initial condition, in terms of which the TDBdG equations (25.52) have to be solved, is that, at the reference time t_0 at which the time-dependent perturbation begins to act on the system, the initial configuration of the system is specified in

[1] Strictly speaking, the completeness condition (25.55) can be invoked by considering also the Kadanoff–Baym equations, which involve $G^>(\mathbf{r}t,\mathbf{r}'t')_{22}$ and $G^>(\mathbf{r}t,\mathbf{r}'t')_{12}$, as well as $G^<(\mathbf{r}t,\mathbf{r}'t')_{22}$ and $G^<(\mathbf{r}t,\mathbf{r}'t')_{12}$.

terms of the solutions $\{u_\nu(\mathbf{r}), v_\nu(\mathbf{r}); \nu = 1, 2, \cdots\}$ of the time-independent BdG equations (25.5) with gap parameter (25.6). For $t \le t_0$, these solutions obey the orthonormality condition (25.7). We are now going to prove that this condition is preserved for later times $t \ge t_0$, that is,

$$\int d\mathbf{r}\, (u_\nu(\mathbf{r}, t)^* u_{\nu'}(\mathbf{r}, t) + v_\nu(\mathbf{r}, t)^* v_{\nu'}(\mathbf{r}, t)) = \delta_{\nu\nu'}\,, \tag{25.53}$$

provided that $\{u_\nu(\mathbf{r}, t), v_\nu(\mathbf{r}, t); \nu = 1, 2, \cdots\}$ are solutions to the TDBdG equations (25.52).

Proof of Eq. (25.53) Consider the time derivative of Eq. (25.53):

$$\begin{aligned}
&\frac{d}{dt}\int d\mathbf{r}\, (u_\nu(\mathbf{r}, t)^* u_{\nu'}(\mathbf{r}, t) + v_\nu(\mathbf{r}, t)^* v_{\nu'}(\mathbf{r}, t)) \\
&= \int d\mathbf{r} \left(\frac{\partial u_\nu(\mathbf{r}, t)^*}{\partial t} u_{\nu'}(\mathbf{r}, t) + u_\nu(\mathbf{r}, t)^* \frac{\partial u_{\nu'}(\mathbf{r}, t)}{\partial t} \right. \\
&\quad \left. + \frac{\partial v_\nu(\mathbf{r}, t)^*}{\partial t} v_{\nu'}(\mathbf{r}, t) + v_\nu(\mathbf{r}, t)^* \frac{\partial v_{\nu'}(\mathbf{r}, t)}{\partial t} \right) \\
&\underset{\substack{\uparrow \\ \text{[Eqs.(25.52)]}}}{=} -i \int d\mathbf{r}\, \{ -[(h(\mathbf{r}, t) - \mu)\, u_\nu(\mathbf{r}, t)^* + \Delta(\mathbf{r}, t)^*\, v_\nu(\mathbf{r}, t)^*]\, u_{\nu'}(\mathbf{r}, t) \\
&\quad + u_\nu(\mathbf{r}, t)^* \, [(h(\mathbf{r}, t) - \mu)\, u_{\nu'}(\mathbf{r}, t) + \Delta(\mathbf{r}, t)\, v_{\nu'}(\mathbf{r}, t)] \\
&\quad + [(h(\mathbf{r}, t) - \mu)\, v_\nu(\mathbf{r}, t)^* - \Delta(\mathbf{r}, t)\, u_\nu(\mathbf{r}, t)^*]\, v_{\nu'}(\mathbf{r}, t) \\
&\quad - v_\nu(\mathbf{r}, t)^* \, [(h(\mathbf{r}, t) - \mu)\, v_{\nu'}(\mathbf{r}, t) - \Delta(\mathbf{r}, t)^*\, u_{\nu'}(\mathbf{r}, t)]\} \\
&= \frac{i}{2m} \int d\mathbf{r}\, \{ u_\nu(\mathbf{r}, t)^* \nabla^2 u_{\nu'}(\mathbf{r}, t) - u_{\nu'}(\mathbf{r}, t) \nabla^2 u_\nu(\mathbf{r}, t)^* \\
&\quad - v_\nu(\mathbf{r}, t)^* \nabla^2 v_{\nu'}(\mathbf{r}, t) + v_{\nu'}(\mathbf{r}, t) \nabla^2 v_\nu(\mathbf{r}, t)^* \} \underset{\substack{\uparrow \\ \text{[by parts]}}}{=} 0\,.
\end{aligned} \tag{25.54}$$

From the result (25.54), we conclude that:

(i) For $\nu = \nu'$, the right-hand side of Eq. (25.53) is unity, which is preserved for all times $t \ge t_0$;
(ii) For $\nu \ne \nu'$, the right-hand side of Eq. (25.53) is zero, which is also preserved for all times $t \ge t_0$.

[QED]

Completeness of the solutions of the TDBdG equations

We have already mentioned that the solutions $\{u_\nu(\mathbf{r}), v_\nu(\mathbf{r}); \nu = 1, 2, \cdots\}$ to the time-independent BdG equations (25.5) satisfy the completeness condition (25.8). We are now going to prove that this condition holds for all times $t \geq t_0$, such that

$$\sum_\nu \left[\begin{pmatrix} u_\nu(\mathbf{r}, t) \\ v_\nu(\mathbf{r}, t) \end{pmatrix} (u_\nu(\mathbf{r}', t)^*, v_\nu(\mathbf{r}', t)^*) + \begin{pmatrix} -v_\nu(\mathbf{r}, t)^* \\ u_\nu(\mathbf{r}, t)^* \end{pmatrix} (-v_\nu(\mathbf{r}', t), u_\nu(\mathbf{r}', t)) \right]$$
$$= \delta(\mathbf{r} - \mathbf{r}') \begin{pmatrix} 1 & 0 \\ 0 & 1 \end{pmatrix}, \tag{25.55}$$

where $\{u_\nu(\mathbf{r}, t), v_\nu(\mathbf{r}, t); \nu = 1, 2, \cdots\}$ are solutions to the TDBdG equations (25.52).

Proof of Eq. (25.55) This condition can be proved by expressing the general relation

$$i\left(G^>(\mathbf{r}t, \mathbf{r}'t)_{ij} - G^<(\mathbf{r}t, \mathbf{r}'t)_{ij}\right) = \left\langle \Psi_i(\mathbf{r}, t)\, \Psi_j^\dagger(\mathbf{r}', t) \right\rangle + \left\langle \Psi_j^\dagger(\mathbf{r}', t)\, \Psi_i(\mathbf{r}, t) \right\rangle$$
$$= \left\langle U(t_0, t) \left(\Psi_i(\mathbf{r})\, \Psi_j^\dagger(\mathbf{r}') + \Psi_j^\dagger(\mathbf{r}')\, \Psi_i(\mathbf{r}) \right) U(t, t_0) \right\rangle = \delta_{ij}\, \delta(\mathbf{r} - \mathbf{r}'), \tag{25.56}$$

which follows from the anti-commutation relations (22.2) in the Nambu representation, in terms of the BdG representation (25.32)–(25.35) for the lesser ($G^<$) and (25.38)–(25.41) for the greater ($G^>$) Green's functions. We then obtain separately for each element of the matrix (25.56):

Element (1,1)

$$\delta(\mathbf{r} - \mathbf{r}') = i\left(G^>(\mathbf{r}t, \mathbf{r}'t)_{11} - G^<(\mathbf{r}t, \mathbf{r}'t)_{11}\right)$$
$$= \sum_\nu \left(u_\nu(\mathbf{r}, t)\, u_\nu(\mathbf{r}', t)^* + v_\nu(\mathbf{r}, t)^* v_\nu(\mathbf{r}', t)\right), \tag{25.57}$$

where Eqs. (25.32) and (25.38) have been used;

Element (1,2)

$$0 = i\left(G^>(\mathbf{r}t, \mathbf{r}'t)_{12} - G^<(\mathbf{r}t, \mathbf{r}'t)_{12}\right) = \sum_\nu \left(u_\nu(\mathbf{r}, t)\, v_\nu(\mathbf{r}', t)^* - v_\nu(\mathbf{r}, t)^* u_\nu(\mathbf{r}', t)\right), \tag{25.58}$$

where Eqs. (25.33) and (25.39) have been used;

Element (2,1)

$$0 = i\left(G^{>}(\mathbf{r}t, \mathbf{r}'t)_{21} - G^{<}(\mathbf{r}t, \mathbf{r}'t)_{21}\right) = \sum_{\nu} \left(v_\nu(\mathbf{r}, t)\, u_\nu(\mathbf{r}', t)^* - u_\nu(\mathbf{r}, t)^* v_\nu(\mathbf{r}', t)\right), \tag{25.59}$$

where Eqs. (25.34) and (25.40) have been used;

Element (2,2)

$$\begin{aligned}\delta(\mathbf{r} - \mathbf{r}') &= i\left(G^{>}(\mathbf{r}t, \mathbf{r}'t)_{22} - G^{<}(\mathbf{r}t, \mathbf{r}'t)_{22}\right) \\ &= \sum_{\nu} \left(v_\nu(\mathbf{r}, t)\, v_\nu(\mathbf{r}', t)^* + u_\nu(\mathbf{r}, t)^* u_\nu(\mathbf{r}', t)\right),\end{aligned} \tag{25.60}$$

where Eqs. (25.35) and (25.41) have been used.

The results (25.57)–(25.60) are seen to coincide with the respective matrix elements of the completeness condition (25.55). [QED]

Remark: A drawback with the solution of the BdG equations

Numerical approaches to the BdG equations (in either their time-independent (25.5) or time-dependent (25.52) versions) require one not only to calculate but also to store a considerable amounts of fine spatial details for a large number of single-particle wave functions (either $\{u_\nu(\mathbf{r}), v_\nu(\mathbf{r})\}$ or $\{u_\nu(\mathbf{r}, t), v_\nu(\mathbf{r}, t)\}$ for given t). These fine details have, however, to be averaged over in order to obtain the physical quantities of interest like the single-particle Green's functions (cf. the expressions from (25.31) to (25.46)), from which the gap parameter (cf. Eqs. (25.30) or (25.36)) and the number density (cf. Eqs. (25.47) and (25.48)), as well as the current density (cf. Eq. (18.4) of Part I) and the average energy (cf. Eq. (31.55)), can eventually be obtained. This double effort represents an inherent inconvenience for the use of the BdG equations, which puts some limits on their practical utility, especially when dealing with problems with nontrivial spatial geometries.

25.3 Connection with Analytic Continuation

We conclude this chapter by showing how the expressions (25.43)–(25.46) for the real-time retarded Green's functions $G^{\mathrm{R}}(\mathbf{r}t, \mathbf{r}'t')_{ij}$ can be *formally* obtained from the Matsubara Green's functions (25.21) by a suitable analytic continuation from imaginary to real time.

To this end, we first perform the analytic continuation $i\omega_n \rightarrow \omega + i\eta$ (where ω is a *real* frequency and $\eta = 0^+$) in the expression (25.21), which thus becomes

$$\overleftrightarrow{\mathcal{G}}(\mathbf{r}, \mathbf{r}'; \omega) = \sum_\nu \left\{ \begin{pmatrix} u_\nu(\mathbf{r}) \\ v_\nu(\mathbf{r}) \end{pmatrix} \frac{1}{\omega + i\eta - \varepsilon_\nu} (u_\nu(\mathbf{r}')^*, v_\nu(\mathbf{r}')^*) \right.$$
$$\left. + \begin{pmatrix} -v_\nu(\mathbf{r})^* \\ u_\nu(\mathbf{r})^* \end{pmatrix} \frac{1}{\omega + i\eta + \varepsilon_\nu} (-v_\nu(\mathbf{r}'), u_\nu(\mathbf{r}')) \right\} . \tag{25.61}$$

Next, we make use of the identity

$$\frac{1}{\omega \pm \varepsilon_\nu + i\eta} = -\, i \int_{-\infty}^{+\infty} dt\, \theta(t)\, e^{i(\omega \pm \varepsilon_\nu + i\eta)t} , \tag{25.62}$$

such that the expression (25.61) becomes

$$\overleftrightarrow{\mathcal{G}}(\mathbf{r}, \mathbf{r}'; \omega) = -\, i \int_{-\infty}^{+\infty} d(t-t')\, \theta(t-t')\, e^{i(\omega+i\eta)(t-t')}$$
$$\times \sum_\nu \left\{ \begin{pmatrix} u_\nu(\mathbf{r}, t) \\ v_\nu(\mathbf{r}, t) \end{pmatrix} (u_\nu(\mathbf{r}', t')^*, v_\nu(\mathbf{r}', t')^*) \right.$$
$$\left. + \begin{pmatrix} -v_\nu(\mathbf{r}, t)^* \\ u_\nu(\mathbf{r}, t)^* \end{pmatrix} (-v_\nu(\mathbf{r}', t'), u_\nu(\mathbf{r}', t')) \right\} \tag{25.63}$$

in terms of the functions (25.28) that hold in the time-independent case. Finally, from the properties of the Fourier transforms, we obtain for the retarded Green's function in real time:

$$\overleftrightarrow{\mathcal{G}}(\mathbf{r}, \mathbf{r}'; t-t') = -\, i\theta(t-t') \sum_\nu \left\{ \begin{pmatrix} u_\nu(\mathbf{r}, t) \\ v_\nu(\mathbf{r}, t) \end{pmatrix} (u_\nu(\mathbf{r}', t')^*, v_\nu(\mathbf{r}', t')^*) \right.$$
$$\left. + \begin{pmatrix} -v_\nu(\mathbf{r}, t)^* \\ u_\nu(\mathbf{r}, t)^* \end{pmatrix} (-v_\nu(\mathbf{r}', t'), u_\nu(\mathbf{r}', t')) \right\} . \tag{25.64}$$

Note that the expression (25.64) *formally* coincides with the expressions (25.43)–(25.46), although the expressions (25.43)–(25.46) take into account the effect of a time-dependent perturbation via the TDBdG equations (25.52), while the expression (25.64), from the way it was derived, applies only to the time-independent case.

In Chapter 27, we will argue that analytic continuation from imaginary to real times can be formulated on a firmer basis, by following the Kadanoff–Baym original arguments [5].

26

A Brief Excursus to the BCS–BEC Crossover

In this chapter, we provide a concise account of salient features of the BCS–BEC crossover, as a prolegomenon to the topics considered in Chapters 28 and 29.

Quite generally, in physics, the term "crossover" refers to a situation when a system evolves from one phase to another by changing a certain parameter, without encountering a phase transition in between. Specifically, for the BCS–BEC crossover, it is the tuning of the interparticle interaction (or the density) that makes the system evolve from a Bardeen–Cooper–Schrieffer (BCS) state, where Cooper pairs of (opposite spin) fermions are described by Fermi statistics, to a Bose–Einstein condensate (BEC) state, where dimers made up of two fermions are described by Bose statistics. In other words, by freezing out the internal degrees of freedom of the fermion pairs, the statistics of fermions smoothly evolves into the statistics of bosons.

The marked similarity between bosonic and fermionic superfluidity has been recognized for some time [91], although these phenomena were originally discovered experimentally in quite different physical systems (namely, ^{4}He and superconductors, respectively). In ultimate analysis, the similarity stems from the fact that these different physical systems share the basic feature of having the same kind of spontaneous broken symmetry associated with the phase of the order parameter.

26.1 Origin and Key Features of the BCS–BEC Crossover

The idea behind the BCS–BEC crossover dates back to just about the same time of the birth of the BCS theory in 1957 [70]. The authors of the BCS theory emphasized the differences between their approach for superconductors based on strongly overlapping Cooper pairs and the Schafroth–Butler–Blatt approach resting on nonoverlapping composite bosons, which undergo Bose–Einstein condensation at low temperature [92]. In this respect, it is interesting to reproduce here footnote 18 of Ref. [70], where it is stated that "Our picture differs from that of Schafroth, Butler, and Blatt, who suggest that pseudo-molecules of pairs of electrons of opposite spin are formed. They show if the size of the pseudo-molecules is less than the average distance between them, and if other conditions are fulfilled, the system has properties similar to that of a charged Bose-Einstein gas, including a Meissner effect and a critical temperature of condensation. Our pairs are not localized in this sense, and our transition is not analogous to a Bose-Einstein condensation."

It was probably for this reason that the interest in the two different (BCS and BEC) situations was kept disjoint for some time, until theoretical interest arose for unifying them as the two limits of *a single theory*, in such a way that the physical system passes through an intermediate situation where the pair size becomes comparable with the average interparticle spacing. (This intermediate situation is nowadays referred to as the "unitary" regime.) In this respect, pioneering work was originally motivated by the exciton condensation in semiconductors [93], as well as by mere intellectual curiosity [94]. Later on, a consistent theory of the BCS–BEC crossover took form through the work by Eagles owing to possible applications to superconducting semiconductors [95]. The formal aspects of the theory were eventually developed at zero temperature by Leggett [96] and above the critical temperature by Nozières and Schmitt-Rink (NSR) [97].

On the experimental side, following up on the realization of Bose–Einstein condensation with ultracold Bose gases (for reviews, see Refs. [98, 99]), the BCS–BEC crossover was soon after realized in all of its aspects with ultracold Fermi gases [20, 100, 101]. In addition, the BCS–BEC crossover was found to be consistent with various aspects of the phenomenology in nuclear systems [20] and is further expected to manifest itself (at least to some extent) in condensed matter [102, 103].

In particular, for ultracold Fermi gases, the parameter that tunes the crossover from the BCS to the BEC limits (or vice versa) is represented by the interparticle coupling $(k_F a_F)^{-1}$, where $k_F = \left(3\pi^2 n\right)^{1/3}$ is the Fermi wave vector for particle density n, and a_F is the scattering length of the two-fermion problem in vacuum. The value of a_F can be varied experimentally by spanning a molecular Fano–Feshbach resonance [104–106]. The coupling parameter $(k_F a_F)^{-1}$ then ranges from $(k_F\, a_F)^{-1} \lesssim -1$ in the BCS regime when $a_F < 0$ to $(k_F\, a_F)^{-1} \gtrsim +1$ in the BEC regime when $a_F > 0$, across unitarity when $|a_F|$ diverges [20]. The flexibility in operating on these systems on the experimental side and the full control of the relevant system degrees of freedom on the theoretical side make the comparison between the experimental data and the corresponding theoretical calculations quite stringent at a fundamental level.

26.2 The BCS Wave Function Gets It (Almost) All

Notwithstanding the circumspection displayed in the original BCS article [70], about the possible connection of the BCS approach with the Bose–Einstein condensation, it is interesting to remark that the BCS wave function for the ground state contains the BEC state of composite bosons as a limiting situation. This can be readily seen as follows:

The BCS ground-state wave function has the form [13, 70]

$$|\Phi\rangle_{\mathrm{BCS}} = \prod_{\mathbf{k}} \left(u_{\mathbf{k}} + v_{\mathbf{k}} c^{\dagger}_{\mathbf{k}\uparrow} c^{\dagger}_{-\mathbf{k}\downarrow} \right) |0\rangle, \tag{26.1}$$

where $|0\rangle$ is the vacuum state, $c^{\dagger}_{\mathbf{k}\sigma}$ is a fermionic creation operator with wave vector $\mathbf{k}$ and spin projection $\sigma = (\uparrow, \downarrow)$, and $u_{\mathbf{k}}$ and $v_{\mathbf{k}}$ are probability amplitudes given by

$$v^2_{\mathbf{k}} = 1 - u^2_{\mathbf{k}} = \frac{1}{2}\left(1 - \frac{\xi_{\mathbf{k}}}{E_{\mathbf{k}}}\right) . \tag{26.2}$$

Here, $\xi_{\mathbf{k}} = \frac{\mathbf{k}^2}{2m} - \mu$ (m being the fermion mass and μ the chemical potential) and $E_{\mathbf{k}} = \sqrt{\xi^2_{\mathbf{k}} + |\Delta_0|^2}$ (Δ_0 being the BCS gap (order) parameter, taken uniform and at zero temperature). Upon setting $g_{\mathbf{k}} = v_{\mathbf{k}}/u_{\mathbf{k}}$, the expression (26.1) can be cast in the alternative form

$$|\Phi\rangle_{\mathrm{BCS}} = \left(\prod_{\mathbf{k}'} u_{\mathbf{k}'} \right) \exp\left[\sum_{\mathbf{k}} g_{\mathbf{k}} c^{\dagger}_{\mathbf{k}\uparrow} c^{\dagger}_{-\mathbf{k}\downarrow} \right] |0\rangle \tag{26.3}$$

since $(c^{\dagger}_{\mathbf{k}\sigma})^2 = 0$ owing to the Pauli principle. The operator $b^{\dagger}_0 \equiv \sum_{\mathbf{k}} g_{\mathbf{k}} c^{\dagger}_{\mathbf{k}\uparrow} c^{\dagger}_{-\mathbf{k}\downarrow}$ corresponds to the creation of fermion pairs, but it is not a truly bosonic operator because the commutator $[b_0, b^{\dagger}_0] = \sum_{\mathbf{k}} |g_{\mathbf{k}}|^2 (1 - n_{\mathbf{k}\uparrow} - n_{-\mathbf{k}\downarrow})$ is not a c-number but explicitly contains the fermionic operators $n_{\mathbf{k}\sigma} = c^{\dagger}_{\mathbf{k}\sigma} c_{\mathbf{k}\sigma}$. However, under some circumstances, it may occur that $[b_0, b^{\dagger}_0] \cong 1$ for all practical purposes, provided ${}_{\mathrm{BCS}}\langle\Phi|n_{\mathbf{k}\sigma}|\Phi\rangle_{\mathrm{BCS}} = v^2_{\mathbf{k}} \ll 1$ for all $\mathbf{k}$ of physical relevance. As a consequence, $|\Phi\rangle_{\mathrm{BCS}} \to \exp\{b^{\dagger}_0\}|0\rangle$ represents a *bosonic condensate* with nonvanishing broken-symmetry average ${}_{\mathrm{BCS}}\langle\Phi|b_0|\Phi\rangle_{\mathrm{BCS}} = \sum_{\mathbf{k}} |g_{\mathbf{k}}|^2 \neq 0$.

It is further interesting to mention that the states

$$|\Phi(\varphi)\rangle_{\mathrm{BCS}} = \prod_{\mathbf{k}} \left(u_{\mathbf{k}} + e^{i\varphi}\, v_{\mathbf{k}} c^{\dagger}_{\mathbf{k}\uparrow} c^{\dagger}_{-\mathbf{k}\downarrow} \right) |0\rangle \tag{26.4}$$

with different values of φ are degenerate. Fixing a particular value of φ corresponds to the relevant *spontaneous symmetry breaking*, whereby states with different values of φ are macroscopically different. In particular, when two superconductors with different values of φ are placed in proximity to each other and separated by a thin barrier, their phase differently manifests itself in the occurrence of a supercurrent of fermion pairs tunneling through the barrier, a phenomenon known as the Josephson effect [107, 108] (which has also been studied throughout the BCS–BEC crossover, both theoretically [109] and experimentally [110, 111]).

26.3 The Special Role Played by the Chemical Potential

The fermionic chemical potential μ entering the expression (26.2) plays the key role of a driving field, which makes the system to evolve from the BCS to the BEC regimes of the BCS–BEC crossover.

It is clear from Eq. (26.2) that the condition $v_{\mathbf{k}}^2 \ll 1$ can be satisfied for *all* $\mathbf{k}$, provided the fermionic chemical potential μ becomes large and negative. This condition is met when a bound state with binding energy $\varepsilon_0 = (ma_F^2)^{-1}$ occurs for the two-body problem *in vacuum* with positive scattering length a_F, and the coupling parameter $(k_F a_F)^{-1}$ becomes large such that $\varepsilon_0/E_F \gg 1$ (where $E_F = k_F^2/(2m)$ is the Fermi energy of the noninteracting system). In this limit, μ approaches the value $-\varepsilon_0/2$, and all fermions are paired up in tight (composite) bosons with a vanishing residual interaction among the bosons. This result for the fermionic chemical potential μ in the BEC limit is obtained by the mean-field gap equation (26.8) below at zero temperature [112].

As an illustration of the evolution from the BCS to the BEC limits driven by the chemical potential, Fig. 26.1 shows the occupation number $n_{\mathbf{k}} = v_{\mathbf{k}}^2$ at zero temperature for different (from positive to negative) values of the chemical potential μ.

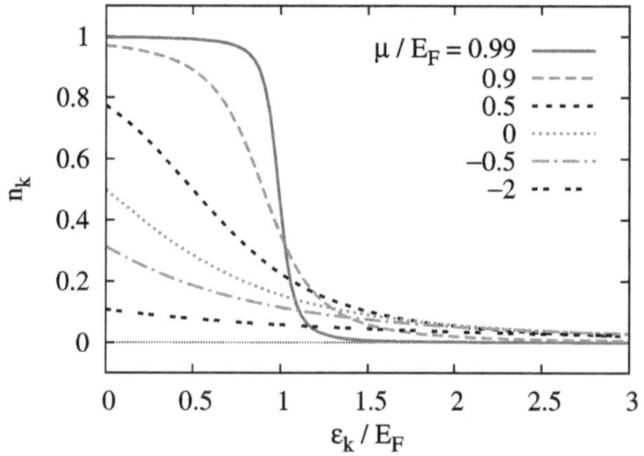

Figure 26.1 Evolution of the occupation number $n_{\mathbf{k}} = v_{\mathbf{k}}^2$ at zero temperature vs $\varepsilon_{\mathbf{k}} = \mathbf{k}^2/(2m)$ (in units of the Fermi energy E_F), for different values of the chemical potential μ. [Reproduced from Fig. 1 of Ref. [20], with permission.]

When $\mu > 0$, the curves have an inflection point at $\varepsilon_{\mathbf{k}} = \mu$, signaling the presence of an underlying Fermi surface even for a system with attractive interparticle interaction. When μ becomes negative, on the other hand, the Fermi sea is completely dissolved and the occupation number becomes small for all $\mathbf{k}$.

26.4 Gap and Density Equations for a Homogeneous System

For a homogeneous system whose interparticle interaction (21.1) is of the contact type, at the mean-field level, the gap equation for a given temperature can be obtained from Eq. (25.6), by replacing therein $\Delta(\mathbf{r}) \to \Delta$, $\nu \to \mathbf{k}$, $\varepsilon_\nu \to E_{\mathbf{k}} = \sqrt{\xi_{\mathbf{k}}^2 + |\Delta|^2}$ (with $\xi_{\mathbf{k}} = \varepsilon_{\mathbf{k}} - \mu$), and $v_\nu(\mathbf{r}) \to v_{\mathbf{k}}$, given by the square root of the expression (26.2). In this way, one obtains

$$-\frac{1}{v_0} = \int \frac{d\mathbf{k}}{(2\pi)^3} \frac{(1 - 2f_F(E_{\mathbf{k}}))}{2E_{\mathbf{k}}}, \tag{26.5}$$

where $v_{\mathbf{k}}$ has been taken real without loss of generality. In addition, with the regularization procedure (25.22), Eq. (26.5) becomes

$$-\frac{m}{4\pi a_F} = \int \frac{d\mathbf{k}}{(2\pi)^3} \left(\frac{1 - 2f_F(E_{\mathbf{k}})}{2E_{\mathbf{k}}} - \frac{m}{\mathbf{k}^2} \right), \tag{26.6}$$

$$n = \int \frac{d\mathbf{k}}{(2\pi)^3} \left(1 - \frac{\xi_{\mathbf{k}}}{E_{\mathbf{k}}} (1 - 2f_F(E_{\mathbf{k}})) \right), \tag{26.7}$$

where we have also reported the corresponding equation for the density obtained from the expressions (25.47) and (25.48) again with the above replacements valid for a homogeneous system.

The two coupled equations (26.6) and (26.7) for the gap parameter Δ and chemical potential μ can be solved for given temperature T and coupling $(k_F a_F)^{-1}$ spanning the BCS–BEC crossover. In general, these equations have to be solved numerically. However, both at zero temperature and close to the critical temperature T_c (where $\Delta \to 0$), useful analytical results can be obtained.

Zero-temperature limit In the zero-temperature limit, we can set $f_F(E_{\mathbf{k}}) = 0$ in Eqs. (26.6) and (26.7) since $E_{\mathbf{k}}$ is always positive for $\Delta(T = 0) \equiv \Delta_0 \neq 0$. The equations (26.6) and (26.7) then become

$$-\frac{m}{4\pi a_F} = \int \frac{d\mathbf{k}}{(2\pi)^3} \left(\frac{1}{2E_{\mathbf{k}}} - \frac{m}{\mathbf{k}^2} \right), \tag{26.8}$$

$$n = \int \frac{d\mathbf{k}}{(2\pi)^3} \left(1 - \frac{\xi_{\mathbf{k}}}{E_{\mathbf{k}}} \right). \tag{26.9}$$

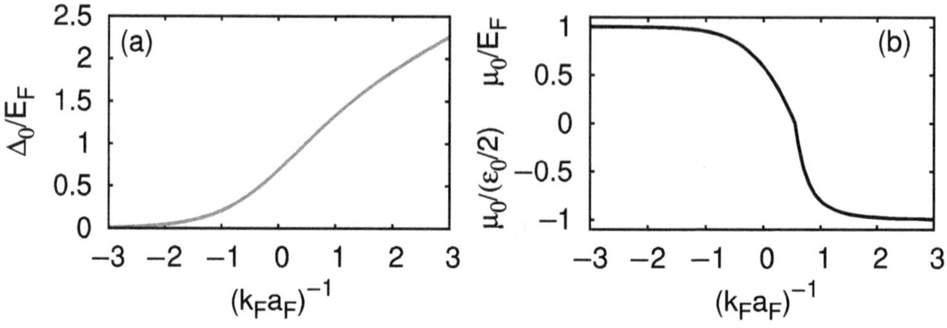

Figure 26.2 Coupling dependence of the gap parameter Δ_0 and chemical potential μ_0 at zero temperature, as obtained from the analytic solution of Ref. [113]. [Reproduced from Fig. 1 of Ref. [109], with permission.]

It turns out that these equations can be solved *analytically* throughout the BCS–BEC crossover in terms of the complete elliptic integrals [113]. Figure 26.2 shows the corresponding results for Δ_0 and $\mu_0 \equiv \mu(T=0)$. Note, in particular, how μ_0 "crosses over," from the value E_F of the BCS regime to the value $-\varepsilon_0/2$ of the BEC regime, in the rather limited coupling range $-1 \lesssim (k_F a_F)^{-1} \lesssim +1$, which embraces the unitary regime about $(k_F a_F)^{-1} = 0$.

Critical temperature At finite temperature, the presence of the Fermi function in Eq. (26.6) and (26.7) makes it impossible to find analytic solutions, and one has thus to resort to numerical solutions of these equations. An exception occurs upon approaching the critical temperature T_c from lower temperatures, where analytic results can be obtained both in BCS and BEC limits.

At the mean-field level that we are considering, the critical temperature T_c is defined by the vanishing of the BCS gap parameter Δ. Equations (26.6) and (26.7) are then replaced by:

$$-\frac{m}{4\pi a_F} = \int \frac{d\mathbf{k}}{(2\pi)^3} \left(\frac{\tanh(\xi_\mathbf{k}/2k_B T_c)}{2\xi_\mathbf{k}} - \frac{m}{\mathbf{k}^2} \right), \tag{26.10}$$

$$n = 2 \int \frac{d\mathbf{k}}{(2\pi)^3} \frac{1}{\exp(\xi_\mathbf{k}/k_B T_c) + 1}. \tag{26.11}$$

In the weak-coupling (BCS) limit where $(k_F a_F)^{-1} \ll -1$, Eq. (26.11) yields the value of the chemical potential of a Fermi gas, which is only slightly smaller than the zero-temperature result $\mu = E_F$ provided $k_B T_c \ll E_F$. Correspondingly, Eq. (26.10) gives the following expression for T_c:

$$k_B T_c = \frac{8 e^\gamma E_F}{\pi e^2} \exp\left(\frac{\pi}{2 k_F a_F} \right), \tag{26.12}$$

where γ is Euler's constant (such that $\Delta_0/(k_BT_c) = \pi/e^\gamma \simeq 1.764$). Since in the weak-coupling regime a_F is negative and $k_F|a_F| \ll 1$, Eq. (26.12) yields $k_BT_c \ll E_F$ consistently with our assumptions.

In the strong-coupling (BEC) limit where $(k_Fa_F)^{-1} \gg +1$, on the other hand, the two equations (26.10) and (26.11) interchange their role. If we assume that $k_BT_c \ll |\mu|$, we may set $\tanh(\xi_{\mathbf{k}}/2k_BT_c) \simeq 1$ in Eq. (26.10), making it to reduce to the bound-state equation and yielding $\mu \simeq -\varepsilon_0/2$ at the leading order [114]. The same kind of approximation, however, cannot be used in Eq. (26.11) since n has to remain finite. In this equation, we instead set $(\exp(\xi_{\mathbf{k}}/k_BT_c)+1)^{-1} \simeq \exp(-\xi_{\mathbf{k}}/k_BT_c)$ and obtain

$$\frac{\mu}{k_BT_c} \simeq \ln\left[\frac{n}{2}\left(\frac{2\pi}{mk_BT_c}\right)^{3/2}\right], \tag{26.13}$$

which coincides with the expression of the classical chemical potential at temperature T_c. Using the value $\mu(T_c) \simeq -\varepsilon_0/2$ determined from Eq. (26.10), we solve the expression (26.13) iteratively for T_c and obtain

$$k_BT_c \simeq \frac{\varepsilon_0}{2\ln\left(\frac{\varepsilon_0}{E_F}\right)^{3/2}} \tag{26.14}$$

at the leading order in $E_F/\varepsilon_0 \ll 1$. This expression for T_c diverges in the BEC limit *at fixed density*, instead of approaching as expected the finite value $k_BT_{\mathrm{BEC}} = \frac{\pi}{m}\left(\frac{n/2}{\zeta(3/2)}\right)^{2/3}$ at which the Bose–Einstein condensation of an ideal gas of composite bosons with mass $2m$ and density $n/2$ would occur ($\zeta(3/2) \approx 2.612$ being the Riemann ζ function of argument $3/2$).

The result (26.14) for the critical temperature has the meaning of the "pair dissociation temperature" of the composite bosons (the factor ln in the denominator originating from entropy effects), which is completely unrelated to the BEC temperature T_{BEC} at which quantum coherence is established among the composite bosons. The reason for this failure is that at the mean-field level of Eqs. (26.10) and (26.11), only the *internal* degrees of freedom of the composite bosons are taken into account, while the *translational* degrees of freedom of the composite bosons are completely neglected. To include these degrees of freedom, consideration of pairing fluctuations beyond the mean field is required. This crucial feature was first pointed out by NSR in their pioneering work on the *t-matrix approach* for the BCS–BEC crossover in the normal phase [97].

26.5 The Need for Pairing Fluctuations beyond Mean Field

It is clear from the above discussion that, to obtain the correct value of the critical temperature T_c in the BEC limit of the BCS–BEC crossover, one needs to modify Eq. (26.11) for the density.

To this end, one can extend the original Galitskii diagrammatic description of a dilute repulsive Fermi gas [54] to the case of an attractive Fermi gas, which undergoes the BCS–BEC crossover, along the lines of the NSR approach of Ref. [97]. In the normal phase above T_c, this amounts to considering the fermionic self-energy

$$\Sigma(k) = - \int \frac{d\mathbf{Q}}{(2\pi)^3} k_B T \sum_{\nu} \Gamma_0(Q)\, G_0(Q-k)\,, \tag{26.15}$$

where $k = (\mathbf{k}, \omega_n)$ is a four-vector with fermionic Matsubara frequency $\omega_n = (2n+1)\pi k_B T$ (n integer), $Q = (\mathbf{Q}, \Omega_\nu)$ is a four-vector with bosonic Matsubara frequency $\Omega_\nu = 2\pi\nu k_B T$ (ν integer), $G_0(k) = (i\omega_n - \xi_{\mathbf{k}})^{-1}$ is the bare fermionic single-particle propagator, and Γ_0 is the effective two-particle interaction (or particle–particle propagator) given by

$$\Gamma_0(Q) = \frac{-v_0}{1 + v_0\, \chi_{\rm pp}(Q)} = - \frac{1}{\frac{m}{4\pi a_F} + R_{\rm pp}(Q)}\,. \tag{26.16}$$

Here,

$$\chi_{\rm pp}(Q) = \int \frac{d\mathbf{k}}{(2\pi)^3} k_B T \sum_n G_0(\mathbf{k}+\mathbf{Q}, \omega_n + \Omega_\nu)\, G_0(-\mathbf{k}, -\omega_n) \tag{26.17}$$

is the particle–particle bubble written in terms of the bare G_0, and

$$R_{\rm pp}(Q) = \chi_{\rm pp}(Q) - \int \frac{d\mathbf{k}}{(2\pi)^3} \frac{m}{\mathbf{k}^2} = \int \frac{d\mathbf{k}}{(2\pi)^3} \left(\frac{1 - f(\xi_{\mathbf{k+Q}}) - f(\xi_{\mathbf{k}})}{\xi_{\mathbf{k+Q}} + \xi_{\mathbf{k}} - i\Omega_\nu} - \frac{m}{\mathbf{k}^2} \right) \tag{26.18}$$

is its regularized version. Note that the expressions (26.15)–(26.18) can be obtained from the results of Chapter 16 of Part I, when the time-dependent external potential is switched off and the time variables are restricted to the vertical track.

In particular, in the strong-coupling (BEC) limit whereby $|\mu|/(k_B T) \gg 1$, the particle–particle propagator Γ_0 can be shown to acquire the polar form [81, 82, 115, 116]

$$\Gamma_0(Q) \simeq - \frac{8\pi}{m^2 a_F} \frac{1}{i\Omega_\nu - \frac{\mathbf{Q}^2}{4m} + \mu_B}\,, \tag{26.19}$$

which (apart from the residue $-8\pi/(m^2 a_F)$) is equivalent to a bare bosonic propagator with mass $2m$ and chemical potential $\mu_B = 2\mu + \varepsilon_0$. In this limit, $|\mu| \simeq \varepsilon_0/2$

is the largest energy scale in the fermionic propagator G_0, such that we can neglect the Q-dependence of G_0 in Eq. (26.15) and obtain:

$$\Sigma(k) \simeq - G_0(-k) \int \frac{d\mathbf{Q}}{(2\pi)^3} k_B T \sum_\nu e^{i\Omega_\nu \eta} \Gamma_0(Q) \qquad (\eta = 0^+) . \tag{26.20}$$

[Note that the convergence factor $e^{i\Omega_\nu \eta}$ in Eq. (26.20) derives from the convention used in Eq. (16.2) of Part I for the expression of the self-energy, where an infinitesimal time shift was added to the second (now imaginary) time argument of the single-particle Green's function.]

We can also consistently expand $G(k) \simeq G_0(k) + G_0(k)\, \Sigma(k)\, G_0(k)$, ending up with the following expression for the density

$$\begin{aligned} n = 2 \int \frac{d\mathbf{k}}{(2\pi)^3} k_B T \sum_n e^{i\omega_n \eta} G(k) &\simeq 2 \int \frac{d\mathbf{k}}{(2\pi)^3} k_B T \sum_n e^{i\omega_n \eta} G_0(k) \\ - 2 \int \frac{d\mathbf{k}}{(2\pi)^3} k_B T \sum_n G_0^2(k)\, G_0(-k) & \int \frac{d\mathbf{Q}}{(2\pi)^3} k_B T \sum_\nu e^{i\Omega_\nu \eta} \Gamma_0(Q) , \end{aligned} \tag{26.21}$$

where the first term on the right-hand side can be neglected since $|\mu|/T \gg 1$ (which holds true provided the value of T_c is finite). When substituting the polar form (26.19) for Γ_0 in the second term on the right-hand side of Eq.(26.21), the factor

$$\int \frac{d\mathbf{k}}{(2\pi)^3} k_B T \sum_n G_0^2(k)\, G_0(-k) \simeq - \frac{m^2}{8\pi\sqrt{2m|\mu|}} \simeq - \frac{m^2 a_F}{8\pi} \tag{26.22}$$

cancels with the residue of the expression (26.19), and we are left with the following Bose-like expression for the density

$$n \simeq 2 \int \frac{d\mathbf{Q}}{(2\pi)^3} \frac{1}{\exp(\xi^B_{\mathbf{Q}}/k_B T) - 1} \equiv 2\, n_B, \tag{26.23}$$

where $\xi^B_{\mathbf{Q}} = \frac{\mathbf{Q}^2}{4m} - \mu_B$. Here, n_B corresponds to the density of a system of noninteracting composite bosons with mass $2m$. The correct value of the Bose–Einstein temperature $k_B T_{\text{BEC}} = \frac{3.31}{2m} n_B^{2/3}$ then results by letting $\mu_B \to 0^-$ in the expression (26.23).

For completeness, the full coupling dependence of the two (mean-field and beyond-mean-field) critical temperatures, whose values in the BEC limit are given by Eq. (26.14) and T_{BEC}, respectively, is reported in Fig. 26.4 as dashed and full lines.

The NSR t-matrix approach at equilibrium and its improved extensions with various degrees of self-consistency, both in the normal phase above T_c [48, 56, 81, 82, 115, 117] and in the superfluid phase below T_c [57, 84, 85], have extensively been utilized for beyond-mean-field descriptions of ultracold Fermi gases (even accounting explicitly for trapping effects relevant to experiments [118, 119]). [For a brief review of recent advances in the theory of the BCS–BEC crossover for fermionic superfluidity based on the t-matrix approach and its variants, see Ref. [120].]

Notwithstanding this large amount of applications of the t-matrix approach throughout the BCS–BEC crossover to equilibrium situations, so far realistic applications of the t-matrix approach to corresponding out-of-equilibrium situations (either in the normal phase as considered in Chapter 16 of Part I or in the superfluid phase as considered in Chapter 24) remain rather limited. And this in spite of the increasing experimental interest in the time-dependent response of a fermionic atomic cloud, such as that measured after an interaction quench is applied to the system, which may abruptly transfer it from the normal to the superfluid phase [24, 121].

26.6 Two Characteristic Lengths

An additional characteristic feature of the BCS–BEC crossover is the occurrence of two (coupling- and temperature-dependent) lengths, which measure the correlation either within a pair of fermions with opposite spins or among different pairs. The first length corresponds to the Cooper pair size (ξ_{pair}) [114], while the second length corresponds to the healing length ξ_{phase} for the spatial variation of the gap parameter [122].

In the superfluid phase below T_c, while ξ_{pair} can already be determined at the mean-field level [63, 114], the calculation of ξ_{phase} requires the inclusion of pairing fluctuations beyond mean field [63, 122]. In the normal phase above T_c, on the other hand, the calculation of both lengths requires the inclusion of pairing fluctuations beyond mean field [63].

Figure 26.3 shows the coupling dependence (from the BCS to the BEC regime) of both the pair size ξ_{pair} and the healing length ξ_{phase} at zero temperature. On the BCS side of unitarity, the two lengths coincide with each other (apart from an irrelevant numerical factor due to their independent definitions) and are much larger than the average interparticle spacing k_F^{-1}, signifying that the Cooper pairs are largely overlapping with each other and justifying a mean-field approach in this case. Progressing toward the BEC side, the pair size ξ_{pair} decreases monotonically, while the healing length ξ_{phase} first reaches a minimum at about unitarity and then starts increasing again, thus signifying that even point-like bosons in the BEC regime correlate with each other.

For increasing temperature, while ξ_{pair} slightly decreases (down to about 80% of its value at zero temperature irrespective of coupling upon reaching T_c [63]), ξ_{phase}

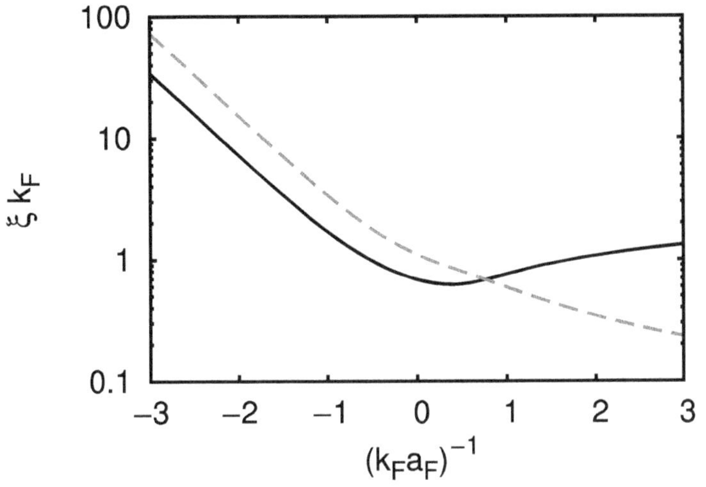

Figure 26.3 Coupling dependence of the pair size ξ_{pair} (dashed line) and the healing length ξ_{phase} (full line) at zero temperature, in units of the average interparticle spacing k_F^{-1}. [Reproduced from Fig. 7 of Ref. [109], with permission.]

diverges at T_c, signaling the occurrence of a phase transition from the superfluid to the normal phase [63]. Consideration of these different behaviors is crucial when addressing the derivation of the Ginzburg–Landau and Gross–Pitaevskii equations, in their respective coupling and temperature regimes, starting from the BdG equations, as discussed in the next section.

26.7 Ginzburg–Landau and Gross–Pitaevskii Equations

There are cases when the (time-independent) BdG equations (25.5) can be replaced by suitable non-linear differential equations for the gap parameter $\Delta(\mathbf{r})$, which are somewhat easier to solve numerically and conceptually more appealing on physical grounds than the BdG equations themselves. Notably, these nonlinear differential equations for $\Delta(\mathbf{r})$ are the Ginzburg–Landau (GL) equation for the Cooper-pair wave function and the Gross–Pitaevskii (GP) equation for the condensate wave function of composite bosons. It turns out that the GL and GP equations can be microscopically derived from the BdG equations in two characteristic situations, namely, the GL equation in the weak-coupling (BCS) limit close to T_c [123] and the GP equation in the strong-coupling (BEC) limit at $T = 0$ [89]. In both cases, the integral form (25.21) of the (single-particle) Gor'kov propagators associated with the solutions of the BdG equations provides the starting point for the microscopic derivation of these nonlinear differential equations for $\Delta(\mathbf{r})$. The domains

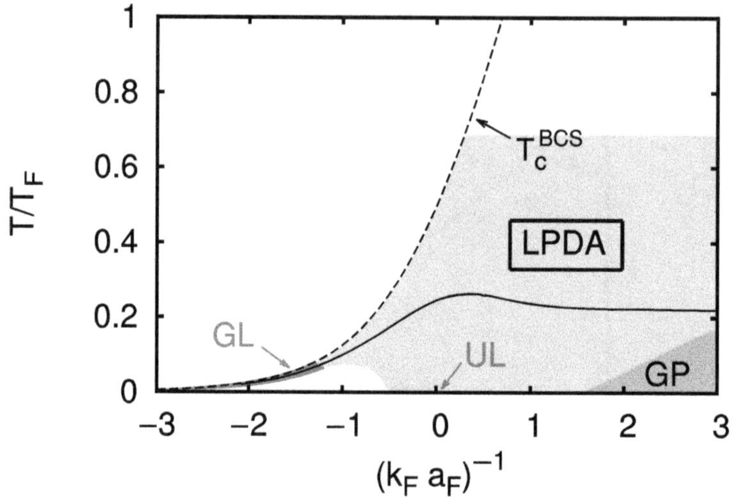

Figure 26.4 Temperature vs coupling phase diagram across the unitary limit (UL), showing the regions of validity of the Ginzburg–Landau (GL) and Gross–Pitaevskii (GP) equations (what is meant here by LPDA will be discussed later). [Reproduced from Fig. 30 of Ref. [20], with permission.]

of validity of the GL and GP equations in the temperature-coupling phase diagram are shown in Fig. 26.4, where the curves of the mean-field (T_c^{BCS} – dashed line) and beyond-mean-field (T_c – full line) critical temperature set the relevant boundaries. Consideration of the Local Phase Density Approximation (LPDA) approach with its domain of validity is deferred to the end of this chapter.

Time-independent GL equation The GL equation for strongly overlapping Cooper pairs is obtained through an expansion of the expressions (25.21) in terms of the small parameter $|\Delta(\mathbf{r})|/(k_B T_c)$ [123], in the form:

$$\left\{ \frac{(i\boldsymbol{\nabla} + 2\mathbf{A}(\mathbf{r}))^2}{4\,m} + \frac{6\pi^2 (k_B T_c)^2}{7\zeta(3) E_F} \left[\left(1 - \frac{\pi}{4 k_F a_F}\right) \frac{V_{\text{ext}}(\mathbf{r})}{E_F} - \left(1 - \frac{T}{T_c}\right) \right] \right\} \Delta(\mathbf{r})$$
$$+ \frac{3}{4E_F} |\Delta(\mathbf{r})|^2 \Delta(\mathbf{r}) = 0 \qquad (26.24)$$

(with $\zeta(3) \approx 1.202$ the Riemann ζ function of argument 3), where the vector potential $\mathbf{A}$ has also been included. In this case, $(k_F a_F)^{-1} \ll -1$ and $T \lesssim T_c$ such that $\xi_{\text{phase}} \gg \xi_{\text{pair}}$, which allows the differential equation for $\Delta(\mathbf{r})$ to be justified to begin with [123]. Note the presence in Eq. (26.24) of an external potential $V_{\text{ext}}(\mathbf{r})$, which is relevant when considering ultracold trapped Fermi gases although it did not appear in the original Gor'kov derivation of the GL equation. This term, which was first considered in Ref. [124], can be readily derived by the methods of Ref. [125] as well as of Chapter 29.

Time-independent GP equation From the BdG equations, one can further obtain the GP equation for a gas of dilute composite bosons of mass $2m$ through an expansion of the expression (25.21) in terms of the small parameter $|\Delta(\mathbf{r})/\mu|$ [89], in the form:

$$\frac{(i\nabla + 2\mathbf{A}(\mathbf{r}))^2}{4\,m}\,\Phi(\mathbf{r}) + 2V_{\rm ext}(\mathbf{r})\,\Phi(\mathbf{r}) + \frac{8\pi a_F}{2\,m}\,|\Phi(\mathbf{r})|^2\Phi(\mathbf{r}) = \mu_B\,\Phi(\mathbf{r})\,. \quad (26.25)$$

Here, $\Phi(\mathbf{r}) = \sqrt{\frac{m^2 a_F}{8\pi}}\,\Delta(\mathbf{r})$ is the condensate wave function expressed in terms of the gap parameter $\Delta(\mathbf{r})$, and μ_B is the residual chemical potential of composite bosons, defined as before by $\mu_B = 2\mu + \varepsilon_0$ with $\mu_B \ll \varepsilon_0$ in the relevant BEC limit (where $a_F > 0$). In this case, $(k_F a_F)^{-1} \gg +1$ and $T \cong 0$, such that $\xi_{\rm phase} \gg \xi_{\rm pair}$, which again allows for the differential equation for $\Delta(\mathbf{r})$ to be justified [89]. Note that the factor of 2 in front of the vector potential $\mathbf{A}(\mathbf{r})$ in both Eqs. (26.24) and (26.25) (as well as in front of the external (trapping) potential $V_{\rm ext}(\mathbf{r})$ in Eq. (26.25)) reflects the nature of Cooper pairs and composite bosons in the two (GL and GP) situations. Note also that the nonlinear term in Eq. (26.25) contains the Born value $a_B = 2a_F$ for the bosonic scattering length a_B in terms of the scattering length a_F of the constituent fermions. This value is consistent with the BdG (mean-field) approach but is larger than the correct value $a_B = 0.6a_F$ obtained by solving directly for the four-fermion problem, either as a scattering problem in vacuum [126] or through a many-body diagrammatic expansion in the zero-density limit [127]. We shall return to this issue in Chapter 28.

In practice, the solutions of the GL (26.24) and GP (26.25) equations may represent useful independent benchmarks on the numerical results obtained by the BdG equations for the gap parameter $\Delta(\mathbf{r})$ and related local quantities, in the respective temperature and coupling regimes of the BCS–BEC crossover [90, 109].

Finally, it is worth mentioning in this context that the theoretical descriptions of BEC for *trapped* Bose gases (which has preceded by a few years the experimental realization of the BCS–BEC crossover for Fermi gases) have heavily relied on the use of the GP equation for bosons treated as point-like entities [128].

Remark: The meaning of the vector potential for neutral fermions

When dealing with the superfluidity of *neutral* particles, the vector potential $\mathbf{A}(\mathbf{r})$ in Eqs. (26.24) and (26.25) is meant to describe a rotating system and is accordingly given by $\mathbf{A}(\mathbf{r}) = m\,\boldsymbol{\Omega} \times \mathbf{r}$, where $\boldsymbol{\Omega}$ is the angular velocity of the rotating trap [129]. In this case, the quadratic term in $\mathbf{A}(\mathbf{r})$ should be omitted from the kinetic energy.

Remark: A brief mention of the LPDA approach
As we have already mentioned, numerical solution of the BdG equations in realistic situations may become computationally prohibitive, owing to exceeding computation time and memory space. To overcome these difficulties, an LPDA to the (time-independent) BdG equations was introduced in Ref. [125], obtained by a suitable double coarse graining (for the phase and amplitude of the local gap parameter $\Delta(\mathbf{r})$) of those equations throughout the BCS–BEC crossover. In this way, the BdG equations were replaced by a single differential equation for $\Delta(\mathbf{r})$, in analogy to what is done with the GL and GP equations. With the difference, however, that the LPDA equation holds over an extended region of the temperature-coupling phase diagram of the BCS–BEC crossover and reduces to the GL and GP equations in appropriate ranges of coupling and temperature, as shown in Fig. 26.4. It was further established that the spatial range of the coarse graining corresponds to the Cooper pair size ξ_{pair} [130], such that the LPDA approach may become less reliable in the BCS regime at low temperature where ξ_{pair} is of the order of the range ξ_{phase} for the spatial variations of $\Delta(\mathbf{r})$. Realistic applications of the LPDA approach have been made to the description of complex arrays of vortices arising in trapped Fermi gases across the BCS–BEC crossover [129], as well as to the proximity [131] and Josephson [132] effects. In addition, the LPDA approach has been extended beyond its initial mean-field perspective, with a suitable inclusion of pairing fluctuations [133–135].

In Chapters 28 and 29, we shall derive, respectively, the time-dependent versions of the GP and GL equations across the BCS–BEC crossover, when a time-dependent disturbance is active. A time-dependent extension of the LPDA equation, valid in wider temperature and coupling ranges than those of the time-dependent GP and GL equations, is however still lacking.

27

Analytic Continuation from the Imaginary to the Real-Time Axis

In Chapter 15 of Part I, the Kadanoff–Baym equations were obtained directly in terms of the real-time Green's functions via the Langreth rules. On the other hand, in the original derivation of the equations that bear their names, Kadanoff and Baym have relied on a procedure of analytic continuation from imaginary to real time, in terms of an "extended" Matsubara approach [5]. Here, for the sake of completeness, we briefly review this procedure (after all, the Kadanoff and Baym work predated by a few years the Keldysh [2] implementation of the original Schwinger contour idea [1]), also because the procedure of analytic continuation turns out to be useful for formal developments, as those considered in Chapters 28 and 29 (although using the Langreth rules as in Chapter 15 of Part I appears preferable for computational and numerical purposes [15]). We will first take into account the case when the system Hamiltonian does not depend on time and then consider the case when it does depend on time. In addition, we shall keep on considering the fermionic (broken-symmetry) superfluid phase, as in the most of Part II.

27.1 Analytic Properties with a Time-Independent Hamiltonian

The analytic properties of the lesser ($G^{<}(t, t')$) and the greater ($G^{>}(t, t')$) Green's functions were first considered in Ref. [5], where it was stated that, in the absence of a time-dependent perturbation, $G^{<}(t, t') = G^{<}(t - t')$ is analytic in the strip $0 < \mathrm{Im}\{t - t'\} < \beta$ and $G^{>}(t, t') = G^{>}(t - t')$ is analytic in the strip $-\beta < \mathrm{Im}\{t - t'\} < 0$ of the complex $(t - t')$-plane. These properties were attributed to the presence of the Gibbs factor $e^{-\beta(H-\mu N)}$, which is sufficient to guarantee the absolute convergence of the trace (present in the definition of these Green's functions) in the above strips of the complex $(t - t')$-plane. In Ref. [5] (as well as later in Ref. [14]), these properties were considered in the normal phase. Here, we extend them to the superfluid phase.

Lesser Green's function This function is defined as follows (cf. Eq. (23.16)):

$$- i\, G^{<}(\mathbf{r}t, \mathbf{r}'t')_{ij} = \left\langle \Psi_j^{\dagger}(\mathbf{r}'t')\, \Psi_i(\mathbf{r}t) \right\rangle_{\eta} \tag{27.1}$$

in terms of the Nambu pseudo-spinor fields (22.1). Note that we have now explicitly indicated that the thermal average is taken over the *η-ensemble* associated with the spontaneous broken-symmetry phase (cf. Section 9.3 of Part I). With this prescription (and dropping any reference to the time t_0 when it is not relevant), the thermal average in Eq. (27.1) can be written as follows:

$$\begin{aligned} - i\, G^{<}(\mathbf{r}t, \mathbf{r}'t')_{ij} &= \mathrm{Tr}\left\{ e^{\beta(\Omega - H + \mu N)}\, \Psi_j^{\dagger}(\mathbf{r}'t')\, \Psi_i(\mathbf{r}t) \right\}_{\eta} \\ &= \sum_{N,n} \sum_{M,m} e^{\beta(\Omega - E_n + \mu N)} \\ &\quad \times \langle N, n| e^{iHt'} \Psi_j^{\dagger}(\mathbf{r}') e^{-iHt'} |M, m\rangle_{\eta} \langle M, m| e^{iHt} \Psi_i(\mathbf{r}) e^{-iHt} |N, n\rangle_{\eta} \\ &= \sum_{N,n,m} e^{\beta(\Omega + \mu N)}\, e^{-E_n[\beta + i(t-t')]}\, e^{iE_m(t-t')} \\ &\quad \times \langle N, n|\Psi_j^{\dagger}(\mathbf{r}')|N-1, m\rangle_{\eta} \langle N-1, m|\Psi_i(\mathbf{r})|N, n\rangle_{\eta}, \end{aligned} \tag{27.2}$$

where Ω is the grand-canonical potential, such that $\mathcal{Z} = e^{-\beta\Omega}$ is the grand-partition function, and the energies E_n (E_m) refer to N ($N-1$) particles (although in the context of the η-ensemble – cf. the remark given later). In the expression (27.2), we set $t - t' \equiv \mathrm{T} = \mathrm{T_R} + i\,\mathrm{T_I}$ and consider T as a complex variable, with real part $\mathrm{T_R}$ and imaginary part $\mathrm{T_I}$. We thus write for the exponential factors in Eq. (27.2):

$$e^{-E_n[\beta + i(t-t')]} = e^{-E_n[\beta + i\mathrm{T}]} = e^{-E_n[\beta + i\mathrm{T_R} - \mathrm{T_I}]} = e^{-E_n[\beta - \mathrm{T_I}]}\, e^{-iE_n \mathrm{T_R}}, \tag{27.3}$$

$$e^{iE_m(t-t')} = e^{iE_m T} = e^{iE_m(\mathrm{T_R} + i\,\mathrm{T_I})} = e^{-E_m \mathrm{T_I}}\, e^{iE_m \mathrm{T_R}}\,. \tag{27.4}$$

Note that the factor (27.3) contains a decaying exponential for $\beta - \mathrm{T_I} > 0$, while the factor (27.4) contains a decaying exponential for $\mathrm{T_I} > 0$. This results in the strip $0 < \mathrm{T_I} < \beta$ in the complex T-plane, as depicted in Fig. 27.1(a).[1] Owing to the presence of these two decaying exponentials, which make the sums over m and n in Eq. (27.2) absolutely converge, $G^{<}(\mathbf{r}t, \mathbf{r}'t')_{ij}$ is an *analytic function of the complex variable* T within the strip $0 < \mathrm{T_I} < \beta$ [136].

[1] Although a single strip could be used for both $G^{<}$ and $G^{>}$ (as remarked in Fig. 27.2), here for didactical purposes, we rather prefer to use initially the two strips of Fig. 27.1(a) for $G^{<}$ and of Fig. 27.1(b) for $G^{>}$.

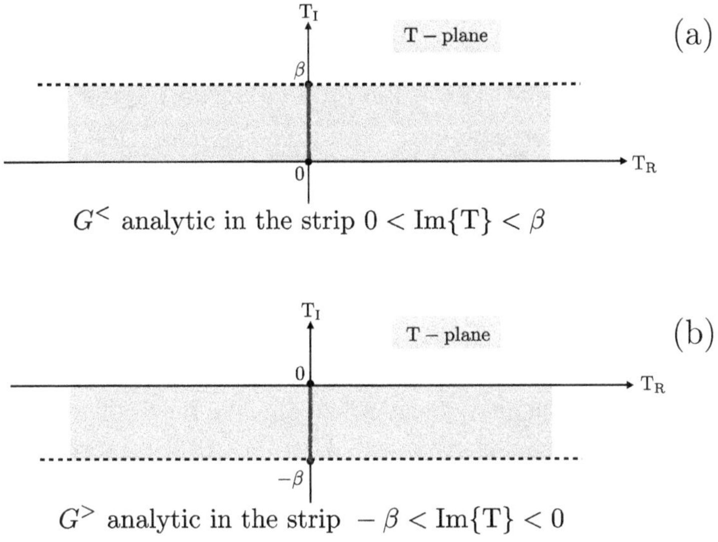

Figure 27.1 Domains of analyticity of (a) the lesser $G^<(t,t')$ and (b) the greater $G^>(t,t')$ Green's functions when the system Hamiltonian does not depend on time. Here, $t-t' \equiv \mathrm{T} = \mathrm{T_R} + i\,\mathrm{T_I}$ is interpreted as a complex variable.

Remark: About shifting the vertical track

Thus far, we have taken the vertical track (or appendix contour) γ^{M} as extending from t_0 to $t_0 - i\beta$ (cf. Fig. 8.1 of Part I). However, there is nothing to prevent us from taking γ^{M} to extend from $t_0 + i\beta$ to t_0 since we may as well write in the place of Eq. (8.2) of Part I

$$e^{-\beta H^{\mathrm{M}}} = e^{-i\int_{t_0+i\beta}^{t_0} dz\, H^{\mathrm{M}}}, \tag{27.5}$$

where $H^{\mathrm{M}} \leftrightarrow H - \mu N$. Accordingly, in the definition of the function $G^<(\mathbf{r}t,\mathbf{r}'t')_{ij}$, we may take the vertical track γ^{M} as in Eq. (27.5), such that it is contained within the strip depicted in Fig. 27.1(a).

Greater Green's function This function is defined as follows (cf. Eq. (23.17)):

$$i\,G^>(\mathbf{r}t,\mathbf{r}'t')_{ij} = \left\langle \Psi_i(\mathbf{r}t)\,\Psi_j^\dagger(\mathbf{r}'t')\right\rangle_\eta, \tag{27.6}$$

where the role of the *η-ensemble* has again been emphasized. Proceeding as in Eq. (27.2), we write:

$$\begin{aligned} i\,G^>(\mathbf{r}t,\mathbf{r}'t')_{ij} &= \mathrm{Tr}\left\{e^{\beta(\Omega-H+\mu N)}\,\Psi_i(\mathbf{r}t)\,\Psi_j^\dagger(\mathbf{r}'t')\right\}_\eta \\ &= \sum_{N,n}\sum_{M,m} e^{\beta(\Omega-E_n+\mu N)} \\ &\quad\times \langle N,n|e^{iHt}\Psi_i(\mathbf{r})e^{-iHt}|M,m\rangle_\eta \langle M,m|e^{iHt'}\Psi_j^\dagger(\mathbf{r}')e^{-iHt'}|N,n\rangle_\eta \end{aligned}$$

$$= \sum_{N,n,m} e^{\beta(\Omega+\mu N)} \, e^{-E_n[\beta - i(t-t')]} \, e^{-iE_m(t-t')}$$
$$\times \langle N, n|\Psi_i(\mathbf{r})|N+1, m\rangle_\eta \langle N+1, m|\Psi_j^\dagger(\mathbf{r}')|N, n\rangle_\eta, \tag{27.7}$$

where now E_n (E_m) refer to N ($N+1$) particles (again in the context of the η-ensemble – cf. the remark given later). In terms of the complex variable T, the exponential factors in Eq. (27.7) read:

$$e^{-E_n[\beta-i(t-t')]} = e^{-E_n[\beta-i\mathrm{T}]} = e^{-E_n[\beta-i\mathrm{T_R}+\mathrm{T_I}]} = e^{-E_n[\beta+\mathrm{T_I}]} \, e^{iE_n\mathrm{T_R}}, \tag{27.8}$$
$$e^{-iE_m(t-t')} = e^{-iE_m T} = e^{-iE_m(\mathrm{T_R}+i\,\mathrm{T_I})} = e^{E_m\mathrm{T_I}} \, e^{-iE_m\mathrm{T_R}} \,. \tag{27.9}$$

Here, the factor (27.8) contains a decaying exponential for $\beta + \mathrm{T_I} > 0$, and the factor (27.9) contains a decaying exponential for $\mathrm{T_I} < 0$. This results in the strip $-\beta < \mathrm{T_I} < 0$ in the complex T-plane, as depicted in Fig. 27.1(b). In this case, the presence of these two decaying exponentials makes $G^>(\mathbf{r}t, \mathbf{r}'t')_{ij}$ an analytic function of the complex variable T within the strip $-\beta < \mathrm{T_I} < 0$ [136].

Remark: Hamiltonian vs grand-canonical Hamiltonian in the time-independent case

In the expressions (27.2) for $G^<$ and (27.7) for $G^>$, we have considered these Green's functions to evolve with the Hamiltonian H. That is to say, the time-evolution operator was taken of the form $U(t, t_0) = e^{-iH(t-t_0)}$ (where we have further set $t_0 = 0$ for simplicity). On the other hand, when deriving the time-dependent BdG equations in Chapter 25, we have considered $G^<$ and $G^>$ to evolve with the grand-canonical Hamiltonian $K = H - \mu N$, thereby measuring the single-particle energies from the value of the chemical potential μ. This was also because μ is expected to play a key role in the BCS–BEC crossover for fermionic superfluidity. The two different definitions of $G^<$ and $G^>$, with respect to either H or K, can be readily related to each other by adapting Eqs. (18.12) and (18.13) of Part I to the Nambu pseudo-spinor field operators (22.1), such that

$$\Psi_{\mathrm{K}i}(\mathbf{r}t) = e^{(-1)^{i+1}i\mu t} \, \Psi_{\mathrm{H}i}(\mathbf{r}t), \tag{27.10}$$
$$\Psi^\dagger_{\mathrm{K}i}(\mathbf{r}'t') = e^{(-1)^{i}i\mu t'} \, \Psi^\dagger_{\mathrm{H}i}(\mathbf{r}'t') \,. \tag{27.11}$$

In this way, for the lesser Green's function (27.1), we obtain

$$-i\,G^<_{\mathrm{K}}(\mathbf{r}t, \mathbf{r}'t')_{ij} = \left\langle \Psi^\dagger_{\mathrm{K}j}(\mathbf{r}'t') \, \Psi_{\mathrm{K}i}(\mathbf{r}t) \right\rangle_\eta = e^{(-1)^{i+1}i\mu t} \, e^{(-1)^{j}i\mu t'} \, (-i) \, G^<_{\mathrm{H}}(\mathbf{r}t, \mathbf{r}'t')_{ij} \,, \tag{27.12}$$

and for the greater Green's function (27.6)

$$i\,G^>_{\mathrm{K}}(\mathbf{r}t, \mathbf{r}'t')_{ij} = \left\langle \Psi_{\mathrm{K}i}(\mathbf{r}t) \, \Psi^\dagger_{\mathrm{K}j}(\mathbf{r}'t') \right\rangle_\eta = e^{(-1)^{i+1}i\mu t} \, e^{(-1)^{j}i\mu t'} \, i\,G^>_{\mathrm{H}}(\mathbf{r}t, \mathbf{r}'t')_{ij} \,. \tag{27.13}$$

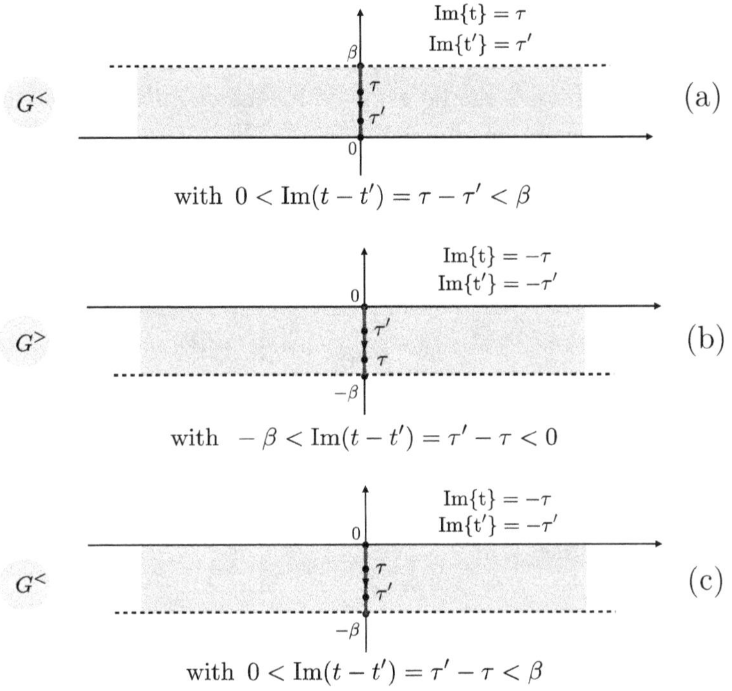

Figure 27.2 Order of the time variables along the vertical track for the (a) lesser $G^<(t.t')$ and (b) the greater $G^>(t, t')$ Green's functions, with reference to the strips depicted in Fig. 27.1. (c) For later convenience, the strip for $G^<$ shown in (a) can be brought to coincide with that for $G^>$ shown in (b).

Remark: Nambu pseudo-spinor fields and the η-ensemble

In arriving at the final expressions in Eq. (27.2) for $G^<$ and in Eq. (27.7) for $G^>$, we have treated the Nambu pseudo-spinor fields $\Psi_i(\mathbf{r})$ as if they were ordinary field operators $\psi_\sigma(\mathbf{r})$. The argument we have implicitly assumed rests on the following reasoning. Consider, for instance, the expression in (27.1) with $i = 1$ and $j = 2$:

$$- i\, G^<(\mathbf{r}t, \mathbf{r}'t')_{12} = \left\langle \Psi_2^\dagger(\mathbf{r}'t')\, \Psi_1(\mathbf{r}t)\right\rangle_\eta \underset{\substack{\uparrow \\ [\text{Eq.(22.1)}]}}{=} \left\langle \psi_\downarrow(\mathbf{r}'t')\, \psi_\uparrow(\mathbf{r}t)\right\rangle_\eta$$

$$= \sum_{N,n,m} e^{\beta(\Omega - E_n + \mu N)}\, e^{iE_n^N t'}\, e^{-iE_m^{N+1} t'}\, e^{iE_m^{N+1} t}\, e^{-iE_n^{N+2} t}$$

$$\times \langle N, n|\psi_\downarrow(\mathbf{r}')|N+1, m\rangle_\eta \langle N+1, m|\psi_\uparrow(\mathbf{r})|N+2, n\rangle_\eta, \qquad (27.14)$$

where the dependence of the eigenvalues on the particle number has been explicitly indicated by a superscript. Asymptotically, for large n, we may consider $E_n^N \simeq E_n^{N+2}$ for large N, which suffices to prove the convergence property of the series in the expression (27.14). Similar considerations apply to the component with $i = 2$ and $j = 1$, whereby we may consider $E_n^N \simeq E_n^{N-2}$.

27.2 Analytic Continuation with a Time-Independent Hamiltonian

We now prove that the lesser $G^{<}(t,t')$ and the greater $G^{>}(t,t')$ Green's functions with *both* real-time variables can be obtained by analytically continuing the Matsubara Green's function for imaginary times. We shall further show that a similar procedure applies to the right $G^{\rceil}(z.z')$ and the left $G^{\lceil}(z.z')$ Keldysh components, for which *only one* variable is moved from the imaginary to the real-time axis.

Quite generally, the contour single-particle Green's function of interest is made up of two pieces (cf. Eq. (14.1) of Part I)

$$G(z,z') = \theta(z,z')\,G^{>}(z,z') + \theta(z',z)\,G^{<}(z,z') \tag{27.15}$$

in terms of the lesser $G^{<}(z,z')$ and the greater $G^{>}(z,z')$ Green's functions, to which it reduces depending on which of the two variables z and z' is past the other along the contour C. This contour is taken here as in Fig. 14.1 of Part I, which includes the vertical track. Keeping with the definition (27.15), we then let the variables z and z' move freely along C, similar to what was done in Chapter 13 of Part I.

In particular, we are interested in letting the variables z and z' land in the final configurations shown in Figs. 27.3 and 27.4, yielding eventually the functions $G^{<}(t,t')$, $G^{>}(t,t')$, $G^{\rceil}(t,\tau)$, and $G^{\lceil}(\tau,t)$. In addition, the Matsubara Green's function, where both time variables lie along the vertical track, will be automatically included. These functions would suffice for a complete description of the

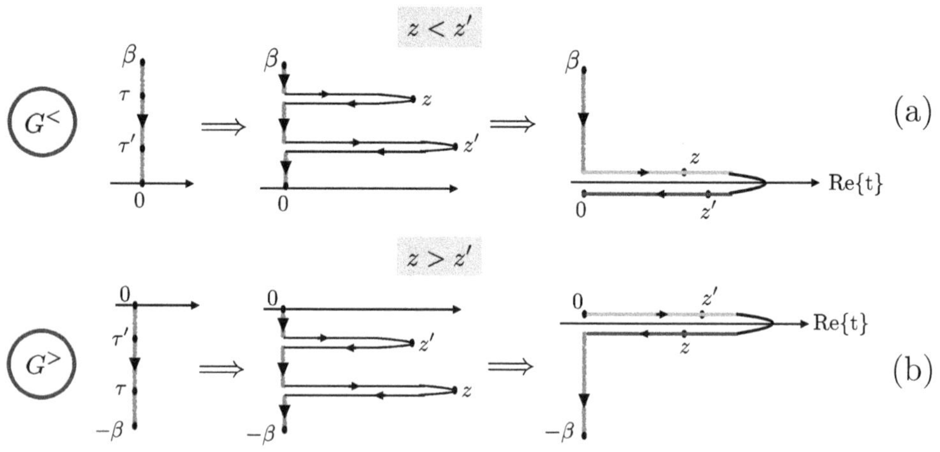

Figure 27.3 Analytic continuation of (a) the lesser $G^{<}(z.z')$ Green's function for which the order $z < z'$ of the time variables along the contour is respected and (b) the greater $G^{>}(z.z')$ Green's function for which the order $z > z'$ of the time variables along the contour is respected.

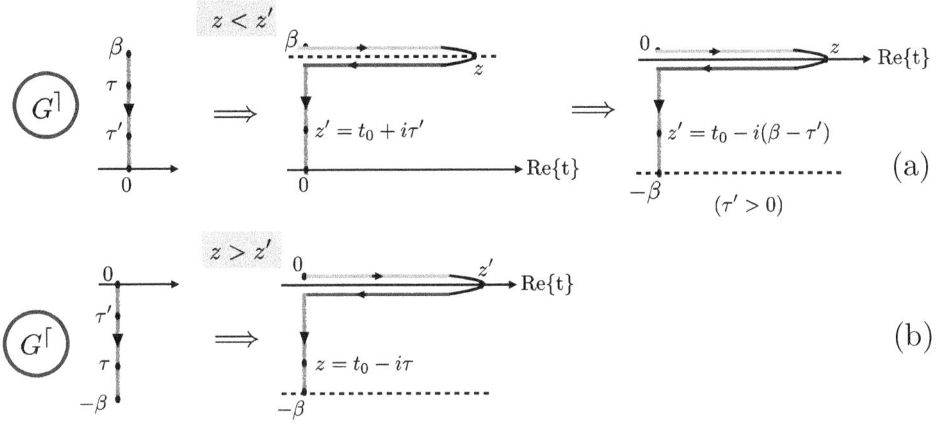

Figure 27.4 Analytic continuation leading to (a) the right Keldysh component $G^{\rceil}(z.z')$ for which the order $z < z'$ of the time variables along the contour is respected and (b) the left Keldysh component $G^{\lceil}(z.z')$ for which the order $z > z'$ of the time variables along the contour is also respected. [Note that in both panels we have momentarily restored the presence of t_0 for clarity.]

independent components of the nonequilibrium single-particle Green's function (cf. Chapter 13).

The case of $G^<$ and $G^>$ We have shown in Section 27.1 that, when the system Hamiltonian H does not depend on time, $G^<(t, t')$ and $G^>(t, t')$ are analytic functions of $(t - t')$ in their respective strips in the complex $(t - t')$-plane (cf. Fig. 27.1), whereby $t \leftrightarrow z$ and $t' \leftrightarrow z'$. For $G^<$, we then start from the settings shown in Fig. 27.2(a), whereby $\tau = \mathrm{Im}\{t\} > \tau' = \mathrm{Im}\{t'\}$ with $\tau - \tau' < \beta$, while for $G^>$, we start from the settings shown in Fig. 27.2(b), whereby $\tau = -\mathrm{Im}\{t\} > \tau' = -\mathrm{Im}\{t'\}$ with $\tau' - \tau > -\beta$. [Note that passing from $G_{\mathrm{H}}^{\lessgtr}$ to $G_{\mathrm{K}}^{\lessgtr}$, according to Eqs. (27.12) and (27.13), does not change the above settings.] In each of the two cases, the relative order of the two time variables *along* the vertical track is respected. Owing to the shown analyticity, the two variables $t \leftrightarrow z$ and $t' \leftrightarrow z'$ can then be moved around these strips, with the provision of maintaining their relative order along the deformed path, as shown in Fig. 27.3(a) for $G^<$ and in Fig. 27.3(b) for $G^>$. At this point, the vertical track in Fig. 27.3(a) can be shifted down by β, in agreement with the remark that has led to Eq. (27.5). Accordingly, in each of the two cases, the appendix contour γ^{M} gets smoothly deformed into the total contour γ_{T} of Fig. 8.1 of Part I (where we set $t_0 = 0$ for simplicity), ending up with the expressions (13.5) for $G^<(z, z')$ and (13.6) for $G^>(z, z')$ of Part I, where both variables z and z' are eventually infinitesimally close to the real-time axis. This proves that the functions $G^<(t, t')$ and $G^>(t, t')$ with real-time variables can be obtained by analytically continuing the Matsubara Green's function with imaginary time variables.

The case of $G^{\rceil}$ and $G^{\lceil}$ Considering again the Matsubara Green's function with both times along the vertical track, we now move only one of these time variables from the imaginary to the real axis, starting again from the settings of Fig. 27.2 and then deforming the path as shown in Fig. 27.4(a) for $G^{\rceil}$ and in Fig. 27.4(b) for $G^{\lceil}$. In this way, we recover the expressions (13.7) for $G^{\rceil}(z.z')$ and (13.9) for $G^{\lceil}(z.z')$ of Part I. Note, however, that Fig. 27.4(a) contains one additional intermediate step with respect to Fig. 27.4(b), which is required to shift down the vertical track by β. This is because in the expression (13.7) for $G^{\rceil}(z.z')$ of Part I, we can write (for fermions)

$$\begin{aligned} G^{\rceil}(\mathbf{x}t, \mathbf{x}'(\beta - \tau')) &= i \langle \psi^{\dagger}_{\mathrm{H}}(\mathbf{x}', t_0 - i(\beta - \tau'))\, \psi_{\mathrm{H}}(\mathbf{x}, t) \rangle \\ &= i \,\mathrm{Tr}\left\{\rho\, U(t_0, t_0 - i(\beta - \tau'))\, \psi^{\dagger}(\mathbf{x}') \right. \\ &\quad \left. \times U(t_0 - i(\beta - \tau'), t_0)\, U(t_0, t)\, \psi(\mathbf{x})\, U(t, t_0)\right\}, \end{aligned} \tag{27.16}$$

where

$$\begin{aligned} U(t_0, t_0 - i(\beta - \tau')) &= U(t_0 + i\beta, t_0 + i\tau'), \\ U(t_0 - i(\beta - \tau'), t_0) &= U(t_0 + i\tau', t_0 + i\beta), \\ U(t_0, t) &= U(t_0 + i\beta, t + i\beta), \\ U(t, t_0) &= U(t + i\beta, t_0 + i\beta), \end{aligned} \tag{27.17}$$

to the extent that $U(z_2, z_1) = e^{-iH(z_2 - z_1)}$ (or, else, for $H \to K$). [The properties (27.17) also hold, more generally, when $U(z_2, z_1) = \mathcal{T}_C \left\{ e^{-i \int_{z_1}^{z_2} dz' \mathcal{H}(z')} \right\}$ provided that $\mathcal{H}(z \pm i\beta) = \mathcal{H}(z)$.]

27.3 Analytic Continuation with a Time-Dependent Hamiltonian

The considerations made so far were limited to the case when the system Hamiltonian does not depend on time. In this case, we were able to show explicitly how all relevant single-particle Green's functions considered in Chapter 13 of Part I can be obtained by analytic continuation from the Matsubara Green's function in the strips of the complex time plane show in Fig. 27.1. In addition, as remarked in Fig. 27.2, the strip from 0 to $-i\beta$ is actually sufficient for the process of analytic continuation, since it contains both the greater (panel (b)) and the lesser (panel (c)) components of the Matsubara Green's function.

In Ref. [5], Kadanoff and Baym extended the aforementioned considerations to the case when the system Hamiltonian $\mathcal{H}(z)$ does depend on time. Their argument was based on the *assumption* that, even in this case, the time-evolution operator $U(z_2, z_1)$ remains an analytic function of its time arguments z_1 and z_2, in such a way that the relevant single-particle Green's functions are also analytic functions of their time arguments, provided these span the strip from 0 to $-i\beta$ in the complex time plane. Through these assumptions, Kadanoff and Baym were able to derive the equations of motion in real time that bear their names (cf. Chapter 15 of Part I) from the equation of motion in imaginary time. Note that, in the present context, the Hamiltonian $\mathcal{H}(z)$ is supposed to depend on time *also* along the imaginary time axis, contrary to what we have assumed so far (barring the formal approach of Chapter 17 of Part I). The Kadanoff and Baym assumption, that the time-evolution operator is an analytic function of its time arguments in the strip from 0 to $-i\beta$ in the complex time plane, is in turn based on the assumption that the external potential, which is responsible for the time dependence to begin with (or the interparticle coupling constant for ultracold gases), is itself an analytic function of time in the same strip. However, according to Kadanoff and Baym, once the analytic continuation has been performed, the requirement of analyticity of the external potential can eventually be relaxed in the final results. We shall take explicit advantage of this formal device of analytic continuation, to derive both the time-dependent Gross–Pitaevskii equation for composite bosons in Chapter 28 and the time-dependent Ginzburg–Landau equation for Cooper pairs in Chapter 29, with reference to the respective restricted regions of the temperature vs coupling phase diagram identified in Fig. 26.4.

Remark: Hamiltonian vs grand-canonical Hamiltonian in the time-dependent case

The results (27.12) and (27.13) remain valid even when the Hamiltonian $\mathcal{H}(t)$ depends on time. This is because

$$\begin{aligned} U_{\mathrm{K}}(t_2, t_1) &= \mathcal{T}\left\{e^{-i\int_{t_1}^{t_2} dt' \mathcal{K}(t')}\right\} = \mathcal{T}\left\{e^{-i\int_{t_1}^{t_2} dt' [\mathcal{H}(t') - \mu N]}\right\} = \mathcal{T}\left\{e^{-i\int_{t_1}^{t_2} dt' \mathcal{H}(t')}\, e^{i\mu N(t_2 - t_1)}\right\} \\ &= e^{i\mu N(t_2 - t_1)}\, \mathcal{T}\left\{e^{-i\int_{t_1}^{t_2} dt' \mathcal{H}(t')}\right\} = e^{i\mu N(t_2 - t_1)}\, U_{\mathrm{H}}(t_2, t_1) \end{aligned} \tag{27.18}$$

provided $\mathcal{H}(t)$ commutes with N. As a consequence, we obtain

$$\begin{aligned} \psi_{\mathrm{K}}(\mathbf{x}t) &= U_{\mathrm{K}}(t_0, t)\, \psi(\mathbf{x})\, U_{\mathrm{K}}(t, t_0) = U_{\mathrm{H}}(t_0, t)\, e^{i\mu N(t_0 - t)}\, \psi(\mathbf{x})\, e^{i\mu N(t - t_0)}\, U_{\mathrm{H}}(t, t_0) \\ &= e^{i\mu(t - t_0)}\, U_{\mathrm{H}}(t_0, t)\, \psi(\mathbf{x})\, U_{\mathrm{H}}(t, t_0) = e^{i\mu(t - t_0)}\, \psi_{\mathrm{H}}(\mathbf{x}t), \end{aligned} \tag{27.19}$$

which is the analog of Eq. (18.12) of Part I, as well as

$$\psi^{\dagger}_{\mathrm{K}}(\mathbf{x}t) = U_{\mathrm{K}}(t_0, t)\, \psi^{\dagger}(\mathbf{x})\, U_{\mathrm{K}}(t, t_0) = e^{-i\mu(t - t_0)}\, \psi^{\dagger}_{\mathrm{H}}(\mathbf{x}t), \tag{27.20}$$

which is the analog of Eq. (18.13) of Part I.

Remark: Boundary conditions and chemical potential

The boundary conditions given by Eqs. (12.2) and (12.3) of Part I differ from those given by Eq. (1.10) of Ref. [5] by the presence of an extra factor $e^{\beta\mu}$ in the latter case. This difference is due to the different convention about the Hamiltonians utilized in the vertical track, such that, for instance, $\psi(\mathrm{x}, t=-i\beta)=e^{\beta(H-\mu N)}\,\psi(\mathrm{x}, t=0)\,e^{-\beta(H-\mu N)}$ by our convention, while $\psi(\mathrm{x}, t=-i\beta)=e^{\beta H}\,\psi(\mathrm{x}, t=0)\,e^{-\beta H}$ by the convention of Ref. [5].

28

Derivation of the Time-Dependent Gross–Pitaevskii Equation for Composite Bosons in the BEC Limit of the BCS–BEC Crossover

The time-dependent Gross–Pitaevskii (TDGP) equation that we are going to derive will describe composite bosons as bound-fermion pairs, which form on the BEC side of the BCS–BEC crossover at sufficiently low temperature, in correspondence to the restricted Gross–Pitaevskii (GP) region of the temperature vs coupling phase diagram shown in Fig. 26.4. In this way, we will extend to the time domain the time-independent GP equation, which was derived in Ref. [89] in the same restricted GP region by relying on the Green's functions method at equilibrium. The key feature, which made that derivation possible, was the fact that, in this context, the fermionic chemical potential μ is the largest energy scale in the problem, such that $|\Delta(\mathbf{r})/\mu| \ll 1$.

The time-dependent GP equation will correspondingly be derived by relying on the Green's functions method for nonequilibrium problems. The assumption that the fermionic chemical potential μ, associated with the initial preparation of the system at thermodynamic equilibrium, is the largest energy scale in the problem will still hold and will duly be explored. Further assumptions to be taken into account will be that:

(i) Both the spatial and temporal variations of the gap parameter $\Delta(\mathbf{r}, t)$ are sufficiently "slow";
(ii) The fluctuating part $\delta\Delta(\mathbf{r}, t)$ of the gap parameter is "small" compared with the underlying static profile $\Delta_e(\mathbf{r})$.

Accordingly, we set

$$\Delta(\mathbf{r}, t) = \Delta_e(\mathbf{r}) + \delta\Delta(\mathbf{r}, t), \tag{28.1}$$

where $|\delta\Delta(\mathbf{r}, t)| \ll |\Delta_e(\mathbf{r})|$ and $\delta\Delta(\mathbf{r}, t)$ is both "slow" and "small." In addition, we shall initially consider $\Delta_e(\mathbf{r})$ to correspond to a homogeneous situation and later introduce its spatial dependence via a "local density" approach, consistently with the assumption that $\Delta(\mathbf{r}, t)$ is slowly varying in space (and similar to what was done in Ref. [89] to derive the time-independent GP equation).

On the basis of all these assumptions, in Section 28.1, we will set up a general framework, which will then be implemented in Section 28.2 to obtain the TDGP equation within the GP region of the temperature vs coupling phase diagram. The same general framework will be further utilized in Chapter 29 to obtain the time-dependent Ginzburg–Landau (TDGL) equation within the GL region of the temperature vs coupling phase diagram shown in Fig. 26.4.

28.1 General Framework

In this section, we will closely parallel the procedure developed in Ref. [137], although in that reference only the weak-coupling (BCS) limit was considered, while here we are interested in the strong-coupling (BEC) limit at low temperature. To this end, we consider the Dyson equation in the differential form (23.5) with Nambu indices, where we now restrict the time variables to the vertical track as we did in Chapter 15 of Part I for the Matsubara (M) component. We also adopt here the convention of Ref. [7] as we did in Eq. (15.5) of Part I (although we now drop the suffix FW for simplicity), to rewrite Eq. (23.5) in the form (where we set $t_0 = 0$ for simplicity)

$$\left[-\frac{d}{d\tau_1} \overleftrightarrow{\mathbb{1}} - \overleftrightarrow{h}^{\mathrm{M}}(\mathbf{r}_1, \tau_1) \right] \overleftrightarrow{G}^{\mathrm{M}}(\mathbf{r}_1\tau_1, \mathbf{r}_2\tau_2)$$
$$- \int_0^{\beta} d\tau_3 \int d\mathbf{r}_3 \, \overleftrightarrow{\Sigma}^{\mathrm{M}}(\mathbf{r}_1\tau_1, \mathbf{r}_3\tau_3) \, \overleftrightarrow{G}^{\mathrm{M}}(\mathbf{r}_3\tau_3, \mathbf{r}_2\tau_2)$$
$$= \delta(\tau_1 - \tau_2) \, \delta(\mathbf{r}_1 - \mathbf{r}_2) \, \overleftrightarrow{\mathbb{1}} \tag{28.2}$$

with $\overleftrightarrow{h}^{\mathrm{M}}(\mathbf{r}, \tau) = \overleftrightarrow{\tau}_3 \, h^{\mathrm{M}}(\mathbf{r}, \tau)$ as in Eq. (23.6). Here, $h^{\mathrm{M}}(\mathbf{r}, \tau)$ is referred to the fermionic chemical potential μ at equilibrium in line with the Matsubara approach (cf. Section 25.1) and is assumed to contain a τ-dependent external potential $V_{\mathrm{ext}}(\mathbf{r}, \tau)$ in line with the approach of Ref. [5] (cf. Section 27.3). [Note that, in the absence of a τ-dependent external potential, all quantities in Eq. (28.2) consistently reduce to those reported in Ref. [7].]

In particular, within a *mean-field decoupling*, the self-energy in Eq. (28.2) is written in the form

$$\overleftrightarrow{\Sigma}^{\mathrm{M}}(\mathbf{r}_1\tau_1, \mathbf{r}_2\tau_2) = \begin{pmatrix} 0 & \Delta(\mathbf{r}_1, \tau_1) \\ \Delta^{\dagger}(\mathbf{r}_1, \tau_1) & 0 \end{pmatrix} \delta(\tau_1 - \tau_2) \, \delta(\mathbf{r}_1 - \mathbf{r}_2) \tag{28.3}$$

(cf. Eq. (23.10)), where self-consistency requires that

$$\Delta(\mathbf{r},\tau) = v_0 \left\langle \psi_\downarrow(\mathbf{r},\tau)\, \psi_\uparrow(\mathbf{r},\tau) \right\rangle = v_0 \left\langle \Psi_2^\dagger(\mathbf{r},\tau)\, \Psi_1(\mathbf{r},\tau) \right\rangle = v_0\, G^{\mathrm{M}}(\mathbf{r}\tau, \mathbf{r}\tau^+)_{12}, \tag{28.4}$$

$$\Delta^\dagger(\mathbf{r},\tau) = v_0 \left\langle \psi_\uparrow^\dagger(\mathbf{r},\tau)\, \psi_\downarrow^\dagger(\mathbf{r},\tau) \right\rangle = v_0 \left\langle \Psi_1^\dagger(\mathbf{r},\tau)\, \Psi_2(\mathbf{r},\tau) \right\rangle = v_0\, G^{\mathrm{M}}(\mathbf{r}\tau, \mathbf{r}\tau^+)_{21}\,. \tag{28.5}$$

Note here that $\Delta^\dagger(\mathbf{r},\tau) \neq \Delta(\mathbf{r},\tau)^*$ since $U(0,-i\tau) \neq U^\dagger(-i\tau,0)$ with $U(-i\tau,0) = \mathcal{T}_\tau \left\{ e^{-\int_0^\tau d\tau' \mathcal{K}(\tau')} \right\}$. In what follows, we will restrict to the form (28.3) of the self-energy, where $\Delta(\mathbf{r},\tau)$ and $\Delta^\dagger(\mathbf{r},\tau)$ are further split according to Eq. (28.1) (now with $t \to \tau$).

On the basis of the above approximations, we next manipulate Eq. (28.2) according to the following steps.

Step # 1: From the differential to the integral equation

We begin by transforming the (integro-)differential equation (28.2) into an integral equation, similar to what was done in Chapter 11 of Part I, where the integral form of the Dyson equation was obtained. To this end, we introduce the noninteracting counterpart $\overleftrightarrow{G}_0^{\mathrm{M}}$ to the Green's function $\overleftrightarrow{G}^{\mathrm{M}}$, such that

$$\left[-\frac{d}{d\tau_1} \overleftrightarrow{\mathbb{1}} - \overleftrightarrow{h}^{\mathrm{M}}(\mathbf{r_1}, \tau_1) \right] \overleftrightarrow{G}_0^{\mathrm{M}}(\mathbf{r}_1\tau_1, \mathbf{r}_2\tau_2) = \delta(\tau_1 - \tau_2)\, \delta(\mathbf{r}_1 - \mathbf{r}_2)\, \overleftrightarrow{\mathbb{1}}, \tag{28.6}$$

which is diagonal in Nambu indices. We thus rewrite the (integro-)differential equation (28.2) in the integral form:

$$\begin{aligned} \overleftrightarrow{G}^{\mathrm{M}}(\mathbf{r}_1\tau_1, \mathbf{r}_2\tau_2) &= \overleftrightarrow{G}_0^{\mathrm{M}}(\mathbf{r}_1\tau_1, \mathbf{r}_2\tau_2) + \int_0^\beta d\tau_3 \int d\mathbf{r}_3 \int_0^\beta d\tau_4 \int d\mathbf{r}_4 \\ &\times \overleftrightarrow{G}_0^{\mathrm{M}}(\mathbf{r}_1\tau_1, \mathbf{r}_3\tau_3)\, \overleftrightarrow{\Sigma}^{\mathrm{M}}(\mathbf{r}_3\tau_3, \mathbf{r}_4\tau_4)\, \overleftrightarrow{G}^{\mathrm{M}}(\mathbf{r}_4\tau_4, \mathbf{r}_2\tau_2)\,. \end{aligned} \tag{28.7}$$

This is because, upon applying the operator $\left[-\frac{d}{d\tau_1} \overleftrightarrow{\mathbb{1}} - \overleftrightarrow{h}^{\mathrm{M}}(\mathbf{r_1}, \tau_1) \right]$ to both sides of Eq. (28.7) and taking into account Eq. (28.6), we obtain

$$\begin{aligned} \left[-\frac{d}{d\tau_1} \overleftrightarrow{\mathbb{1}} - \overleftrightarrow{h}^{\mathrm{M}}(\mathbf{r_1}, \tau_1) \right] \overleftrightarrow{G}^{\mathrm{M}}(\mathbf{r}_1\tau_1, \mathbf{r}_2\tau_2) &= \delta(\tau_1 - \tau_2)\, \delta(\mathbf{r}_1 - \mathbf{r}_2)\, \overleftrightarrow{\mathbb{1}} \\ &+ \int_0^\beta d\tau_4 \int d\mathbf{r}_4\, \overleftrightarrow{\Sigma}^{\mathrm{M}}(\mathbf{r}_1\tau_1, \mathbf{r}_4\tau_4)\, \overleftrightarrow{G}^{\mathrm{M}}(\mathbf{r}_4\tau_4, \mathbf{r}_2\tau_2)\,, \end{aligned} \tag{28.8}$$

which coincides with Eq. (28.2). [QED]

Step # 2: Splitting into two coupled integral equations

Next, we split the self-energy $\overleftrightarrow{\Sigma}^{\mathrm{M}}$ of the mean-field form (28.3), by writing $\Delta(\mathbf{r}\tau)$ and $\Delta^{\dagger}(\mathbf{r}\tau)$ therein as in Eq. (28.1). We thus set:

$$\overleftrightarrow{\Sigma}^{\mathrm{M}}(\mathbf{r}_1\tau_1, \mathbf{r}_2\tau_2) = \overleftrightarrow{\Sigma}_e^{\mathrm{M}}(\mathbf{r}_1\tau_1, \mathbf{r}_2\tau_2) + \delta\overleftrightarrow{\Sigma}^{\mathrm{M}}(\mathbf{r}_1\tau_1, \mathbf{r}_2\tau_2), \tag{28.9}$$

where

$$\overleftrightarrow{\Sigma}_e^{\mathrm{M}}(\mathbf{r}_1\tau_1, \mathbf{r}_2\tau_2) = \begin{pmatrix} 0 & \Delta_e(\mathbf{r}_1) \\ \Delta_e^*(\mathbf{r}_1) & 0 \end{pmatrix} \delta(\tau_1 - \tau_2)\,\delta(\mathbf{r}_1 - \mathbf{r}_2) \tag{28.10}$$

and

$$\delta\overleftrightarrow{\Sigma}^{\mathrm{M}}(\mathbf{r}_1\tau_1, \mathbf{r}_2\tau_2) = \begin{pmatrix} 0 & \delta\Delta(\mathbf{r}_1, \tau_1) \\ \delta\Delta^{\dagger}(\mathbf{r}_1, \tau_1) & 0 \end{pmatrix} \delta(\tau_1 - \tau_2)\,\delta(\mathbf{r}_1 - \mathbf{r}_2)\,. \tag{28.11}$$

In this way, we can employ the general method discussed in Section 20.1 of Part I and split the integral equation (28.7) in the following way. For the time being, we do not consider the presence of a (space- and τ-dependent) external potential $V_{\mathrm{ext}}(\mathbf{r}, \tau)$ in the single-particle Hamiltonian $\overleftrightarrow{h}^{\mathrm{M}}(\mathbf{r}, \tau)$ of Eq. (28.6), which however still contains the chemical potential μ associated with the initial thermodynamic equilibrium. Only at a later stage, in fact, $V_{\mathrm{ext}}(\mathbf{r}, \tau)$ will be reintroduced into the problem via a "local" space and (imaginary) time approximation, in a similar way to what was done in Ref. [89] for the derivation of the time-independent GP equation. Accordingly, we rewrite the integral equation (28.7) in the form of two coupled integral equations

$$\overleftrightarrow{G}_e^{\mathrm{M}}(\mathbf{r}_1\tau_1, \mathbf{r}_2\tau_2) = \overleftrightarrow{G}_0^{\mathrm{M}}(\mathbf{r}_1\tau_1, \mathbf{r}_2\tau_2) + \int_0^{\beta} d\tau_3 \int d\mathbf{r}_3 \int_0^{\beta} d\tau_4 \int d\mathbf{r}_4 \\ \times \overleftrightarrow{G}_0^{\mathrm{M}}(\mathbf{r}_1\tau_1, \mathbf{r}_3\tau_3)\, \overleftrightarrow{\Sigma}_e^{\mathrm{M}}(\mathbf{r}_3\tau_3, \mathbf{r}_4\tau_4)\, \overleftrightarrow{G}_e^{\mathrm{M}}(\mathbf{r}_4\tau_4, \mathbf{r}_2\tau_2) \tag{28.12}$$

$$\overleftrightarrow{G}^{\mathrm{M}}(\mathbf{r}_1\tau_1, \mathbf{r}_2\tau_2) = \overleftrightarrow{G}_e^{\mathrm{M}}(\mathbf{r}_1\tau_1, \mathbf{r}_2\tau_2) + \int_0^{\beta} d\tau_3 \int d\mathbf{r}_3 \int_0^{\beta} d\tau_4 \int d\mathbf{r}_4 \\ \times \overleftrightarrow{G}_e^{\mathrm{M}}(\mathbf{r}_1\tau_1, \mathbf{r}_3\tau_3)\, \delta\overleftrightarrow{\Sigma}^{\mathrm{M}}(\mathbf{r}_3\tau_3, \mathbf{r}_4\tau_4)\, \overleftrightarrow{G}^{\mathrm{M}}(\mathbf{r}_4\tau_4, \mathbf{r}_2\tau_2)\,. \tag{28.13}$$

which we now pass to solve separately.

Step # 3: Solving the first integral equation

By our (temporary) assumption, that the single-particle Hamiltonian $\overleftrightarrow{h}^{\mathrm{M}}(\mathbf{r}, \tau)$ does not contain the external potential $V_{\mathrm{ext}}(\mathbf{r}, \tau)$, the gap parameter $\Delta_e(\mathbf{r})$ in

Eq. (28.10) does not depend on the spatial coordinate $\mathbf{r}$, such that we can consider $\Delta_e(\mathbf{r}) = \Delta_e$ as constant. We then solve Eq. (28.6) by setting

$$\overleftrightarrow{\mathcal{G}}_0^{\mathrm{M}}(\mathbf{r}_1\tau_1, \mathbf{r}_2\tau_2) = \overleftrightarrow{\mathcal{G}}_0^{\mathrm{M}}(\mathbf{r}_1 - \mathbf{r}_2, \tau_1 - \tau_2)$$
$$= \int \frac{d\mathbf{k}}{(2\pi)^3} e^{i\mathbf{k}\cdot(\mathbf{r}_1-\mathbf{r}_2)} \frac{1}{\beta} \sum_n e^{-i\omega_n(\tau_1-\tau_2)} \overleftrightarrow{\mathcal{G}}_0^{\mathrm{M}}(\mathbf{k}, \omega_n), \quad (28.14)$$

where $\mathbf{k}$ is a wave vector and $\omega_n = (2n+1)\pi/\beta$ (n integer) a fermionic Matsubara frequency. Equation (28.6) then becomes

$$\left[i\,\omega_n \overleftrightarrow{\mathbb{1}} - \begin{pmatrix} \xi_{\mathbf{k}} & 0 \\ 0 & -\xi_{\mathbf{k}} \end{pmatrix} \right] \overleftrightarrow{\mathcal{G}}_0^{\mathrm{M}}(\mathbf{k}, \omega_n) = \overleftrightarrow{\mathbb{1}} \quad (28.15)$$

with $\xi_{\mathbf{k}} = \frac{\mathbf{k}^2}{2m} - \mu$, whose solution is

$$\overleftrightarrow{\mathcal{G}}_0^{\mathrm{M}}(\mathbf{k}, \omega_n) = \begin{pmatrix} \frac{1}{i\omega_n - \xi_{\mathbf{k}}} & 0 \\ 0 & \frac{1}{i\omega_n + \xi_{\mathbf{k}}} \end{pmatrix}. \quad (28.16)$$

In addition, Eq. (28.12) becomes an algebraic equation

$$\overleftrightarrow{\mathcal{G}}_e^{\mathrm{M}}(\mathbf{k}, \omega_n) = \overleftrightarrow{\mathcal{G}}_0^{\mathrm{M}}(\mathbf{k}, \omega_n) + \overleftrightarrow{\mathcal{G}}_0^{\mathrm{M}}(\mathbf{k}, \omega_n)\, \overleftrightarrow{\Delta}_e\, \overleftrightarrow{\mathcal{G}}_e^{\mathrm{M}}(\mathbf{k}, \omega_n) \quad (28.17)$$

where now (cf. Eq. (28.10))

$$\overleftrightarrow{\Delta}_e = \begin{pmatrix} 0 & \Delta_e \\ \Delta_e^* & 0 \end{pmatrix}. \quad (28.18)$$

The solution to Eq. (28.17) is readily obtained by casting it in the form

$$\overleftrightarrow{\mathcal{G}}_e^{\mathrm{M}}(\mathbf{k}, \omega_n)^{-1} = \overleftrightarrow{\mathcal{G}}_0^{\mathrm{M}}(\mathbf{k}, \omega_n)^{-1} - \overleftrightarrow{\Delta}_e = \begin{pmatrix} i\omega_n - \xi_{\mathbf{k}} & -\Delta_e \\ -\Delta_e^* & i\omega_n + \xi_{\mathbf{k}} \end{pmatrix}, \quad (28.19)$$

such that

$$\overleftrightarrow{\mathcal{G}}_e^{\mathrm{M}}(\mathbf{k}, \omega_n) = \frac{1}{(i\omega_n)^2 - E_{\mathbf{k}}^2} \begin{pmatrix} i\omega_n + \xi_{\mathbf{k}} & \Delta_e \\ \Delta_e^* & i\omega_n - \xi_{\mathbf{k}} \end{pmatrix} \quad (28.20)$$

with $E_{\mathbf{k}} = \sqrt{\xi_{\mathbf{k}}^2 + |\Delta_e|^2}$ and $\Delta_e = |\Delta_e| e^{i\varphi_e}$. The result (28.20) is consistent with the BCS form of the normal and anomalous single-particle Green's functions within the BCS theory [7]. Note, however, that, in the expressions (28.15)–(28.20), the value of Δ_e is left unspecified and will eventually be determined together with the fluctuating part $\delta\Delta$ in the form (28.1) by solving the final TDGP equation.

Step # 4: Manipulating the second integral equation

Owing to the assumptions leading to the expression (28.1), according to which the quantity $\delta\overleftrightarrow{\Sigma}^{\mathrm{M}}$ of Eq. (28.11) can be considered sufficiently "small," we retain the integral equation (28.13) at the lowest order in $\delta \overleftrightarrow{\Sigma}^{\mathrm{M}}$ and approximate it as follows:

$$\begin{aligned}
\overleftrightarrow{G}^{\mathrm{M}}(\mathbf{r}_1\tau_1, \mathbf{r}_2\tau_2) &= \overleftrightarrow{G}^{\mathrm{M}}_e(\mathbf{r}_1\tau_1, \mathbf{r}_2\tau_2) + \int_0^\beta d\tau_3 \int d\mathbf{r}_3 \int_0^\beta d\tau_4 \int d\mathbf{r}_4 \\
&\quad \times \overleftrightarrow{G}^{\mathrm{M}}_e(\mathbf{r}_1\tau_1, \mathbf{r}_3\tau_3)\, \delta \overleftrightarrow{\Sigma}^{\mathrm{M}}(\mathbf{r}_3\tau_3, \mathbf{r}_4\tau_4)\, \overleftrightarrow{G}^{\mathrm{M}}_e(\mathbf{r}_4\tau_4, \mathbf{r}_2\tau_2) + \cdots \\
&\simeq \overleftrightarrow{G}^{\mathrm{M}}_e(\mathbf{r}_1\tau_1, \mathbf{r}_2\tau_2) + \int_0^\beta d\tau_3 \int d\mathbf{r}_3 \\
&\quad \times \overleftrightarrow{G}^{\mathrm{M}}_e(\mathbf{r}_1\tau_1, \mathbf{r}_3\tau_3) \begin{pmatrix} 0 & \delta\Delta(\mathbf{r}_3, \tau_3) \\ \delta\Delta^\dagger(\mathbf{r}_3, \tau_3) & 0 \end{pmatrix} \overleftrightarrow{G}^{\mathrm{M}}_e(\mathbf{r}_3\tau_3, \mathbf{r}_2\tau_2)\,.
\end{aligned} \tag{28.21}$$

For our purposes, it will be sufficient to consider the element $(1, 2)$ of the above expression, where we further take the limits $\mathbf{r}_2 \to \mathbf{r}_1$ and $\tau_2 \to \tau_1^+$, such that (cf. also Eq. (23.21))

$$G^{\mathrm{M}}(\mathbf{r}_1\tau_1, \mathbf{r}_1\tau_1^+)_{12} = \frac{1}{v_0}\, \Delta(\mathbf{r}_1, \tau_1) \tag{28.22}$$

owing to Eq. (28.4), and

$$\begin{aligned}
G^{\mathrm{M}}_e(\mathbf{r}_1\tau_1, \mathbf{r}_1\tau_1^+)_{12} &= \int \frac{d\mathbf{k}}{(2\pi)^3} \frac{1}{\beta} \sum_n e^{i\omega_n\eta} \frac{\Delta_e}{(i\omega_n)^2 - E^2_{\mathbf{k}}} \\
&= -\,\Delta_e \int \frac{d\mathbf{k}}{(2\pi)^3} \frac{[1 - 2f_F(E_{\mathbf{k}})]}{2E_{\mathbf{k}}},
\end{aligned} \tag{28.23}$$

where [7]

$$\frac{1}{\beta} \sum_n \frac{e^{i\omega_n\eta}}{i\omega_n - \varepsilon} = \frac{1}{e^{\beta\varepsilon} + 1} = f_F(\varepsilon) \tag{28.24}$$

is the Fermi–Dirac distribution function (such that $f_F(-\varepsilon) = 1 - f_F(\varepsilon)$). Correspondingly, the element $(1, 2)$ in the last term of Eq. (28.21) reads

$$\begin{aligned}
\int_0^\beta d\tau_3 \int d\mathbf{r}_3 \Big(& G^{\mathrm{M}}_e(\mathbf{r}_1\tau_1, \mathbf{r}_3\tau_3)_{11}\, \delta\Delta(\mathbf{r}_3, \tau_3)\, G^{\mathrm{M}}_e(\mathbf{r}_3\tau_3, \mathbf{r}_1\tau_1)_{22} \\
& + G^{\mathrm{M}}_e(\mathbf{r}_1\tau_1, \mathbf{r}_3\tau_3)_{12}\, \delta\Delta^\dagger(\mathbf{r}_3, \tau_3)\, G^{\mathrm{M}}_e(\mathbf{r}_3\tau_3, \mathbf{r}_1\tau_1)_{12} \Big)\,,
\end{aligned} \tag{28.25}$$

where it is convenient to represent the mean-field Green's functions $G^{\mathrm{M}}_e(\mathbf{r}\tau, \mathbf{r}'\tau')_{ij}$ as in Eq. (28.14), as well as to set

$$\delta\Delta(\mathbf{r}, \tau) = \int \frac{d\mathbf{q}}{(2\pi)^3} \frac{1}{\beta} \sum_\nu e^{i\mathbf{q}\cdot\mathbf{r} - i\Omega_\nu\tau}\, \delta\Delta(\mathbf{q}, \Omega_\nu) \tag{28.26}$$

$$\delta\Delta^\dagger(\mathbf{r}, \tau) = \int \frac{d\mathbf{q}}{(2\pi)^3} \frac{1}{\beta} \sum_\nu e^{i\mathbf{q}\cdot\mathbf{r} - i\Omega_\nu\tau}\, \delta\Delta^\dagger(\mathbf{q}, \Omega_\nu) \tag{28.27}$$

with wave vector $\mathbf{q}$ and bosonic Matsubara frequency $\Omega_\nu = 2\pi\nu/\beta$ (ν integer). In this way, the expression (28.25) becomes:

$$\int \frac{d\mathbf{q}}{(2\pi)^3} \frac{1}{\beta} \sum_\nu e^{i\mathbf{q}\cdot\mathbf{r}_1 - i\Omega_\nu \tau_1}$$
$$\times \left(\delta\Delta(\mathbf{q}, \Omega_\nu) \int \frac{d\mathbf{k}}{(2\pi)^3} \frac{1}{\beta} \sum_n \mathcal{G}_e(\mathbf{k}, \omega_n)_{11}\, \mathcal{G}_e(\mathbf{k}\text{-}\mathbf{q}, \omega_n - \Omega_\nu)_{22} \right.$$
$$\left. + \, \delta\Delta^\dagger(\mathbf{q}, \Omega_\nu) \int \frac{d\mathbf{k}}{(2\pi)^3} \frac{1}{\beta} \sum_n \mathcal{G}_e(\mathbf{k}, \omega_n)_{12}\, \mathcal{G}_e(\mathbf{k}\text{-}\mathbf{q}, \omega_n - \Omega_\nu)_{12} \right) . \tag{28.28}$$

Step # 5: Focusing on the particle–particle bubbles

In the expression (28.28), we identify the "normal" particle–particle bubble

$$\int \frac{d\mathbf{k}}{(2\pi)^3} \frac{1}{\beta} \sum_n \mathcal{G}_e(\mathbf{k}, \omega_n)_{11}\, \mathcal{G}_e(\mathbf{k}\text{-}\mathbf{q}, \omega_n - \Omega_\nu)_{22}$$
$$= - \int \frac{d\mathbf{k}}{(2\pi)^3} \frac{1}{\beta} \sum_n \mathcal{G}_e(-\mathbf{k}, -\omega_n)_{11}\, \mathcal{G}_e(\mathbf{k}\text{+}\mathbf{q}, \omega_n + \Omega_\nu)_{11}$$
$$\equiv A_e(\mathbf{q}, \Omega_\nu) + \frac{1}{v_0} , \tag{28.29}$$

as well as the "anomalous" particle–particle bubble

$$\int \frac{d\mathbf{k}}{(2\pi)^3} \frac{1}{\beta} \sum_n \mathcal{G}_e(\mathbf{k}, \omega_n)_{12}\, \mathcal{G}_e(\mathbf{k}\text{-}\mathbf{q}, \omega_n - \Omega_\nu)_{12}$$
$$= \int \frac{d\mathbf{k}}{(2\pi)^3} \frac{1}{\beta} \sum_n \mathcal{G}_e(-\mathbf{k}, -\omega_n)_{12}\, \mathcal{G}_e(\mathbf{k}\text{+}\mathbf{q}, \omega_n + \Omega_\nu)_{12}$$
$$\equiv B_e(\mathbf{q}, \Omega_\nu) , \tag{28.30}$$

which both hold in the superfluid phase and where the notation of Ref. [83] has been utilized. [Care should be taken of the fact that, in general, Δ_e, which enters the definition (28.20) of $(\mathcal{G}_e)_{12}$, is a complex factor.] The expression (28.28) can thus be cast in the compact form

$$\int \frac{d\mathbf{q}}{(2\pi)^3} \frac{1}{\beta} \sum_\nu e^{i\mathbf{q}\cdot\mathbf{r}_1 - i\Omega_\nu \tau_1} \left[\delta\Delta(\mathbf{q}, \Omega_\nu) \left(A_e(\mathbf{q}, \Omega_\nu) + \frac{1}{v_0} \right) + \delta\Delta^\dagger(\mathbf{q}, \Omega_\nu) B_e(\mathbf{q}, \Omega_\nu) \right] . \tag{28.31}$$

There remains to enter the results (28.22), (28.23), and (28.31) into the approximate equation (28.21). Before doing that, however, it is convenient to make use of the *identity*

$$A_e(\mathbf{q}=0,\Omega_\nu=0) - |B_e(\mathbf{q}=0,\Omega_\nu=0)| + \frac{1}{v_0} = -\int \frac{d\mathbf{k}}{(2\pi)^3}\frac{[1-2f_F(E_\mathbf{k})]}{2E_\mathbf{k}}\,, \tag{28.32}$$

which can be readily proved from the explicit expressions of the bubbles (28.29) and (28.30), as reported in Section 29.1 (and which was also utilized in Ref. [57] for rephrasing the gap equation at the mean-field level in terms of these bubbles). In this way, Eq. (28.21) becomes:

$$\begin{aligned}
\frac{1}{v_0}\Delta(\mathbf{r},\tau) &= \Delta_e\left(A_e(\mathbf{q}=0,\Omega_\nu=0) - |B_e(\mathbf{q}=0,\Omega_\nu=0)| + \frac{1}{v_0}\right) \\
&\quad + \int \frac{d\mathbf{q}}{(2\pi)^3}\frac{1}{\beta}\sum_\nu e^{i\mathbf{q}\cdot\mathbf{r}-i\Omega_\nu\tau} \\
&\quad \times \left[\delta\Delta(\mathbf{q},\Omega_\nu)\left(A_e(\mathbf{q},\Omega_\nu) + \frac{1}{v_0}\right) + \delta\Delta^\dagger(\mathbf{q},\Omega_\nu)\,B_e(\mathbf{q},\Omega_\nu)\right] \\
&= \frac{1}{v_0}\left(\Delta_e + \delta\Delta(\mathbf{r},\tau)\right) + \Delta_e\left(A_e(\mathbf{q}=0,\Omega_\nu=0) - |B_e(\mathbf{q}=0,\Omega_\nu=0)|\right) \\
&\quad + \int \frac{d\mathbf{q}}{(2\pi)^3}\frac{1}{\beta}\sum_\nu e^{i\mathbf{q}\cdot\mathbf{r}-i\Omega_\nu\tau} \\
&\quad \times \left[\delta\Delta(\mathbf{q},\Omega_\nu)\,A_e(\mathbf{q},\Omega_\nu) + \delta\Delta^\dagger(\mathbf{q},\Omega_\nu)\,B_e(\mathbf{q},\Omega_\nu)\right]\,.
\end{aligned} \tag{28.33}$$

By recalling that we have assumed that $\Delta(\mathbf{r},\tau) = \Delta_e + \delta\Delta(\mathbf{r},\tau)$, Eq. (28.33) simplifies eventually to the final form:

$$\boxed{\begin{aligned}
&\Delta_e\left(A_e(\mathbf{q}=0,\Omega_\nu=0) - |B_e(\mathbf{q}=0,\Omega_\nu=0)|\right) \\
&+\int \tfrac{d\mathbf{q}}{(2\pi)^3}\tfrac{1}{\beta}\textstyle\sum_\nu e^{i\mathbf{q}\cdot\mathbf{r}-i\Omega_\nu\tau}\left[\delta\Delta(\mathbf{q},\Omega_\nu)A_e(\mathbf{q},\Omega_\nu) + \delta\Delta^\dagger(\mathbf{q},\Omega_\nu)B_e(\mathbf{q},\Omega_\nu)\right] = 0\,.
\end{aligned}} \tag{28.34}$$

Starting from this equation, we will obtain both the TDGP equation in Section 28.2 and the TDGL equation in Chapter 29, depending on the expressions of the particle–particle bubbles A_e and B_e that hold specifically in the two (GP and GL) cases.

Remark: Inhomogeneous mean-field decoupling vs homogeneous pairing fluctuations

It is worth emphasizing that, although Eq. (28.34) was obtained starting from a "mean-field decoupling" for a space- and time-*dependent* situation, this equation inherently contains the particle–particle bubbles (28.29) and (28.30) that would arise when including "beyond-mean-field pairing fluctuations" for a space- and time-*independent* situation [83, 84]. On physical grounds, this is because, when a physical system is subject to space and time variations of its environment, fluctuations over and above the mean field unavoidably build up in the system, to which the system in turn reacts. Specifically, we note that the above particle–particle bubbles correspond to those identified in Fig. 24.5 in the context of the T-matrix approximation. This is because to the BCS self-energy in the equilibrium case (as given by the expression (28.3) with the τ-dependence suppressed), there corresponds the following expression of the kernel Ξ of the Bethe–Salpeter equation

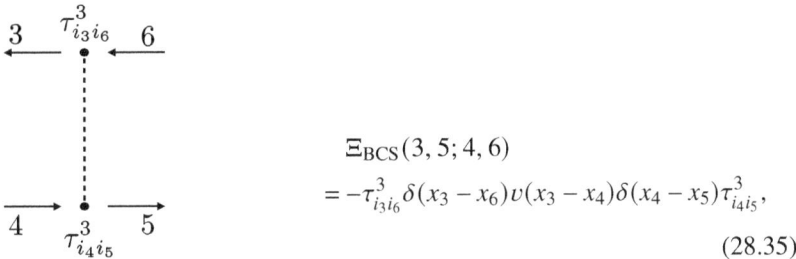

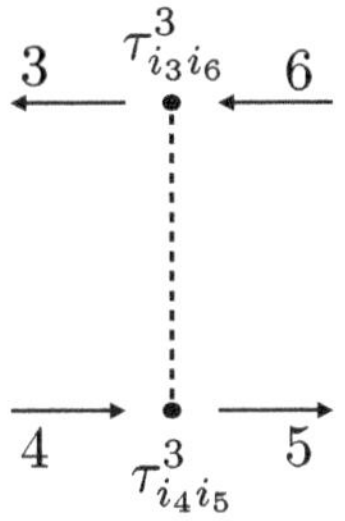

Figure 28.1 Graphical view of Eq. (28.35).

$$\Xi_{\mathrm{BCS}}(3,5;4,6) = -\tau^3_{i_3 i_6}\delta(x_3 - x_6)v(x_3 - x_4)\delta(x_4 - x_5)\tau^3_{i_4 i_5}, \tag{28.35}$$

(as given by Eq. (17.27) of Part I, now translated into Nambu indices), where $x = (\mathbf{r}, \tau)$ and with the further provision that $i_3 = i_6 \neq i_4 = i_5$ for a contact potential. The kernel (28.35) then gives rise to a whole sequence of ladder diagrams (collectively referred to as the BCS–RPA approximation), which are built on the above normal and anomalous particle–particle bubbles. These diagrams, in turn, are required to guarantee gauge invariance in the response functions when starting from a mean-field (BCS) configuration [80, 138].

28.2 TDGP Equation Arising in the BEC Limit at Low Temperature

It remains to specify Eq. (28.34) in the BEC limit of the BCS–BEC crossover at low temperature, which is characterized by the condition $|\Delta_e|/|\mu| \ll 1$ with $\mu < 0$. We postpone to Chapter 29, giving the general expressions for $A_e(\mathbf{q}, \Omega_\nu)$ of Eq. (28.29) and $B_e(\mathbf{q}, \Omega_\nu)$ of Eq. (28.30), which are valid at any temperature in the superfluid phase throughout the BCS–BEC crossover. Specifically, at zero temperature, these expressions can be calculated analytically throughout the BCS–BEC crossover by the methods of Ref. [113] in terms of complete elliptic integrals, for small values of $\mathbf{q}$ and $|\Omega_\nu|$. In particular, in the BEC limit of interest here, we obtain by the methods of Ref. [113] (cf. also Ref. [57]):

$$A_e(\mathbf{q}, \Omega_\nu) = \frac{m^2 a_F}{8\pi}\left(m a_F^2 |\Delta_e|^2 - \mu_B - i\Omega_\nu + \frac{\mathbf{q}^2}{4m} + \cdots\right), \tag{28.36}$$

$$B_e(\mathbf{q}, \Omega_\nu) = \frac{m^2 a_F}{8\pi}\left(\frac{1}{2} m a_F^2 \Delta_e^2 + \cdots\right), \tag{28.37}$$

where μ_B is the *bosonic chemical potential* for the composite bosons made up of fermion pairs in the BEC limit, obtained from the fermionic chemical potential μ via the relation [20]

$$2\mu = -\varepsilon_0 + \mu_B \,, \tag{28.38}$$

$\varepsilon_0 = (m a_F^2)^{-1}$ being the binding energy of the two-fermion problem in vacuum (with $\varepsilon_0 \gg \mu_B$). Note that, while $|\Delta_e|^2$ is present in the expression (28.36), Δ_e^2 is instead present in the expression (28.37), which thus contains twice the phase factor $e^{i\varphi_e}$ of the gap parameter Δ_e.

To arrive at the TDGP equation, we now manipulate Eq. (28.34) through the following steps.

Step # 6: Analytic continuation from imaginary to real time

Entering the expressions (28.36) and (28.37) into Eq. (28.34), we can perform the analytic continuation of the resulting expression from imaginary time $-i\tau$ to real time t (cf. also the illustration at the end of Chapter 12 in Part I).

We first consider the contribution of the constant terms in the expressions (28.36) and (28.37) to the second term on the right-hand side of Eq. (28.34), for which we obtain

$$\begin{aligned}
&\frac{m^2 a_F}{8\pi} \int \frac{d\mathbf{q}}{(2\pi)^3} \frac{1}{\beta} \sum_\nu e^{i(\mathbf{q}\cdot\mathbf{r} - i\Omega_\nu \tau)} \\
&\times \left[\delta\Delta(\mathbf{q}, \Omega_\nu) \left(m a_F^2 |\Delta_e|^2 - \mu_B \right) + \delta\Delta^\dagger(\mathbf{q}, \Omega_\nu) \, \frac{1}{2} m a_F^2 \Delta_e^2 \right] \\
&\underset{\text{[Eqs.(28.26) and (28.27)]}}{\overset{\uparrow}{=}} \frac{m^2 a_F}{8\pi} \left[\delta\Delta(\mathbf{r}, \tau) \left(m a_F^2 |\Delta_e|^2 - \mu_B \right) + \delta\Delta^\dagger(\mathbf{r}, \tau) \, \frac{1}{2} m a_F^2 \Delta_e^2 \right] \\
&\xrightarrow{(-i\tau \to t)} \frac{m^2 a_F}{8\pi} \left[\delta\Delta(\mathbf{r}, t) \left(m a_F^2 |\Delta_e|^2 - \mu_B \right) + \delta\Delta^*(\mathbf{r}, t) \, \frac{1}{2} m a_F^2 \Delta_e^2 \right] .
\end{aligned} \tag{28.39}$$

Adding to this expression the first term on the right-hand side of Eq. (28.34) yields

$$\begin{aligned}
&\frac{m^2 a_F}{8\pi} \left\{ \Delta_e \left[m a_F^2 |\Delta_e|^2 - \mu_B - \frac{1}{2} m a_F^2 |\Delta_e|^2 \right] \right. \\
&\qquad \left. + \, \delta\Delta(\mathbf{r}, t) \left(m a_F^2 |\Delta_e|^2 - \mu_B \right) + \delta\Delta^*(\mathbf{r}, t) \, \frac{1}{2} m a_F^2 \Delta_e^2 \right\} \\
&= \frac{m^2 a_F}{8\pi} \left\{ (\Delta_e + \delta\Delta(\mathbf{r}, t)) \left(\frac{1}{2} m a_F^2 |\Delta_e|^2 - \mu_B \right) \right. \\
&\qquad \left. + \, \frac{1}{2} m a_F^2 \left(\delta\Delta(\mathbf{r}, t) |\Delta_e|^2 + \delta\Delta^*(\mathbf{r}, t) \Delta_e^2 \right) \right\} ,
\end{aligned} \tag{28.40}$$

where the factor in the last term can be rewritten up to linear order in $\delta\Delta(\mathbf{r}, t)$

$$\delta\Delta(\mathbf{r}, t)|\Delta_e|^2 + \delta\Delta^*(\mathbf{r}, t)\Delta_e^2 \simeq \left(|\Delta(\mathbf{r}, t)|^2 - |\Delta_e|^2\right)\Delta(\mathbf{r}, t) \tag{28.41}$$

with $\Delta(\mathbf{r}, t) = \Delta_e + \delta\Delta(\mathbf{r}, t)$. In this way, the expression (28.40) simplifies to the form

$$\frac{m^2 a_F}{8\pi}\left(\frac{1}{2} m a_F^2 \, |\Delta(\mathbf{r}, t)|^2 - \mu_B\right)\Delta(\mathbf{r}, t) \, . \tag{28.42}$$

There remains to consider the wave-vector- and frequency-dependent terms in the expression (28.36), once inserted in the second term on the right-hand side of Eq. (28.34). We obtain for these terms:

$$\begin{aligned}
&\frac{m^2 a_F}{8\pi}\int\frac{d\mathbf{q}}{(2\pi)^3}\frac{1}{\beta}\sum_\nu e^{i\mathbf{q}\cdot\mathbf{r} - i\Omega_\nu\tau}\,\delta\Delta(\mathbf{q}, \Omega_\nu)\left(-i\Omega_\nu + \frac{\mathbf{q}^2}{4m}\right)\\
&= \frac{m^2 a_F}{8\pi}\left(\frac{\partial}{\partial\tau} - \frac{\nabla^2}{4m}\right)\int\frac{d\mathbf{q}}{(2\pi)^3}\frac{1}{\beta}\sum_\nu e^{i\mathbf{q}\cdot\mathbf{r} - i\Omega_\nu\tau}\,\delta\Delta(\mathbf{q}, \Omega_\nu)\\
&= \frac{m^2 a_F}{8\pi}\left(\frac{\partial}{\partial\tau} - \frac{\nabla^2}{4m}\right)\delta\Delta(\mathbf{r}, \tau) \xrightarrow{(-i\tau\to t)} \frac{m^2 a_F}{8\pi}\left(-i\frac{\partial}{\partial t} - \frac{\nabla^2}{4m}\right)\Delta(\mathbf{r}, t) \, .
\end{aligned} \tag{28.43}$$

Here, as anticipated at the beginning of this chapter and explicitly implemented in Step 7, we have eventually endowed Δ_e with a "slow" spatial variation (whereby $\Delta_e \to \Delta_e(\mathbf{r})$), in such a way that the assumption $|\nabla^2\Delta_e(\mathbf{r})| \ll |\nabla^2\delta\Delta(\mathbf{r}, t)|$ holds.

Putting together the results (28.42) and (28.43) into Eq. (28.34), we obtain eventually the following space- and time-dependent nonlinear differential equation for $\Delta(\mathbf{r}, t)$:

$$\left(-i\frac{\partial}{\partial t} - \frac{\nabla^2}{4m} + \frac{1}{2} m a_F^2 \, |\Delta(\mathbf{r}, t)|^2 - \mu_B\right)\Delta(\mathbf{r}, t) = 0 \, . \tag{28.44}$$

This is not yet the final result we are after, and a few more steps are still required to arrive at it.

Step # 7: A "local" approximation in space and time

In Step # 3, we have temporarily assumed to neglect the external potential $V_{\text{ext}}(\mathbf{r}, t)$, which would be the source of the space and time inhomogeneities to begin with. We are now going to reintroduce $V_{\text{ext}}(\mathbf{r}, t)$ via a *local* approximation in space and time, in a similar way to what was done in Refs. [89, 125] in a time-independent context. Accordingly, we replace the fermionic chemical potential μ, whenever it occurs in the single-particle Green's functions entering the expressions (28.29) and (28.30), by the "local" value $\mu - V_{\text{ext}}(\mathbf{r}, t)$ in space and time. In this way, the system is meant to be broken up into a set of homogeneous subsystems on which $V_{\text{ext}}(\mathbf{r}, t)$ acts locally, the different subsystems being connected to each other by sharing a common overall chemical potential μ. Owing to the definition (28.38) for the bosonic chemical potential μ_B, this local replacement on μ implies the following replacement on μ_B:

$$\mu_B \longrightarrow \mu_B - 2\, V_{\text{ext}}(\mathbf{r}, t)\ . \tag{28.45}$$

In this way, Eq. (28.44) becomes:

$$\left(-i\frac{\partial}{\partial t} - \frac{\nabla^2}{4m} + 2\, V_{\text{ext}}(\mathbf{r}, t) + \frac{1}{2} m a_F^2\, |\Delta(\mathbf{r}, t)|^2 - \mu_B \right) \Delta(\mathbf{r}, t) = 0\, . \tag{28.46}$$

Note here that the factor of 2, which multiplies the potential $V_{\text{ext}}(\mathbf{r}, t)$ as well as the fermionic mass m in the kinetic energy term, signals the composite nature of the bosons (as it was already pointed out in Section 26.7).

Step # 8: Reabsorbing the bosonic chemical potential

By our assumption, that the single-particle Hamiltonian $h^{\mathrm{M}}(\mathbf{r}, \tau)$ in Eq. (28.2) contains the fermionic chemical potential μ, we are also assuming that $\Delta(\mathbf{r}, \tau)$ given by Eq. (28.4) and $\Delta(\mathbf{r}, t)$ given by Eq. (23.21) are calculated with the grand-canonical Hamiltonian $K = H - \mu N$. This implies that $\Delta(\mathbf{r}, t)$ in Eq. (28.46) should be correctly interpreted as $\Delta_{\mathrm{K}}(\mathbf{r}, t)$. However, the TDGP equation for point-like bosons (to be distinguished from the composite bosons of interest here) is usually obtained from the Hamiltonian H (cf., e.g., Ref. [98]). To map the two equations into each other, we have thus to connect $\Delta_{\mathrm{K}}(\mathbf{r}, t)$ with $\Delta_{\mathrm{H}}(\mathbf{r}, t)$. This is readily done by making use of the definition (23.21) and of the transformation (27.12) with $i = 1$ and $j = 2$, yielding

$$\begin{aligned}
\Delta(\mathbf{r}, t) \leftrightarrow \Delta_{\mathrm{K}}(\mathbf{r}, t) &= e^{i\,2\mu t} \Delta_{\mathrm{H}}(\mathbf{r}, t) \underset{\substack{\uparrow \\ [\text{Eq.}(28.38)]}}{=} e^{i(-\varepsilon_0 + \mu_B)t} \Delta_{\mathrm{H}}(\mathbf{r}, t) \\
&= e^{i\mu_B t} \left(e^{i(-\varepsilon_0)t} \Delta_{\mathrm{H}}(\mathbf{r}, t) \right) \equiv e^{i\mu_B t} \bar{\Delta}(\mathbf{r}, t)\, .
\end{aligned} \tag{28.47}$$

Here, the definition of $\bar{\Delta}(\mathbf{r}, t) = e^{i(-\varepsilon_0)t} \Delta_{\mathrm{H}}(\mathbf{r}, t)$ emphasizes the fact that $(-\varepsilon_0)$ is the bound eigenvalue of the two-fermion eigen-problem in vacuum, to which the mean-field gap equation reduces in the (extreme) BEC limit, as mentioned in Section 26.4. In this way, $\bar{\Delta}(\mathbf{r}, t)$ is meant to reabsorb the fast (and uninteresting) time oscillations due to the trivial phase factor $e^{i\varepsilon_0 t}$ related to the internal degrees of freedom of the two-body problem (thereby concentrating on the translational degrees of freedom relevant for the residual interaction among composite bosons). We then obtain

$$-\,i\frac{\partial}{\partial t} \Delta(\mathbf{r}, t) = e^{i\mu_B t} \left(\mu_B - i\frac{\partial}{\partial t} \right) \bar{\Delta}(\mathbf{r}, t)\, , \tag{28.48}$$

such that Eq. (28.46) becomes

$$\left(-i\frac{\partial}{\partial t}-\frac{\nabla^2}{4m}+2\,V_{\text{ext}}(\mathbf{r},t)+\frac{1}{2}ma_F^2\,|\bar{\Delta}(\mathbf{r},t)|^2\right)\bar{\Delta}(\mathbf{r},t)=0\,. \tag{28.49}$$

Step # 9: Rescaling the gap parameter

It remains to rescale the gap parameter $\bar{\Delta}(\mathbf{r},t)$, in analogy to what was done in Ref. [89] in the time-independent case. This is done by introducing the time-dependent *condensate wave function*

$$\Phi(\mathbf{r},t)=\sqrt{\frac{m^2a_F}{8\pi}}\,\bar{\Delta}(\mathbf{r},t)\,. \tag{28.50}$$

In this way, the self-interaction term in Eq. (28.49) becomes

$$\frac{1}{2}ma_F^2\,|\bar{\Delta}(\mathbf{r},t)|^2=\frac{1}{2}ma_F^2\,\frac{8\pi}{m^2a_F}\,|\Phi(\mathbf{r},t)|^2=\frac{4\pi\,(2a_F)}{(2m)}\,|\Phi(\mathbf{r},t)|^2\,. \tag{28.51}$$

Note that in the last term we have highlighted the structure $\frac{4\pi a_B}{m_B}$ of the coefficient of the self-interaction term in the GP equation for point-like bosons with mass m_B and scattering length a_B, which Eq. (28.51) identifies as $m_B \leftrightarrow 2m$ and $a_B \leftrightarrow 2a_F$.

Accordingly, Eq. (28.49) reduces eventually the *TDGP equation* for composite bosons we were after:

$$\boxed{\left(-i\frac{\partial}{\partial t}-\frac{\nabla^2}{4m}+2\,V_{\text{ext}}(\mathbf{r},t)+\frac{4\pi\,(2a_F)}{(2m)}\,|\Phi(\mathbf{r},t)|^2\right)\Phi(\mathbf{r},t)=0\,.} \tag{28.52}$$

In Chapter 31, we shall further verify that the time derivative of first order appearing in Eq. (28.52) is indeed the dominant one in the BEC limit of the BCS–BEC crossover.

Remark: Scattering length a_B for composite bosons vs scattering length a_F for constituent fermions

As it is mentioned in Chapter 26, the value $a_B=2a_F$ that was obtained in Eq. (28.52) for the scattering length a_B of the low-energy scattering processes between two composite bosons (or dimers) is not the exact value, but it corresponds to the so-called Born approximation, which includes only the lowest-order scattering processes [87]. Consideration of higher-order scattering processes would, in fact, require one to include additional diagrammatic structures over and above those utilized to derive Eq. (28.52). This can be done by including an infinite sequence of dimer–dimer scattering processes, for which the "internal" structure of fermion pairs is consistently taken into account, so that, in the intermediate (virtual) states of the scattering processes, one of the dimers is broken into its fermionic constituents. These processes are sketched in Fig. 28.2. In particular, Fig. 28.2(a) shows the particle–particle propagators Γ_0 (given by Eq. (26.19) in the BEC limit), which can be connected with each other through four-leg structures. This is what is shown in Fig. 28.2(b), where the collection of all these diagrammatic structures is represented by a block that connects two incoming and two outgoing dimers. Specifically, the first two terms corresponding to the aforementioned Born approximation for the dimer–dimer scattering are here explicitly reported, which when evaluated in the limit of vanishing four-vectors yield the value $a_B = 2a_F$ [116], consistently with the result of Eq. (28.52). This result can be improved by considering the dimer–dimer T-matrix, which includes an infinite sequence of dimer–dimer scattering processes, as shown in Fig. 28.2(c). In this way, the value $a_B \simeq 0.75a_F$ was first obtained in Ref. [116]. This result was later improved in Ref. [127], where a complete diagrammatic treatment was given by summing up not only the series for the dimer–dimer T-matrix of Fig. 28.2(c) but also an additional infinite series for the residual dimer–dimer interaction (indicated simply by dots in Fig. 28.2(b)), yielding eventually the "exact" value $a_B \simeq 0.60a_F$.

Finally, we mention that the above results for the dimer–dimer scattering obtained in the BEC limit of the BCS–BEC crossover can be considered as a unique one of a kind, to the extent that it has been possible to obtain from first principles the relevant scattering properties of composite objects in terms of the scattering properties of their constituent fermions. An analogous information is not available, for instance, for ^{4}He, whose inter-particle potential is commonly modeled by a simplified phenomenological form (like a hard-sphere model), while the bosons are considered to be effectively point-like objects.

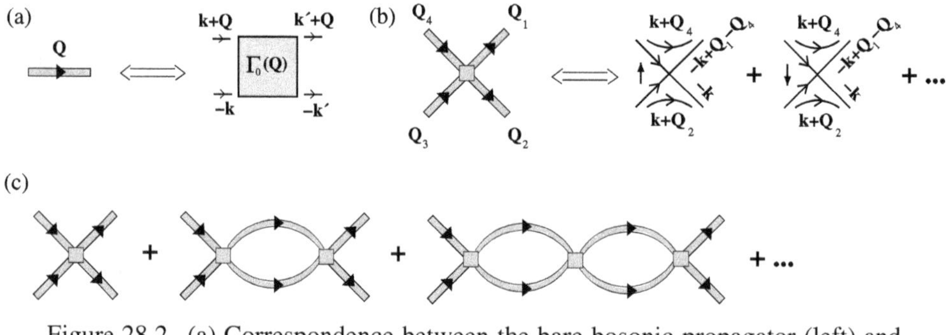

Figure 28.2 (a) Correspondence between the bare bosonic propagator (left) and the particle–particle propagator Γ_0 (right). (b) The dimer–dimer residual interaction is represented by a block (with $Q_1 + Q_2 = Q_3 + Q_4$ in four-vector notation), for which the two lowest-order terms are explicitly shown (right). Here, thin lines represent bare fermionic single-particle propagators G_0, whose spin labels are indicated in the internal lines. (c) T-matrix for the dimer–dimer scattering, where successive blocks of panel (b) are connected through a sequence of dimer–dimer bubbles. [Reproduced from Fig. 16 of Ref. [20], with permission.]

In practice, the TDGP equation for composite bosons has consistently been utilized in the BEC limit of the BCS–BEC crossover (reasonably, with the correct value of the bosonic scattering length a_B), to account, for instance, for the experimental results on the Josephson critical current when a barrier moves through an elongated trap containing a gas of Fermi atoms [139].

29

Derivation of the Time-Dependent Ginzburg–Landau Equation for Cooper Pairs in the BCS Limit of the BCS–BEC Crossover

Similar to the time-independent Ginzburg–Landau (GL) equation mentioned in Section 26.7, the time-dependent Ginzburg–Landau (TDGL) equation also holds close to the critical temperature of the initial equilibrium preparation and in the (extreme) BCS limit of the BCS–BEC crossover when Cooper pairs are largely overlapping with each other. These conditions correspond to the GL region shown in the temperature vs coupling phase diagram of Fig. 26.4.

In the present chapter, we will derive the TDGL equation from first principles by relying on the same approach that had led us to the TDGP equation in Chapter 28. Accordingly, we will start from Eq. (28.34) obtained therein, which we rewrite here for convenience

$$\Delta_e \left(A_e(\mathbf{q}=0,\Omega_\nu=0) - |B_e(\mathbf{q}=0,\Omega_\nu=0)|\right) + \int \frac{d\mathbf{q}}{(2\pi)^3} \frac{1}{\beta} \sum_\nu e^{i(\mathbf{q}\cdot\mathbf{r}-i\Omega_\nu\tau)}$$
$$\times \left[\delta\Delta(\mathbf{q},\Omega_\nu)\, A_e(\mathbf{q},\Omega_\nu) + \delta\Delta^\dagger(\mathbf{q},\Omega_\nu)\, B_e(\mathbf{q},\Omega_\nu)\right] = 0\,, \qquad (29.1)$$

where the normal and anomalous particle–particle bubbles are defined by Eqs. (28.29) and (28.30), respectively. In this way, the derivations of the TDGP and TDGL equations are placed on equal footing, to the extent that they differ by the explicit expressions of the particle–particle bubbles relevant to the two cases. In addition, besides the generic condition $|\delta\Delta(\mathbf{q},\Omega_\nu)| \ll |\Delta_e|$, we will now require that $|\Delta_e| \ll k_B T_c$ consistently with an initial preparation at a temperature T, such that $(T_c - T) \ll T_c$. In this limit, the expressions of $A_e(\mathbf{q},\Omega_\nu)$ and $B_e(\mathbf{q},\Omega_\nu)$ considerably simplify.

29.1 Generic Expressions of the Normal and Anomalous Particle–Particle Bubbles

Quite generally, the explicit expressions of the particle–particle bubbles $A_e(\mathbf{q},\Omega_\nu)$ and $B_e(\mathbf{q},\Omega_\nu)$ read

$$-A_e(\mathbf{q},\Omega_\nu)-\frac{1}{v_0}=\int\frac{d\mathbf{k}}{(2\pi)^3}\frac{1}{\beta}\sum_n G_e(\mathbf{k}+\mathbf{q},\omega_n+\Omega_\nu)_{11}G_e(-\mathbf{k},-\omega_n)_{11}$$
$$=\int\frac{d\mathbf{k}}{(2\pi)^3}\left\{u^2_{\mathbf{k}+\mathbf{q}}u^2_{\mathbf{k}}\frac{\left[1-f_F\left(E_{\mathbf{k}+\mathbf{q}}\right)-f_F\left(E_{\mathbf{k}}\right)\right]}{E_{\mathbf{k}+\mathbf{q}}+E_{\mathbf{k}}-i\Omega_\nu}+v^2_{\mathbf{k}+\mathbf{q}}v^2_{\mathbf{k}}\frac{\left[1-f_F\left(E_{\mathbf{k}+\mathbf{q}}\right)-f_F\left(E_{\mathbf{k}}\right)\right]}{E_{\mathbf{k}+\mathbf{q}}+E_{\mathbf{k}}+i\Omega_\nu}\right.$$
$$\left.-u^2_{\mathbf{k}+\mathbf{q}}v^2_{\mathbf{k}}\frac{\left[f_F\left(E_{\mathbf{k}+\mathbf{q}}\right)-f_F\left(E_{\mathbf{k}}\right)\right]}{E_{\mathbf{k}+\mathbf{q}}-E_{\mathbf{k}}-i\Omega_\nu}-v^2_{\mathbf{k}+\mathbf{q}}u^2_{\mathbf{k}}\frac{\left[f_F\left(E_{\mathbf{k}+\mathbf{q}}\right)-f_F\left(E_{\mathbf{k}}\right)\right]}{E_{\mathbf{k}+\mathbf{q}}-E_{\mathbf{k}}+i\Omega_\nu}\right\} \tag{29.2}$$

and

$$B_e(\mathbf{q},\Omega_\nu)=\int\frac{d\mathbf{k}}{(2\pi)^3}\frac{1}{\beta}\sum_n G_e(\mathbf{k}+\mathbf{q},\omega_n+\Omega_\nu)_{12}G_e(-\mathbf{k},-\omega_n)_{12}$$
$$=\int\frac{d\mathbf{k}}{(2\pi)^3}u_{\mathbf{k}+\mathbf{q}}u_{\mathbf{k}}v_{\mathbf{k}+\mathbf{q}}v_{\mathbf{k}}\left\{\frac{\left[1-f_F\left(E_{\mathbf{k}+\mathbf{q}}\right)-f_F\left(E_{\mathbf{k}}\right)\right]}{E_{\mathbf{k}+\mathbf{q}}+E_{\mathbf{k}}-i\Omega_\nu}\right.$$
$$\left.+\frac{\left[1-f_F\left(E_{\mathbf{k}+\mathbf{q}}\right)-f_F\left(E_{\mathbf{k}}\right)\right]}{E_{\mathbf{k}+\mathbf{q}}+E_{\mathbf{k}}+i\Omega_\nu}+\frac{\left[f_F\left(E_{\mathbf{k}+\mathbf{q}}\right)-f_F\left(E_{\mathbf{k}}\right)\right]}{E_{\mathbf{k}+\mathbf{q}}-E_{\mathbf{k}}-i\Omega_\nu}+\frac{\left[f_F\left(E_{\mathbf{k}+\mathbf{q}}\right)-f_F\left(E_{\mathbf{k}}\right)\right]}{E_{\mathbf{k}+\mathbf{q}}-E_{\mathbf{k}}+i\Omega_\nu}\right\}, \tag{29.3}$$

where $v^2_{\mathbf{k}}=1-u^2_{\mathbf{k}}=\frac{1}{2}\left(1-\frac{\xi_{\mathbf{k}}}{E_{\mathbf{k}}}\right)$ with $\xi_{\mathbf{k}}=\frac{\mathbf{k}^2}{2m}-\mu$ and $E_{\mathbf{k}}=\sqrt{\xi^2_{\mathbf{k}}+|\Delta_e|^2}$ (and where the right-hand side of Eq. (29.3) is meant to be multiplied by twice the phase factor $e^{i\varphi_e}$ originating from Δ_e).

Remark: Proof of Eq. (28.32)

In Chapter 28, we made use of the identity (28.32) to arrive at the equation (28.34). This identity can now be explicitly verified in terms of the expressions (29.2) and (29.3), from which we obtain

$$A_e(\mathbf{q}=0,\Omega_\nu=0)+\frac{1}{v_0}-|B_e(\mathbf{q}=0,\Omega_\nu=0)|=-\int\frac{d\mathbf{k}}{(2\pi)^3}\left(u^4_{\mathbf{k}}+v^4_{\mathbf{k}}+2u^2_{\mathbf{k}}v^2_{\mathbf{k}}\right)\frac{[1-2f_F(E_{\mathbf{k}})]}{2E_{\mathbf{k}}}$$
$$=-\int\frac{d\mathbf{k}}{(2\pi)^3}\frac{[1-2f_F(E_{\mathbf{k}})]}{2E_{\mathbf{k}}} \tag{29.4}$$

since $\left(u^4_{\mathbf{k}}+v^4_{\mathbf{k}}+2u^2_{\mathbf{k}}v^2_{\mathbf{k}}\right)=1$. Note that the result (29.4) holds irrespective of how the limit $\mathbf{q}=0$ and $\Omega_\nu=0$ is reached since in this case the particle–hole contributions to A_e and B_e cancel out with each other.

29.2 Static and Dynamic Limits of Their Particle–Hole Subunits

As already mentioned while commenting on Fig. 24.5, we emphasize again that the name "particle–particle" bubbles used for A_e and B_e is only conventional owing to the particle–hole mixing of the pairing theory. This is explicitly verified from the expressions (29.2) and (29.3), which contain, at the same time, *particle–particle contributions* (corresponding to the factor $\left[1 - f_F(E_{\mathbf{k+q}}) - f_F(E_{\mathbf{k}})\right]$ in the numerator and the energy sum $E_{\mathbf{k+q}} + E_{\mathbf{k}}$ in the denominator), as well as *particle–hole contributions* (corresponding to the factor $\left[f_F(E_{\mathbf{k+q}}) - f_F(E_{\mathbf{k}})\right]$ in the numerator and the energy difference $E_{\mathbf{k+q}} - E_{\mathbf{k}}$ in the denominator). The crucial point here is that, while the particle–particle contributions admit an analytic expansion about $\mathbf{q} = 0$ and $\Omega_\nu = 0$, the particle–hole contributions do not. This is because the ratio

$$\frac{\left[1 - f_F(E_{\mathbf{k+q}}) - f_F(E_{\mathbf{k}})\right]}{E_{\mathbf{k+q}} + E_{\mathbf{k}} \pm i\Omega_\nu} \xrightarrow{(\Omega_\nu \to 0 \text{ and } \mathbf{q} \to 0)} \frac{\left[1 - 2 f_F(E_{\mathbf{k}})\right]}{2\, E_{\mathbf{k}}} \tag{29.5}$$

approaches a well-defined limit when $\Omega_\nu \to 0$ and $\mathbf{q} \to 0$ owing to the presence of the energy sum in the denominator (which thus never vanishes), while the ratio

$$\frac{\left[f_F(E_{\mathbf{k+q}}) - f_F(E_{\mathbf{k}})\right]}{E_{\mathbf{k+q}} - E_{\mathbf{k}} \pm i\Omega_\nu} \tag{29.6}$$

admits instead different limits when *either* $\Omega_\nu = 0$ and $\mathbf{q} \to 0$ *or* $\mathbf{q} = 0$ and $\Omega_\nu \to 0$ (which are referred to as the *static limit* and *dynamic limit*, respectively). This can be seen by expanding for small $\mathbf{q}$

$$f_F(E_{\mathbf{k+q}}) \simeq f_F(E_{\mathbf{k}} + \nabla E_{\mathbf{k}} \cdot \mathbf{q}) \simeq f_F(E_{\mathbf{k}}) + \frac{\partial f_F(E_{\mathbf{k}})}{\partial E_{\mathbf{k}}} \nabla E_{\mathbf{k}} \cdot \mathbf{q}, \tag{29.7}$$

such that

$$f_F(E_{\mathbf{k+q}}) - f_F(E_{\mathbf{k}}) \simeq \frac{\partial f_F(E_{\mathbf{k}})}{\partial E_{\mathbf{k}}} \left[E_{\mathbf{k+q}} - E_{\mathbf{k}}\right], \tag{29.8}$$

from which

$$\frac{\left[f_F(E_{\mathbf{k+q}}) - f_F(E_{\mathbf{k}})\right]}{E_{\mathbf{k+q}} - E_{\mathbf{k}} \pm i\Omega_\nu} \simeq \frac{\frac{\partial f_F(E_{\mathbf{k}})}{\partial E_{\mathbf{k}}} \left[E_{\mathbf{k+q}} - E_{\mathbf{k}}\right]}{E_{\mathbf{k+q}} - E_{\mathbf{k}} \pm i\Omega_\nu} = \begin{cases} \frac{\partial f_F(E_{\mathbf{k}})}{\partial E_{\mathbf{k}}} & \text{for } \Omega_\nu = 0 \text{ and } \mathbf{q} \to 0 \\ 0 & \text{for } \mathbf{q} = 0 \text{ and } \Omega_\nu \to 0\,. \end{cases} \tag{29.9}$$

The relevance of the difference between static and dynamic limits is well known in the context of the response functions of many-body physics. At its simplest level, this difference is already contained in the density–density correlation (or Lindhard) function for free fermions [7]. In more elaborated contexts, taking due account of this difference is crucial, for instance, in the microscopic derivation of the transport equation in the Landau theory of normal Fermi liquids [6, 9, 12]. In the next section, we shall explicitly highlight only those physical regimes for which the particle–hole terms do not contribute to the expressions (29.2) and (29.3), avoiding in this way taking any side (as it was done in Ref. [137]) about which of the two limits would apply to more general physical situations one may like to consider otherwise.

29.3 Three Relevant Cases

The particle–hole terms do not contribute to the expressions (29.2) and (29.3) in the following cases.

CASE 1: The limit of zero temperature for any coupling In this case,

$$f_F(E_{\mathbf{k}}) = \frac{1}{e^{\beta E_{\mathbf{k}}} + 1} \xrightarrow{(\beta \to \infty)} 0 \tag{29.10}$$

provided $E_{\mathbf{k}} = \sqrt{\xi_{\mathbf{k}}^2 + |\Delta_e|^2} \neq 0$, which is the case when $|\Delta_e| \neq 0$ irrespective of coupling. As a consequence, in the expressions (29.2) and (29.3), the particle–hole contributions drop out, and only the terms containing the unity survive in the particle–particle contributions. We have already utilized this result in Chapter 28 to arrive at the results (28.36) and (28.37) in the BEC limit.

CASE 2: The BEC limit The BEC limit is slightly more general, being achieved even at finite β for $\mu < 0$ with $|\mu|$ sufficiently large, such that $\beta\mu \longrightarrow -\infty$. In this case, $E_{\mathbf{k}} = \sqrt{\left(\frac{\mathbf{k}^2}{2m} + |\mu|\right)^2 + |\Delta_e|^2}$ becomes large irrespective of the value of $|\Delta_e|$, such that $f_F(E_{\mathbf{k}}) \to 0$. The final results for the expressions (29.2) and (29.3) then coincide with those of Case 1.

CASE 3: The BCS limit when $T \to T_c^-$ In this case, $|\Delta_e| \to 0$, such that the term $B_e(\mathbf{q}, \Omega_\nu)$ given by Eq. (29.3) can be neglected being proportional to $|\Delta_e|^2$.[1] This is because $u_{\mathbf{k}} v_{\mathbf{k}} = \frac{1}{2}\sqrt{1 - \frac{\xi_{\mathbf{k}}^2}{E_{\mathbf{k}}^2}} = \frac{|\Delta_e|}{2E_{\mathbf{k}}} \propto |\Delta_e|$ for each of the two factors in Eq. (29.3). On the other hand, that in the limit $|\Delta_e| \to 0$, the particle–hole contributions disappear also from the term $A_e(\mathbf{q}, \Omega_\nu)$ given by Eq. (29.2) is not a priori evident and can explicitly be verified as follows.

We begin by changing the integration variables in the expression (29.2), by setting $\mathbf{k} + \mathbf{q} = \mathbf{k}' + \frac{\mathbf{q}}{2}$ and $\mathbf{k} = \mathbf{k}' - \frac{\mathbf{q}}{2}$, such that we write

$$u_{\mathbf{k}+\mathbf{q}} = u_{\mathbf{k}'+\frac{\mathbf{q}}{2}} \equiv u_{\mathbf{k}'}^{(+)} = \sqrt{\frac{1}{2}\left(1 + \frac{\xi_{\mathbf{k}'}^{(+)}}{E_{\mathbf{k}'}^{(+)}}\right)}, \; u_{\mathbf{k}} = u_{\mathbf{k}'-\frac{\mathbf{q}}{2}} \equiv u_{\mathbf{k}'}^{(-)} = \sqrt{\frac{1}{2}\left(1 + \frac{\xi_{\mathbf{k}'}^{(-)}}{E_{\mathbf{k}'}^{(-)}}\right)} \tag{29.11}$$

$$v_{\mathbf{k}+\mathbf{q}} = v_{\mathbf{k}'+\frac{\mathbf{q}}{2}} \equiv v_{\mathbf{k}'}^{(+)} = \sqrt{\frac{1}{2}\left(1 - \frac{\xi_{\mathbf{k}'}^{(+)}}{E_{\mathbf{k}'}^{(+)}}\right)}, \; v_{\mathbf{k}} = v_{\mathbf{k}'-\frac{\mathbf{q}}{2}} \equiv v_{\mathbf{k}'}^{(-)} = \sqrt{\frac{1}{2}\left(1 - \frac{\xi_{\mathbf{k}'}^{(-)}}{E_{\mathbf{k}'}^{(-)}}\right)} \tag{29.12}$$

1 What one actually neglects is the term $\delta\Delta^\dagger(\mathbf{q}, \Omega_\nu) B_e(\mathbf{q}, \Omega_\nu)$ in Eq. (29.1).

with $\xi_{\mathbf{k}'}^{(\pm)} = \xi_{\mathbf{k}' \pm \frac{\mathbf{q}}{2}}$ and $E_{\mathbf{k}'}^{(\pm)} = E_{\mathbf{k}' \pm \frac{\mathbf{q}}{2}}$. In this way, the expression (29.2) becomes

$$
\begin{aligned}
-A_e(\mathbf{q}, \Omega_\nu) - \frac{1}{v_0} = \frac{1}{4} \int \frac{d\mathbf{k}'}{(2\pi)^3} \Bigg\{ & \left(1 + \frac{\xi_{\mathbf{k}'}^{(+)}}{E_{\mathbf{k}'}^{(+)}}\right)\left(1 + \frac{\xi_{\mathbf{k}'}^{(-)}}{E_{\mathbf{k}'}^{(-)}}\right) \frac{\left[1 - f_F\left(E_{\mathbf{k}'}^{(+)}\right) - f_F\left(E_{\mathbf{k}'}^{(-)}\right)\right]}{E_{\mathbf{k}'}^{(+)} + E_{\mathbf{k}'}^{(-)} - i\Omega_\nu} \\
& + \left(1 - \frac{\xi_{\mathbf{k}'}^{(+)}}{E_{\mathbf{k}'}^{(+)}}\right)\left(1 - \frac{\xi_{\mathbf{k}'}^{(-)}}{E_{\mathbf{k}'}^{(-)}}\right) \frac{\left[1 - f_F\left(E_{\mathbf{k}'}^{(+)}\right) - f_F\left(E_{\mathbf{k}'}^{(-)}\right)\right]}{E_{\mathbf{k}'}^{(+)} + E_{\mathbf{k}'}^{(-)} + i\Omega_\nu} \\
& - \left(1 + \frac{\xi_{\mathbf{k}'}^{(+)}}{E_{\mathbf{k}'}^{(+)}}\right)\left(1 - \frac{\xi_{\mathbf{k}'}^{(-)}}{E_{\mathbf{k}'}^{(-)}}\right) \frac{\left[f_F\left(E_{\mathbf{k}'}^{(+)}\right) - f_F\left(E_{\mathbf{k}'}^{(-)}\right)\right]}{E_{\mathbf{k}'}^{(+)} - E_{\mathbf{k}'}^{(-)} - i\Omega_\nu} \\
& - \left(1 - \frac{\xi_{\mathbf{k}'}^{(+)}}{E_{\mathbf{k}'}^{(+)}}\right)\left(1 + \frac{\xi_{\mathbf{k}'}^{(-)}}{E_{\mathbf{k}'}^{(-)}}\right) \frac{\left[f_F\left(E_{\mathbf{k}'}^{(+)}\right) - f_F\left(E_{\mathbf{k}'}^{(-)}\right)\right]}{E_{\mathbf{k}'}^{(+)} - E_{\mathbf{k}'}^{(-)} + i\Omega_\nu} \Bigg\},
\end{aligned}
\tag{29.13}
$$

such that when $|\Delta_e| \longrightarrow 0$, we may replace $E_{\mathbf{k}'}^{(\pm)} \longrightarrow |\xi_{\mathbf{k}'}^{(\pm)}|$ in all terms within braces. Depending on the signs of $\xi_{\mathbf{k}'}^{(+)}$ and $\xi_{\mathbf{k}'}^{(-)}$, four cases are then possible for the expression within braces in Eq. (29.13). [Recall that, in the BCS limit, we are considering, $\mu \simeq E_F$ such that $\xi_{\mathbf{k}'}$ can take either sign.]

$\xi_{\mathbf{k}'}^{(+)} > 0$ and $\xi_{\mathbf{k}'}^{(-)} > 0$

$$
\{\cdots\} \longrightarrow 4 \frac{\left[1 - f_F\left(\xi_{\mathbf{k}'}^{(+)}\right) - f_F\left(\xi_{\mathbf{k}'}^{(-)}\right)\right]}{\xi_{\mathbf{k}'}^{(+)} + \xi_{\mathbf{k}'}^{(-)} - i\Omega_\nu}. \tag{29.14}
$$

$\xi_{\mathbf{k}'}^{(+)} < 0$ and $\xi_{\mathbf{k}'}^{(-)} < 0$

$$
\{\cdots\} \longrightarrow 4 \frac{\left[1 - f_F\left(-\xi_{\mathbf{k}'}^{(+)}\right) - f_F\left(-\xi_{\mathbf{k}'}^{(-)}\right)\right]}{-\xi_{\mathbf{k}'}^{(+)} - \xi_{\mathbf{k}'}^{(-)} + i\Omega_\nu} = 4 \frac{\left[1 - f_F\left(\xi_{\mathbf{k}'}^{(+)}\right) - f_F\left(\xi_{\mathbf{k}'}^{(-)}\right)\right]}{\xi_{\mathbf{k}'}^{(+)} + \xi_{\mathbf{k}'}^{(-)} - i\Omega_\nu} \tag{29.15}
$$

since $f_F(-\varepsilon) = 1 - f_F(\varepsilon)$.

$\xi_{\mathbf{k}'}^{(+)} > 0$ and $\xi_{\mathbf{k}'}^{(-)} < 0$

$$\{\cdots\} \longrightarrow -4\,\frac{\left[f_F\left(\xi_{\mathbf{k}'}^{(+)}\right)-f_F\left(-\xi_{\mathbf{k}'}^{(-)}\right)\right]}{\xi_{\mathbf{k}'}^{(+)}+\xi_{\mathbf{k}'}^{(-)}-i\Omega_\nu} = 4\,\frac{\left[1-f_F\left(\xi_{\mathbf{k}'}^{(+)}\right)-f_F\left(\xi_{\mathbf{k}'}^{(-)}\right)\right]}{\xi_{\mathbf{k}'}^{(+)}+\xi_{\mathbf{k}'}^{(-)}-i\Omega_\nu}. \tag{29.16}$$

$\xi_{\mathbf{k}'}^{(+)} < 0$ and $\xi_{\mathbf{k}'}^{(-)} > 0$

$$\{\cdots\} \longrightarrow -4\,\frac{\left[f_F\left(-\xi_{\mathbf{k}'}^{(+)}\right)-f_F\left(\xi_{\mathbf{k}'}^{(-)}\right)\right]}{-\xi_{\mathbf{k}'}^{(+)}-\xi_{\mathbf{k}'}^{(-)}+i\Omega_\nu} = 4\,\frac{\left[1-f_F\left(\xi_{\mathbf{k}'}^{(+)}\right)-f_F\left(\xi_{\mathbf{k}'}^{(-)}\right)\right]}{\xi_{\mathbf{k}'}^{(+)}+\xi_{\mathbf{k}'}^{(-)}-i\Omega_\nu}. \tag{29.17}$$

We then conclude that, in the limit $|\Delta_e| \longrightarrow 0$, the same result is obtained for the expression within braces in Eq. (29.13) in all four cases. We thus write in this limit,

$$\begin{aligned} -A_e(\mathbf{q},\Omega_\nu) - \frac{1}{v_0} &\xrightarrow{(|\Delta_e|\to 0)} \int \frac{d\mathbf{k}'}{(2\pi)^3}\,\frac{\left[1-f_F\left(\xi_{\mathbf{k}'}^{(+)}\right)-f_F\left(\xi_{\mathbf{k}'}^{(-)}\right)\right]}{\xi_{\mathbf{k}'}^{(+)}+\xi_{\mathbf{k}'}^{(-)}-i\Omega_\nu} \\ &\xrightarrow{(\mathbf{q}=0 \text{ and } \Omega_\nu=0)} \int \frac{d\mathbf{k}}{(2\pi)^3}\,\frac{[1-2f_F(\xi_{\mathbf{k}})]}{2\,\xi_{\mathbf{k}}}, \end{aligned} \tag{29.18}$$

where the last line holds no matter how the limit $\mathbf{q}=0$ and $\Omega_\nu=0$ is reached. The first line of Eq. (29.18) can be further manipulated as follows. From $\xi_{\mathbf{k}}^{(\pm)} = \xi_{\mathbf{k}\pm\frac{\mathbf{q}}{2}} = \xi_{\mathbf{k}} + \frac{\mathbf{q}^2}{8m} \pm \frac{\mathbf{k}\cdot\mathbf{q}}{2m}$, it follows that $\xi_{\mathbf{k}}^{(+)} + \xi_{\mathbf{k}}^{(-)} = 2\left(\xi_{\mathbf{k}} + \frac{\mathbf{q}^2}{8m}\right) = 2\left(\xi_{\mathbf{k}}^{(-)} + \frac{\mathbf{k}\cdot\mathbf{q}}{2m}\right)$. In this way, the expression (29.18) becomes:

$$\begin{aligned} -A_e(\mathbf{q},\Omega_\nu) - \frac{1}{v_0} &\xrightarrow{(|\Delta_e|\to 0)} \int \frac{d\mathbf{k}}{(2\pi)^3}\,\frac{\left[1-f_F\left(\xi_{\mathbf{k}}^{(+)}\right)-f_F\left(\xi_{\mathbf{k}}^{(-)}\right)\right]}{2\xi_{\mathbf{k}}+\frac{\mathbf{q}^2}{4m}-i\Omega_\nu} \\ &= \int \frac{d\mathbf{k}}{(2\pi)^3}\,\frac{\left[1-2f_F\left(\xi_{\mathbf{k}}^{(-)}\right)\right]}{2\xi_{\mathbf{k}}^{(-)}+\frac{\mathbf{k}\cdot\mathbf{q}}{m}-i\Omega_\nu} \underset{[\mathbf{p}=\mathbf{k}-\mathbf{q}/2]}{\overset{\uparrow}{=}} \int \frac{d\mathbf{p}}{(2\pi)^3}\,\frac{[1-2f_F(\xi_{\mathbf{p}})]}{2\xi_{\mathbf{p}}+\frac{\mathbf{p}\cdot\mathbf{q}}{m}+\frac{\mathbf{q}^2}{2m}-i\Omega_\nu} \\ &= \int \frac{d\mathbf{p}}{(2\pi)^3}\,\frac{\tanh\left(\frac{\beta\xi_{\mathbf{p}}}{2}\right)}{2\xi_{\mathbf{p}}+\frac{\mathbf{p}\cdot\mathbf{q}}{m}+\frac{\mathbf{q}^2}{2m}-i\Omega_\nu}. \end{aligned} \tag{29.19}$$

This is the approximate expression for $A_e(\mathbf{q},\Omega_\nu)$ that we shall use in Eq. (29.1) (together with the other approximate expression $B_e(\mathbf{q},\Omega_\nu) \simeq 0$), to arrive eventually at the TDGL equation.

29.4 Anticipating the Analytic Continuation from Imaginary to Real Time

The integral over the wave vector in Eq. (29.19) needs to be regularized because the denominator in the integrand may vanish when $\Omega_\nu = 0$. This regularization can consistently be achieved by anticipating at this point the planned analytic continuation from imaginary to real time, thereby letting $i\Omega_\nu \rightarrow \omega + i\eta$, where ω is a real frequency and $\eta = 0^+$. We thus write:

$$\frac{1}{2\,\xi_{\mathbf{p}} + \frac{\mathbf{p}\cdot\mathbf{q}}{m} + \frac{\mathbf{q}^2}{2m} - i\Omega_\nu} \longrightarrow \frac{1}{2\,\xi_{\mathbf{p}} + \frac{\mathbf{p}\cdot\mathbf{q}}{m} + \frac{\mathbf{q}^2}{2m} - \omega - i\eta}\,. \tag{29.20}$$

On physical grounds, this replacement is consistent with the causal (or retarded) condition relevant to the physical situation of interest. In this way, the right-hand side of Eq. (29.20) can be expanded for small enough values of $\mathbf{q}$ and ω, yielding

$$\begin{aligned}\frac{1}{2\xi_{\mathbf{p}} + \frac{\mathbf{p}\cdot\mathbf{q}}{m} + \frac{\mathbf{q}^2}{2m} - \omega - i\eta} &= \frac{1}{2\xi_{\mathbf{p}} - i\eta}\,\frac{1}{1 + \left(\frac{\frac{\mathbf{p}\cdot\mathbf{q}}{m} + \frac{\mathbf{q}^2}{2m} - \omega}{2\xi_{\mathbf{p}} - i\eta}\right)} \\ &\simeq \frac{1}{2\xi_{\mathbf{p}} - i\eta}\left[1 - \left(\frac{\frac{\mathbf{p}\cdot\mathbf{q}}{m} + \frac{\mathbf{q}^2}{2m} - \omega}{2\xi_{\mathbf{p}} - i\eta}\right) + \left(\frac{\frac{\mathbf{p}\cdot\mathbf{q}}{m} + \frac{\mathbf{q}^2}{2m} - \omega}{2\xi_{\mathbf{p}} - i\eta}\right)^2 + \cdots\right].\end{aligned} \tag{29.21}$$

Accordingly, the relevant expression for $A_e(\mathbf{q}, \omega)$ becomes:

$$\begin{aligned}- A_e(\mathbf{q}, \Omega_\nu) - \frac{1}{v_0} &\xrightarrow{(|\Delta_e| \to 0)} \int \frac{d\mathbf{p}}{(2\pi)^3}\,\frac{\tanh\left(\frac{\beta\xi_{\mathbf{p}}}{2}\right)}{2\xi_{\mathbf{p}}} - \left(\frac{\mathbf{q}^2}{2m} - \omega\right)\int \frac{d\mathbf{p}}{(2\pi)^3}\,\frac{\tanh\left(\frac{\beta\xi_{\mathbf{p}}}{2}\right)}{(2\xi_{\mathbf{p}} - i\eta)^2} \\ &+ \int \frac{d\mathbf{p}}{(2\pi)^3}\,\frac{\tanh\left(\frac{\beta\xi_{\mathbf{p}}}{2}\right)}{(2\xi_{\mathbf{p}} - i\eta)^3}\left[\left(\frac{\mathbf{p}\cdot\mathbf{q}}{m}\right)^2 + \omega^2 + \cdots\right].\end{aligned} \tag{29.22}$$

[Note that the irrelevant term $-i\eta$ has been dropped from the denominator in the first integral on the right-hand side of Eq. (29.22).]

To proceed further, we recall that the present calculation applies in the weak-coupling (BCS) limit and close to T_c. We can then resort to a standard approximation adopted in the BCS limit, whereby

$$\int \frac{d\mathbf{k}}{(2\pi)^3}\,F(\xi_{\mathbf{p}}) \simeq N_0 \int_{-\infty}^{+\infty} d\xi\, F(\xi)\;, \tag{29.23}$$

that holds for a smooth function $F(\xi_{\mathbf{p}})$ of $\xi_{\mathbf{p}} = \frac{\mathbf{p}^2}{2m} - \mu$ for which the integral on the right-hand side is convergent in the ultraviolet, where $\mu \simeq E_F$ and $N_0 = \frac{mk_F}{2\pi^2}$ is the density of states at the Fermi level (per spin component). This approximation thus applies to the integrals in the second and third terms on the right-hand side of Eq. (29.22), while the integral in the first term will be regularized below in a different way. In addition, consistent with the above approximation, in the last term on the right-hand side of Eq. (29.22), we replace $\left(\frac{\mathbf{p}\cdot\mathbf{q}}{m}\right)^2 \rightarrow \frac{\mathbf{p}^2\,\mathbf{q}^2}{3\,m^2} \rightarrow \frac{k_F^2\,\mathbf{q}^2}{3\,m^2}$. Finally, in the second and third terms on the right-hand side of Eq. (29.22), we can set $\mu = E_F$ and $\beta = \beta_c$ since these terms are proportional to powers of $\mathbf{q}$ and ω. On the other hand, we shall keep μ and β as they stand in the first term on the right-hand side of Eq. (29.22), in conjunction with an analogous term originating from $(A_e(\mathbf{q}=0,\Omega_\nu=0) - |B_e(\mathbf{q}=0,\Omega_\nu=0)|)$ in Eq. (29.1) (where we shall also keep $|\Delta_e| \neq 0$ albeit with a small value).

29.5 Evaluation of a Main Integral

There remains to evaluate the following integral

$$K_n \equiv \int_{-\infty}^{+\infty} d\xi \; \frac{\tanh\left(\frac{\beta\xi}{2}\right)}{(2\,\xi + i\eta)^n} \tag{29.24}$$

for $n=2$ and $n=3$, in terms of which the expression (29.22) can be cast in the compact form

$$\begin{aligned} &- A_e(\mathbf{q},\Omega_\nu) - \frac{1}{v_0} \xrightarrow{(|\Delta_e|\to 0)} \int \frac{d\mathbf{p}}{(2\pi)^3}\, \frac{\tanh\left(\frac{\beta\xi_{\mathbf{p}}}{2}\right)}{2\xi_{\mathbf{p}}} \\ &+ N_0 \left(\frac{\mathbf{q}^2}{2m} - \omega\right) K_2 + N_0 \left(\frac{v_F^2\mathbf{q}^2}{3} + \omega^2\right) K_3 + \cdots \end{aligned} \tag{29.25}$$

where $v_F = k_F/m$ is the Fermi velocity. We are going to prove that

$$K_n = 2\left(\frac{\beta}{2\,\pi\, i}\right)^{n-1} (1 - 2^{-n})\;\zeta(n) \qquad (n > 1), \tag{29.26}$$

where $\zeta(n)$ is the Riemann zeta function of integer argument n [140].

Proof of Eq. (29.26)

We recall that the function $\tanh(\pi z)$ of the complex variable z has simple poles along the imaginary axis at $z_\ell = i(\ell + 1/2)$ (ℓ integer) with residue $1/\pi$. On the other hand, the function $(2z + i\eta)^n$ has no singularities in the upper-half of the complex z-plane (including the real axis) owing to the presence of the $i\eta$ term. As a consequence, the integral (29.24) with $n > 1$ can be evaluated with the help of Jordan lemma, by deforming the integration contour as indicated in Fig. 29.1:

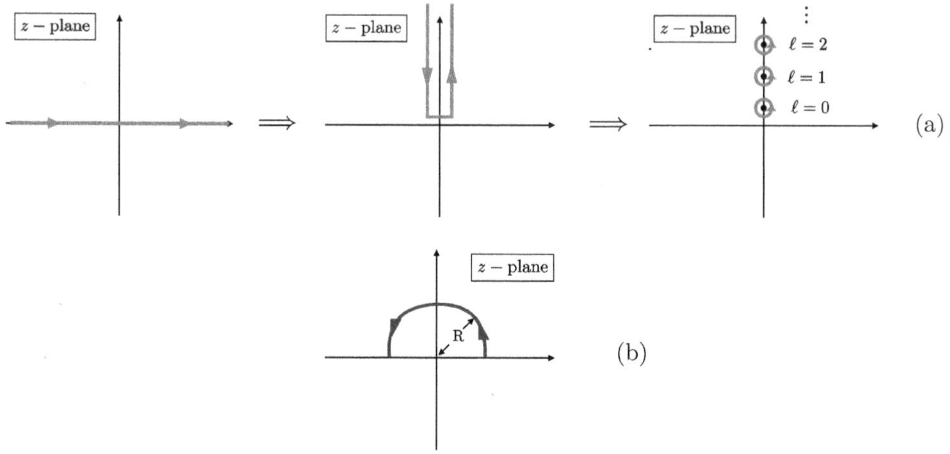

Figure 29.1 (a) The original contour along the horizontal axis of the integral (29.24) (left) is first deformed into a contour surrounding the upper part of the vertical axis (center) and then broken up into a sequence of small circles (right). (b) This is possible because the contribution of the half-circle of radius R in the upper-half of the complex z-plane is negligible in the limit $\mathrm{R} \to \infty$.

$$
\begin{aligned}
K_n &= \int_{-\infty}^{+\infty} dz \frac{\tanh\left(\frac{\beta z}{2}\right)}{(2z+i\eta)^n} \underset{[z'=\frac{\beta z}{2\pi}]}{\overset{=}{\uparrow}} \frac{2\pi}{\beta} \sum_{\ell=0}^{+\infty} \oint_{C_\ell} dz' \frac{\tanh(\pi z')}{\left(2z'\frac{2\pi}{\beta}\right)^n} \\
&= \frac{2\pi}{\beta} \frac{2i}{i^n \left(\frac{2\pi}{\beta}\right)^n} \sum_{\ell=0}^{+\infty} \frac{1}{(2\ell+1)^n} = 2\left(\frac{\beta}{2\pi i}\right)^{n-1} \sum_{\ell=0}^{+\infty} \frac{1}{(2\ell+1)^n} \\
&= 2\left(\frac{\beta}{2\pi i}\right)^{n-1} \left(1-2^{-n}\right)\zeta(n)
\end{aligned}
\tag{29.27}
$$

from the definition of the function $\zeta(n)$ [140], where C_ℓ are small circular contours centered about z_ℓ (cf. Fig. 29.1(a)). [QED]

In particular, for $n=2$ and $n=3$, the expression (29.26) yields:

$$
K_2 = 2\left(\frac{\beta}{2\pi i}\right)\left(1-\frac{1}{4}\right)\zeta(2) = -i\frac{\beta}{\pi}\frac{3}{4}\zeta(2), \tag{29.28}
$$

$$
K_3 = 2\left(\frac{\beta}{2\pi i}\right)^2\left(1-\frac{1}{8}\right)\zeta(3) = -\frac{\beta^2}{2\pi^2}\frac{7}{8}\zeta(3), \tag{29.29}
$$

where $\zeta(2) = \frac{\pi^2}{6} \simeq 1.645$ and $\zeta(3) \simeq 1.202$. Recall also that we are allowed to set $\beta = \beta_c = \frac{1}{k_B T_c}$ in Eqs. (29.28) and (29.29), consistently with our approximations.

There remains to regularize the first integral on the right-hand side of Eq. (29.25), which is divergent in the ultraviolet (but convergent in the infrared).

This regularization is simply achieved by moving the constant term $1/v_0$ from the left- to the right-hand side of Eq. (29.25) and recalling Eq. (25.22), which relates v_0 to the scattering length a_F of the two-fermion problem. In this way, we end up with the expression

$$\int \frac{d\mathbf{p}}{(2\pi)^3} \frac{\tanh\left(\frac{\beta\xi_\mathbf{p}}{2}\right)}{2\,\xi_\mathbf{p}} + \frac{1}{v_0} = \frac{m}{4\,\pi\,a_F} + \int \frac{d\mathbf{p}}{(2\pi)^3} \left[\frac{\tanh\left(\frac{\beta\xi_\mathbf{p}}{2}\right)}{2\,\xi_\mathbf{p}} - \frac{m}{\mathbf{p}^2}\right], \tag{29.30}$$

where the integral over the wave vector $\mathbf{p}$ is now convergent in the ultraviolet.

At this point, we enter the results (29.28), (29.29), and (29.30) into the expression (29.25), to obtain eventually:

$$\begin{aligned} -\,A_e(\mathbf{q},\Omega_\nu) \xrightarrow{(|\Delta_e|\to 0)} & \frac{m}{4\,\pi\,a_F} + \int \frac{d\mathbf{p}}{(2\pi)^3} \left[\frac{\tanh\left(\frac{\beta\xi_\mathbf{p}}{2}\right)}{2\,\xi_\mathbf{p}} - \frac{m}{\mathbf{p}^2}\right] \\ & -\,i\,N_0\,\frac{\beta_c}{\pi}\,\frac{3}{4}\,\zeta(2)\left(\frac{\mathbf{q}^2}{2m} - \omega\right) - N_0\,\frac{\beta_c^2}{2\,\pi^2}\,\frac{7}{8}\,\zeta(3)\left(\frac{v_F^2\,\mathbf{q}^2}{3} + \omega^2\right). \end{aligned} \tag{29.31}$$

29.6 Contribution of Higher-Order Terms in Space and Time

The result (29.31) contains terms of order ω and ω^2 as well as two terms of order $\mathbf{q}^2$. Before proceeding further, it is convenient to estimate the relative importance of these terms, with reference to the (weak-coupling (BCS) and near-T_c) limits that we are considering.

ω vs ω^2 terms: From the last two terms of Eq. (29.31), we get:

$$\frac{N_0\,\frac{\beta_c}{\pi}\,\frac{3}{4}\,\zeta(2)\,\omega}{N_0\,\frac{\beta_c^2}{2\,\pi^2}\,\frac{7}{8}\,\zeta(3)\,\omega^2} = \frac{12\,\pi\,\zeta(2)}{7\,\zeta(3)}\,\frac{k_B T_c}{\omega} \simeq 7.37\,\frac{k_B T_c}{\omega} \gg 1 \tag{29.32}$$

since $\omega \ll k_B T_c$ by our assumption. Accordingly, in Eq. (29.31), we shall retain the term $\propto \omega$ and disregard the term $\propto \omega^2$.

$\mathbf{q}^2$ vs $E_F\mathbf{q}^2$ terms: From the last two terms of Eq. (29.31), we also get:

$$\frac{N_0\,\frac{\beta_c}{\pi}\,\frac{3}{4}\,\zeta(2)\,\frac{\mathbf{q}^2}{2m}}{N_0\,\frac{\beta_c^2}{2\,\pi^2}\,\frac{7}{8}\,\zeta(3)\,\frac{v_F^2\,\mathbf{q}^2}{3}} = \frac{9\,\pi\,\zeta(2)}{7\,\zeta(3)}\,\frac{k_B T_c}{E_F} \simeq 5.53\,\frac{k_B T_c}{E_F} \ll 1 \tag{29.33}$$

since $k_B T_c \ll E_F$ by our assumption. Accordingly, out of the two terms in Eq. (29.31) containing $\mathbf{q}^2$, we shall retain only that $\propto v_F^2$.

In this way, Eq. (29.31) reduces to:

$$- A_e(\mathbf{q}, \Omega_\nu) \xrightarrow{(|\Delta_e|\to 0)} \frac{m}{4\pi a_F} + \int \frac{d\mathbf{p}}{(2\pi)^3} \left[\frac{\tanh\left(\frac{\beta \xi_\mathbf{p}}{2}\right)}{2\xi_\mathbf{p}} - \frac{m}{\mathbf{p}^2} \right]$$
$$+ N_0 \left(i\omega \frac{\beta_c}{\pi} \frac{3}{4} \zeta(2) - \frac{\mathbf{q}^2}{4m} \frac{\beta_c^2 E_F}{\pi^2} \frac{7}{6} \zeta(3) \right) . \quad (29.34)$$

29.7 Effect of the Deviations from the Critical Temperature and the Thermodynamic Chemical Potential

It remains to specify the temperature and the chemical potential entering the integral over the wave vector $\mathbf{p}$ in Eq. (29.34). Specifically, we shall set therein

$$\beta = \beta_c + \delta\beta \quad \text{and} \quad \mu = E_F - V_{\text{ext}}(\mathbf{r}, t) \quad (29.35)$$

consistently with our assumptions. We consider these two deviations separately.

Effect of the deviation $\beta = \beta_c + \delta\beta$: Here, $\delta\beta = \beta - \beta_c \simeq \frac{(T_c - T)}{k_B T_c^2}$ with $T \le T_c$ by construction. We then have:

$$\tanh\left(\frac{\beta\xi_\mathbf{p}}{2}\right) = \tanh\left[\frac{(\beta_c + \delta\beta)\,\xi_\mathbf{p}}{2}\right] \simeq \tanh\left(\frac{\beta_c\xi_\mathbf{p}}{2}\right) + \frac{\delta\beta\xi_\mathbf{p}}{2} \frac{1}{\cosh^2\left(\frac{\beta_c\xi_\mathbf{p}}{2}\right)} + \cdots , \quad (29.36)$$

such that

$$\int \frac{d\mathbf{p}}{(2\pi)^3} \left[\frac{\tanh\left(\frac{\beta\xi_\mathbf{p}}{2}\right)}{2\xi_\mathbf{p}} - \frac{m}{\mathbf{p}^2} \right] \simeq \int \frac{d\mathbf{p}}{(2\pi)^3} \left[\frac{\tanh\left(\frac{\beta_c\xi_\mathbf{p}}{2}\right)}{2\xi_\mathbf{p}} - \frac{m}{\mathbf{p}^2} \right]$$
$$+ \frac{\delta\beta}{4} \int \frac{d\mathbf{p}}{(2\pi)^3} \frac{1}{\cosh^2\left(\frac{\beta_c\xi_\mathbf{p}}{2}\right)}, \quad (29.37)$$

where

$$\int \frac{d\mathbf{p}}{(2\pi)^3} \frac{1}{\cosh^2\left(\frac{\beta_c\xi_\mathbf{p}}{2}\right)} \simeq N_0 \int_{-\infty}^{+\infty} d\xi \, \frac{1}{\cosh^2\left(\frac{\beta_c\xi}{2}\right)} = \frac{4\,N_0}{\beta_c} \int_0^{+\infty} dx \, \frac{1}{\cosh^2(x)} = \frac{4\,N_0}{\beta_c} . \quad (29.38)$$

The expression (29.37) thus becomes

$$\int \frac{d\mathbf{p}}{(2\pi)^3} \left[\frac{\tanh\left(\frac{\beta\xi_\mathbf{p}}{2}\right)}{2\,\xi_\mathbf{p}} - \frac{m}{\mathbf{p}^2} \right] \simeq \int \frac{d\mathbf{p}}{(2\pi)^3} \left[\frac{\tanh\left(\frac{\beta_c\xi_\mathbf{p}}{2}\right)}{2\,\xi_\mathbf{p}} - \frac{m}{\mathbf{p}^2} \right] + N_0 \frac{\delta\beta}{\beta_c} \quad (29.39)$$

with $\frac{\delta\beta}{\beta_c} \simeq \frac{(T_c - T)}{T_c}$.

Effect of the deviation $\mu = E_F - V_{\text{ext}}(\mathbf{r}, t)$: The integral over the wave vector $\mathbf{p}$ on the right-hand side of Eq. (29.39) can be evaluated in a closed form, as it will be shown in Section 31.3. We report here the result (cf. Eq. (31.20))

$$\int \frac{d\mathbf{p}}{(2\pi)^3} \left[\frac{\tanh\left(\frac{\beta_c \xi_{\mathbf{p}}}{2}\right)}{2\,\xi_{\mathbf{p}}} - \frac{m}{\mathbf{p}^2} \right] = N(\mu)\, \ln\left(\frac{8\, e^{\gamma}\, \beta_c\, \mu}{\pi\, e^2} \right), \tag{29.40}$$

where γ is the Euler constant (such that $e^{\gamma} \simeq 1.781$) and (cf. Eq. (31.21))

$$N(\mu) = \frac{m^{3/2}\,\sqrt{\mu}}{\sqrt{2}\,\pi^2} \tag{29.41}$$

is the free-particle density of states (per spin component) for $\varepsilon = \mu$. In particular, for $\varepsilon = \mu = E_F$, we have $N(\mu = E_F) = \frac{m\, k_F}{2\,\pi^2} = N_0$, as expected. If we now set $\mu = E_F - V_{\text{ext}}(\mathbf{r}, t)$ as in Eq. (29.35) and expand in the small quantity $V_{\text{ext}}(\mathbf{r}, t)/E_F$ both the square root and the ln factor that contain μ in the expression (29.40), we obtain approximately

$$N(\mu)\, \ln\left(\frac{8\, e^{\gamma}\, \beta_c\, \mu}{\pi\, e^2} \right) \simeq N_0\, \ln\left(\frac{8\, e^{\gamma}\, \beta_c\, E_F}{\pi\, e^2} \right) - \frac{V_{\text{ext}}(\mathbf{r}, t)}{E_F}\, N_0 \left[1 + \frac{1}{2}\, \ln\left(\frac{8\, e^{\gamma}\, \beta_c\, E_F}{\pi\, e^2} \right) \right], \tag{29.42}$$

where the condition

$$N_0\, \ln\left(\frac{8\, e^{\gamma}\, \beta_c\, E_F}{\pi\, e^2} \right) = - \frac{m}{4\,\pi\, a_F} \tag{29.43}$$

or else

$$k_B T_c = \frac{8\, e^{\gamma}\, E_F}{\pi\, e^2}\; e^{\frac{\pi}{2 k_F a_F}} \qquad (a_F < 0) \tag{29.44}$$

determines the critical temperature in the BCS regime (cf. Eq. (26.12)). In this way, the expression (29.40) becomes eventually:

$$\begin{aligned}
\int \frac{d\mathbf{p}}{(2\pi)^3} \left[\frac{\tanh\left(\frac{\beta_c \xi_{\mathbf{p}}}{2}\right)}{2\,\xi_{\mathbf{p}}} - \frac{m}{\mathbf{p}^2} \right] &\simeq - \frac{m}{4\,\pi\, a_F} - \frac{V_{\text{ext}}(\mathbf{r}, t)}{E_F}\, N_0 \left(1 - \frac{1}{2\, N_0}\, \frac{m}{4\,\pi\, a_F} \right) \\
&= - \frac{m}{4\,\pi\, a_F} - \frac{V_{\text{ext}}(\mathbf{r}, t)}{E_F}\, N_0 \left(1 - \frac{\pi}{4\, k_F\, a_F} \right).
\end{aligned} \tag{29.45}$$

Entering the results (29.39) and (29.45) into Eq. (29.34) yields the final expression:

$$\begin{aligned}
- A_e(\mathbf{q}, \Omega_\nu) \xrightarrow{(|\Delta_e| \to 0)} N_0 \Bigg\{ & \frac{(T_c - T)}{T_c} - \frac{V_{\text{ext}}(\mathbf{r}, t)}{E_F} \left(1 - \frac{\pi}{4 k_F\, a_F} \right) \\
& + i\, \omega \frac{\beta_c}{\pi} \frac{3}{4} \zeta(2) - \frac{\mathbf{q}^2}{4m} \frac{\beta_c^2 E_F}{\pi^2} \frac{7}{6} \zeta(3) \Bigg\}.
\end{aligned} \tag{29.46}$$

29.8 Contributions Arising at the Mean-Field Level

We are left with evaluating the factor $(A_e(\mathbf{q}=0,\Omega_\nu=0) - |B_e(\mathbf{q}=0,\Omega_\nu=0)|)$, which enters Eq. (29.1). This can be done using the expression (29.4) where we utilize Eq. (25.22) for regularizing the integral over the wave vector $\mathbf{p}$. We thus write

$$- A_e(\mathbf{q}=0,\Omega_\nu=0) + |B_e(\mathbf{q}=0,\Omega_\nu=0)| = \frac{m}{4\pi a_F} + \int \frac{d\mathbf{p}}{(2\pi)^3} \left\{ \frac{[1-2f_F(E_\mathbf{p})]}{2E_\mathbf{p}} - \frac{m}{\mathbf{p}^2} \right\}, \tag{29.47}$$

where the integral over $\mathbf{p}$ differs from that entering Eq. (29.34) by the replacement $\xi_\mathbf{p} \to E_\mathbf{p}$. It can be shown that these two integrals can be related to each other in the following way

$$\int \frac{d\mathbf{p}}{(2\pi)^3} \left\{ \frac{[1-2f_F(E_\mathbf{p})]}{2E_\mathbf{p}} - \frac{m}{\mathbf{p}^2} \right\} = \int \frac{d\mathbf{p}}{(2\pi)^3} \left\{ \frac{[1-2f_F(\xi_\mathbf{p})]}{2\xi_\mathbf{p}} - \frac{m}{\mathbf{p}^2} \right\} - \int \frac{d\mathbf{p}}{(2\pi)^3} \frac{1}{\beta} \sum_n \frac{|\Delta_e|^2}{\left[\omega_n^2 + \xi_\mathbf{p}^2\right]^2}, \tag{29.48}$$

which holds near T_c, and where $\omega_n = (2n+1)\pi/\beta$ (n integer) are fermionic Matsubara frequencies.

Proof of Eq. (29.48)

We begin by considering the expression ($\eta = 0^+$)

$$\frac{1}{\beta}\sum_n \frac{e^{i\omega_n\eta}}{(i\omega_n - E_\mathbf{p})(i\omega_n + E_\mathbf{p})} = \frac{1}{2E_\mathbf{p}}\frac{1}{\beta}\sum_n \left[\frac{e^{i\omega_n\eta}}{i\omega_n - E_\mathbf{p}} - \frac{e^{i\omega_n\eta}}{i\omega_n + E_\mathbf{p}} \right] = -\frac{[1-2f_F(E_\mathbf{p})]}{2E_\mathbf{p}}, \tag{29.49}$$

where use has been made of the result (28.24). On the other hand, the denominators on the left-hand side of Eq. (29.49) can be written as

$$(i\omega_n - E_\mathbf{p})\,(i\omega_n + E_\mathbf{p}) = \left[(i\omega_n)^2 - \xi_\mathbf{p}^2\right] \left[1 - \frac{|\Delta_e|^2}{(i\omega_n)^2 - \xi_\mathbf{p}^2} \right], \tag{29.50}$$

such that

$$\frac{1}{\beta}\sum_n \frac{e^{i\omega_n\eta}}{(i\omega_n - E_\mathbf{p})\,(i\omega_n + E_\mathbf{p})} \simeq \frac{1}{\beta}\sum_n e^{i\omega_n\eta} \left\{ \frac{1}{(i\omega_n)^2 - \xi_\mathbf{p}^2} + \frac{|\Delta_e|^2}{\left[(i\omega_n)^2 - \xi_\mathbf{p}^2\right]^2} \right\}$$

$$= -\frac{\left[1 - 2f_F(\xi_{\mathbf{p}})\right]}{2\xi_{\mathbf{p}}} + \frac{1}{\beta}\sum_n \frac{|\Delta_e|^2}{\left[\omega_n^2 + \xi_{\mathbf{p}}^2\right]^2} \tag{29.51}$$

where the convergence factor $e^{i\omega_n \eta}$ has safely been dropped from the last sum. Combining together Eqs. (29.49) and (29.51) yields the result (29.48). [QED]

Remark: A small expansion parameter
The approximation made in Eq. (29.51) relies on the fact that $\min\{\omega_n^2 + \xi_{\mathbf{p}}^2\} = \omega_{n=0}^2 = (\pi k_B T)^2$, since $\xi_{\mathbf{p}}^2$ can vanish when $\mu > 0$. The expansion parameter is then $|\Delta_e|^2/(\pi k_B T)^2$, which is small for T near T_c when $|\Delta_e| \ll k_B T_c$. This was the key assumption used by Gor'kov for the derivation from first principles of the time-independent GL equation [123].

There remains to calculate explicitly the last term on the right-hand side of Eq. (29.48). By interchanging the integral over $\mathbf{p}$ with the sum over n, we obtain:

$$\begin{aligned}
&\int \frac{d\mathbf{p}}{(2\pi)^3} \frac{1}{\beta}\sum_n \frac{|\Delta_e|^2}{\left[\omega_n^2 + \xi_{\mathbf{p}}^2\right]^2} = |\Delta_e|^2 \frac{1}{\beta}\sum_n \int \frac{d\mathbf{p}}{(2\pi)^3} \frac{1}{\left[\omega_n^2 + \xi_{\mathbf{p}}^2\right]^2} \\
&\simeq N_0 |\Delta_e|^2 \frac{1}{\beta}\sum_n \int_{-\infty}^{+\infty} d\xi \frac{1}{\left[\omega_n^2 + \xi^2\right]^2} = N_0 |\Delta_e|^2 \frac{1}{\beta}\sum_n \int_{-\infty}^{+\infty} d\xi \frac{1}{\left[(\xi - i\omega_n)(\xi + i\omega_n)\right]^2} \\
&= N_0 |\Delta_e|^2 \frac{1}{\beta}\sum_n \begin{cases} \dfrac{\pi}{2\,\omega_n^3} & \text{for } \omega_n > 0 \\[2ex] -\dfrac{\pi}{2\,\omega_n^3} & \text{for } \omega_n < 0 \end{cases}
\end{aligned} \tag{29.52}$$

with

$$\frac{1}{\beta}\sum_n \frac{\pi}{2\,|\omega_n|^3} = \frac{\beta^2}{\pi^2}\sum_{n=0}^{+\infty} \frac{1}{(2n+1)^3} = \frac{\beta^2}{\pi^2}\left(1 - 2^{-3}\right)\zeta(3) = \frac{\beta^2}{\pi^2}\frac{7}{8}\zeta(3), \tag{29.53}$$

where the last line of Eq. (29.27) has been utilized. Entering the results (29.48), (29.52), and (29.53) into Eq. (29.4) yields eventually

$$\begin{aligned}
-A_e(\mathbf{q}=0, \Omega_\nu = 0) + |B_e(\mathbf{q}=0, \Omega_\nu = 0)| = \frac{m}{4\pi a_F} + \int \frac{d\mathbf{p}}{(2\pi)^3}\left\{\frac{\left[1 - 2f_F(\xi_{\mathbf{p}})\right]}{2\xi_{\mathbf{p}}} - \frac{m}{\mathbf{p}^2}\right\} \\
- N_0 |\Delta_e|^2 \frac{\beta_c^2}{\pi^2}\frac{7}{8}\zeta(3),
\end{aligned} \tag{29.54}$$

where, owing to the results (29.39) and (29.45),

$$\int \frac{d\mathbf{p}}{(2\pi)^3} \left\{ \frac{[1 - 2f_F(\xi_\mathbf{p})]}{2\xi_\mathbf{p}} - \frac{m}{\mathbf{p}^2} \right\} = - \frac{m}{4\,\pi\,a_F} - \frac{V_{\text{ext}}(\mathbf{r},t)}{E_F} N_0 \left(1 - \frac{\pi}{4\,k_F\,a_F}\right) + N_0 \frac{(T_c - T)}{T_c}\,, \tag{29.55}$$

such that the constant term $\frac{m}{4\,\pi\,a_F}$ cancels out from Eqs. (29.54) and (29.55).

29.9 The Final Form of the TDGL Equation

To obtain the desired TDGL equation, we proceed at this point by entering the results (29.46) for $A_e(\mathbf{q},\Omega_\nu)$ and (29.54) and (29.55) for $(A_e(\mathbf{q}=0,\Omega_\nu=0) - |B_e(\mathbf{q}=0,\Omega_\nu=0)|)$ into the general expression (29.1), where we set $B_e(\mathbf{q},\Omega_\nu)=0$ in line with our assumptions. In the process, we also replace $\mathbf{q}^2 \to -\nabla^2$ and $i\omega \to -\frac{\partial}{\partial t}$, consistently with the anticipated analytic continuation from imaginary to real times, such that $e^{-i\Omega_\nu\tau} \xrightarrow{(i\Omega_\nu \to \omega+i\eta)} e^{-(\omega+i\eta)\tau} \xrightarrow{(\tau\to it)} e^{-i(\omega+i\eta)t}$ in Eq. (29.1). In this way, we end up with the expression

$$\begin{aligned}
&\left\{ \frac{(T_c - T)}{T_c} - \frac{V_{\text{ext}}(\mathbf{r},t)}{E_F}\left(1 - \frac{\pi}{4k_F a_F}\right) - \frac{\beta_c^2}{\pi^2}\frac{7}{8}\zeta(3)|\Delta_e(\mathbf{r})|^2 \right\} \Delta_e(\mathbf{r}) \\
&+ \left\{ \frac{(T_c - T)}{T_c} - \frac{V_{\text{ext}}(\mathbf{r},t)}{E_F}\left(1 - \frac{\pi}{4k_F a_F}\right) - \frac{\beta_c}{\pi}\frac{3}{4}\zeta(2)\frac{\partial}{\partial t} + \frac{\beta_c^2}{\pi^2}\frac{7}{6}\zeta(3)E_F\frac{\nabla^2}{4m} \right\} \delta\Delta(\mathbf{r},t) \\
&\simeq \left\{ \frac{(T_c - T)}{T_c} - \frac{V_{\text{ext}}(\mathbf{r},t)}{E_F}\left(1 - \frac{\pi}{4k_F a_F}\right)\right\} (\Delta_e(\mathbf{r}) + \delta\Delta(\mathbf{r},t)) - \frac{\beta_c^2}{\pi^2}\frac{7}{8}\zeta(3)|\Delta_e(\mathbf{r})|^2\Delta_e(\mathbf{r}) \\
&- \frac{\beta_c}{\pi}\frac{3}{4}\zeta(2)\frac{\partial}{\partial t}(\Delta_e(\mathbf{r}) + \delta\Delta(\mathbf{r},t)) + \frac{\beta_c^2}{\pi^2}\frac{7}{6}\zeta(3)E_F\frac{\nabla^2}{4m}(\Delta_e(\mathbf{r}) + \delta\Delta(\mathbf{r},t)) = 0,
\end{aligned} \tag{29.56}$$

where we have eventually endowed Δ_e with a sufficiently "slow" spatial variation (whereby $\Delta_e \to \Delta_e(\mathbf{r})$), consistently with what we did also in Chapter 28. In addition, in the self-interaction term of Eq. (29.56), we can replace $|\Delta_e(\mathbf{r})|^2\Delta_e(\mathbf{r}) \to |(\Delta_e(\mathbf{r}) + \delta\Delta(\mathbf{r},t))|^2(\Delta_e(\mathbf{r}) + \delta\Delta(\mathbf{r},t)) = |\Delta(\mathbf{r},t)|^2\Delta(\mathbf{r},t)$, the difference being assumed to be small. With these provisions, Eq. (29.56) becomes

$$\begin{aligned}
&\left\{ \frac{(T_c - T)}{T_c} - \frac{V_{\text{ext}}(\mathbf{r},t)}{E_F}\left(1 - \frac{\pi}{4\,k_F\,a_F}\right) - \frac{\beta_c^2}{\pi^2}\frac{7\,\zeta(3)}{8}|\Delta(\mathbf{r},t)|^2 - \frac{\beta_c}{\pi}\frac{3\,\zeta(2)}{4}\frac{\partial}{\partial t} \right. \\
&\left. + \frac{\beta_c^2\,E_F}{\pi^2}\frac{7\,\zeta(3)}{6}\frac{\nabla^2}{4\,m} \right\} \Delta(\mathbf{r},t) = 0\,,
\end{aligned} \tag{29.57}$$

where the coefficient of the term $\frac{\partial}{\partial t}$ coincides with that obtained in Ref. [141] by using a different method.

Although this is already our final result, it is convenient to cast it into a form that allows for a direct comparison with the time-independent GL equation (26.24), which is achieved by bringing the coefficient of the kinetic term $\frac{\nabla^2}{4m}$ to be unity. In this way, we arrive at the final form of the *TDGL equation*

$$\boxed{\begin{aligned}&\left\{-\frac{\nabla^2}{4m}+\frac{3\pi^3}{28\zeta(3)E_F\beta_c}\frac{\partial}{\partial t}+\frac{6\pi^2}{7\zeta(3)E_F\beta_c^2}\left[\frac{V_{\mathrm{ext}}(\mathbf{r},t)}{E_F}\left(1-\frac{\pi}{4k_Fa_F}\right)-\frac{(T_c-T)}{T_c}\right]\right.\\&\left.\quad+\frac{3}{4E_F}|\Delta(\mathbf{r},t)|^2\right\}\Delta(\mathbf{r},t)=0\end{aligned}} \tag{29.58}$$

where the coefficients of the terms $\frac{(T_c-T)}{T_c}$, $\frac{V_{\mathrm{ext}}(\mathbf{r},t)}{E_F}$, and $|\Delta(\mathbf{r},t)|^2$ have the same form of those entering Eq. (26.24). Note that the TDGL equation (29.58) is of the "diffusion type," with *diffusion coefficient D* given by

$$D^{-1}=\frac{3\,\pi^3}{28\,\zeta(3)\,E_F\,\beta_c}\simeq\frac{2.764}{E_F\,\beta_c} \tag{29.59}$$

(apart from the factor $(4m)^{-1}$ present in the kinetic term). This physical behavior has to be contrasted with that of the TDGP equation (28.52), whose time dependence is instead of the "quantum type." Note also that the presence of a vector potential $\mathbf{A}$ could as well be included when needed, as it was done in the time-independent version (26.24).

Remark: Grand-canonical Hamiltonian *K* vs Hamiltonian *H*, again

Similar to the derivation of the TDGP equation made in Chapter 28, we have here considered the time-dependent gap function $\Delta(\mathbf{r},t)$ to develop in time with the grand-canonical Hamiltonian $K=H-\mu N$ and not with the Hamiltonian H. In both cases, this is due to the special role played by the chemical potential μ in the context of the BCS–BEC crossover, where μ refers to the initial equilibrium state before the time-dependent perturbation begins to act on the system. In this respect, the first equality in Eq. (28.47) still gives the relationship between the two time-dependent gap functions $\Delta_{\mathrm{K}}(\mathbf{r},t)$ and $\Delta_{\mathrm{H}}(\mathbf{r},t)$, developing with either K or H, respectively.

In practice, also due to its simplicity compared to alternative microscopic theories, the TDGL equation has been instrumental in assessing numerous experimental results in type-II superconductors, especially in exploring nonequilibrium vortex configurations at the microscopic scale. [See Ref. [142] for a review on recent progress in TDGL simulations, which also focuses on key aspects of superconductor applications.] In addition, although it can only be derived near critical temperature T_c, substantial evidence suggests that the TDGL equation is surprisingly accurate even for temperatures far below T_c. This conclusion has also been independently reached for the time-independent GL equation, which was tested against the BdG calculations in a less extreme BCS coupling regime, showing that the GL results smoothly merge into the BdG results and that this merging appears to be essentially complete in the extended sector $(k_F a_F)^{-1} \lesssim -0.7$ and $T/T_c \gtrsim 0.85$ of the temperature-coupling phase diagram [132].

30

Real-Frequency Green's Functions from the Kadanoff–Baym Equations in the Equilibrium Case

For a system at equilibrium, the external potential $V_{\text{ext}}(\mathbf{r}, t) \rightarrow V_{\text{ext}}(\mathbf{r})$ does not depend on time, such that the single-particle Hamiltonian $h(\mathbf{r}, t) \rightarrow h(\mathbf{r})$ of Eq. (11.1) of Part I also does not depend on time. Under these circumstances, it is customary to resort to the Matsubara formalism to obtain not only the thermodynamic properties of the system at equilibrium but also its dynamical properties (at least, within linear-response theory) [6, 7]. For the latter properties, it is also necessary to perform an *analytic continuation* from Matsubara ($i\Omega_\nu$) to real (ω) frequencies. Only in a few simple cases, however, this analytic continuation can be readily implemented directly in the diagrammatic structure (such as in the non-self-consistent t-matrix approach for the single-particle self-energy [115]), while more elaborate approximations (such as the corresponding fully self-consistent t-matrix approach [48]) require recourse to Padé approximants [143]. The problem with this numerical procedure, however, is that it is mathematically ill defined and accompanied by large systematic uncertainties.

The closed-time-path Green's functions approach, once eased from the nonequilibrium tasks for which it was originally conceived and specified instead to equilibrium situations whereby $h(\mathbf{r}, t) \rightarrow h(\mathbf{r})$, offers an alternative to the more standard Matsubara plus analytic continuation procedure, for obtaining physical quantities *directly* in real frequency ω. In this case, the price to pay is having to solve for additional components of the single-particle Green's function, while, in the Matsubara procedure, the Green's function has only one component. Nevertheless, at the time of writing, the use of the closed-time-path Green's functions approach for calculating dynamical excitations in equilibrium situations appears to show increasing interest in the literature [144]–[147].

With reference to the Kadanoff–Baym equations of Chapter 15 of Part I, in this chapter, we will show how, in the equilibrium case we are interested in here, the number of independent components of the single-particle Green's function (as well as of the related self-energy) reduces considerably, thereby making it easier to solve for the Kadanoff–Baym equations. In the present context, we shall keep the reference time t_0 at a formal level and postpone to Section 31.4 the proof that t_0 can be dismissed from the Kadanoff–Baym equations in the equilibrium case, making it possible to solve them directly in real frequencies via Fourier transforms. When dealing with these formal arguments, we shall not put too much attention in distinguishing whether we are dealing with fermions or bosons nor whether we are considering the normal or the fermionic superfluid phase (for which, however, we will make an interesting exception in Section 30.4). In addition, the physical system is assumed to evolve in time with the Hamiltonian H (and not with the grand-canonical Hamiltonian K) to the extent that no specific emphasis will be given here to the BCS–BEC crossover.

30.1 Identities Holding between the Components of the Single-Particle Green's Functions in the Equilibrium Case

The desired identities can be obtained directly from the definitions of the components of the single-particle Green's functions introduced in Chapter 13 of Part I, once they are specified to the equilibrium case. In this section, we shall consider the various identities separately.

Left Keldysh component $G^{\lceil}$ vs greater Green's function $G^{>}$

From the definition (13.9) of Part I, we obtain

$$G^{\lceil}(\mathbf{x}\tau, \mathbf{x}'t') = -\, i\, \langle \psi_{\mathrm{H}}(\mathbf{x}t_0 - i\tau)\, \psi^{\dagger}_{\mathrm{H}}(\mathbf{x}'t') \rangle$$
$$= -\, i\, \langle e^{\tau(H-\mu N)} \psi(\mathbf{x}) e^{-\tau(H-\mu N)}\, e^{iH(t'-t_0)} \psi^{\dagger}(\mathbf{x}') e^{-iH(t'-t_0)} \rangle \tag{30.1}$$

with the usual notation $\mathbf{x}$ for space and spin variables, and where we have used the expression (12.6) of Part I for $\psi_{\mathrm{H}}(\mathbf{x}t_0 - i\tau)$ and the form of $\psi^{\dagger}_{\mathrm{H}}(\mathbf{x}'t')$ at equilibrium. We also recall that $\langle \cdots \rangle = \sum_k \rho_k \langle \Phi_k | \cdots | \Phi_k \rangle$ stands for the thermal average (cf. Chapter 4 of Part I), where the index k refers to the energy eigenfunctions $H|\Phi_k\rangle = |\Phi_k\rangle E_k$ with the given particle number N_k (such that a sum over all possible values of N_k is implied), and take into account the identity (27.19), where now $i(t - t_0) \longrightarrow \tau$

$$e^{\tau(H-\mu N)} \psi(\mathbf{x}) e^{-\tau(H-\mu N)} = e^{\mu\tau}\, e^{\tau H} \psi(\mathbf{x}) e^{-\tau H}\, . \tag{30.2}$$

The expression (30.1) then becomes:

$$G^{\lceil}(\mathbf{x}\tau, \mathbf{x}'t') = -\, i\, e^{\mu\tau} \langle e^{\tau H} \psi(\mathbf{x}) e^{-\tau H} e^{iH(t'-t_0)} \psi^{\dagger}(\mathbf{x}') e^{-iH(t'-t_0)} \rangle\, . \tag{30.3}$$

This expression has to be compared with (cf. Eq. (13.6) of Part I)

$$G^{>}(\mathbf{x}t_0 - i\tau, \mathbf{x}'t') = -i \langle \psi_{\mathrm{H}}(\mathbf{x}t_0 - i\tau)\, \psi^{\dagger}_{\mathrm{H}}(\mathbf{x}'t') \rangle$$
$$= -i \langle e^{\tau H} \psi(\mathbf{x}) e^{-\tau H} e^{iH(t'-t_0)} \psi^{\dagger}(\mathbf{x}') e^{-iH(t'-t_0)} \rangle , \qquad (30.4)$$

where we have analytically continued the operator $\psi_{\mathrm{H}}(\mathbf{x}t)$ (as it originally appears in the definition (13.6) of Part I for $G^{>}(\mathbf{x}t, \mathbf{x}'t')$) from real time t to imaginary time $t_0 - i\tau$, in line with the considerations made in Chapter 27. The comparison yields:

$$\boxed{G^{\ulcorner}(\mathbf{x}\tau, \mathbf{x}'t') = e^{\mu\tau}\, G^{>}(\mathbf{x}t_0 - i\tau, \mathbf{x}'t') .} \qquad (30.5)$$

Right Keldysh component $G^{\urcorner}$ vs lesser Green's function $G^{<}$

From the definition (13.7) of Part I, we obtain

$$G^{\urcorner}(\mathbf{x}t, \mathbf{x}'\tau') = \mp i \langle \psi^{\dagger}_{\mathrm{H}}(\mathbf{x}'t_0 - i\tau')\, \psi_{\mathrm{H}}(\mathbf{x}t) \rangle$$
$$= \mp i\, e^{-\mu\tau'} \langle e^{\tau' H} \psi^{\dagger}(\mathbf{x}') e^{-\tau' H} e^{iH(t-t_0)} \psi(\mathbf{x}) e^{-iH(t-t_0)} \rangle, \qquad (30.6)$$

where the upper (lower) sign holds for bosons (fermions), and the identity (27.20) with $i(t - t_0) \to \tau$ has been utilized. This expression has to be compared with (cf. Eq. (13.5) of Part I)

$$G^{<}(\mathbf{x}t, \mathbf{x}'t_0 - i\tau') = \mp i \langle \psi^{\dagger}_{\mathrm{H}}(\mathbf{x}'t_0 - i\tau')\, \psi_{\mathrm{H}}(\mathbf{x}t) \rangle$$
$$= \mp i \langle e^{\tau' H} \psi^{\dagger}_{\mathrm{H}}(\mathbf{x}') e^{-\tau' H}\, e^{iH(t-t_0)} \psi_{\mathrm{H}}(\mathbf{x}) e^{-iH(t-t_0)} \rangle, \qquad (30.7)$$

where we have analytically continued the operator $\psi^{\dagger}_{\mathrm{H}}(\mathbf{x}'t')$ (as it originally appears in the definition (13.5) of Part I for $G^{<}(\mathbf{x}t, \mathbf{x}'t')$) from real time t' to imaginary time $t_0 - i\tau'$. The comparison yields:

$$\boxed{G^{\urcorner}(\mathbf{x}\tau, \mathbf{x}'t') = e^{-\mu\tau'}\, G^{<}(\mathbf{x}t, \mathbf{x}'t_0 - i\tau') .} \qquad (30.8)$$

Retarded G^{R} vs greater $G^{>}$ and lesser $G^{<}$ Green's functions

In Chapter 13 of Part I, we have introduced the retarded Green's function given by Eq. (13.15), which can also be written in terms of the lesser and greater Green's functions given by Eqs. (13.5) and (13.6), respectively, in the form:

$$G^{\mathrm{R}}(\mathbf{x}t, \mathbf{x}'t') = \theta(t - t') \left(G^{>}(\mathbf{x}t, \mathbf{x}'t') - G^{<}(\mathbf{x}t, \mathbf{x}'t') \right) . \qquad (30.9)$$

This expression holds, quite generally, in nonequilibrium situations. At equilibrium, on the other hand, all Green's functions depend on the time difference $t - t'$, such that the above expression can be Fourier transformed to real frequency ω. To this end, we recall that the Fourier transform of the product $h(t) = f(t)g(t)$

is given by $h(\omega) = \int_{-\infty}^{+\infty} \frac{d\omega'}{2\pi} f(\omega - \omega') g(\omega')$. In the present case, $h(t) \to \theta(t)$ with Fourier transform $\frac{i}{\omega + i\eta}$ (with $\eta = 0^+$). We thus obtain for the Fourier transform of $G^{\mathrm{R}}(\mathbf{x}t, \mathbf{x}'t')$:

$$\boxed{G^{\mathrm{R}}(\mathbf{x}, \mathbf{x}'; \omega) = \int_{-\infty}^{+\infty} \frac{d\omega'}{2\pi} \frac{i}{\omega - \omega' + i\eta} \left(G^{>}(\mathbf{x}, \mathbf{x}'; \omega') - G^{<}(\mathbf{x}, \mathbf{x}'; \omega') \right) ,} \tag{30.10}$$

where the term $+i\eta$ is consistent with the "outgoing wave" boundary condition of scattering theory [87].

Advanced G^{A} vs greater $G^{>}$ and lesser $G^{<}$ Green's functions

Similarly, the advanced Green's function given by Eq. (13.17) of Part I can be written in terms of the lesser and greater Green's functions, in the form:

$$G^{\mathrm{A}}(\mathbf{x}t, \mathbf{x}'t') = -\theta(t' - t) \left(G^{>}(\mathbf{x}t, \mathbf{x}'t') - G^{<}(\mathbf{x}t, \mathbf{x}'t') \right) . \tag{30.11}$$

Since the Fourier transform of $\theta(-t)$ is $\frac{i}{-\omega + i\eta}$ in the equilibrium case, we write:

$$\boxed{G^{\mathrm{A}}(\mathbf{x}, \mathbf{x}'; \omega) = \int_{-\infty}^{+\infty} \frac{d\omega'}{2\pi} \frac{i}{\omega - \omega' - i\eta} \left(G^{>}(\mathbf{x}, \mathbf{x}'; \omega') - G^{<}(\mathbf{x}, \mathbf{x}'; \omega') \right) ,} \tag{30.12}$$

where the term $-i\eta$ is consistent with the "ingoing wave" boundary condition of scattering theory [87].

Remark: Combining Eqs. (30.10) and (30.12)
Upon subtracting Eq. (30.12) from Eq. (30.10) and recalling the identity $\frac{1}{\omega \pm i\eta} = \mathcal{P}\frac{1}{\omega} \mp i\pi\delta(\omega)$ for real ω, where $\mathcal{P}$ stands for Cauchy principal value, we obtain

$$G^{\mathrm{R}}(\mathbf{x}, \mathbf{x}'; \omega) - G^{\mathrm{A}}(\mathbf{x}, \mathbf{x}'; \omega) = G^{>}(\mathbf{x}, \mathbf{x}'; \omega) - G^{<}(\mathbf{x}, \mathbf{x}'; \omega) , \tag{30.13}$$

which is just the Fourier transform of the expression (13.19) of Part I, once specified to the equilibrium case.

Greater $G^{>}$ vs lesser $G^{<}$ Green's functions

At equilibrium, $G^{>}(\mathbf{x}, \mathbf{x}'; \omega)$ and $G^{<}(\mathbf{x}, \mathbf{x}'; \omega)$ are not independent but are related via the expression

$$\boxed{G^{<}(\mathbf{x}, \mathbf{x}'; \omega) = \pm\, e^{\beta(\mu - \omega)}\, G^{>}(\mathbf{x}, \mathbf{x}'; \omega),} \tag{30.14}$$

where the plus (minus) sign holds for bosons (fermions).

Proof of Eq. (30.14). From the definition (13.6) of Part I, we obtain in the equilibrium case

$$\begin{aligned} G^{>}(\mathbf{x}t, \mathbf{x}'t') &= -i\langle e^{iH(t-t_0)}\psi(\mathbf{x})e^{-iH(t-t')}\psi^{\dagger}(\mathbf{x}')e^{-iH(t'-t_0)}\rangle \\ &= -i\sum_{kp}\rho_k\langle\Phi_k|\psi(\mathbf{x})|\Phi_p\rangle\langle\Phi_p|\psi^{\dagger}(\mathbf{x}')|\Phi_k\rangle e^{-i(E_p-E_k)(t-t')}, \end{aligned} \tag{30.15}$$

such that its Lehmann representation reads:

$$\begin{aligned} G^{>}(\mathbf{x}, \mathbf{x}'; \omega) &= \int_{-\infty}^{+\infty} d(t-t')e^{i\omega(t-t')}G^{>}(\mathbf{x}t, \mathbf{x}'t') \\ &= -2\pi i\sum_{kp}\rho_k\langle\Phi_k|\psi(\mathbf{x})|\Phi_p\rangle\langle\Phi_p|\psi^{\dagger}(\mathbf{x}')|\Phi_k\rangle\delta(\omega - E_p + E_k). \end{aligned} \tag{30.16}$$

By a similar token, we obtain from the definition (13.5) of Part I in the equilibrium case

$$\begin{aligned} G^{<}(\mathbf{x}t, \mathbf{x}'t') &= \mp i\langle e^{iH(t'-t_0)}\psi^{\dagger}(\mathbf{x}')e^{-iH(t'-t)}\psi(\mathbf{x})e^{-iH(t-t_0)}\rangle \\ &= \mp i\sum_{kp}\rho_k\langle\Phi_k|\psi^{\dagger}(\mathbf{x}')|\Phi_p\rangle\langle\Phi_p|\psi(\mathbf{x})|\Phi_k\rangle e^{-i(E_p-E_k)(t'-t)}, \end{aligned} \tag{30.17}$$

such that its Lehmann representation reads:

$$\begin{aligned} G^{<}(\mathbf{x}, \mathbf{x}'; \omega) &= \int_{-\infty}^{+\infty} d(t-t')e^{i\omega(t-t')}G^{<}(\mathbf{x}t, \mathbf{x}'t') \\ &= \mp 2\pi i\sum_{kp}\rho_k\langle\Phi_k|\psi^{\dagger}(\mathbf{x}')|\Phi_p\rangle\langle\Phi_p|\psi(\mathbf{x})|\Phi_k\rangle\delta(\omega + E_p - E_k). \end{aligned} \tag{30.18}$$

To compare the two results (30.16) and (30.18) with each other, we recall the following relations that hold at thermal equilibrium

$$\rho_k = \frac{e^{-\beta(E_k-\mu N_k)}}{\mathrm{Tr}\{e^{-\beta(H-\mu N)}\}} \quad \text{and} \quad \rho_p = \frac{e^{-\beta(E_p-\mu N_p)}}{\mathrm{Tr}\{e^{-\beta(H-\mu N)}\}}, \tag{30.19}$$

such that

$$\frac{\rho_k}{e^{-\beta(E_k-\mu N_k)}} = \frac{\rho_p}{e^{-\beta(E_p-\mu N_p)}}\,. \tag{30.20}$$

The expression (30.18) can then be rewritten as follows:

$$G^<(\mathbf{x},\mathbf{x}';\omega) = \mp\, 2\pi i \sum_{kp} \rho_p \frac{e^{-\beta(E_k-\mu N_k)}}{e^{-\beta(E_p-\mu N_p)}} \langle\Phi_k|\psi^\dagger(\mathbf{x}')|\Phi_p\rangle\, \langle\Phi_p|\psi(\mathbf{x})|\Phi_k\rangle \delta(\omega+E_p-E_k)$$

$$\underset{[k \leftrightarrows p]}{\overset{\uparrow}{=}} \mp\, 2\pi i \sum_{kp} \rho_k \frac{e^{-\beta(E_p-\mu N_p)}}{e^{-\beta(E_k-\mu N_k)}} \langle\Phi_k|\psi(\mathbf{x})|\Phi_p\rangle\, \langle\Phi_p|\psi^\dagger(\mathbf{x}')|\Phi_k\rangle\, \delta(\omega+E_k-E_p)$$

$$\underset{[N_p = N_k+1]}{\overset{\uparrow}{=}} \mp\, 2\pi i\, e^{\beta\mu} \sum_{kp} \rho_k\, e^{-\beta(E_p-E_k)}\, \langle\Phi_k|\psi(\mathbf{x})|\Phi_p\rangle\, \langle\Phi_p|\psi^\dagger(\mathbf{x}')|\Phi_k\rangle\, \delta(\omega-E_p+E_k)$$

$$\underset{[\text{Eq.}(30.16)]}{\overset{\uparrow}{=}} \pm\, e^{\beta(\mu-\omega)}\, G^>(\mathbf{x},\mathbf{x}';\omega)\,. \qquad [\text{QED}] \qquad (30.21)$$

Remark: An alternative route to arrive at the result (30.14)

The result (30.14) can alternatively be obtained from the Kubo–Martin–Schwinger boundary conditions of Chapter 12 in Part I, following the steps indicated schematically in Fig. 30.1 that hold at equilibrium.

One starts from the definition (13.5) of Part I (where for simplicity the space and spin labels $\mathbf{x}$ and $\mathbf{x}'$ are suppressed) for $G^<(t_0,t')$, with $t=t_0$ an t' in the backward branch – cf. panel (a). One then moves t' to the forward branch with no harm and shifts t_0 down to $t_0 - i\beta$ by making use of the boundary condition (12.2) of Part I with reference to the Konstantinov–Perel contour, ending up in this way with $\pm\, G(t_0 - i\beta, t')$, where the upper (lower) sign holds for bosons (fermions) – cf. panel (b). This setup corresponds to $\pm$ times the left Keldysh component $G^{\lceil}(\tau=\beta,t')$ of Eq. (13.9)

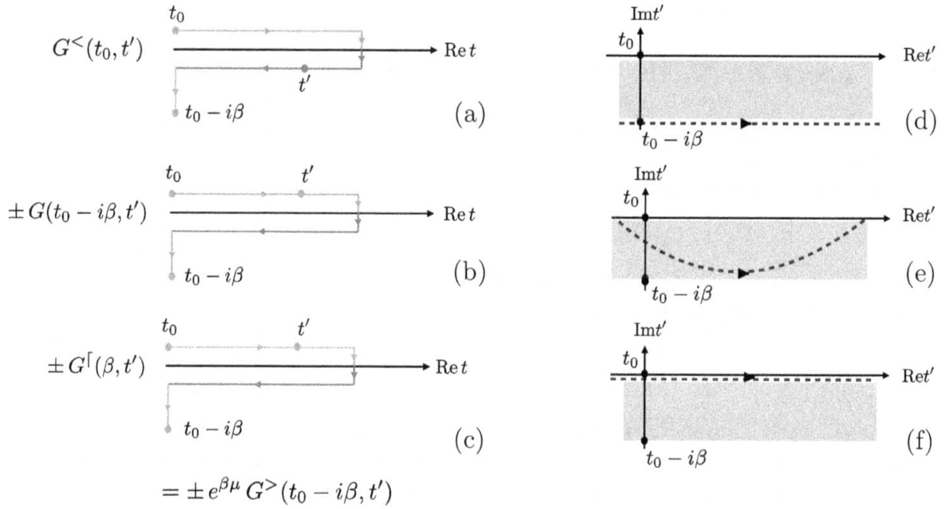

Figure 30.1 Schematic steps (a)→(f) as described in detail in the text, for arriving at the result (30.14) through an alternative route.

of Part I, where $\tau = \beta$ – cf. panel (c). Finally, taking into account the result (30.5), one ends up with the function $\pm\, e^{\beta\mu}\, G^{>}(t_0 - i\beta, t')$, as reported in the lower part of panel (c).

There remains to Fourier transform this result in the relative time $(t_0 - t')$, thereby obtaining

$$\begin{aligned} G^{<}(\omega) &= \int_{-\infty}^{+\infty} d(t_0 - t')\, e^{i\omega\,(t_0 - t')}\, G^{<}(t_0 - t') \\ &= \pm\, e^{\beta\mu} \int_{-\infty}^{+\infty} d(t_0 - t')\, e^{i\omega\,(t_0 - t')}\, G^{>}(t_0 - i\beta - t') \\ &= \pm\, e^{\beta(\mu - \omega)} \int_{-\infty}^{+\infty} d(t_0 - t')\, e^{i\omega\,(t_0 - i\beta - t')}\, G^{>}(t_0 - i\beta - t')\,, \end{aligned} \tag{30.22}$$

where the last integral is meant to run over the straight (dashed) line in complex t'-plane passing through $t_0 - i\beta$, as shown in panel (d). One can then modify the integration contour from the straight to the bent (dashed) curve shown in panel (e), using the fact that the integrand vanishes for sufficiently large values of $\mathrm{Re}t'$ away from t_0. As a last step, one can exploit the analytic properties of $G^{>}(t, t')$ as a function of the complex variables t and t' discussed in Chapter 27 (cf., in particular, Fig. 27.2 therein), so as to modify the integration contour into the straight (dashed) line in complex t'-plane passing through t_0, as shown in panel (f). In this way, the integral in the last line of Eq. (30.22) recovers $G^{>}(\omega)$, and the result (30.14) is eventually reproduced. [QED]

Note that the reference to the time t_0 is here purely conventional, in the sense that t_0 is meant to select a time interval when measurements may occur. As anticipated in Section 27.1, where t_0 was dropped at the outset, in Section 31.4, we shall show that, in the equilibrium case, t_0 can explicitly be eliminated from any consideration.

30.2 Related Identities for the Components of the Self-Energy

Quite generally, identities that hold among the components of the single-particle Green's function hold for the corresponding self-energies as well. This statement stems from the fact that diagrammatically the self-energy is the main building block of the Green's function itself. Here, we shall explicitly verify that identities related to those of Section 30.1 hold for the self-energies in terms of the specific example of the second-order Born (2B) approximation depicted graphically in Fig. 30.2, which, for definiteness, we specify to the case of spin-$\frac{1}{2}$ fermions in the normal phase. This may be considered in itself as an exercise for applying the Langreth rules for particle–hole-type (p–h) and particle–particle-type (p–p) products, treated, respectively, in Sections 14.3 and 14.4 of Part I.

According to the Feynman rules of Chapter 10 of Part I, the analytic expression of $\Sigma_{2\mathrm{B}}$ reads:

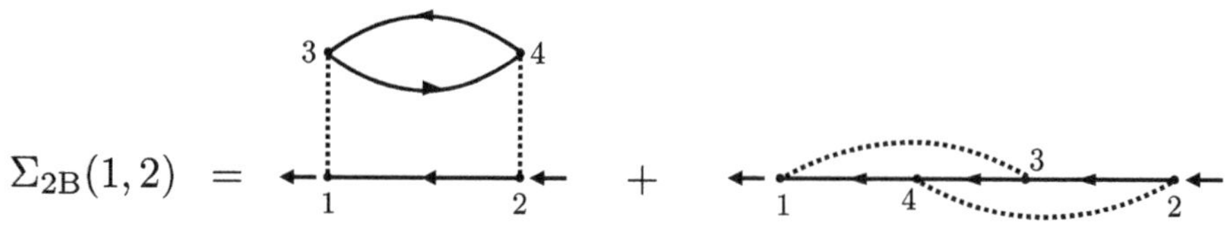

Figure 30.2 Graphical representation of the self-energy within the second-order Born (2B) approximation. Note the presence of a closed loop in the left term.

$$\Sigma_{2B}(1,2) = -i^2 \int d3d4v(1,3)v(2,4)\,[G(1,2)G(3,4)G(4,3) \\ -G(1,4)G(4,3)G(3,2)] \tag{30.23}$$

with the usual notation $1 \leftrightarrow (\mathbf{r}_1, \sigma_1, z_1)$ for the set of space, spin, and contour time variables, and with a genetic interparticle interaction v. From the graphical representation of Σ_{2B} shown in Fig. 30.2, both contributions are seen to correspond to "conserving" approximations in the Baym–Kadanoff sense (cf., e.g., Fig. 3 of Ref. [32]).
The expression (30.23) can be further simplified as follows:

(i) Each interaction potential contains a Dirac delta function along the time contour (cf. Eq. (11.20) of Part I), such that both time integrations drop out from Eq. (30.23), yielding:

$$\Sigma_{2B}(\mathbf{r}_1\sigma_1 t_1, \mathbf{r}_2\sigma_2 t_2) = -\,i^2 \sum_{\sigma_3\sigma_4} \int d\mathbf{r}_3\, d\mathbf{r}_4\, v(\mathbf{r}_1 - \mathbf{r}_3)\, v(\mathbf{r}_2 - \mathbf{r}_4) \\ \times [\, G(\mathbf{r}_1\sigma_1 t_1, \mathbf{r}_2\sigma_2 t_2)\, G(\mathbf{r}_3\sigma_3 t_1, \mathbf{r}_4\sigma_4 t_2)\, G(\mathbf{r}_4\sigma_4 t_2, \mathbf{r}_3\sigma_3 t_1) \\ -\, G(\mathbf{r}_1\sigma_1 t_1, \mathbf{r}_4\sigma_4 t_2)\, G(\mathbf{r}_4\sigma_4 t_2, \mathbf{r}_3\sigma_3 t_1)\, G(\mathbf{r}_3\sigma_3 t_1, \mathbf{r}_2\sigma_2 t_2)\,]\,. \tag{30.24}$$

(ii) In the absence of magnetic effects, spin is conserved along each fermion line and does not change at the interaction vertices. The sums over the spin indices then yields

$$\Sigma_{2B}(\mathbf{r}_1\sigma_1 t_1, \mathbf{r}_2\sigma_2 t_2) = -\,i^2\, \delta_{\sigma_1,\sigma_2} \int d\mathbf{r}_3\, d\mathbf{r}_4\, v(\mathbf{r}_1 - \mathbf{r}_3)\, v(\mathbf{r}_2 - \mathbf{r}_4) \\ \times [\, 2\, G(\mathbf{r}_1 t_1, \mathbf{r}_2 t_2)\, G(\mathbf{r}_3 t_1, \mathbf{r}_4 t_2)\, G(\mathbf{r}_4 t_2, \mathbf{r}_3 t_1) \\ -\, G(\mathbf{r}_1 t_1, \mathbf{r}_4 t_2)\, G(\mathbf{r}_4 t_2, \mathbf{r}_3 t_1)\, G(\mathbf{r}_3 t_1, \mathbf{r}_2 t_2)\,]\,. \tag{30.25}$$

Here, for the expression within brackets, we adopt the following short-hand notation

$$[\cdots] \longrightarrow [2G_{12}(t_1,t_2)G_{34}(t_1,t_2)G_{43}(t_2,t_1) - G_{14}(t_1,t_2)G_{43}(t_2,t_1)\, G_{32}(t_1,t_2)\,]\ , \tag{30.26}$$

which gives emphasis to the time variables (t_1, t_2), while the space variables are schematically displaced to the suffix. Note that this expression contains both particle–hole-type (p–h) and particle–particle-type (p–p) products (cf. Sections 14.3 and 14.4 of Part I).

To relate the various components of the self-energy (30.25) to each other, it is sufficient to consider the various components of the expression (30.26). We shall then assume that the results we will obtain in this way can as well be extended to other choices of the self-energy besides Σ_{2B}.

Left Keldysh self-energy $\Sigma^{\lceil}$ vs greater self-energy $\Sigma^{>}$

An identity as that given in Eq. (30.5) holds also for the corresponding self-energies, namely,

$$\boxed{\Sigma^{\lceil}(\mathbf{x}_1\tau_1, \mathbf{x}_2 t_2) = e^{\mu\tau_1}\, \Sigma^{>}(\mathbf{x}_1 t_0 - i\tau_1, \mathbf{x}_2 t_2)\ .} \tag{30.27}$$

In particular, within the second-order Born approximation, we obtain for the greater component of the expression (30.26) according to the Langreth rules of Sections 14.3 and 14.4 of Part I:

$$\begin{aligned}
&\left[2\, \underbrace{G_{12}(t_1,t_2)\, \overbrace{G_{34}(t_1,t_2)\, G_{43}(t_2,t_1)}^{\text{p–h}}}_{\text{p–p}} - \underbrace{\overbrace{G_{14}(t_1,t_2)\, G_{43}(t_2,t_1)}^{\text{p–h}}\, G_{32}(t_1,t_2)}_{\text{p–p}} \right]^{>} \\
&= 2\, G^{>}_{12}(t_1,t_2)\, G^{>}_{34}(t_1,t_2)\, G^{<}_{43}(t_2,t_1) - G^{>}_{14}(t_1,t_2)\, G^{<}_{43}(t_2,t_1)\, G^{>}_{32}(t_1,t_2) \\
&\xrightarrow{(t_1 \to t_0 - i\tau_1)} e^{-\mu\tau_1} \left[G^{\lceil}_{12}(\tau_1,t_2)\, G^{\lceil}_{34}(\tau_1,t_2)\, G^{\rceil}_{43}(t_2,\tau_1) - G^{\lceil}_{14}(\tau_1,t_2) G^{\rceil}_{43}(t_2,\tau_1)\, G^{\lceil}_{32}(\tau_1,t_2) \right] \\
&= e^{-\mu\tau_1}\, [2\, G_{12}(\tau_1,t_2)\, G_{34}(\tau_1,t_2)\, G_{43}(t_2,\tau_1) - G_{14}(\tau_1,t_2) G_{43}(t_2,\tau_1)\, G_{32}(\tau_1,t_2)]^{\lceil}\ ,
\end{aligned} \tag{30.28}$$

where the properties (30.5) and (30.8) for the Green's functions have been utilized. This result is consistent with the general property (30.27).

Right Keldysh self-energy $\Sigma^{\rceil}$ vs lesser self-energy $\Sigma^{<}$

Similarly, an identity as that given by Eq. (30.8) holds for the corresponding self-energies, namely,

$$\boxed{\Sigma^{\rceil}(\mathbf{x}_1\tau_1, \mathbf{x}_2 t_2) = e^{-\mu\tau_2}\, \Sigma^{<}(\mathbf{x}_2 t_1, \mathbf{x}_2 t_0 - i\tau_2)\ .} \tag{30.29}$$

Considering again the second-order Born approximation, we obtain for the lesser component of the expression (30.26):

$$
\begin{aligned}
&\left[2\,\underbrace{G_{12}(t_1,t_2)\,\overbrace{G_{34}(t_1,t_2)G_{43}(t_2,t_1)}^{\text{p–h}}}_{\text{p–p}} - \underbrace{\overbrace{G_{14}(t_1,t_2)G_{43}(t_2,t_1)}^{\text{p–h}}\,G_{32}(t_1,t_2)}_{\text{p–p}}\right]^{<} \\
&= 2G^{<}_{12}(t_1,t_2)G^{<}_{34}(t_1,t_2)G^{>}_{43}(t_2,t_1) - G^{<}_{14}(t_1,t_2)G^{>}_{43}(t_2,t_1)G^{<}_{32}(t_1,t_2) \\
&\xrightarrow{(t_2\to t_0-i\tau_2)} e^{\mu\tau_2}\left[G^{\rceil}_{12}(t_1,\tau_2)G^{\rceil}_{34}(t_1,\tau_2)G^{\lceil}_{43}(\tau_2,t_1) - G^{\rceil}_{14}(t_1,\tau_2)G^{\lceil}_{43}(\tau_2,t_1)G^{\rceil}_{32}(t_1,\tau_2)\right] \\
&= e^{\mu\tau_2}\left[2G_{12}(\tau_1,t_2)G_{34}(\tau_1,t_2)G_{43}(t_2,\tau_1) - G_{14}(\tau_1,t_2)G_{43}(t_2,\tau_1)G_{32}(\tau_1,t_2)\right]^{\rceil}, \qquad (30.30)
\end{aligned}
$$

where the properties (30.5) and (30.8) for the Green's functions have again been utilized. This result is consistent with the general property (30.29).

Retarded Σ^{R} vs greater $\Sigma^{>}$ and lesser $\Sigma^{<}$ self-energies

An expression like Eq. (30.9) holds for the "regular" (r) part of the retarded self-energy, namely,

$$
\boxed{\Sigma^{\mathrm{R}}_{\mathrm{r}}(\mathbf{x}_1 t_1, \mathbf{x}_2 t_2) = \theta(t_1 - t_2)\left(\Sigma^{>}(\mathbf{x}_1 t_1, \mathbf{x}_2 t_2) - \Sigma^{<}(\mathbf{x}_1 t_1, \mathbf{x}_2 t_2)\right)\,.} \qquad (30.31)
$$

As we did before, this identity can be verified within the second-order Born approximation, whereby the retarded component of the expression (30.26) is given by:

$$
\begin{aligned}
&\left[2\,\underbrace{G_{12}(t_1,t_2)\,\overbrace{G_{34}(t_1,t_2)\,G_{43}(t_2,t_1)}^{\text{p–h}}}_{\text{p–p}} - \underbrace{\overbrace{G_{14}(t_1,t_2)\,G_{43}(t_2,t_1)}^{\text{p–h}}\,G_{32}(t_1,t_2)}_{\text{p–p}}\right]^{\mathrm{R}} \\
&= 2\left[G^{\mathrm{R}}_{12}(t_1,t_2)\,G^{>}_{34}(t_1,t_2)\,G^{<}_{43}(t_2,t_1) + G^{<}_{12}(t_1,t_2)\,G^{\mathrm{R}}_{34}(t_1,t_2)\,G^{<}_{43}(t_2,t_1)\right. \\
&\qquad \left. + G^{<}_{12}(t_1,t_2)\,G^{<}_{34}(t_1,t_2)\,G^{\mathrm{A}}_{43}(t_2,t_1)\right] \\
&\quad - \left[G^{\mathrm{R}}_{14}(t_1,t_2)\,G^{<}_{43}(t_2,t_1)\,G^{>}_{32}(t_1,t_2) + G^{<}_{14}(t_1,t_2)\,G^{\mathrm{A}}_{43}(t_2,t_1)\,G^{>}_{32}(t_1,t_2)\right. \\
&\qquad \left. + G^{<}_{14}(t_1,t_2)\,G^{>}_{43}(t_2,t_1)\,G^{\mathrm{R}}_{32}(t_1,t_2)\right] \\
&= \theta(t_1-t_2)\left[2G^{>}_{12}(t_1,t_2)\,G^{>}_{34}(t_1,t_2)\,G^{<}_{43}(t_2,t_1) - G^{>}_{14}(t_1,t_2)G^{<}_{43}(t_2,t_1)\,G^{>}_{32}(t_1,t_2)\right. \\
&\quad \left. -\ 2\,G^{<}_{12}(t_1,t_2)\,G^{<}_{34}(t_1,t_2)\,G^{>}_{43}(t_2,t_1) + G^{<}_{14}(t_1,t_2)\,G^{>}_{43}(t_2,t_1)\,G^{<}_{32}(t_1,t_2)\right], \qquad (30.32)
\end{aligned}
$$

where Eqs. (30.9) and (30.11) have been utilized. With the help of the second lines of the expressions (30.28) and (30.30), which correspond to the greater and lesser components of Σ_{2B}, respectively, the result (30.31) then follows within the present approximation.

In addition, at equilibrium, a result like (30.10) follows from Eq. (30.31), namely,

$$\Sigma_r^R(\mathbf{x},\mathbf{x}';\omega)=\int_{-\infty}^{+\infty}\frac{d\omega'}{2\pi}\,\frac{i}{\omega-\omega'+i\eta}\left(\Sigma^>(\mathbf{x},\mathbf{x}';\omega')-\Sigma^<(\mathbf{x},\mathbf{x}';\omega')\right)\ . \tag{30.33}$$

Advanced Σ^A vs greater $\Sigma^>$ and lesser $\Sigma^<$ self-energies

By a similar token, an identity like that given by Eq. (30.11) holds for the "regular" (r) part of the advanced self-energy, namely,

$$\boxed{\Sigma_r^A(\mathbf{x}_1t_1,\mathbf{x}_2t_2)=-\,\theta(t_2-t_1)\left(\Sigma^>(\mathbf{x}_1t_1,\mathbf{x}_2t_2)-\Sigma^<(\mathbf{x}_1t_1,\mathbf{x}_2t_2)\right)\ .} \tag{30.34}$$

Within the second-order Born approximation of Fig. 30.2, this result can be verified in analogy to Eq. (30.32), where now

$$\begin{aligned}
&\Bigg[\,2\,\underbrace{G_{12}(t_1,t_2)\,\overbrace{G_{34}(t_1,t_2)\,G_{43}(t_2,t_1)}^{\text{p–h}}}_{\text{p–p}}\;-\;\underbrace{\overbrace{G_{14}(t_1,t_2)\,G_{43}(t_2,t_1)}^{\text{p–h}}\;G_{32}(t_1,t_2)}_{\text{p–p}}\,\Bigg]^A\\
&=2\left[G_{12}^A(t_1,t_2)\,G_{34}^>(t_1,t_2)\,G_{43}^<(t_2,t_1)+G_{12}^<(t_1,t_2)\,G_{34}^A(t_1,t_2)\,G_{43}^<(t_2,t_1)\right.\\
&\qquad\left.+\,G_{12}^<(t_1,t_2)\,G_{34}^<(t_1,t_2)\,G_{43}^R(t_2,t_1)\right]\\
&\quad-\left[G_{14}^A(t_1,t_2)\,G_{43}^<(t_2,t_1)\,G_{32}^>(t_1,t_2)+G_{14}^<(t_1,t_2)\,G_{43}^R(t_2,t_1)\,G_{32}^>(t_1,t_2)\right.\\
&\qquad\left.+\,G_{14}^<(t_1,t_2)\,G_{43}^>(t_2,t_1)\,G_{32}^A(t_1,t_2)\right]\\
&=-\,\theta(t_2-t_1)\left[2\,G_{12}^>(t_1,t_2)\,G_{34}^>(t_1,t_2)\,G_{43}^<(t_2,t_1)\right.\\
&\qquad\qquad-G_{14}^>(t_1,t_2)G_{43}^<(t_2,t_1)\,G_{32}^>(t_1,t_2)\\
&\quad\left.-\,2\,G_{12}^<(t_1,t_2)\,G_{34}^<(t_1,t_2)\,G_{43}^>(t_2,t_1)+G_{14}^<(t_1,t_2)\,G_{43}^>(t_2,t_1)\,G_{32}^<(t_1,t_2)\right]\ .
\end{aligned} \tag{30.35}$$

In addition, at equilibrium, we can write

$$\Sigma_r^A(\mathbf{x},\mathbf{x}';\omega)=\int_{-\infty}^{+\infty}\frac{d\omega'}{2\pi}\,\frac{i}{\omega-\omega'-i\eta}\left(\Sigma^>(\mathbf{x},\mathbf{x}';\omega')-\Sigma^<(\mathbf{x},\mathbf{x}';\omega')\right) \tag{30.36}$$

in analogy to the result (30.12).

Note further that, quite generally, from Eqs. (30.32) and (30.35), it follows that

$$\boxed{\Sigma^R(\mathbf{x}_1t_1,\mathbf{x}_2t_2)-\Sigma^A(\mathbf{x}_1t_1,\mathbf{x}_2t_2)=\Sigma^>(\mathbf{x}_1t_1,\mathbf{x}_2t_2)-\Sigma^<(\mathbf{x}_1t_1,\mathbf{x}_2t_2),} \tag{30.37}$$

where the "singular" parts of Σ^{R} and Σ^{A} have formally been reincluded since they cancel out in the difference.

Greater $\Sigma^{>}$ vs lesser $\Sigma^{<}$ self-energies

An identity like Eq. (30.14) holds also for the corresponding self-energies, namely,

$$\boxed{\Sigma^{<}(\mathbf{x}_2, \mathbf{x}_2; \omega) = \pm\, e^{\beta(\mu-\omega)}\, \Sigma^{>}(\mathbf{x}_1, \mathbf{x}_2; \omega)} \tag{30.38}$$

with the plus (minus) sign holding for bosons (fermions). This identity can be proved by following a procedure similar to that illustrated in Fig. 30.1 for the single-particle Green's functions.

Still within the second-order Born approximation, the result (30.38) readily follows by comparing the second lines of the expressions (30.28) and (30.30) with each other, where at equilibrium one further takes the Fourier transform with respect to the time difference $(t_1 - t_2)$ and makes repeated use of the property (30.14) for the Green's functions.

30.3 Fluctuation–Dissipation Theorem

From the identities (30.13) and (30.14) one can prove the so-called *fluctuation–dissipation theorem*, which holds at equilibrium, in the form

$$\boxed{\begin{cases} G^{<}(\mathbf{x}, \mathbf{x}'; \omega) = \pm f_{\mp}(\omega - \mu)\left[G^{\mathrm{R}}(\mathbf{x}, \mathbf{x}'; \omega) - G^{\mathrm{A}}(\mathbf{x}, \mathbf{x}'; \omega)\right], \\[2ex] G^{>}(\mathbf{x}, \mathbf{x}'; \omega) = (1 \pm f_{\mp}(\omega - \mu))\left[G^{\mathrm{R}}(\mathbf{x}, \mathbf{x}'; \omega) - G^{\mathrm{A}}(\mathbf{x}, \mathbf{x}'; \omega)\right], \end{cases}} \tag{30.39}$$

where the upper (lower) sign refers to bosons (fermions) and $f_{\mp}(\varepsilon) = \left(e^{\beta\varepsilon} \mp 1\right)^{-1}$, which encompasses both the Bose–Einstein (f_{-}) and Fermi–Dirac (f_{+}) distribution functions.[1]

Proof of the first of Eqs. (30.39). By combining Eqs. (30.13) and (30.14), we obtain

$$\begin{aligned} G^{>}(\mathbf{x}, \mathbf{x}'; \omega) - G^{<}(\mathbf{x}, \mathbf{x}'; \omega) &\underset{\substack{\uparrow \\ [\text{Eq.(30.14)}]}}{=} \left(\pm\, e^{\beta(\omega-\mu)} - 1\right) G^{<}(\mathbf{x}, \mathbf{x}'; \omega) \\ &\underset{\substack{\uparrow \\ [\text{Eq.(30.13)}]}}{=} G^{\mathrm{R}}(\mathbf{x}, \mathbf{x}'; \omega) - G^{\mathrm{A}}(\mathbf{x}, \mathbf{x}'; \omega), \end{aligned} \tag{30.40}$$

[1] Note that, in the expressions (30.39), the frequency ω is explicitly referred to the chemical potential μ, while this dependence would have been implicit if the grand-canonical Hamiltonian K had been used in place of the Hamiltonian H.

from which

$$G^<(\mathbf{x},\mathbf{x}';\omega) = \frac{(\pm 1)}{e^{\beta(\omega-\mu)} \mp 1}\left[G^{\mathrm{R}}(\mathbf{x},\mathbf{x}';\omega) - G^{\mathrm{A}}(\mathbf{x},\mathbf{x}';\omega)\right] . \tag{30.41}$$

[QED]

Proof of the second of Eqs. (30.39). By combining again Eqs. (30.13) and (30.14), we obtain

$$\begin{aligned} G^>(\mathbf{x},\mathbf{x}';\omega) - G^<(\mathbf{x},\mathbf{x}';\omega) &\underset{[\text{Eq.}(30.14)]}{\overset{\uparrow}{=}} \left(1 \mp e^{\beta(\mu-\omega)}\right) G^>(\mathbf{x},\mathbf{x}';\omega) \\ &\underset{[\text{Eq.}(30.13)]}{\overset{\uparrow}{=}} G^{\mathrm{R}}(\mathbf{x},\mathbf{x}';\omega) - G^{\mathrm{A}}(\mathbf{x},\mathbf{x}';\omega) , \end{aligned} \tag{30.42}$$

from which

$$G^>(\mathbf{x},\mathbf{x}';\omega) = \frac{1}{1 \mp e^{\beta(\mu-\omega)}}\left[G^{\mathrm{R}}(\mathbf{x},\mathbf{x}';\omega) - G^{\mathrm{A}}(\mathbf{x},\mathbf{x}';\omega)\right] , \tag{30.43}$$

where $\left(1 \mp e^{\beta(\mu-\omega)}\right)^{-1} = 1 \pm f_{\mp}(\omega-\mu)$. [QED]

Remark: Keldysh Green's function vs fluctuation–dissipation theorem
By combining the results (30.39) with the expression (13.18) of Part I for the Keldysh Green's function, we further obtain

$$\begin{aligned} G^{\mathrm{K}}(\mathbf{x},\mathbf{x}';\omega) &= G^>(\mathbf{x},\mathbf{x}';\omega) + G^<(\mathbf{x},\mathbf{x}';\omega) \\ &= (1 \pm 2f_{\mp}(\omega-\mu))\left[G^{\mathrm{R}}(\mathbf{x},\mathbf{x}';\omega) - G^{\mathrm{A}}(\mathbf{x},\mathbf{x}';\omega)\right] \\ &= \left[G^{\mathrm{R}}(\mathbf{x},\mathbf{x}';\omega) - G^{\mathrm{A}}(\mathbf{x},\mathbf{x}';\omega)\right] \begin{cases} \coth\left[\frac{\beta(\omega-\mu)}{2}\right] & \text{(bosons)} \\ \tanh\left[\frac{\beta(\omega-\mu)}{2}\right] & \text{(fermions),} \end{cases} \end{aligned} \tag{30.44}$$

which generalizes to an arbitrary interacting system the result (20.11) of Part I valid for a noninteracting system.

Note finally that, from the identity (30.37) specified at equilibrium and the result (30.38), it follows that relations such as (30.39) hold for the corresponding self-energies, namely,

$$\boxed{\begin{cases} \Sigma^<(\mathbf{x},\mathbf{x}';\omega) = \pm f_{\mp}(\omega-\mu)\left[\Sigma^{\mathrm{R}}(\mathbf{x},\mathbf{x}';\omega) - \Sigma^{\mathrm{A}}(\mathbf{x},\mathbf{x}';\omega)\right] \\ \Sigma^>(\mathbf{x},\mathbf{x}';\omega) = (1 \pm f_{\mp}(\omega-\mu))\left[\Sigma^{\mathrm{R}}(\mathbf{x},\mathbf{x}';\omega) - \Sigma^{\mathrm{A}}(\mathbf{x},\mathbf{x}';\omega)\right] . \end{cases}} \tag{30.45}$$

30.4 The Single-Particle Spectral Function

A quantity of physical interest is the *single-particle spectral function*, defined at equilibrium by

$$A(\mathbf{x}, \mathbf{x}'; \omega) = i\left[G^{\mathrm{R}}(\mathbf{x}, \mathbf{x}'; \omega) - G^{\mathrm{A}}(\mathbf{x}, \mathbf{x}'; \omega)\right], \tag{30.46}$$

where $G^{\mathrm{A}}(\mathbf{x}, \mathbf{x}'; \omega) = G^{\mathrm{R}}(\mathbf{x}', \mathbf{x}; \omega)^*$ owing to the symmetry property (13.25) of Part I. The definition (30.46) can be combined with the expressions (30.10) and (30.12), as well as with the result (30.13), to yield

$$G^{\mathrm{R}}(\mathbf{x}, \mathbf{x}'; \omega) = \int_{-\infty}^{+\infty} \frac{d\omega'}{2\pi} \frac{A(\mathbf{x}, \mathbf{x}'; \omega')}{\omega - \omega' + i\eta} \tag{30.47}$$

$$G^{\mathrm{A}}(\mathbf{x}, \mathbf{x}'; \omega) = \int_{-\infty}^{+\infty} \frac{d\omega'}{2\pi} \frac{A(\mathbf{x}, \mathbf{x}'; \omega')}{\omega - \omega' - i\eta}, \tag{30.48}$$

which justify referring to $A(\mathbf{x}, \mathbf{x}'; \omega)$ as the "spectral" function.

Remark: A more general expression for the single-particle spectral function

More generally, we may define

$$A(\mathbf{x}t, \mathbf{x}'t') = i\left[G^{\mathrm{R}}(\mathbf{x}t, \mathbf{x}'t') - G^{\mathrm{A}}(\mathbf{x}t, \mathbf{x}'t')\right] \tag{30.49}$$

to hold even in nonequilibrium situations. In this case, we may further introduce the relative $t_{\mathrm{rel}} = t - t'$ and total $t_{\mathrm{tot}} = (t + t')/2$ times, such that $t = t_{\mathrm{tot}} + t_{\mathrm{rel}}/2$ and $t' = t_{\mathrm{tot}} - t_{\mathrm{rel}}/2$, and then Fourier transform $A(\mathbf{x}, \mathbf{x}'; t_{\mathrm{rel}}, t_{\mathrm{tot}})$ with respect to the relative time t_{rel}, yielding

$$A(\mathbf{x}, \mathbf{x}'; t_{\mathrm{tot}}|\omega) = i\int_{-\infty}^{+\infty} dt_{\mathrm{rel}}\, e^{i\,\omega\, t_{\mathrm{rel}}} \left[G^{\mathrm{R}}(\mathbf{x}, \mathbf{x}'; t_{\mathrm{rel}}, t_{\mathrm{tot}}) - G^{\mathrm{A}}(\mathbf{x}, \mathbf{x}'; t_{\mathrm{rel}}, t_{\mathrm{tot}})\right]. \tag{30.50}$$

An explicit expression for $A(\mathbf{x}, \mathbf{x}'; \omega)$ at equilibrium can be obtained from the Lehmann representations (30.16) for $G^{>}(\mathbf{x}, \mathbf{x}'; \omega)$ and the relation (30.14) for $G^{<}(\mathbf{x}, \mathbf{x}'; \omega)$, such that

$$A(\mathbf{x}, \mathbf{x}'; \omega) = 2\pi\left(1 \mp e^{\beta(\mu-\omega)}\right) \times \sum_{kp} \rho_k \langle\Phi_k|\psi(\mathbf{x})|\Phi_p\rangle \langle\Phi_p|\psi^\dagger(\mathbf{x}')|\Phi_k\rangle\, \delta(\omega - E_p + E_k). \tag{30.51}$$

From this expression, it follows that $A(\mathbf{x}, \mathbf{x}'; \omega)$ satisfies the *sum rule*

$$\boxed{\int_{-\infty}^{+\infty} \frac{d\omega}{2\pi} A(\mathbf{x}, \mathbf{x}'; \omega) = \delta(\mathbf{x}, \mathbf{x}').} \tag{30.52}$$

Proof of Eq. (30.52). By integrating the expression (30.51) over ω, we obtain

$$\begin{aligned}
\int_{-\infty}^{+\infty} \frac{d\omega}{2\pi} A(\mathbf{x}, \mathbf{x}'; \omega) &= \Sigma_{kp}\, \rho_k \, \langle \Phi_k | \psi(\mathbf{x}) | \Phi_p \rangle \, \langle \Phi_p | \psi^\dagger(\mathbf{x}') | \Phi_k \rangle \, \left(1 \mp e^{\beta(\mu - E_p + E_k)}\right) \\
&\underset{\substack{\uparrow \\ [\text{Eq.(30.20) with } N_p = N_k + 1]}}{=} \Sigma_{kp} \, \langle \Phi_k | \psi(\mathbf{x}) | \Phi_p \rangle \, \langle \Phi_p | \psi^\dagger(\mathbf{x}') | \Phi_k \rangle \, \left(\rho_k \mp \rho_p\right) \\
&\underset{\substack{\uparrow \\ [k \leftrightarrows p \text{ in the second term and } \Sigma_p |\Phi_p\rangle \langle \Phi_p| = \mathbb{1}]}}{=} \Sigma_k \, \rho_k \, \langle \Phi_k | \left(\psi(\mathbf{x}) \, \psi^\dagger(\mathbf{x}') \mp \psi^\dagger(\mathbf{x}') \, \psi(\mathbf{x})\right) | \Phi_k \rangle \\
&= \delta(\mathbf{x}, \mathbf{x}') \, \Sigma_k \, \rho_k = \delta(\mathbf{x}, \mathbf{x}') \,,
\end{aligned} \tag{30.53}$$

where the commutation (anti-commutation) relations (2.2) and the normalization condition of the density matrix given by Eq. (4.10) of Part I have been utilized in the last steps. [QED]

Remark: Sum rule in the nonequilibrium case

The sum rule (30.52) holds more generally even in nonequilibrium situations for the expression (30.50) of the single-particle spectral function. This statement can be readily proved by using the relation (13.19) of Part I, according to which $G^{\mathrm{R}}(\mathbf{x}t, \mathbf{x}'t') - G^{\mathrm{A}}(\mathbf{x}t, \mathbf{x}'t') = G^{>}(\mathbf{x}t, \mathbf{x}'t') - G^{<}(\mathbf{x}t, \mathbf{x}'t')$, as well as the definitions (13.5) for $G^{<}$ and (13.6) for $G^{>}$ of Part I. This is because,

$$\begin{aligned}
&\int_{-\infty}^{+\infty} \frac{d\omega}{2\pi} A(\mathbf{x}, \mathbf{x}'; t_{\text{tot}} | \omega) = \int_{-\infty}^{+\infty} \frac{d\omega}{2\pi} i \int_{-\infty}^{+\infty} d(t - t') e^{i\omega(t-t')} \left[G^{>}(\mathbf{x}t, \mathbf{x}'t') - G^{<}(\mathbf{x}t, \mathbf{x}'t')\right] \\
&= i \left[G^{>}(\mathbf{x}t, \mathbf{x}'t) - G^{<}(\mathbf{x}t, \mathbf{x}'t)\right] = \langle U(t_0, t) \left(\psi(\mathbf{x}) \psi^\dagger(\mathbf{x}') \mp \psi^\dagger(\mathbf{x}') \psi(\mathbf{x})\right) U(t, t_0) \rangle = \delta(\mathbf{x}, \mathbf{x}'),
\end{aligned} \tag{30.54}$$

which holds quite generally without invoking the Lehman representation (an argument that can be used also in the equilibrium case for which the dependence on t_{tot} can safely be dropped).

In particular, for a spin-independent and spatially homogeneous system, the single-particle spectral function becomes

$$A(\mathbf{x}, \mathbf{x}'; \omega) = \delta_{\sigma,\sigma'} \, A(\mathbf{r} - \mathbf{r}'; \omega) \,. \tag{30.55}$$

As a consequence, upon taking the spatial Fourier transform with wave vector $\mathbf{k}$ of this expression and recalling that $\delta(\mathbf{x}, \mathbf{x}') = \delta(\mathbf{r} - \mathbf{r}') \, \delta_{\sigma,\sigma'}$, the sum rule (30.52) simplifies to the form

$$\int_{-\infty}^{+\infty} \frac{d\omega}{2\pi} A(\mathbf{k}; \omega) = 1 \,. \tag{30.56}$$

In addition, upon taking the spatial Fourier transform of the definition (30.46), the standard result

$$A(\mathbf{k};\omega) = i\int d\mathbf{r}\, e^{-i\mathbf{k}\cdot\mathbf{r}} \left[G^{\mathrm{R}}(\mathbf{r};\omega) - G^{\mathrm{A}}(\mathbf{r};\omega)\right]$$
$$= i\int d\mathbf{r}\, e^{-i\mathbf{k}\cdot\mathbf{r}} \left[G^{\mathrm{R}}(\mathbf{r};\omega) - G^{\mathrm{R}}(-\mathbf{r};\omega)^*\right]$$
$$= i\left[G^{\mathrm{R}}(\mathbf{k};\omega) - G^{\mathrm{R}}(\mathbf{k};\omega)^*\right] = -\,2\,\mathrm{Im}\left\{G^{\mathrm{R}}(\mathbf{k};\omega)\right\} \qquad (30.57)$$

follows. Finally, for a noninteracting system $G^{\mathrm{R}}(\mathbf{k},\omega) = \frac{1}{\omega - \epsilon(\mathbf{k}) + i\eta}$ (cf. Eq. (20.6) of Part I), such that $A(\mathbf{k};\omega) = 2\,\pi\,\delta(\omega - \epsilon(\mathbf{k}))$ with the sum rule (30.56) being trivially satisfied.

Remark: An additional sum rule valid in the fermionic superfluid phase
When dealing specifically with the superfluid phase of a Fermi system with a point-contact interparticle interaction, an interesting sum rule arises *beyond* the lowest-order moment represented by Eq. (30.54) (and even for nonequilibrium situations). To this end, let us consider the off-diagonal element (in Nambu space)

$$A(\mathbf{r}t,\mathbf{r}'t')_{12} = i\left[G^{>}(\mathbf{r}t,\mathbf{r}'t')_{12} - G^{<}(\mathbf{r}t,\mathbf{r}'t')_{12}\right] = \langle \Psi_1(\mathbf{r}t)\,\Psi_2^\dagger(\mathbf{r}'t') + \Psi_2^\dagger(\mathbf{r}'t')\,\Psi_1(\mathbf{r}t)\rangle$$
$$= \langle \psi_\uparrow(\mathbf{r}t)\,\psi_\downarrow(\mathbf{r}'t') + \psi_\downarrow(\mathbf{r}'t')\,\psi_\uparrow(\mathbf{r}t)\rangle, \qquad (30.58)$$

where Eqs. (22.1), (23.16), and (23.17) have been utilized. We then obtain along the lines of the expression (30.54) for the lowest-order moment

$$\int_{-\infty}^{+\infty}\frac{d\omega}{2\pi}A(\mathbf{r},\mathbf{r}';t_{\mathrm{tot}}|\omega)_{12} = \int_{-\infty}^{+\infty}\frac{d\omega}{2\pi}\,i\int_{-\infty}^{+\infty}d(t-t')\,e^{i\omega(t-t')}\left[G^{>}(\mathbf{r}t,\mathbf{r}'t')_{12} - G^{<}(\mathbf{r}t,\mathbf{r}'t')_{12}\right]$$
$$= i\left[G^{>}(\mathbf{r}t,\mathbf{r}'t)_{12} - G^{<}(\mathbf{r}t,\mathbf{r}'t)_{12}\right]$$
$$= \langle U(t_0,t)\left(\psi_\uparrow(\mathbf{r})\,\psi_\downarrow(\mathbf{r}') + \psi_\downarrow(\mathbf{r}')\,\psi_\uparrow(\mathbf{r})\right)U(t,t_0)\rangle = 0\,, \qquad (30.59)$$

which does not provide any useful information (apart from the fact that the integrand over ω is not positive definite). Considering the first-order moment, we instead obtain

$$\int_{-\infty}^{+\infty}\frac{d\omega}{2\pi}\,\omega\,A(\mathbf{r},\mathbf{r}';t_{\mathrm{tot}}|\omega)_{12}$$
$$= \int_{-\infty}^{+\infty}\frac{d\omega}{2\pi}\,\omega\, i\int_{-\infty}^{+\infty}d(t-t')\,e^{i\omega(t-t')}\left[G^{>}(\mathbf{r}t,\mathbf{r}'t')_{12} - G^{<}(\mathbf{r}t,\mathbf{r}'t')_{12}\right]$$
$$\underset{\substack{\uparrow\\ \text{[by parts]}}}{=} \int_{-\infty}^{+\infty}\frac{d\omega}{2\pi}\int_{-\infty}^{+\infty}d(t-t')\,e^{i\omega(t-t')}\left[-\frac{\partial}{\partial t}G^{>}(\mathbf{r}t,\mathbf{r}'t')_{12} + \frac{\partial}{\partial t}G^{<}(\mathbf{r}t,\mathbf{r}'t')_{12}\right]$$

$$\underset{\substack{\uparrow\\ \text{[Eq. (30.58)]}}}{=} \int_{-\infty}^{+\infty} \frac{d\omega}{2\pi} \int_{-\infty}^{+\infty} d(t-t')\, e^{i\omega(t-t')} \left\langle \left(i\frac{\partial}{\partial t}\psi_\uparrow(\mathbf{r}t)\right) \psi_\downarrow(\mathbf{r}'t') + \psi_\downarrow(\mathbf{r}'t') \left(i\frac{\partial}{\partial t}\psi_\uparrow(\mathbf{r}t)\right) \right\rangle$$

$$\underset{\substack{\uparrow\\ \text{[Eq. (21.4)]}}}{=} v_0\, \delta(\mathbf{r}-\mathbf{r}')\, \langle \psi_\downarrow(\mathbf{r}t)\, \psi_\uparrow(\mathbf{r}t)\rangle \underset{\substack{\uparrow\\ \text{[Eq. (21.6)]}}}{=} \delta(\mathbf{r}-\mathbf{r}')\, \Delta(\mathbf{r},t) \tag{30.60}$$

with a time-dependent gap parameter. The results (30.59) and (30.60) generalize to a time- and space-dependent case the results obtained in Appendix A of Ref. [84] for a homogeneous superfluid Fermi system at equilibrium.

30.5 From Four to Two Independent Green's Functions

In Section 13.3 of Part I, we concluded that, in the general nonequilibrium case, it is sufficient to consider only *four* independent physical Green's functions, say, G^{M}, $G^{\rceil}$, $G^{<}$, and G^{R}. In the equilibrium case, on the other hand, we have just proved that it is sufficient to consider only *two* independent Green's functions, say, G^{M} and $G^{<}$. This is because

(i) $G^{\rceil}$ can be expressed in terms of $G^{<}$ (cf. Eq. (30.8)) and further becomes irrelevant when the limit $t_0 \longrightarrow -\infty$ is eventually considered (cf. also Section 31.4) and
(ii) G^{R} can also be expressed in terms of $G^{<}$ via Eqs. (30.10) and (30.14).

The progressive reduction in the number of independent components of the single-particle contour Green's function, when passing from the nonequilibrium "full" (Konstantinov–Perel) formalism to the Keldysh formalism, and considering eventually the equilibrium case, is schematically indicated in Table 30.1 for the sake of completeness.

We have also shown that similar conclusions hold for the components of the self-energy, among which Σ^{M} is needed to obtain G^{M} via the Dyson equation

Table 30.1 *Components of the single-particle contour Green's function occurring in the various formalisms.*

Formalism	G^{M}	$G^{\rceil}$	$G^{<}$	$G^{>}$
Full	Yes	Yes	Yes	Yes
Keldysh	Yes*	No	Yes	Yes
Equilibrium	Yes	No	Yes	No

[The asterisk (∗) signifies that, in this case, $G^{\mathrm{M}} \rightarrow G_0^{\mathrm{M}}$.]

(15.10) of Part I. In turn, G^{M} is needed to fix the relation between the thermodynamic chemical potential μ and the (inverse) temperature β (which also specifies the relation (30.14) between $G^<$ and $G^>$), via the density equation

$$n = \pm \int \frac{d\mathbf{k}}{(2\pi)^3} \frac{1}{\beta} \sum_n e^{i\omega_n \eta} \, \mathrm{tr}\left\{ iG^{\mathrm{M}}(\mathbf{k}, \omega_n) \right\}, \tag{30.61}$$

where the upper (lower) sign holds for bosons (fermions), the trace is over the spin components, and

$$\omega_n = \begin{cases} \frac{2n\pi}{\beta} & \text{(bosons)} \\ \\ \frac{(2n+1)\pi}{\beta} & \text{(fermions)} \end{cases} \tag{30.62}$$

(n integer) are Matsubara frequencies.

There still remains to show that, at equilibrium, the reference time t_0 drops out from the Kadanoff–Baym equation for $G^<$. Although this may appear evident on physical grounds, it will be explicitly shown mathematically in Section 31.4.

31

Miscellany and Addenda to Part II

Similar to what we did at the end of Part I, we add here the detailed treatment of a few issues that complement some of the topics treated in Chapters 23 and 28–30 of Part II.

31.1 Dyson Equations for the Contour Single-Particle Green's Function in the Nambu Representation

For completeness, here, we provide the proof of the differential equation (23.5) in the variables $(\mathbf{r}_1, z_1)$, as well as its companion in the variables $(\mathbf{r}_2, z_2)$, for the contour single-particle Green's function $G(\mathbf{r}_1 z_1, \mathbf{r}_2 z_2)_{i_1 i_2}$ in the Nambu representation.

To this end, we begin to calculate the commutator of the field operators $\Psi_k(\mathbf{r})$ and $\Psi_k^\dagger(\mathbf{r})$ (where k is a Nambu pseudo-spinor index), with the Hamiltonian (23.3) in the Nambu representation. Taking into account the anti-commutator relations (22.2), we obtain:

$$
\begin{aligned}
[\Psi_k(\mathbf{r}), H(t)] &= \sum_j \tau^3_{kj}\, h(\mathrm{r}, t)\, \Psi_j(\mathbf{r}) \\
&+ \frac{1}{2} \int d\mathbf{r}' \sum_{ij} \Psi_i^\dagger(\mathbf{r}')\, \tau^3_{ij}\, \Psi_j(\mathbf{r}')\, v(\mathbf{r}' - \mathbf{r}) \sum_{j'} \tau^3_{kj'}\, \Psi_{j'}(\mathbf{r}) \\
&+ \frac{1}{2} \int d\mathbf{r}' \sum_{j'} \tau^3_{kj'}\, \Psi_{j'}(\mathbf{r})\, v(\mathbf{r} - \mathbf{r}') \sum_{ij} \Psi_i^\dagger(\mathbf{r}')\, \tau^3_{ij}\, \Psi_j(\mathbf{r}'), \qquad (31.1)
\end{aligned}
$$

as well as

$$
\begin{aligned}
[\Psi_k^\dagger(\mathbf{r}), H(t)] &= -\, [\Psi_k(\mathbf{r}), H(t)]^\dagger = -\sum_j h(\mathrm{r}, t)\, \Psi_j^\dagger(\mathbf{r})\, \tau^3_{jk} \\
&- \frac{1}{2} \int d\mathbf{r}' \sum_{ij} \Psi_i^\dagger(\mathbf{r}')\, \tau^3_{ij}\, \Psi_j(\mathbf{r}')\, v(\mathbf{r}' - \mathbf{r}) \sum_{j'} \Psi_{j'}^\dagger(\mathbf{r})\, \tau^3_{j'k} \\
&- \frac{1}{2} \int d\mathbf{r}' \sum_{j'} \Psi_{j'}^\dagger(\mathbf{r})\, \tau^3_{j'k}\, v(\mathbf{r} - \mathbf{r}') \sum_{ij} \Psi_i^\dagger(\mathbf{r}')\, \tau^3_{ij}\, \Psi_j(\mathbf{r}'). \qquad (31.2)
\end{aligned}
$$

The last two terms on the right-hand side of both Eqs. (31.1) and (31.2), which originate from the presence of the interparticle interaction, will give the same contribution to the equations of motion for the contour single-particle Green's function, such that we can combine them in a single term.

In this way, we obtain the equations of motion for the contour single-particle Green's function:

(i) In the *first* (left) variable $(\mathbf{r}_1 z_1)$,

$$\frac{d}{dz_1} G(\mathbf{r}_1 z_1, \mathbf{r}_2 z_2)_{i_1 i_2} = -i\,\delta(z_1, z_2)\,\delta(\mathbf{r}_1 - \mathbf{r}_2)\,\delta_{i_1 i_2}$$
$$- \sum_j \tau^3_{i_1 j}\, h(\mathbf{r}_1, z_1) \left\langle \mathcal{T}_C \left\{ \Psi_j(\mathbf{r}_1 z_1)\, \Psi^\dagger_{i_2}(\mathbf{r}_2 z_2) \right\} \right\rangle - \int_C dz_3 \int d\mathbf{r}_3$$
$$\times \left\langle \mathcal{T}_C \left\{ \sum_{ij} \Psi^\dagger_i(\mathbf{r}_3 z_3)\, \tau^3_{ij}\, \Psi_j(\mathbf{r}_3 z_3)\, v(\mathbf{r}_3 z_3, \mathbf{r}_1 z_1) \sum_{j'} \tau^3_{i_1 j'}\, \Psi_{j'}(\mathbf{r}_1 z_1)\, \Psi^\dagger_{i_2}(\mathbf{r}_2 z_2) \right\} \right\rangle, \tag{31.3}$$

(ii) In the *second* (right) variable $(\mathbf{r}_2 z_2)$,

$$\frac{d}{dz_2} G(\mathbf{r}_1 z_1, \mathbf{r}_2 z_2)_{i_1 i_2} = i\,\delta(z_1, z_2)\,\delta(\mathbf{r}_1 - \mathbf{r}_2)\,\delta_{i_1 i_2}$$
$$+ \sum_j h(\mathbf{r}_2, z_2) \left\langle \mathcal{T}_C \left\{ \Psi_{i_1}(\mathbf{r}_1 z_1)\, \Psi^\dagger_j(\mathbf{r}_2 z_2) \right\} \right\rangle \tau^3_{j i_2} + \int_C dz_3 \int d\mathbf{r}_3$$
$$\times \left\langle \mathcal{T}_C \left\{ \Psi_{i_1}(\mathbf{r}_1 z_1) \sum_{ij} \Psi^\dagger_i(\mathbf{r}_3 z_3)\, \tau^3_{ij}\, \Psi_j(\mathbf{r}_3 z_3)\, v(\mathbf{r}_3 z_3, \mathbf{r}_2 z_2) \sum_{j'} \Psi^\dagger_{j'}(\mathbf{r}_2 z_2)\, \tau^3_{j' i_2} \right\} \right\rangle, \tag{31.4}$$

where C is the closed contour along which the single-particle Green's function is defined to begin with and $v(\mathbf{r}_1 z_1, \mathbf{r}_2 z_2) = \delta(z_1, z_2)\, v(\mathbf{r}_1 - \mathbf{r}_2)$ as in Eq. (11.20) (cf. also Eq. (11.21)) of Part I.

We note that in Eq. (31.3) the interparticle interaction $v(\mathbf{r}_3 z_3, \mathbf{r}_1 z_1)$ relates to the variables $(\mathbf{r}_1 z_1)$ and $(\mathbf{r}_3 z_3)$, but not to $(\mathbf{r}_2 z_2)$. With reference to the diagrammatic rules in the Nambu representation, ($-i$ times) the last term on the right-hand side of Eq. (31.3) then corresponds to the diagrammatic structure depicted in Fig. 31.1(a), with the self-energy Σ at the left and the Green's function G at the right. On the other hand, in Eq. (31.4), the interparticle interaction $v(\mathbf{r}_3 z_3, \mathbf{r}_2 z_2)$ relates to the variables $(\mathbf{r}_2 z_2)$ and $(\mathbf{r}_3 z_3)$, but not to $(\mathbf{r}_1 z_1)$. Again with reference to the diagrammatic rules in the Nambu representation, ($-i$ times) the last term on the right-hand side of Eq. (31.4) then corresponds to the diagrammatic structure depicted in

Figure 31.1 Compact diagrammatic representation of ($-\,i$ times) the last term on the right-hand side of the expressions (a) (31.3) and (b) (31.4). Here, filled dots stand for τ^3 matrices with Nambu indices, and interrupted dashes lines signal the presence of an interaction line at the given vertex.

Fig. 31.1(b), with the Green's function G at the left and the *same* self-energy Σ at the right. Accordingly, Eqs. (31.3) and (31.4) can be cast in the compact form

$$\sum_{i_3}\left[i\,\delta_{i_1 i_3}\frac{d}{dz_1} - \tau^3_{i_1 i_3} h(\mathbf{r}_1, z_1)\right] G(\mathbf{r}_3 z_3, \mathbf{r}_2 z_2)_{i_1 i_2}$$
$$- \int_C dz_3 \int d\mathbf{r}_3 \sum_{i_3} \Sigma(\mathbf{r}_1 z_1, \mathbf{r}_3 z_3)_{i_1 i_3}\, G(\mathbf{r}_3 z_3, \mathbf{r}_2 z_2)_{i_3 i_2}$$
$$= \delta(z_1, z_2)\,\delta(\mathbf{r}_1 - \mathbf{r}_2)\,\delta_{i_1 i_2}, \tag{31.5}$$

which coincides with Eq. (23.5), and

$$\sum_{i_3}\left[-\,i\,\delta_{i_1 i_3}\frac{d}{dz_2} - \tau^3_{i_3 i_2} h(\mathbf{r}_2, z_2)\right] G(\mathbf{r}_3 z_3, \mathbf{r}_2 z_2)_{i_1 i_3}$$
$$- \int_C dz_3 \int d\mathbf{r}_3 \sum_{i_3} G(\mathbf{r}_1 z_1, \mathbf{r}_3 z_3)_{i_1 i_3}\, \Sigma(\mathbf{r}_3 z_3, \mathbf{r}_2 z_2)_{i_3 i_2}$$
$$= \delta(z_1, z_2)\,\delta(\mathbf{r}_1 - \mathbf{r}_2)\,\delta_{i_1 i_2}\,, \tag{31.6}$$

respectively. We emphasize that, on the basis of the above direct recourse to a diagrammatic analysis, no distinction had to be made between the two possible versions Σ and $\bar{\Sigma}$ of the self-energy, as it was (at least initially) done in Chapter 11 in Part I. Finally, we recall in this context that the Hartree–Fock self-energy has to be dealt with in the ordinary spin representation rather than in the Nambu representation, as it was discussed in Section 22.2.

31.2 The ω vs ω^2 Terms in the Derivation of the TDGP Equation

When deriving the time-dependent Gross–Pitaevskii equation for composite bosons in Chapter 28, we have retained only the term of linear order in the time derivative, leaving it open the issue of the contribution of terms of higher-than-linear orders in the time derivative. On the other hand, a term of second order in the time derivative was explicitly considered (although not eventually retained) in the derivation of the time-dependent Ginzburg–Landau equation of Chapter 29. Here, we remedy the above omission and consider also the term of second order in the time derivative in the context of the time-dependent Gross–Pitaevskii equation, arriving eventually to the conclusion that this term is subleading with respect to the linear term in the relevant BEC limit of the BCS–BEC crossover.

To this end, we consider the results reported in section 3 of Ref. [113], where the expressions of the particle–particle bubbles $A_e(\mathbf{q}, \omega)$ and $B_e(\mathbf{q}, \omega)$ for small $\mathbf{q}$ and ω are reported, including the terms of order ω^2 that were omitted in Chapter 28. We thus need to evaluate explicitly the coefficients a_3 in Eq. (43) and b_3 in Eq. (44) of Ref. [113], in the BEC limit of the BCS–BEC crossover at zero temperature. These coefficients are given, respectively, by the expressions

$$a_3 = -\int \frac{d\mathbf{k}}{(2\pi)^3} \frac{\left(2\,\xi_{\mathbf{k}}^2 + |\Delta_{\mathrm{e}}|^2\right)}{16\,E_{\mathbf{k}}^5} \tag{31.7}$$

[cf. Eq. (48) of Ref. [113]], and

$$b_3 = \int \frac{d\mathbf{k}}{(2\pi)^3} \frac{|\Delta_{\mathrm{e}}|^2}{16\,E_{\mathbf{k}}^5} \tag{31.8}$$

[cf. Eq. (51) of Ref. [113]]. For the sake of comparison between the terms of order ω^2 and the term of order ω, we will also need to evaluate the coefficient

$$a_1 = -\int \frac{d\mathbf{k}}{(2\pi)^3} \frac{\xi_{\mathbf{k}}}{4\,E_{\mathbf{k}}^3} \tag{31.9}$$

[cf. Eq. (46) of Ref. [113]]. In the expressions (31.7)–(31.9),

$$\xi_{\mathbf{k}} = \frac{\mathbf{k}^2}{2m} - \mu \quad \text{and} \quad E_{\mathbf{k}} = \sqrt{\xi_{\mathbf{k}}^2 + |\Delta_{\mathrm{e}}|^2} \tag{31.10}$$

where the notation of Chapter 28 has been maintained for the magnitude $|\Delta_{\mathrm{e}}|$ of the gap parameter.

Recall further from Chapter 26 that, in the BEC limit of the BCS–BEC crossover, $|\mu| \simeq \frac{\varepsilon_0}{2} = \frac{1}{2ma_F^2}$ (with $a_F > 0$) is the largest energy scale in the problem, such that $|\mu| \gg |\Delta_{\mathrm{e}}|$, and that $\xi_{\mathbf{k}} > 0$ for all $\mathbf{k}$. We can then approximate in the expressions (31.7)–(31.9):

(i)

$$\int \frac{d\mathbf{k}}{(2\pi)^3} \frac{\xi_{\mathbf{k}}^2}{E_{\mathbf{k}}^5} \simeq \int \frac{d\mathbf{k}}{(2\pi)^3} \frac{1}{\xi_{\mathbf{k}}^3} = \frac{\left(2\,m\,a_F^2\right)^3}{2\,\pi^2} \int_0^\infty dk \, \frac{k^2}{\left(k^2 a_F^2 + 1\right)^3}$$

$$= \frac{(2\,m)^3\,a_F^3}{2\,\pi^2} \int_0^\infty dx \, \frac{x^2}{\left(x^2+1\right)^3} = \frac{(m\,a_F)^3}{4\,\pi}\,; \tag{31.11}$$

(ii)

$$\int \frac{d\mathbf{k}}{(2\pi)^3} \frac{1}{E_{\mathbf{k}}^5} \simeq \int \frac{d\mathbf{k}}{(2\pi)^3} \frac{1}{\xi_{\mathbf{k}}^5} = \frac{\left(2\,m\,a_F^2\right)^5}{2\,\pi^2} \int_0^\infty dk \, \frac{k^2}{\left(k^2 a_F^2 + 1\right)^5}$$

$$= \frac{(2\,m)^5\,a_F^7}{2\,\pi^2} \int_0^\infty dx \, \frac{x^2}{\left(x^2+1\right)^5} = \frac{5\,m^5\,a_F^7}{16\,\pi}\,; \tag{31.12}$$

(iii)

$$\int \frac{d\mathbf{k}}{(2\pi)^3} \frac{\xi_{\mathbf{k}}}{E_{\mathbf{k}}^3} \simeq \int \frac{d\mathbf{k}}{(2\pi)^3} \frac{1}{\xi_{\mathbf{k}}^2} = \frac{\left(2\,m\,a_F^2\right)^2}{2\,\pi^2} \int_0^\infty dk \, \frac{k^2}{\left(k^2 a_F^2 + 1\right)^2}$$

$$= \frac{(2\,m)^2\,a_F}{2\,\pi^2} \int_0^\infty dx \, \frac{x^2}{\left(x^2+1\right)^2} = \frac{m^2\,a_F}{2\,\pi}\,. \tag{31.13}$$

To obtain the last line of each of the expressions (31.11)–(31.13), we, have made use of the result [148]

$$\int_0^\infty dx \, \frac{x^{\alpha-1}}{\left(q\,x^\beta + p\right)^{n+1}} = \frac{(p/q)^{\alpha/\beta}}{\beta\,p^{n+1}} \, \frac{\Gamma\left(\alpha/\beta\right)\,\Gamma\left(n+1-\alpha/\beta\right)}{\Gamma\left(n+1\right)}\,, \tag{31.14}$$

where $0 < \frac{\alpha}{\beta} < n+1$, $p \neq 0$ and $q \neq 0$, and $\Gamma\,(z)$ is the Euler Gamma function of argument z.

In terms of the approximate expressions (31.11)–(31.13) valid in the BEC limit of the BCS–BEC crossover, the coefficients (31.7)–(31.9) become eventually:

$$a_3 = -\,\frac{1}{8} \int \frac{d\mathbf{k}}{(2\pi)^3} \frac{\xi_{\mathbf{k}}^2}{E_{\mathbf{k}}^5} - \frac{|\Delta_{\mathrm{e}}|^2}{16} \int \frac{d\mathbf{k}}{(2\pi)^3} \frac{1}{16\,E_{\mathbf{k}}^5} \simeq -\,\frac{1}{8}\,\frac{(m\,a_F)^3}{4\,\pi} - \frac{|\Delta_{\mathrm{e}}|^2}{16}\,\frac{5\,m^5\,a_F^7}{16\,\pi}$$

$$= -\left(\frac{m^2\,a_F}{8\,\pi}\right) \frac{1}{4\,\varepsilon_0} \left(1 + \frac{5\,|\Delta_{\mathrm{e}}|^2}{8\,\varepsilon_0^2}\right), \tag{31.15}$$

$$b_3 = \frac{|\Delta_{\mathrm{e}}|^2}{16} \int \frac{d\mathbf{k}}{(2\pi)^3} \frac{1}{E_{\mathbf{k}}^5} \simeq \frac{|\Delta_{\mathrm{e}}|^2}{16}\,\frac{5\,m^5\,a_F^7}{16\,\pi} = \left(\frac{m^2\,a_F}{8\,\pi}\right) \frac{1}{4\,\varepsilon_0}\,\frac{5\,|\Delta_{\mathrm{e}}|^2}{8\,\varepsilon_0^2}\,, \tag{31.16}$$

and

$$a_1 = -\frac{1}{4} \int \frac{d\mathbf{k}}{(2\pi)^3} \frac{\xi_{\mathbf{k}}}{E_{\mathbf{k}}^3} \simeq -\frac{1}{4} \int \frac{d\mathbf{k}}{(2\pi)^3} \frac{1}{\xi_{\mathbf{k}}^2} = -\left(\frac{m^2 \, a_F}{8\,\pi} \right) . \tag{31.17}$$

Finally, upon comparing the terms $a_3\,\omega^2$ and $b_3\,\omega^2$ with the term $a_1\,\omega$, we obtain the results

$$\frac{|a_1|\,\omega}{|a_3|\,\omega^2} = \frac{4\,\epsilon_0}{\omega} \frac{1}{1 + \frac{5\,|\Delta_e|^2}{8\,\varepsilon_0^2}} \simeq \frac{4\,\epsilon_0}{\omega} \gg 1, \tag{31.18}$$

$$\frac{|a_1|\,\omega}{|b_3|\,\omega^2} = \frac{4\,\epsilon_0}{\omega} \frac{8\,\varepsilon_0^2}{5\,|\Delta_e|^2} \gg 1\,, \tag{31.19}$$

which hold not only because $\varepsilon_0 \gg |\Delta_e|$ as appropriate to the BEC limit of the BCS–BEC crossover but also because $\varepsilon_0 \gg \omega$ in line with the derivation of the TDGP equation for composite bosons of Chapter 28.

In conclusion, we have verified that the expressions (28.36) and (28.37), which have been utilized in Chapter 28 to derive the TDGP equation for composite bosons, contain the leading terms in the expansions of $A_e(\mathbf{q}, \omega)$ and $B_e(\mathbf{q}, \omega)$ for sufficiently small values of $\mathbf{q}$ and ω.

31.3 Calculation of an Integral Occurring in the Derivation of the TDGL Equation

In Section 29.7, we left open the calculation of the expression (29.40), valid in the BCS limit of the BCS–BEC crossover at the critical temperature, which we rewrite here for convenience in the form

$$\int \frac{d\mathbf{p}}{(2\pi)^3} \left[\frac{(1 - 2f_F(\xi_{\mathbf{p}}))}{2\,\xi_{\mathbf{p}}} - \frac{m}{\mathbf{p}^2} \right] = N(\mu) \, \ln \left(\frac{8\,e^{\gamma}\,\beta_c\,\mu}{\pi\,e^2} \right) , \tag{31.20}$$

where f_F is the Fermi–Dirac distribution function taken at the critical temperature T_c, and $N(\mu)$ is the free-particle density of states (per spin component)

$$N(\varepsilon) = \frac{m^{3/2}\,\sqrt{\varepsilon}}{\sqrt{2}\,\pi^2}, \tag{31.21}$$

with ε taken at the chemical potential μ, and γ is the Euler constant (such that $e^{\gamma} \simeq 1.781$). Here, we explicitly provide an analytic evaluation of the expression (31.20) through the following steps.

Step # 1 We begin by making use of the expression (31.21) for the density of states , to rewrite the left-hand side of Eq. (31.20) in the form

$$\mathcal{I}(\beta_c, \mu) \equiv \int \frac{d\mathbf{p}}{(2\pi)^3} \left[\frac{(1 - 2f_F(\xi_{\mathbf{p}}))}{2\,\xi_{\mathbf{p}}} - \frac{m}{\mathbf{p}^2} \right] = \int_0^\infty d\varepsilon\, N(\varepsilon) \left[\frac{(1 - 2f_F(\varepsilon - \mu))}{2\,(\varepsilon - \mu)} - \frac{1}{2\,\varepsilon} \right]$$

$$= \fint_0^\infty d\varepsilon\, N(\varepsilon) \left[-\frac{f_F(\varepsilon - \mu)}{\varepsilon - \mu} + \frac{1}{2} \left(\frac{1}{\varepsilon - \mu} - \frac{1}{\varepsilon} \right) \right], \tag{31.22}$$

where we have introduced the prescription of taking the principal part of the integral since the original well-defined integral has been split into two pieces, which are both ill-defined when $\varepsilon = \mu$.

Step # 2 We next introduce the variables $x = \varepsilon/\mu$ and $y = \beta_c \mu$, such that $f_F(\varepsilon - \mu) = \left(e^{y(x-1)} + 1\right)^{-1}$. The expression (31.22) then becomes:

$$\mathcal{I}(\beta_c, \mu) = N(\mu) \fint_0^\infty dx\, \sqrt{x} \left[-\frac{1}{e^{y(x-1)} + 1} \frac{1}{x - 1} + \frac{1}{2} \left(\frac{1}{x - 1} - \frac{1}{x} \right) \right]. \tag{31.23}$$

We are now going to evaluate separately the two integrals corresponding to the two terms within brackets in Eq. (31.23).

Step # 3 In the first integral of Eq. (31.23), we write $\frac{\sqrt{x}}{x-1} = \frac{1}{\sqrt{x}+1} + \frac{1}{x-1}$, such that

$$\fint_0^\infty dx\, \sqrt{x} \frac{1}{e^{y(x-1)} + 1} \frac{1}{x - 1} = \int_0^\infty dx\, \frac{1}{e^{y(x-1)} + 1} \frac{1}{\sqrt{x} + 1} + \fint_0^\infty dx\, \frac{1}{e^{y(x-1)} + 1} \frac{1}{x - 1}. \tag{31.24}$$

The first integral on the right-hand side of Eq. (31.24) has no singularity when $x = 1$. In addition, since $y = \beta_c \mu \gg 1$, this integral can be approximated by its value for $y = \infty$ in such a way that the variable x extends in practice up to $\max\{x\} = 1$, yielding

$$\int_0^1 dx\, \frac{1}{\sqrt{x} + 1} = 2 - 2\, \ln 2\,. \tag{31.25}$$

The second integral on the right-hand side of Eq. (31.24), on the other hand, has a singularity when $x = 1$, which is treated by the principal part. Accordingly, we rewrite this integral as follows:

$$\fint_0^\infty dx\, \frac{1}{e^{y(x-1)} + 1} \frac{1}{x - 1} \underset{[\,z=x-1\,]}{\overset{\uparrow}{=}} \int_1^\eta dz\, \frac{1}{e^{-yz} + 1} \frac{1}{z} + \int_\eta^\infty dz\, \frac{1}{e^{yz} + 1} \frac{1}{z} \qquad (\eta = 0^+)$$

$$
\underset{[\text{ by parts }]}{\overset{\uparrow}{=}} y \int_0^1 dz \, \ln z \, \frac{e^{-yz}}{(e^{-yz}+1)^2} + y \int_0^\infty dz \, \ln z \, \frac{e^{yz}}{(e^{yz}+1)^2} \underset{[\, y \gg 1 \,]}{\overset{\uparrow}{\simeq}} \frac{y}{2} \int_0^\infty dz \, \frac{\ln z}{\cosh^2\left(\frac{yz}{2}\right)}
$$

$$
= \int_0^\infty dx \, \frac{1}{\cosh^2 x} \left[\ln\left(\frac{2}{y}\right) + \ln x \right] = - \ln\left(\frac{2\, y\, e^\gamma}{\pi}\right), \tag{31.26}
$$

where the last result has been obtained in terms of standard definite integrals [148].

Grouping together the results (31.25) and (31.26), the first integral in Eq. (31.23) becomes:

$$
\int_0^\infty dx \, \sqrt{x} \frac{1}{e^{y(x-1)}+1} \, \frac{1}{x-1} = 2 - 2 \, \ln 2 - \ln\left(\frac{2\, y\, e^\gamma}{\pi}\right) = - \ln\left(\frac{8\, y\, e^\gamma}{\pi\, e^2}\right), \tag{31.27}
$$

where we recall that $y = \beta_c \mu$.

Step # 4 Finally, the second integral in Eq. (31.23) can be evaluated as follows:

$$
\frac{1}{2} \fint_0^\infty dx \, \sqrt{x} \left(\frac{1}{x-1} - \frac{1}{x} \right) \underset{[x=z^2]}{\overset{\uparrow}{=}} \frac{1}{2} \fint_0^\infty dz \left(\frac{1}{z-1} - \frac{1}{z+1} \right)
$$

$$
= \frac{1}{2} \lim_{\bar{z} \to \infty} \fint_0^{\bar{z}} dx \left(\frac{1}{z-1} - \frac{1}{z+1} \right) = \frac{1}{2} \lim_{\bar{z} \to \infty} \ln\left(\frac{\bar{z}-1}{\bar{z}+1} \right) = 0 \, , \tag{31.28}
$$

a result that can also be independently verified by a numerical calculation.

In conclusion, entering the results (31.27) and (31.28) into Eq. (31.23), the expression (31.20) readily follows.

31.4 Irrelevance of the Reference Time t_0 for the Convolutions Entering the Kadanoff–Baym Equations at Equilibrium

A necessary step, to verify that the nonequilibrium approach appropriately reduces to the equilibrium one when $h(\mathbf{r}, t) \to h(\mathbf{r})$, consists in mastering the fate of the time t_0, which enters the Kadanoff–Baym equations of Chapter 15 in Part I. Recall that the "reference time" t_0 is, by definition, the time at which a time dependence is turned on in the external potential $V_{\text{ext}}(\mathbf{r})$ (or else, in the interparticle interaction $v(\mathbf{r} - \mathbf{r}'))$. It is for this reason that, in the nonequilibrium case, one has to specify how the system was initially prepared before a time-dependent perturbance begins to act on the system. For the same reason, in the equilibrium case, when no time-dependent perturbance acts on the system, the time t_0 should *formally* be removed from the relevant expressions, and in particular from the Kadanoff–Baym equations where it enters the convolution integrals. Here, the results from Chapter 30 will be utilized for the purpose, following closely the treatment given in section 9.6 of Ref. [15].

For definiteness, we consider the convolution integral

$$I^<(t,t'|t_0) = \int_{t_0}^{t} d\bar{t}\; \Sigma^{\mathrm{R}}(t,\bar{t})\, G^<(\bar{t},t') + \int_{t_0}^{t'} d\bar{t}\; \Sigma^<(t,\bar{t})\, G^{\mathrm{A}}(\bar{t},t')$$
$$- i \int_0^{\beta} d\bar{\tau}\, \Sigma^{\rceil}(t,\bar{\tau})\, G^{\lceil}(\bar{\tau},t'), \tag{31.29}$$

which enters the equation of motion (15.13) of Part I for $G^<$, where the dependence on the spatial and spin variables has been omitted for simplicity, and manipulate it through the following steps.

Step # 1 We begin by noting that the "singular" part $\sigma^{\mathrm{R}}(t)$ of the retarded self-energy $\Sigma^{\mathrm{R}}(t,t')$

$$\Sigma^{\mathrm{R}}(t,t') = \sigma^{\mathrm{R}}(t)\,\delta(t-t') + \Sigma^{\mathrm{R}}_{\mathrm{r}}(t,t') \tag{31.30}$$

with "regular" part $\Sigma^{\mathrm{R}}_{\mathrm{r}}(t,t')$ would give the following contribution to $I^<(t,t'|t_0)$ of Eq. (31.29)

$$\int_{t_0}^{t} d\bar{t}\; \sigma^{\mathrm{R}}(t)\,\delta(t-\bar{t})\, G^<(\bar{t},t') = \sigma^{\mathrm{R}}(t)\, G^<(t,t') \tag{31.31}$$

for $t > t_0$ and zero otherwise (where at equilibrium $\sigma^{\mathrm{R}}(t)$ should not depend on t). This contribution would thus appear to depend on t_0 through the step function $\theta(t-t_0)$. However, this dependence on t_0 does not explicitly appear if one retraces back to Eq. (14.13) of Part I the origin of the singular contribution to Eq. (14.25) of Part I, in which t_0 can be made to recede to $-\infty$ with no harm. We thus concentrate on $I^<(t,t'|t_0)|_{\mathrm{r}}$, which contains only the contribution of the "regular" part $\Sigma^{\mathrm{R}}_{\mathrm{r}}(t,t')$.

Step # 2 Next, we Fourier transform all single-particle Green's functions and self-energies entering the expression (31.29) and make use of Eqs. (30.5) and (30.29), in conjunction with the analytic properties discussed in Chapter 27, to obtain:

$$I^<(t,t'|t_0)|_{\mathrm{r}} = \int_{-\infty}^{+\infty} \frac{d\omega_1}{2\pi} \int_{-\infty}^{+\infty} \frac{d\omega_2}{2\pi}\, e^{-i\omega_1 t}\, e^{i\omega_2 t'} \Big\{ \Sigma^{\mathrm{R}}_{\mathrm{r}}(\omega_1)\, G^<(\omega_2) \int_{t_0}^{t} d\bar{t}\, e^{i(\omega_1-\omega_2)\,\bar{t}}$$
$$+ \Sigma^<(\omega_1)\, G^{\mathrm{A}}(\omega_2) \int_{t_0}^{t'} d\bar{t}\, e^{i(\omega_1-\omega_2)\,\bar{t}} - \Sigma^<(\omega_1)\, G^>(\omega_2)\; i \int_0^{\beta} d\bar{\tau}\, e^{i(\omega_1-\omega_2)\,(t_0 - i\bar{\tau})} \Big\}. \tag{31.32}$$

Step # 3 The two integrals over $\bar{t}$ in expression (31.32) would be ill-defined in the case where t_0 is made to recede to $-\infty$. To avoid any divergence, we then regularize these integrals by replacing therein $\omega_1 - \omega_2 \to \omega_1 - \omega_2 - i\eta$ (with $\eta = 0^+$). Physically, this corresponds to introducing an asymmetry between past and future, thereby assuming that the system was prepared at equilibrium far in the past and that it remains undisturbed until it is subject to measurements. Accordingly, we write

$$\int_{t_0}^{t} d\bar{t}\, e^{i(\omega_1-\omega_2)\,\bar{t}} \longrightarrow \int_{t_0}^{t} d\bar{t}\, e^{i(\omega_1-\omega_2-i\eta)\,\bar{t}} = \frac{e^{i(\omega_1-\omega_2)\,t} - e^{i(\omega_1-\omega_2)\,t_0}}{i(\omega_1-\omega_2-i\eta)} \tag{31.33}$$

and similarly for the integral over $\bar{t}$ from t_0 to t'. Although not strictly necessary, to maintain a coherent notation, we make the replacement $\omega_1-\omega_2 \longrightarrow \omega_1-\omega_2-i\eta$ also in the integral over $\bar{\tau}$ from 0 to β and write

$$\int_{0}^{\beta} d\bar{\tau}\, e^{(\omega_1-\omega_2-i\eta)\,\bar{\tau}} = \frac{e^{(\omega_1-\omega_2)\,\beta}-1}{(\omega_1-\omega_2-i\eta)}\,. \tag{31.34}$$

In this way, the expression (31.32) becomes:

$$\begin{aligned}
I^{<}(t,t'|t_0)|_{\mathrm{r}} = &\int_{-\infty}^{+\infty}\frac{d\omega_1}{2\pi}\int_{-\infty}^{+\infty}\frac{d\omega_2}{2\pi}\,\frac{e^{-i\omega_1 t}\,e^{i\omega_2 t'}}{i(\omega_1-\omega_2-i\eta)}\Big\{\Sigma_{\mathrm{r}}^{\mathrm{R}}(\omega_1)\,G^{<}(\omega_2)\\
&\times\left(e^{i(\omega_1-\omega_2)t}-e^{i(\omega_1-\omega_2)t_0}\right)+\Sigma^{<}(\omega_1)\,G^{\mathrm{A}}(\omega_2)\left(e^{i(\omega_1-\omega_2)t'}-e^{i(\omega_1-\omega_2)t_0}\right)\\
&+\left(e^{(\omega_1-\omega_2)\beta}-1\right)\Sigma^{<}(\omega_1)\,G^{>}(\omega_2)\,e^{i(\omega_1-\omega_2)t_0}\Big\}\,.
\end{aligned} \tag{31.35}$$

Step # 4 To get rid of the factor $\left(e^{(\omega_1-\omega_2)\beta}-1\right)$ entering the last term within braces in Eq. (31.35), we make use of the properties (30.39) for $G^{>}(\omega_2)$ and (30.45) for $\Sigma^{<}(\omega_1)$, thus writing

$$\begin{aligned}
\left(e^{(\omega_1-\omega_2)\beta}-1\right)\Sigma^{<}(\omega_1)\,G^{>}(\omega_2) &= \pm\left(e^{(\omega_1-\omega_2)\beta}-1\right)f_{\mp}(\omega_1-\mu)\,(1\pm f_{\mp}(\omega_2-\mu))\\
&\quad\times\left[\Sigma^{\mathrm{R}}(\omega_1)-\Sigma^{\mathrm{A}}(\omega_1)\right]\left[G^{R}(\omega_2)-G^{\mathrm{A}}(\omega_2)\right]\\
&=\pm\,(f_{\mp}(\omega_2-\mu)-f_{\mp}(\omega_1-\mu))\left[\Sigma^{\mathrm{R}}(\omega_1)-\Sigma^{\mathrm{A}}(\omega_1)\right]\left[G^{R}(\omega_2)-G^{\mathrm{A}}(\omega_2)\right]\\
&=\left[\Sigma^{\mathrm{R}}(\omega_1)-\Sigma^{\mathrm{A}}(\omega_1)\right]G^{<}(\omega_2)-\Sigma^{<}(\omega_1)\left[G^{R}(\omega_2)-G^{\mathrm{A}}(\omega_2)\right],
\end{aligned} \tag{31.36}$$

where, in the last line, the properties (30.39) for $G^{<}(\omega_2)$ and (30.45) for $\Sigma^{<}(\omega_1)$ have been used in reverse. In this way, some of the terms containing the factor $e^{i(\omega_1-\omega_2)t_0}$ cancel out with each other in Eq. (31.35), and we are left with

$$\begin{aligned}
I^{<}(t,t'|t_0)|_{\mathrm{r}} = &\int_{-\infty}^{+\infty}\frac{d\omega_1}{2\pi}\int_{-\infty}^{+\infty}\frac{d\omega_2}{2\pi}\,\frac{e^{-i\omega_1 t}\,e^{i\omega_2 t'}}{i(\omega_1-\omega_2-i\eta)}\Big\{e^{i(\omega_1-\omega_2)t}\,\Sigma_{\mathrm{r}}^{\mathrm{R}}(\omega_1)\,G^{<}(\omega_2)\\
&+e^{i(\omega_1-\omega_2)t'}\Sigma^{<}(\omega_1)\,G^{\mathrm{A}}(\omega_2)-e^{i(\omega_1-\omega_2)t_0}\left[\Sigma_{\mathrm{r}}^{\mathrm{A}}(\omega_1)\,G^{<}(\omega_2)+\Sigma^{<}(\omega_1)\,G^{R}(\omega_2)\right]\Big\}.
\end{aligned} \tag{31.37}$$

Step # 5 Finally, the factor $e^{i(\omega_1-\omega_2)t_0}$ can be eliminated from Eq. (31.37) by exploiting the properties

$$\begin{cases} \int_{t_0}^{t} d\bar{t}\ \Sigma_{\rm r}^{\rm A}(t,\bar{t})\ G^<(\bar{t},t') = 0 & \text{since } \bar{t} > t, \\ \int_{t_0}^{t'} d\bar{t}\ \Sigma_{\rm r}^<(t,\bar{t})\ G^{\rm R}(\bar{t},t') = 0 & \text{since } \bar{t} > t'. \end{cases} \tag{31.38}$$

In Fourier space, Eqs. (31.38) read

$$\int_{-\infty}^{+\infty}\frac{d\omega_1}{2\pi}\int_{-\infty}^{+\infty}\frac{d\omega_2}{2\pi}\frac{e^{-i\omega_1 t}\,e^{i\omega_2 t'}}{i(\omega_1-\omega_2-i\eta)}\left(e^{i(\omega_1-\omega_2)t}-e^{i(\omega_1-\omega_2)t_0}\right)\Sigma_{\rm r}^{\rm A}(\omega_1)\,G^<(\omega_2)=0 \tag{31.39}$$

and

$$\int_{-\infty}^{+\infty}\frac{d\omega_1}{2\pi}\int_{-\infty}^{+\infty}\frac{d\omega_2}{2\pi}\frac{e^{-i\omega_1 t}\,e^{i\omega_2 t'}}{i(\omega_1-\omega_2-i\eta)}\left(e^{i(\omega_1-\omega_2)t'}-e^{i(\omega_1-\omega_2)t_0}\right)\Sigma^<(\omega_1)\,G^{\rm R}(\omega_2)=0, \tag{31.40}$$

respectively, where the infinitesimal frequency shift $-i\eta$ has been introduced as in Eq. (31.33). In this way, the expression (31.37) becomes

$$\begin{aligned} I^<(t,t'|t_0)|_{\rm r} &= \int_{-\infty}^{+\infty}\frac{d\omega_1}{2\pi}\int_{-\infty}^{+\infty}\frac{d\omega_2}{2\pi}\frac{e^{-i\omega_1 t}\,e^{i\omega_2 t'}}{i(\omega_1-\omega_2-i\eta)} \\ &\quad\times\left\{e^{i(\omega_1-\omega_2)t}\left[\Sigma_{\rm r}^{\rm R}(\omega_1)-\Sigma_{\rm r}^{\rm A}(\omega_1)\right]G^<(\omega_2)\right. \\ &\quad\left.+\,e^{i(\omega_1-\omega_2)t'}\Sigma^<(\omega_1)\left[G^{\rm A}(\omega_2)-G^{\rm R}(\omega_2)\right]\right\} \\ &= \int_{-\infty}^{+\infty}\frac{d\omega_2}{2\pi}\,e^{-i\omega_2(t-t')}\,G^<(\omega_2)\int_{-\infty}^{+\infty}\frac{d\omega_1}{2\pi}\,i\,\frac{\left[\Sigma_{\rm r}^{\rm R}(\omega_1)-\Sigma_{\rm r}^{\rm A}(\omega_1)\right]}{\omega_2-\omega_1+i\eta} \\ &\quad+\int_{-\infty}^{+\infty}\frac{d\omega_1}{2\pi}\,e^{-i\omega_1(t-t')}\,\Sigma^<(\omega_1)\int_{-\infty}^{+\infty}\frac{d\omega_2}{2\pi}\,i\,\frac{\left[G^{\rm R}(\omega_2)-G^{\rm A}(\omega_2)\right]}{\omega_1-\omega_2-i\eta} \\ &= \int_{-\infty}^{+\infty}\frac{d\omega}{2\pi}\,e^{-i\omega(t-t')}\left(\Sigma_{\rm r}^{\rm R}(\omega)\,G^<(\omega)+\Sigma^<(\omega)\,G^{\rm A}(\omega)\right), \end{aligned} \tag{31.41}$$

where the spectral representations (30.12) for $G^{\rm A}(\omega)$ and (30.33) for $\Sigma_{\rm r}^{\rm R}(\omega)$ have been utilized. This is the final result we were looking for, from which any explicit dependence on t_0 has been removed.

[QED]

Remark: A direct way to arrive at the result (31.41)
The final result (31.41) could have been anticipated directly from the original expression (31.29), by letting therein

$$\begin{cases} t_0 \longrightarrow -\infty \\ \Sigma^{\rceil}(t,\tau) \xrightarrow{t_0 \longrightarrow -\infty} 0 \,. \end{cases} \tag{31.42}$$

On physical grounds, the above condition on $\Sigma^{\rceil}$ is expected to hold for a macroscopic system in the thermodynamic limit, where at equilibrium the initial configuration specified by the Matsubara self-energy Σ^{M} does not evolve further through $\Sigma^{\rceil}$. Under this perspective, the mathematical derivation that has led to Eq. (31.41) appears somewhat superfluous.

It may be interesting to mention that reference to t_0 can be removed also from the convolution integral

$$I^{\mathrm{R}}(t,t'|t_0) = \int_{t_0}^{t} d\bar{t}\; \Sigma^{\mathrm{R}}(t,\bar{t})\, G^{\mathrm{R}}(\bar{t},t') \tag{31.43}$$

that enters the Dyson equation (15.19) of Part I for the retarded Green's function G^{R}. This is readily seen from the property

$$\int_{t_0}^{t'} d\bar{t}\; \Sigma_{\mathrm{r}}^{\mathrm{R}}(t,\bar{t})\;\, G^{\mathrm{R}}(\bar{t},t') = 0 \quad \text{since } \bar{t} > t' \,, \tag{31.44}$$

from which we obtain

$$\begin{aligned} I^{\mathrm{R}}(t,t'|t_0) &= \int_{t_0}^{t} d\bar{t}\; \Sigma^{\mathrm{R}}(t,\bar{t})\, G^{\mathrm{R}}(\bar{t},t') = \int_{t_0}^{t'} d\bar{t}\; \Sigma^{\mathrm{R}}(t,\bar{t})\, G^{\mathrm{R}}(\bar{t},t') \\ &\quad + \int_{t'}^{t} d\bar{t}\; \Sigma^{\mathrm{R}}(t,\bar{t})\, G^{\mathrm{R}}(\bar{t},t') \\ &= \int_{t'}^{t} d\bar{t}\; \Sigma^{\mathrm{R}}(t,\bar{t})\, G^{\mathrm{R}}(\bar{t},t') = \int_{-\infty}^{+\infty} d\bar{t}\; \Sigma^{\mathrm{R}}(t,\bar{t})\, G^{\mathrm{R}}(\bar{t},t') \,. \end{aligned} \tag{31.45}$$

31.5 Average Energy of the System Expressed in Terms of $G^{<}$

In Chapter 18 of Part I, we have seen that the number and current densities can be expressed in terms of the lesser Green's function $G^{<}$. In addition, in Section 30.4, we have selected $G^{<}$ as the only relevant dynamical Green's function at equilibrium. Here, the special role played by $G^{<}$ is further highlighted by showing that the average energy of the system can also be expressed in terms of $G^{<}$.

To this end, we consider the equation of motion of the field operator in the Heisenberg representation for real time with a generic two-body interaction (cf. Eq. (11.19) of Part I)

$$i \frac{\partial}{\partial t} \psi_{\mathrm{H}}(\mathbf{r}, \sigma, t) = h(\mathbf{r}, t)\, \psi_{\mathrm{H}}(\mathbf{r}, \sigma, t) + \int d\mathbf{r}' \sum_{\sigma'} \psi^{\dagger}_{\mathrm{H}}(\mathbf{r}', \sigma', t)\, \psi_{\mathrm{H}}(\mathbf{r}', \sigma', t)\, v(\mathbf{r} - \mathbf{r}')\, \psi_{\mathrm{H}}(\mathbf{r}, \sigma, t), \tag{31.46}$$

where the spin label σ refers to spin-$\frac{1}{2}$ fermions. Next, we multiply both sides of Eq. (31.46) from the left by $\psi^{\dagger}_{\mathrm{H}}(\mathbf{r}, \sigma, t)$ and integrate the result over $\mathbf{r}$ and sum over σ, to obtain:

$$\int d\mathbf{r} \sum_{\sigma} \psi^{\dagger}_{\mathrm{H}}(\mathbf{r}, \sigma, t)\, i \frac{\partial}{\partial t} \psi_{\mathrm{H}}(\mathbf{r}, \sigma, t) = \int d\mathbf{r} \sum_{\sigma} \psi^{\dagger}_{\mathrm{H}}(\mathbf{r}, \sigma, t)\, h(\mathbf{r}, t)\, \psi_{\mathrm{H}}(\mathbf{r}, \sigma, t) + \int d\mathbf{r} d\mathbf{r}' \sum_{\sigma\sigma'} \psi^{\dagger}_{\mathrm{H}}(\mathbf{r}, \sigma, t)\, \psi^{\dagger}_{\mathrm{H}}(\mathbf{r}', \sigma', t)\, v(\mathbf{r} - \mathbf{r}')\, \psi_{\mathrm{H}}(\mathbf{r}', \sigma', t)\, \psi_{\mathrm{H}}(\mathbf{r}, \sigma, t) \tag{31.47}$$

(which holds even when $v(\mathbf{r} - \mathbf{r}') \rightarrow v(\mathbf{r} - \mathbf{r}', t)$ would explicitly depend on time). From this expression, we can identify the interaction part of the system Hamiltonian *at any time t* as given by

$$\begin{aligned} H^{\mathrm{int}}(t) &\equiv \frac{1}{2} \int d\mathbf{r} d\mathbf{r}' \sum_{\sigma\sigma'} \psi^{\dagger}_{\mathrm{H}}(\mathbf{r}, \sigma, t)\, \psi^{\dagger}_{\mathrm{H}}(\mathbf{r}', \sigma', t)\, v(\mathbf{r} - \mathbf{r}')\, \psi_{\mathrm{H}}(\mathbf{r}', \sigma', t)\, \psi_{\mathrm{H}}(\mathbf{r}, \sigma, t) \\ &= \frac{1}{2} \int d\mathbf{r} \sum_{\sigma} \psi^{\dagger}_{\mathrm{H}}(\mathbf{r}, \sigma, t) \left(i \frac{\partial}{\partial t} - h(\mathbf{r}, t) \right) \psi_{\mathrm{H}}(\mathbf{r}, \sigma, t). \end{aligned} \tag{31.48}$$

This gives for the total Hamiltonian:

$$\begin{aligned} \mathcal{H}(t) &= \int d\mathbf{r} \sum_{\sigma} \psi^{\dagger}_{\mathrm{H}}(\mathbf{r}, \sigma, t)\, h(\mathbf{r}, t)\, \psi_{\mathrm{H}}(\mathbf{r}, \sigma, t) + H^{\mathrm{int}}(t) \\ &= \frac{1}{2} \int d\mathbf{r} \sum_{\sigma} \psi^{\dagger}_{\mathrm{H}}(\mathbf{r}, \sigma, t) \left(i \frac{\partial}{\partial t} + h(\mathbf{r}, t) \right) \psi_{\mathrm{H}}(\mathbf{r}, \sigma, t) \,. \end{aligned} \tag{31.49}$$

In addition, from Eq. (13.5) of Part I applied to fermions, we identify

$$G^{<}_{\sigma}(\mathbf{r}t, \mathbf{r}'t') = i \left\langle \psi^{\dagger}_{\mathrm{H}}(\mathbf{r}', \sigma, t')\, \psi_{\mathrm{H}}(\mathbf{r}, \sigma, t) \right\rangle \tag{31.50}$$

as the lesser Green's function, such that, upon averaging over the initial configuration, the average of the Hamiltonian (31.49) can be cast in the form:

$$\langle \mathcal{H}(t) \rangle = \frac{1}{2} \int d\mathbf{r} \sum_{\sigma} \left(i \frac{\partial}{\partial t} \left\langle \psi^{\dagger}_{\mathrm{H}}(\mathbf{r}, \sigma, t')\, \psi_{\mathrm{H}}(\mathbf{r}, \sigma, t) \right\rangle \Big|_{t'=t^+} + h(\mathbf{r}, t) \left\langle \psi^{\dagger}_{\mathrm{H}}(\mathbf{r}', \sigma, t^+)\, \psi_{\mathrm{H}}(\mathbf{r}, \sigma, t) \right\rangle \Big|_{\mathbf{r}'=\mathbf{r}} \right)$$

$$= \frac{1}{2} \int d\mathbf{r} \sum_{\sigma} \left(\frac{\partial}{\partial t} G^{<}_{\sigma}(\mathbf{r}t, \mathbf{r}t')\Big|_{t'=t^+} - h(\mathbf{r}, t)\, i\, G^{<}_{\sigma}(\mathbf{r}t, \mathbf{r}'t^+)\Big|_{\mathbf{r}'=\mathbf{r}} \right) . \quad (31.51)$$

We now consider two examples of the application of the expression (31.51).

Example # 1: TDBdG equations in the superfluid phase

Recall that the single-particle Hamiltonian $h(\mathbf{r}, t)$ entering the TDBdG equations is referred to the chemical potential μ. When this provision is consistently applied to the right-hand side of Eq. (31.51), we have to interpret the ensuing expression as the time-dependent average $\langle \mathcal{K}(t) \rangle$ of the grand-canonical Hamiltonian. Recall also from Eq. (25.32) that

$$\begin{aligned} G^{<}_{\uparrow}(\mathbf{r}t, \mathbf{r}'t') &= i \left\langle \psi^{\dagger}_{\mathrm{K}}(\mathbf{r}', \sigma = \uparrow, t')\, \psi_{\mathrm{K}}(\mathbf{r}, \sigma = \uparrow, t) \right\rangle \\ &= i \sum_{\nu} \left[u_{\nu}(\mathbf{r}', t')^{*}\, u_{\nu}(\mathbf{r}, t) f_F(\varepsilon_{\nu}) + v_{\nu}(\mathbf{r}', t')\, v_{\nu}(\mathbf{r}, t)^{*}\, (1 - f_F(\varepsilon_{\nu})) \right] , \end{aligned} \quad (31.52)$$

where we assume that there is no difference from the case with $\sigma = \sigma' = \downarrow$. Accordingly, the two terms on the right-hand side of the expression (31.51) become, in the order,

$$\begin{aligned} &\frac{\partial}{\partial t} G^{<}_{\sigma}(\mathbf{r}t, \mathbf{r}t')|_{t'=t^+} \\ &= i \sum_{\nu} \left[u_{\nu}(\mathbf{r}, t)^{*}\, \frac{\partial u_{\nu}(\mathbf{r}, t)}{\partial t} f_F(\varepsilon_{\nu}) + v_{\nu}(\mathbf{r}, t)\, \frac{\partial v_{\nu}(\mathbf{r}, t)^{*}}{\partial t}\, (1 - f_F(\varepsilon_{\nu})) \right] \\ &= \sum_{\nu} \{ u_{\nu}(\mathbf{r}, t)^{*}\, [(h(\mathbf{r}, t) - \mu)\, u_{\nu}(\mathbf{r}, t) + \Delta(\mathbf{r}, t)\, v_{\nu}(\mathbf{r}, t)] f_F(\varepsilon_{\nu}) \\ &+ v_{\nu}(\mathbf{r}, t)\, [(h(\mathbf{r}, t) - \mu)\, v_{\nu}(\mathbf{r}, t)^{*} - \Delta(\mathbf{r}, t)\, u_{\nu}(\mathbf{r}, t)^{*}]\, (1 - f_F(\varepsilon_{\nu})) \} , \end{aligned} \quad (31.53)$$

where use has been made of the TDBdG equations (25.52), and

$$\begin{aligned} -\, (h(\mathbf{r}, t) - \mu)\; i\, G^{<}_{\sigma}(\mathbf{r}t, \mathbf{r}'t^+)|_{\mathbf{r}'=\mathbf{r}} = \sum_{\nu} &\left[u_{\nu}(\mathbf{r}, t)^{*}\, (h(\mathbf{r}, t) - \mu)\, u_{\nu}(\mathbf{r}, t) f_F(\varepsilon_{\nu}) \right. \\ &\left. + v_{\nu}(\mathbf{r}, t)\, (h(\mathbf{r}, t) - \mu)\, v_{\nu}(\mathbf{r}, t)^{*}\, (1 - f_F(\varepsilon_{\nu})) \right] . \end{aligned} \quad (31.54)$$

Entering these results into the expression (31.51), where $\langle \mathcal{H}(t) \rangle \to \langle \mathcal{K}(t) \rangle$, and recalling the expression (25.30) for time-dependent gap parameter $\Delta(\mathbf{r}, t)$, we obtain eventually:

$$\langle \mathcal{K}(t)\rangle = \int d\mathbf{r} \left\{ \sum_\sigma \sum_\nu \left[u_\nu(\mathbf{r},t)^* \, (h(\mathbf{r},t)-\mu) \, u_\nu(\mathbf{r},t) f_F(\varepsilon_\nu) \right.\right.$$
$$\left.\left. + \, v_\nu(\mathbf{r},t) \, (h(\mathbf{r},t)-\mu) \, v_\nu(\mathbf{r},t)^* \, (1-f_F(\varepsilon_\nu)) \right] + \frac{1}{v_0} \, |\Delta(\mathbf{r},t)|^2 \right\} . \tag{31.55}$$

In addition, upon writing the expression for the total number of particles N in the form

$$N = \int d\mathbf{r} \sum_\sigma (-i) \, G^<_\sigma(\mathbf{r}t, \mathbf{r}t^+)$$
$$= \int d\mathbf{r} \sum_\sigma \sum_\nu \left[|u_\nu(\mathbf{r},t)|^2 f_F(\varepsilon_\nu) + |v_\nu(\mathbf{r},t)|^2 \, (1-f_F(\varepsilon_\nu)) \right] , \tag{31.56}$$

in the present context to obtain the expression for $\langle \mathcal{H}(t)\rangle$, we can simply eliminate the chemical potential from Eq. (31.55).

Remark: An explicit expression for the ground-state (grand-canonical) energy in the BdG approach

The above expression for $\langle \mathcal{K}(t)\rangle$, once applied to the time-independent case, allows us to identify an explicit expression for the constant C in Eq. (25.10). By making use of the time-independent BdG equations (25.5) and the orthonormality condition (25.7), and recalling the expression (25.6) for the gap parameter, the time-independent counterpart of Eq. (31.55) reads

$$\langle \mathcal{K}\rangle = 2 \sum_\nu \varepsilon_\nu f_F(\varepsilon_\nu) - 2 \sum_\nu \varepsilon_\nu \int d\mathbf{r} \, |v_\nu(\mathbf{r})|^2 - \frac{1}{v_0} \int d\mathbf{r} \, |\Delta(\mathbf{r})|^2, \tag{31.57}$$

where the factor of 2 accounts for the spin degeneracy. The last two terms in Eq. (31.57) identify the ground-state (grand-canonical) energy K_0 (and thus the constant C in Eq. (25.10)).

In particular, for a homogeneous system whereby $\int d\mathbf{r} \, |v_\nu(\mathbf{r})|^2 \to \frac{1}{2}\left(1 - \frac{\xi_\mathbf{k}}{E_\mathbf{k}}\right)$ and $\int d\mathbf{r} \, |\Delta(\mathbf{r})|^2 \to V|\Delta|^2$ with V the volume occupied by the system, K_0 reduces to

$$\frac{K_0}{V} \longrightarrow \int \frac{d\mathbf{k}}{(2\pi)^3} \, (\xi_\mathbf{k} - E_\mathbf{k}) - \frac{|\Delta|^2}{v_0}, \tag{31.58}$$

which identifies the constant C in Eq. (25.10) in the homogeneous case. Note here that (i) the term with $\xi_\mathbf{k}$ corresponds to the second term on the left-hand side of Eq. (22.9) when specified to the time-independent case, (ii) the term with $E_\mathbf{k}$ stems from having interchanged $\gamma_{\nu 2}\gamma^\dagger_{\nu 2} \to 1 - \gamma^\dagger_{\nu 2}\gamma_{\nu 2}$ to get the first term in Eq. (25.10), and (iii) the term containing $|\Delta|^2/v_0$ is required to avoid double counting when calculating the average of the (grand-canonical) Hamiltonian within a mean-field (BCS) decoupling. In addition, upon recalling the gap equation (26.5) in the zero-temperature limit, the expression (31.58) can be cast in the alternative form

$$\frac{K_0}{V} = 2 \int \frac{d\mathbf{k}}{(2\pi)^3} \xi_{\mathbf{k}} v_{\mathbf{k}}^2 + \frac{|\Delta|^2}{v_0}, \tag{31.59}$$

which coincides with the standard expression of the BCS ground-state (grand-canonical) energy [13, 70].

Example # 2: Homogeneous normal Fermi system at equilibrium

With references to Eqs. (30.39) and (30.40), as well as Eq. (30.46), we write in this case

$$G_{\sigma}^{<}(\mathbf{k}, \omega) = -f_{+}(\omega - \mu) \left[G_{\sigma}^{>}(\mathbf{k}, \omega) - G_{\sigma}^{<}(\mathbf{k}, \omega)\right] = -i f_{+}(\omega - \mu) A_{\sigma}(\mathbf{k}, \omega) . \tag{31.60}$$

Here,

$$G_{\sigma}^{<}(\mathbf{r}t, \mathbf{r}'t') = \int \frac{d\mathbf{k}}{(2\pi)^3} \int_{-\infty}^{+\infty} \frac{d\omega}{2\pi} e^{i\mathbf{k}\cdot(\mathbf{r}-\mathbf{r}')} e^{-i\omega(t-t')} G_{\sigma}^{<}(\mathbf{k}, \omega) , \tag{31.61}$$

so that for the two terms on the right-hand side of Eq. (31.51), we obtain, respectively,

$$\frac{\partial}{\partial t} G_{\sigma}^{<}(\mathbf{r}t, \mathbf{r}t')|_{t'=t^{+}} = \int \frac{d\mathbf{k}}{(2\pi)^3} \int_{-\infty}^{+\infty} \frac{d\omega}{2\pi} (-i\omega)\, G_{\sigma}^{<}(\mathbf{k}, \omega) \tag{31.62}$$

and

$$-h(\mathbf{r}, t)\, i\, G_{\sigma}^{<}(\mathbf{r}t, \mathbf{r}t^{+})|_{\mathbf{r}'=\mathbf{r}} = \int \frac{d\mathbf{k}}{(2\pi)^3} \int_{-\infty}^{+\infty} \frac{d\omega}{2\pi} \left(-i\frac{\mathbf{k}^2}{2m}\right) G_{\sigma}^{<}(\mathbf{k}, \omega) \tag{31.63}$$

since $h(\mathbf{r}, t) = -\frac{\nabla^2}{2m}$ in the present case. The expression (31.51) then reduces to

$$\begin{aligned} \langle \mathcal{H} \rangle &= \frac{1}{2} \int d\mathbf{r} \sum_{\sigma} \int \frac{d\mathbf{k}}{(2\pi)^3} \int_{-\infty}^{+\infty} \frac{d\omega}{2\pi} \left(-i\omega - i\frac{\mathbf{k}^2}{2m}\right) G_{\sigma}^{<}(\mathbf{k}, \omega) \\ &= \frac{V}{2} \sum_{\sigma} \int \frac{d\mathbf{k}}{(2\pi)^3} \int_{-\infty}^{+\infty} \frac{d\omega}{2\pi} \left(\omega + \frac{\mathbf{k}^2}{2m}\right) f_F(\omega) A_{\sigma}(\mathbf{k}, \omega), \end{aligned} \tag{31.64}$$

where $f_F(\omega) = \left(e^{\beta(\omega-\mu)} + 1\right)^{-1}$ is the Fermi–Dirac distribution function. Note how the above result generalizes to the real-frequency domain a standard expression obtained at equilibrium in the context of the Matsubara formalism (cf., e.g., sections 25 and 31 of Ref. [7]). In particular, for noninteracting fermions, whereby

$$A_{\sigma}(\mathbf{k}, \omega) = 2\pi\, \delta\left(\omega - \frac{\mathbf{k}^2}{2m}\right), \tag{31.65}$$

Eq. (31.64) becomes as expected

$$\langle \mathcal{H} \rangle = \frac{V}{2} \sum_{\sigma} \int \frac{d\mathbf{k}}{(2\pi)^3} \int_{-\infty}^{+\infty} \frac{d\omega}{2\pi} \left(\omega + \frac{\mathbf{k}^2}{2m} \right) f_F(\omega)\, 2\,\pi\, \delta \left(\omega - \frac{\mathbf{k}^2}{2m} \right)$$
$$= V \sum_{\sigma} \int \frac{d\mathbf{k}}{(2\pi)^3} \frac{\mathbf{k}^2}{2m} f_F \left(\frac{\mathbf{k}^2}{2m} \right) . \tag{31.66}$$

Part III

Applications

32

An Overview of Applications: Yesterday, Today, and Tomorrow

Why, sir, there is every probability that you will soon be able to tax it
Faraday's reply to UK's finance minister, asking about the practical value of electricity

The present chapter differs in spirit and style from those that preceded it in Parts I and II and from those that will follow it in Part III. This is because this chapter is not meant to select a single topic related to the nonequilibrium Schwinger–Keldysh Green's functions technique and to expand on it in detail. Rather, it is devoted to giving a brief overview of a number of specific physical problems, which are of recent, current, and possibly future interest (with reference to the time of writing), problems that can be ideally dealt with in terms of the nonequilibrium Schwinger–Keldysh Green's functions technique developed at a formal level in Parts I and II. This is also because discussing specific applications of general theoretical methods is essential to appraise their overall usefulness to begin with. However, while one may expect that the general methodological aspects should survive the course of time and remain of use even much later than they were introduced, specific applications that are fashionable at the time of their publication may be quickly (and, in some sense, unavoidably) replaced by more updated ones. For these reasons, although the material discussed here bears on a rather large selection of original as well as review articles available in the literature, this chapter should by no means be expected to update those articles but only to provide a synthetic demonstration of the versatility of the Schwinger–Keldysh formalism, especially in the view of possible future applications to scientific problems as well as to technological issues. In this respect, it is hopefully anticipated that the readers may eventually select the physical problems meeting their own interests and to which it is appropriate to apply the nonequilibrium Schwinger–Keldysh Green's function technique or its variants.

32.1 General Considerations

With the above premises, the claim here is that the nonequilibrium Schwinger–Keldysh Green's functions technique should be considered as one of the most versatile theoretical approaches to study a variety of time-dependent phenomena that occur in quantum many-body correlated systems, especially at the nanoscale. This is because no special requirement has to be set in advance about the system dimensionality, the strength of the external perturbations, zero or elevated temperatures, or the transient regime versus the stationary state. This technique can then be used to raise and possibly answer questions ranging from fundamental to practical levels.

On the one hand, numerical simulations can be set up for addressing theoretically out-of (and even far from) equilibrium phenomena arising at ultrashort timescales after the action of a time-dependent strong drive, whereby neither the transient behavior (possibly with an associated relaxation time) nor the asymptotic behavior (possibly, but not necessarily, leading to thermalization) is known in advance. In this respect, there is actually no a priori guarantee for the system to relax toward stationary-like states on reasonable timescales.

On the other hand, combined experimental advances in the miniaturization of electronic devices and in the development of ultra-fast time-resolved spectroscopies in solids would need support from computationally demanding theoretical simulations of time-dependent transport properties of correlated systems at that timescale and in line with the relevant experimental setups. In this context, it is worth recalling that light pulses can probe the dynamics of charges, spins, and atoms with an experimental resolution of picosecond ($= 10^{-12}$ s), femtosecond ($= 10^{-15}$ s), and even down to attosecond ($= 10^{-18}$ s) timescales, while the fastest electronic devices work at most at a frequency of $10^{15} s^{-1}$ (cf. Ref. [149]).

Within this framework, it may be useful to begin by summarizing the terminology mostly used in the literature when referring to the quantum physical systems we shall be concerned with and to the methods used to excite them out-of-equilibrium from an initially prepared (possibly, at equilibrium) configuration.

32.1.1 About the Most Commonly Utilized Terminology

Quantum physical systems can, quite generally, be classified as being "closed" or "open" according to the following criteria.

(a) A *closed quantum system* is assumed to be completely decoupled from its surrounding "environment." In particular, the environment is referred to as the "reservoir" when it has an infinite set of degrees of freedom with a *continuous* spectrum (a property that is essential to yield a truly irreversible and dissipative dynamics for an open quantum system). In addition, when the reservoir is in a thermal equilibrium state, it is referred to as a "bath" (or heat bath). The unitary dynamic of a closed system is formulated in terms of a time-dependent Hamiltonian $H(t)$ (if a closed system is further "isolated," its Hamiltonian H is time independent). Sometimes, however, it is important to take into account the environment when it is crucial for describing the behavior of the system. Under these circumstances, one considers the system-plus-reservoir as a closed system.

(b) An *open quantum system* interacts with an external quantum system (environment, reservoir, or bath), with which it can exchange particles and energy. This interaction modifies the dynamics of the system, which may occur either without or with quantum dissipation (meaning that, in the latter case, the information contained in the system is lost to its environment). For an open system, when the interaction between the system and its environment leads to dissipative processes, the dynamics of the system cannot be accurately described by unitary operators alone. Nonetheless, all physical observations refer to the open quantum system itself and not to its environment, whose degrees of freedom are unobservable. In practice, open quantum systems play a particularly important role because no quantum system is completely isolated. In most cases, it is again assumed that the combined system-plus-environment is closed.

A notable example of closed quantum systems is represented by *ultracold gases*, which are kept in site by the action of magnetic and/or optical traps and are accordingly completely isolated from their surroundings. On the other hand, *condensed-matter samples* are, in practice, hard to isolate from their surrounding and are thus typical open quantum systems.

These (open and closed) systems differ also in the way they can be brought out-of-equilibrium, or initially prepared in a state that subsequently leads to a nontrivial time evolution.

(i) A closed quantum system can be acted upon by a Hamiltonian *quench* represented by a sudden change in some Hamiltonian parameters (say, at time t_0). In this way, the initial quantum state (just before the quench) is not an eigenstate of the Hamiltonian that governs the subsequent time evolution after the quench. What is most interesting in this case is following the *transient dynamics* of the system. In ultracold gases, this situation can be achieved by an abrupt change of the strength of the interparticle interaction and/or of the spatial profiles of the trapping potentials. These systems offer also the advantage that their time evolution can be controlled over typical timescales of millisecond ($= 10^{-3}$ s), instead of femtosecond ($= 10^{-15}$ s) like in condensed matter. Apart from a quench, a closed quantum system can also be affected by a *periodically driven term* occurring in its Hamiltonian.

(ii) An open quantum system can be *driven* out of equilibrium, for example, by suddenly modifying the coupling to its environment (again, at time t_0). This is the case of a typical quantum transport setup, whereby (at least) two macroscopic electronic "leads" are coupled to a microscopic system like a quantum dot. As before, what is most interesting in this case is following the transient dynamics of the system. In addition, when *dissipation* toward the environment becomes of importance, to compensate for losses of particles and energy the system may be suitably driven by external forces, up to the point of establishing a dynamical balance between losses and driving forces, which may lead to a nonequilibrium stationary state. In this case, one refers to a *driven-dissipative system* and may not be concerned with transient dynamics.

(iii) In condensed matter, ultrafast time-resolved (*pump and probe*) spectroscopies have recently become quite popular, whereby an intense "pump" laser pulse is used to drive the system strongly out-of-equilibrium into highly excited states, while "probe" (possibly not so intense) pulses with some delay are then used to track the subsequent temporal evolution of the system. Even more recently, the newly developed X-ray free-electron lasers (XFELs) have been utilized as probe pulses with unprecedented brightness, pulse duration, and coherence, which make them ideal for performing time-resolved experiments on various materials (cf. Ref. [150]). In this case, the system can be regarded as either closed or open, depending on the relevance of dissipation. From a physical point of view, ultrafast pump–probe spectroscopies not only serve to disclose the response to strong external fields but can also be used to manipulate phases of correlated electron systems.

In all the above cases, the *perturbations* that bring the system out-of-equilibrium are *strong* enough that they have access to the nonlinear regime of the system. Accordingly, more standard equilibrium quantum many-body techniques (like the Matsubara formalism), which have access only to the linear-response regime, cannot be utilized to account for the effects of these perturbations. In addition, for mesoscopic systems whose spatial size may decrease down to the nanoscale (like in ultracold atomic clouds and electronic devices), it is not only easier to drive them out of equilibrium, but it also becomes more relevant to follow in detail the explicit time dependence they acquire over their intrinsic timescales in response to the rapid action of perturbations.

32.1.2 About Transient Effects in Quantum Transport

As mentioned earlier, transient effects following a quench are relevant to both closed and open quantum systems. What is typically quenched at a given time t_0, for a closed system (like a trapped ultracold gas), is the interparticle interaction strength, while for an open system (like a quantum dot coupled to two leads) is the dot-lead coupling.

32.1.3 About Reaching Equilibrium

Highly relevant questions are *"if and how"* an isolated quantum system reaches thermal equilibrium following the action of a (strong) external perturbation, losing at the same time memory of its initial preparation. More specifically, the questions are whether the system reaches directly a new state of *quasi-thermal* equilibrium, which is close to being indistinguishable from a thermal equilibrium state; or else, on a short timescale, the system reaches first a *pre-thermalization* condition toward an apparent *meta-stable state*, before it eventually relaxes on a longer timescale to a state indistinguishable from a genuine thermal state. And this in spite of the quantum analog of the recurrence theorem by Poincarè (cf. Ref. [151]), which for a finite system with *discrete* energy eigenvalues, would predict a recursive return close enough to the initial configuration (although possibly after an extremely long time). Note that this feature is altogether avoided for an open system in contact with a reservoir whose energy spectrum is *continuous* (cf. Ref. [151]). In any case, the main problem regards the timescales over which many-body systems equilibrate.

32.1.4 About the Following Compendium

Within the above framework, we expect the nonequilibrium Schwinger–Keldysh Green's functions technique, which we have discussed in detail in Parts I and II, to be an appropriate theoretical tool for addressing *all* the above physical issues. In the following sections from 32.2 to 32.7, we shall briefly consider a selection of experimental and theoretical works, which deal with transient dynamics in closed (cf. Section 32.2) and open (cf. Section 32.3) quantum systems, with pump and probes spectroscopies (cf. Section 32.4), with metastable photo-induced superconductivity (cf. Section 32.5), with quench-induced dynamics in closed quantum systems with emphasis on thermalization (cf. Section 32.6), and with driven open quantum systems with emphasis on dissipation (cf. Section 32.7). In this way, all previously mentioned key words will be suitably illustrated in a concise way. We shall further discuss these problems in the perspective of the nonequilibrium Schwinger–Keldysh Green's functions technique by considering relevant theoretical works from the current literature, and we shall also refer to specific experiments and/or computer simulations on these topics when available. Later on, we shall consider in fuller depth the general description of driven open quantum systems (cf. Chapter 33), also with applications to superfluid Fermi systems (cf. Chapter 34), as well as of dissipation effects in connection with the Lindblad nonequilibrium approach (cf. Chapters 35 and 37).

32.2 Closed Quantum Systems

As already mentioned, ultracold Bose and Fermi gases provide an ideal platform to study out-of-equilibrium phenomena in closed quantum systems with a clean and well-controlled environment. In this context, they offer a rather unique possibility to follow in detail the nonequilibrium dynamics developing after a quench, by allowing for a full real-time control of all relevant parameters. In particular, quantum *quenches* can be readily implemented in the system Hamiltonian, by changing instantaneously either the strength of the interparticle interaction or the shape of the trapping potential.

In this respect, a pioneering experiment was realized in Ref. [152], where the expansion of an initially localized fermionic cloud of a balanced spin mixture of the two lowest hyperfine states of ^{40}K atoms embedded in a homogeneous optical lattice (mimicking a Hubbard-model setup) was studied, following a rapid quench that completely eliminates the harmonic trapping potential. This experiment was performed both in the absence and in the presence of the interparticle interaction, whose control through Fano–Feshbach resonances represents a peculiar feature of ultracold gases (as recalled in Section 26.1 of Part II). A crossover was then observed, from a ballistic expansion for vanishing interaction (or small density) to a bimodal expansion in both the attractive and repulsive interacting cases, by examining in situ absorption images with a single-site resolution. The main experimental results for noninteracting and interacting fermions, respectively, are reported in Figs. 2 and 3 of Ref. [152], which are reproduced here for convenience in Fig. 32.1.

On the theoretical side, it is worth mentioning the work of Ref. [38], which has applied the nonequilibrium Schwinger–Keldysh Green's functions technique within the t-matrix self-energy to finite inhomogeneous fermionic lattice systems of dimension D = (1, 2, 3) in terms of the Hubbard model, by implementing both the full two-time propagation of the Green's functions and its simplified version in terms of the generalized Kadanoff–Baym ansatz (to be discussed later on in Section 36.2). This work was able to show that the t-matrix self-energy captures the interaction effects governing the experiment of Ref. [152], in spite of the unavoidable limited particle number used in the simulations. In this context,

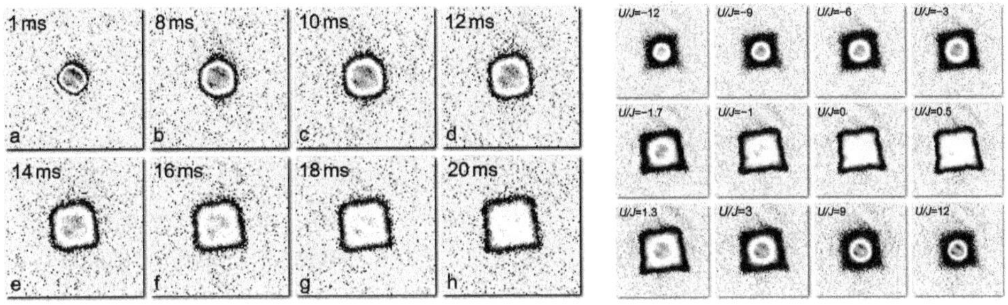

(a) (b)

Figure 32.1 (a) In situ absorption images of an expanding noninteracting cloud. [Adapted from Fig. 2 of Ref. [152], with permission.] (b) In situ absorption images for different interactions after 25 ms expansion. [Adapted from Fig. 3 of Ref. [152], with permission.]

the relevant (and most important) quantity for the study of nonequilibrium transport processes is confirmed to be the time-dependent single-particle density matrix expressed in terms of the lesser Green's function $G^<$. The diffusion process is set up starting from an initial state in which all lattice sites within a limited spatial region are doubly occupied and then removing abruptly the confining potential. In addition, the expansion is studied for different values of the Hubbard interaction parameter U. For instance, Fig. 32 of Ref. [38] displays numerical results for the spreading of the particle density in the 2D Hubbard model obtained from the two-time t-matrix approximation, for different particle numbers N and two values of U. The actual comparison with the experiment of Ref. [152] is reported in Fig. 42 of Ref. [38], where the results for the core expansion velocity versus the interaction strength are shown for various lattice depths in 2D. These results show quite a good agreement between theory (Hubbard model) and experiment (optical lattices).

In Section 32.6, we shall further discuss closed quantum systems, with specific emphasis on the issue of thermalization.

32.3 Driven Open Quantum Systems

A typical problem of interest for condensed matter is that of a time-dependent current flowing through a driven open quantum system (for simplicity, assimilated to a junction made up of noninteracting terminals [leads] and of an interacting intermediate [central] region of finite size), as a result of a time-dependent bias superposed on the terminals. In most cases, no dissipation effects are taken into account for this case. Here, we briefly summarize the main theoretical works that have applied the Schwinger–Keldysh perturbation theory for nonequilibrium systems to study this problem.

A pioneering work on driven open quantum systems, which made use of the Schwinger–Keldysh perturbation theory for nonequilibrium systems, was provided by Ref. [153], where the tunneling current flowing through the junction was calculated in a stationary situation to all orders in the applied bias voltage. In this work, the terminals were considered initially unconnected and in equilibrium at different chemical potentials, while a more realistic approach should consider the whole system initially contacted and in equilibrium with a common chemical potential and later on driven out of equilibrium by an applied bias voltage between the terminals. In addition, in Ref. [153], the many-body effects of interactions in the intermediate region were altogether neglected.

These effects were later included in Ref. [154], but still restricting to a steady-state flow without considering transient effects. Although this restriction was later removed in Ref. [155], this work again relies on an initial configuration with three uncoupled regions, making it questionable the nature of transient effects that could in practice develop from this configuration. Later on, in Ref. [156], the current response to an arbitrary time-dependent bias in a multi-lead system was obtained with a proper inclusion of the switch-on effect by extending the approach of Ref. [157], but similarly neglecting many-body interactions in the intermediate region. These interactions were eventually included in Ref. [158] within the second-order Born approximation for the correlation self-energy (which, as we have seen in Section 30.2 of Part II, has the advantage of not requiring integrations over time due to the delta-like structure of the two-body interaction potential). However, to drastically reduce the computational time, in this case, the generalized Kadanoff–Baym ansatz was further adopted. As already mentioned, in Section 36.2, we will discuss in detail this ansatz as a rather efficient and reliable approach to solve numerically the Kadanoff–Baym equations.

We also mention that Ref. [155] assumes that the decoupled leads are prepared at $t_0 = -\infty$ with an appropriate time-dependent bias and that the coupling between the leads and the central system is enabled immediately afterwards. This approach has the problem that the bias and the coupling are established at the same time in the distant past. Therefore, the applicability of this approach rests on the assumption that the system arrives at the correct nonequilibrium state at times t of interest after a semi-infinite development from $-\infty$, which cannot be a priori guaranteed as the perturbations due to the coupling and the bias are added together. However, in a real experiment, one would switch on the bias and not the coupling between regions. Typical results obtained by the approach of Ref. [155] are reproduced in Fig. 32.2.

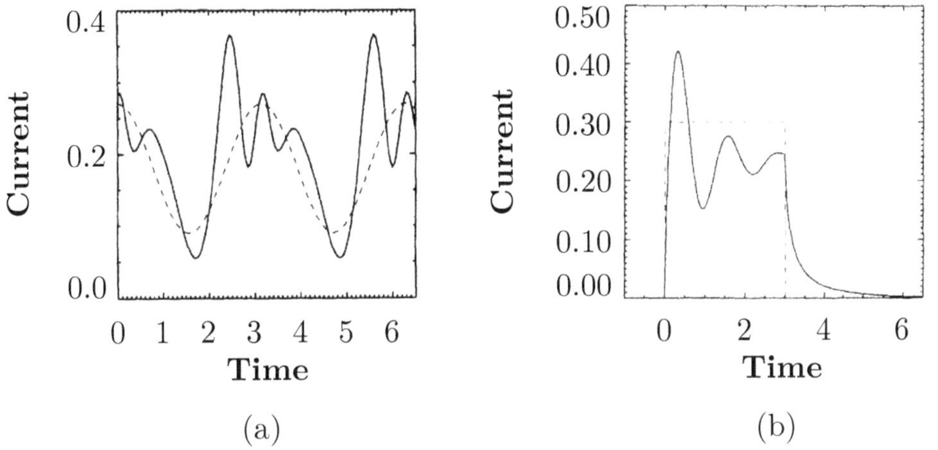

Figure 32.2 (a) Time-dependent current (full line) for a harmonic modulation (dotted line). [Reproduced from Fig. 4 of Ref. [155], with permission.] (b) Time-dependent current (full line) in response to a rectangular bias pulse (dotted line). [Reproduced from Fig. 7 of Ref. [155], with permission.]

In Chapter 33, we shall consider in more detail the way how the nonequilibrium Schwinger–Keldysh approach addresses the problem of a current flowing across a junction as representative of driven open quantum systems. For the sake of definiteness as well as for didactical purposes, we shall consider in particular the approach of Ref. [155] as *a typical case study* (although it may not deal in the best possible way with the initial configuration of the junction). In any case, we shall complement this treatment with the more modern and complete one as summarized in the review article of Ref. [159].

32.4 Spectroscopic Problems: Pump and Probe Photoemission

As mentioned in Section 32.1, ultrafast time-resolved *pump and probe* spectroscopies utilize ultra-short pulse lasers for the study of dynamics on extremely short timescales. In general, dynamics on the picosecond (= 10^{-12} s) to the attosecond (= 10^{-18} s) timescale are too fast to be measured electronically. Most measurements are performed by employing a sequence of ultrashort light pulses to initiate a process and record its dynamics. The temporal width (duration) of the light pulses has to be on the same scale as the dynamics that are to be measured or even shorter. In addition, X-ray free-electron lasers (XFELs) add to a time resolution at the femtosecond (= 10^{-15} s) scale a spatial resolution at the angstrom (=10^{-10} m) scale and generate electromagnetic radiation with very high brightness, larger than any other existing source (cf. Ref. [160]).

In this context, time-resolved *and* angle-resolved photoemission spectroscopy (tr-ARPES) has received much attention. In these experiments, an intense pulse of radiation "pumps" the system into a highly excited nonequilibrium state. After a variable time delay, the system is subject to a weak "probe" pulse of higher-energy photons, ejecting photoelectrons that are detected with energy (and angle) resolution. In many circumstances, it is reasonable to assume that these highly excited photoelectrons interact weakly with the other degrees of freedom in the sample (typically, with the hole left behind), such that final-state effects can be neglected in the highly excited state.

A theoretical description of conventional photoemission processes (conventionally referred to as equilibrium ARPES) based on the Schwinger–Keldysh formalism was originally formulated by the pioneering work of Ref. [161], in the spirit of the approach considered in Chapter 30 in Part II. Here, a general expression for the photoelectron current was obtained for real frequencies up to second order in the interaction with an external light field, by addressing the time-dependent perturbation theory directly in terms of the lesser Green's function while starting from an equilibrium configuration and considering the ensuing stationary (dc) response. This approach was later resumed in Ref. [162], with the main focus of providing a method for experimentalists that allows for fast simulations in the interpretation and analysis of ARPES spectra without the need for mastering the electronic structure theory. This work also presents a comparison of theoretical simulations and experimental data for Ag(111) and Au(111).

A theoretical approach to derive practical expressions to analyze specifically the novel ultra-fast time-resolved, pump–probe, photoemission spectroscopy (as before, referred to as time-resolved ARPES or tr-ARPES) was instead developed in Ref. [163] again in terms of the Schwinger–Keldysh formalism, which allows the strong pump pulse to be treated nonperturbatively to all-order perturbation theory, while the weak probe pulse can be treated perturbatively to first order in perturbation theory. This difference is what distinguishes the approach of Ref. [163] from the previous treatments of photoemission developed in Refs. [161] and [162] mentioned earlier. Later on, Ref. [164] has provided an alternative derivation [cf. Eq. (22) therein] of the main result of Ref. [163] [cf. Eq. (7) therein]. In both approaches, the problem consists in computing the lesser Green's function, which describes the temporal evolution of the electronic degrees of freedom after the "pump" pulse. [It should be mentioned in this context that Ref. [165] [cf. Eq. (7) therein], inspired by the work of Ref. [166], has corrected an inaccuracy contained in Ref. [163].]

In all the above works, no final-state effect was included, in the sense that the photoelectron was assumed to be uncorrelated with the system left behind. Final-state (like excitonic) effects have been instead included in Ref. [167], which

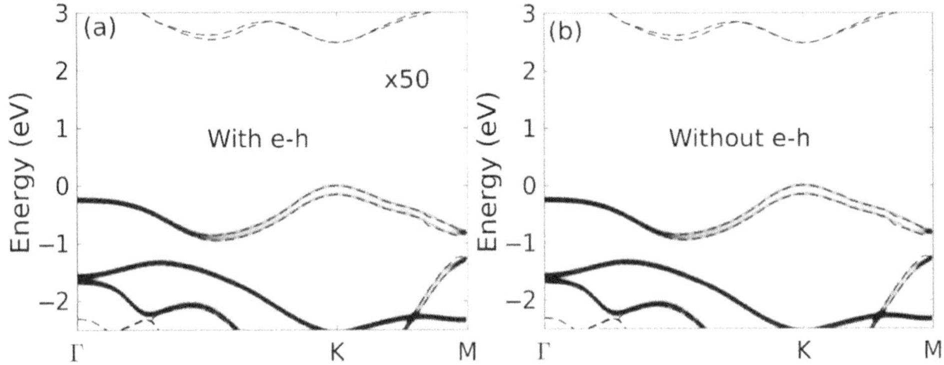

Figure 32.3 Computed tr-ARPES intensity of monolayer MoS_2 (a) with electron–hole interaction and (b) without electron–hole interaction. [Adapted from Fig. 2 of Ref. [168], under the Creative Commons Attribution 4.0 International license (https://creativecommons.org/licenses/by/4.0/).]

also provides a more general form [cf. Eq. (10) therein] of Eq. (7) of Ref. [163]. In addition, Ref. [167] performs explicit numerical calculations using the generalized Kadanoff–Baym ansatz [cf. Eq. (13) therein], with a specific choice for the self-energy [cf. Fig. 2(a) therein]. Nevertheless, a time-dependent first-principles simulation of tr-ARPES for real materials that includes electron–hole interactions still remains a major challenge, with only limited comparison with experiments available thus far.

An attempt along these lines was made in Ref. [168], where a time-domain GW approach to tr-ARPES was implemented to probe excitons in momentum space and applied to monolayer MoS_2. In this specific context, tr-ARPES allows for studying the dynamics of the electron response, whereby the pump light is first used to dress selected energy levels and the probe light is then used to detect the energy level shifts of the dressed exciton states. Figure 32.3 shows typical results of the computed tr-ARPES, with a 1.9 eV pump light for the valence band dispersion, where a feature appears around the K point at 0.50 eV below the conduction band minimum when including final-state effects (a), while this feature is absent when final-state effects are neglected (b).

In this context, it is worth witnessing the raising of the experimental interest in characterizing the excitonic dynamics in photo-excited semiconductors, either by using tr-ARPES at the picosecond scale to capture the timing of the early-stage exciton dynamics, thus providing a full characterization of the exciton formation [169], or by using ultrafast X-ray transient absorption spectroscopy (XTA) also at the picosecond scale to demonstrate the dynamic Coulomb screening of core excitons induced by photoexcited carriers [170]. These experimental works follow the path traced by the early theoretical work in the dynamical shift and broadening of core excitons in semiconductors [171], where the associated dynamical screening was originally accounted for in terms of the Bethe–Salpeter equation [172].

32.5 Metastable Photo-Induced Superconductivity

Superconductors probably represent the physical context where pump–probe experiments lead to results that are most exciting and a forerunner of even more intriguing discoveries in the context of the nonequilibrium control of emergent phenomena in solids. This is because recent pump–probe experiments on underdoped cuprates and similar systems have suggested the possible existence of a transient photo-induced superconducting phase, which can exist at elevated temperatures, even well above the equilibrium superconducting critical temperature T_c. This phenomenon has been referred to as *"light-enhanced superconductivity."*

These experiments have been performed in a systematic way over the last several years, primarily at the Centre for Free-Electron-Laser Science in Hamburg (Germany). In this context, it is in particular worth mentioning that: (i) in Ref. [21], the appearance of long-lived metastable photo-induced superconductivity in $LESCO_{1/8}$ was reported, lasting longer than 100 ps, while the timescale needed to form the superconducting phase was about 1 ps; (ii) in Ref. [22], a large increase in carrier mobility in metallic K_3C_{60} was induced, accompanied by the opening of a gap in the optical conductivity even well above T_c; (iii) in Ref. [23], the above experimental findings were confirmed in organic conductors; (iv) in Ref. [173], the same superconducting-like optical properties, observed previously for femtosecond ($= 10^{-15}$s) excitations, were shown to become metastable under sustained optical driving with lifetime in excess of 10 nanoseconds ($= 10^{-9}$s) up to a temperature of 100 K (with $T_c = 20.0$ K) and also to survive far longer than the drive pulse; (v) in Ref. [174], a dominant energy scale was searched underlying all previous observations in alkali-doped fullerides. Typical experimental results relevant to point (iv) are shown in Fig. 32.4.

On the theoretical side, the nonequilibrium dynamics of superconductors has been addressed in terms of the nonequilibrium Schwinger–Keldysh technique in several articles, although no direct comparison with experiments has yet been attempted at the time of writing, which thus still awaits for future more dedicated and extensive work. In the following, we shall limit ourselves to mentioning a few of these preliminary theoretical works along these lines, which may be of interest to the readers.

In particular, Ref. [175] is concerned with tr-ARPES in superconductors. Here, the claim is that tr-ARPES can be used to directly observe the amplitude Higgs mode, whereby the nonlinear coupling associated with the strong pump pulse is

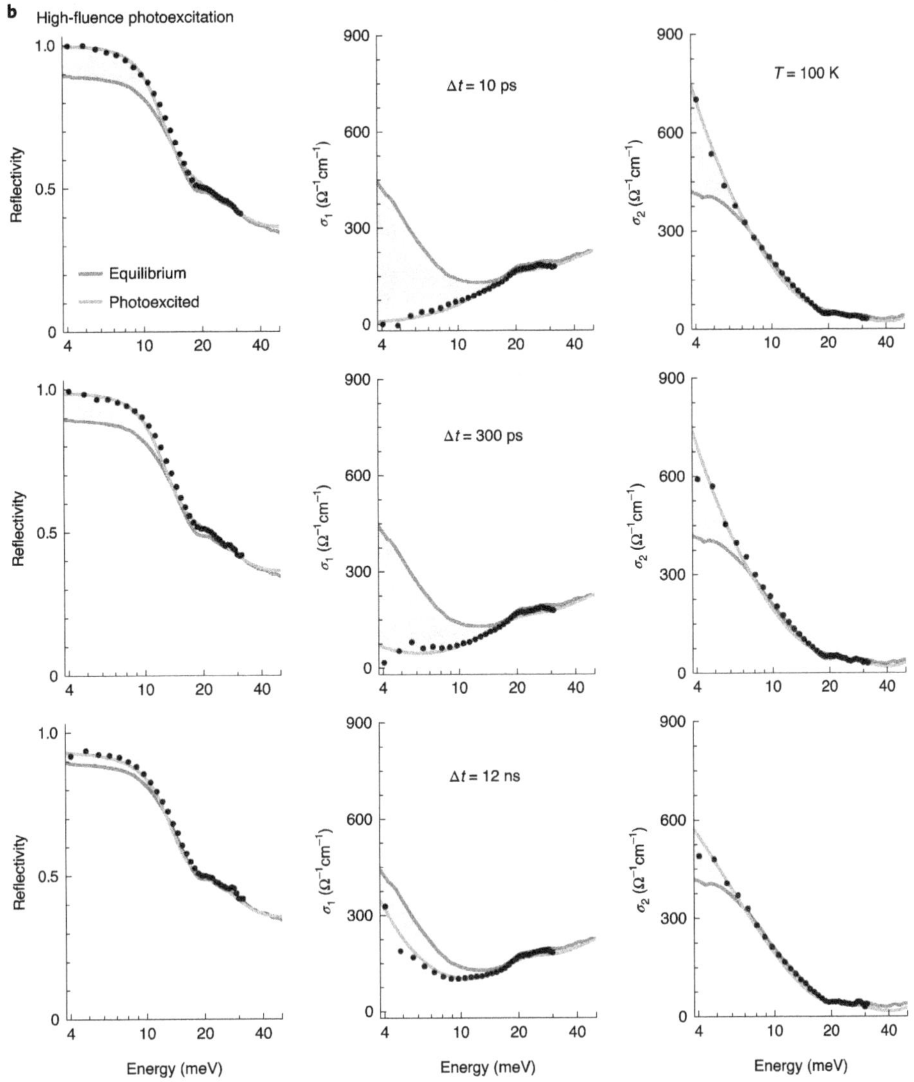

Figure 32.4 Typical results for long-lived "light-enhanced superconductivity," generated in K_3C_{60} with 1-ps-long excitation pulses and acquired at a temperature $T = 100$ K. Shown are the reflectivity and the real and imaginary parts of the optical conductivity, measured at equilibrium (lines), and 10 ps, 300 ps, and 12 ns (filled symbols) after photoexcitation. [Adapted from Fig. 2 of Ref. [173], under the Creative Commons Attribution 4.0 International license (https://creativecommons.org/licenses/by/4.0/).]

crucial for the excitation of this mode. The calculation of the time-resolved single-particle spectral function (as given in Eq. (7) therein) is formally similar to that of Ref. [163] mentioned in Section 32.4, and is now explicitly implemented for the superconducting phase in terms of the Holstein model, whereby noninteracting fermions interact with a gas of Einstein phonons (which is further assumed not to be influenced by the feedback of the fermions) at the level of the Born approximation. The main results of Ref. [175], obtained by solving the Kadanoff–Baym equations starting from a thermal equilibrium configuration before a time-varying vector potential is applied, are shown in Fig. 3 therein, where the "anomalous density" characteristic of a superconductor is also shown.

In addition, Ref. [176] reviews the theory for pump–probe photoemission spectroscopy (tr-ARPES) of electron–phonon–mediated superconductors based on the previous Ref. [175], but now for both normal and superconducting phases and further considering the Hubbard–Holdstein model, which allows for fermionic interaction. This article also mentions the physical effects revealed experimentally in Refs. [21] and [22], for which, by addressing the dynamics of the time-dependent order parameter, it provides a possible explanation in terms of nonlinear phononics, whereby a resonantly excited phonon mode causes a nonoscillatory displacement in another mode.

Finally, Ref. [177] is instead concerned with time-resolved optical conductivity (and not with tr-ARPES), although it does not refer to the previous experiments reported in Refs. [21] and [22]. To this end, Ref. [177] solves the nonequilibrium Dyson equation within the Schwinger–Keldysh formalism for the superconducting state, in a similar way to what was done in Ref. [175]. Three separate timescales are here explicitly considered in this context: (i) the pump time, when the system is brought out of equilibrium; (ii) the probe time, when an electric field is applied to this system brought out of equilibrium; and (iii) the gate time, when the ensuing temperature is measured.

32.6 Dynamics Induced by Quenches and Rumps in "Closed" Quantum Systems, with Emphasis on Thermalization

A closed quantum system can be brought out of equilibrium in several ways. One of these ways is a "quantum quench" corresponding to a change in one of the system parameters, which can happen either slowly or suddenly. For a *slow* quench, it may happen that during the ensuing time evolution the system crosses a phase transition, such that the initial and final phases are macroscopically different. For a *sudden* quench, on the other hand, a relevant question is whether and how the system eventually thermalizes by reaching asymptotically a steady state. Additional ways to bring a system out of equilibrium are ramps and periodic driving.

In this context, Ref. [178] provides an overview on dynamical equilibration and thermalization of "closed" quantum many-body systems out-of-equilibrium due to quenches, ramps, and periodic driving. In particular, the physical systems of interest are assumed to have short-range interparticle interaction and thus to be described by Bose or Fermi Hubbard models. This is a rapidly developing field of research, whose progresses are due to the accessibility of experiments probing relevant physical questions under specifically controlled conditions, as well as to the associated availability of theoretical and computational methods. What is usually observed in these cases is that after a sudden *"global" quench* (affecting, for instance, the strength of the interparticle coupling), the system relaxes and equilibrates, in the sense that the expectation value of (especially "local") observables equilibrates. An alternative scenario rests on activating sudden *"local" quenches*, when it is not the entire system to be uniformly modified but rather the system is suddenly driven out of equilibrium only locally (for instance, by modifying the external potential in a spatial region of limited extent). In a related context, the dynamics of (thermal or quantum) phase transitions can be studied in the course of "ramps" when the Hamiltonian is changed according to some schedule.

As already mentioned in Section 32.2, ultracold trapped gases represent an ideal platform for implementing the dynamics following quenches and ramps in closed quantum systems, owing to both their tunable relevant parameters (like the interparticle interaction) and the experimentally resolved intrinsic timescales (which are much longer than comparable scales in solids). Significant experiments in this context were reported in Ref. [179] for a Bose gas (of ^{39}K atoms) and in Ref. [24] for a two-component Fermi gas (of ^{6} Li atoms).

In Ref. [179], momentum- and time-resolved measurements were used to explore a degenerate and thermal homogeneous Bose gas of ^{39}K atoms confined in an optical-box trap, initially prepared in a weak-coupling regime and then rapidly (within $2\ \mu s = 2\ 10^{-6}$ s) quenched to unitarity (when the interactions are as strong as allowed by quantum mechanics). In addition, after a controlled hold time t_{hold}, the system was quenched back to the weak-coupling regime and released from the trap, so as to measure its momentum distribution n_k for different values of t_{hold}. In spite of the fact that in Bose gases quenches to the strongly interacting regime are accompanied by rapid inelastic losses via three-body collisions, a universal post-quench dynamics was observed in agreement with the emergence of a pre-thermal state. In this case, the intrinsic system timescale is also of the order of microseconds. The key experimental observation for this degenerate Bose gas is reported in Fig. 32.5, where after a rapid initial growth n_k reaches a (quasi-) steady-state plateau, before the longtime heating takes over due to three-body recombinations.

The dynamics of a two-component Fermi gas of ^{6}Li atoms was instead detected in Ref. [24], following a quench from weak to unitarity-limited interactions, which involves crossing the normal to superfluid phase transition. This experiment was able to distinguish pairing governed by local properties of the gas and taking place

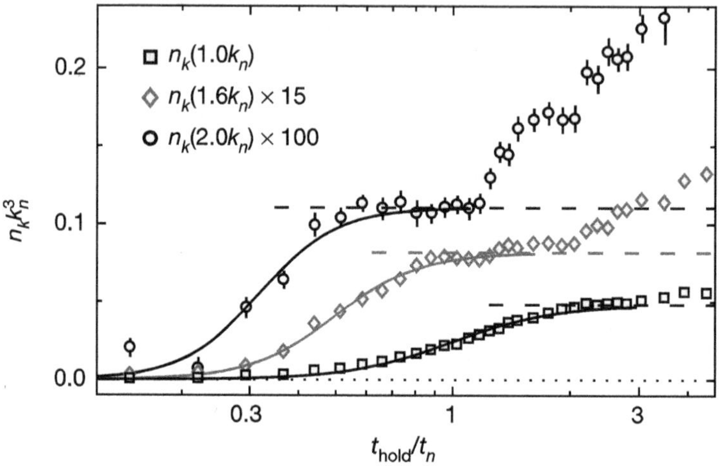

Figure 32.5 Experimental dynamics of a degenerate Bose gas quenched to unitarity, for which the populations of individual k states show a rapid initial growth, an intermediate saturation at quasi-steady-state values, and a final longtime heating. Here, k_n is the analog of the Fermi wave vector for a Bose gas with associated Fermi energy E_n, and $t_n = \hbar/E_n$. [Reproduced from Fig. 1(c) of Ref. [179], with permission.]

on short timescales, from condensation and equilibration of the momentum distribution, which can take much longer depending on the adiabaticity of the quench. Specifically, in the experiment of Ref. [24], the Fermi gas was confined in an oblate harmonic potential (resulting from a combination of optical and magnetic fields) and prepared in the normal phase for a weak attractive interaction at a temperature well above the corresponding superfluid transition temperature. The interaction was then ramped linearly to unitarity in a quench time t_q and, after a further hold time t_h, the dynamics as the gas approaches equilibrium was measured by commencing an imaging sequence. Contrary to Bose gases considered in Ref. [179], two-component Fermi gases are virtually immune to three-body losses, thereby allowing access to dynamics across a broad range of timescales. Since in this case the quench involves crossing the normal to superfluid phase transition, the pair momentum distribution was measured to track both the formation and the condensation of fermion pairs, which take place on very different timescales (of the order of ms), depending on the adiabaticity of the quench. Correlations are found to evolve at vastly different rates, depending on the corresponding length scale. In particular, short-range correlations, based on the occupation of high-momentum modes, evolve far more rapidly than the correlations in low-momentum modes necessary for pair condensation. Typical results obtained by Ref. [24] are reproduced in Fig. 32.6, where in the left figure a progressively sharper central peak

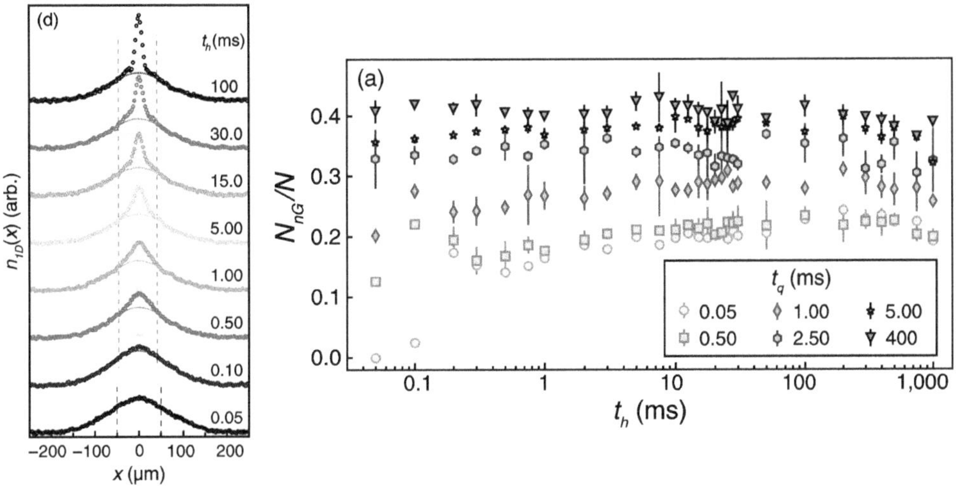

(a) (b)

Figure 32.6 (a) One-dimensional optical density profiles following a 50 μs quench are shown at different hold times t_h. [Reproduced from Fig. 1(d) of Ref. 24, with permission.] (b) Evolution of the non-Gaussian fraction N_{nG}/N in the pair momentum distribution, as a function of the hold time t_h for different quench times t_q. [Reproduced from Fig. 2(a) of Ref. 24, with permission.]

is seen to develop (shaded regions) corresponding to the non-Gaussian component N_{nG} associated with the formation of pairs in modes with low center-of-mass momentum, while in the right figure longer quenches show very stable levels of N_{nG}, indicating that pairing is essentially complete at the end of the quench.

In spite of their importance, the experiments reported for bosons in Ref. [179] and for fermions in Ref. [24] (as well as those later reported for fermions in Ref. [25]) still await detailed theoretical interpretations at the time of writing. Pending this achievement, it is nonetheless worth mentioning two independent theoretical works that, although not directly connected with the above experimental works, employ the nonequilibrium Schwinger–Keldysh approach to study thermalization after sudden and slow quenches, in interacting models for bosons [180] and fermions [181].

Specifically, Ref. [180] considers the dynamics following a sudden (global) quench in the interaction parameter $U(t)$ of the weakly interacting repulsive lattice Bose–Hubbard model. In this context, a bosonic version of the ladder (t-matrix) approximation discussed in Chapter 16 in Part I was adopted, and the integral form of the Dyson equation was solved numerically in a self-consistent way, with a quench in $U(t)$ (starting from an initial configuration with $U = 0$) applied both in 1D and 2D, and at zero as well as finite temperature. To characterize the

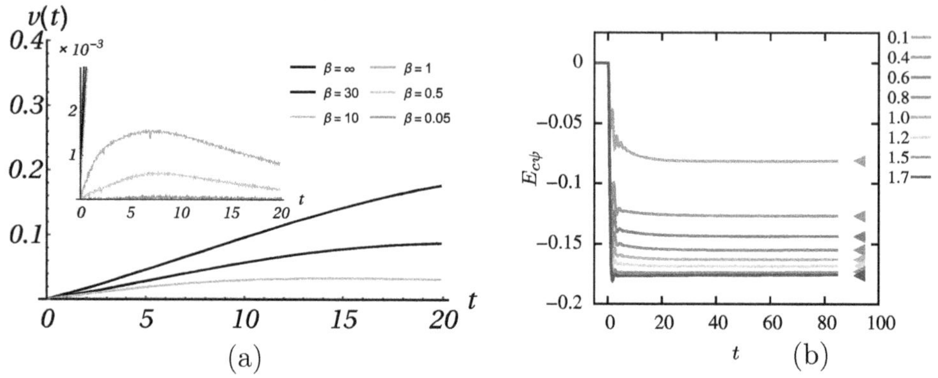

Figure 32.7 (a) Expansion velocity of a Bose gas of 21 particles with final interaction $U = 0.2$ in a 1D lattice of 63 sites, for several initial (inverse) temperatures. [Reproduced from Fig. 6 of Ref. [180], with permission.] (b) Time dependence of the energy $E_{c\psi}$ of c–ψ bonds for site fractions $p = 0.1 - 1.7$. [Adapted from Fig. S3(a) of Ref. [182], under the Creative Commons Attribution 4.0 International license (https://creativecommons.org/licenses/by/4.0/).]

post-quench dynamics, both the expansion of the density of bosons and the spreading of correlations were analyzed. In 1D, the numerical observations for the expansion of the density are consistent with the existence of a transient time, before the system reaches its stationary state, in which the expansion of the density is accelerated up until some characteristic time after which the expansion slows down and the system starts to equilibrate. In this context, the effect of temperature is to decrease the expansion velocity as shown in Fig. 6 of Ref. [180], which is reproduced in Fig. 32.7(a). In 2D, it is also interesting to visualize the spreading of correlations, which turns from ballistic to diffusive, as the final interaction strength increases (cf. Fig. 14 of Ref. [180] in the isotropic case).

In Ref. [181] (see also Ref. [182]), a time-dependent version of the Banerjee–Altman model was instead considered for two species of spinless fermions (c on N sites and ψ on M sites), which are mutually coupled although otherwise isolated. This model (whose time-independent version has an exact solution in the limit of an infinite number of N and M sites with fixed ratio $p = M/N$ [cf. Ref. [183]]) bears a quantum phase transition from non-Fermi liquid (NFL) to a Fermi liquid (FL). The authors of Ref. [181] consider both a sudden quench and a slow ramp for the coupling between the two (c and ψ) species of fermions and study the ensuing thermalization process. Before the quench, the two disconnected subsystems are in their own thermal equilibria at (or close to) zero temperature. The Kadanoff–Baym equations for both species of fermions are then solved using a form for the self-energies obtained in the large N limit (cf. Eqs. (A5) of Ref. [181] and Fig. 3 of Ref. [183], where $f(t)$ is the coupling strength in Eq. (1c) of Ref. [181]). In this context, the "fluctuation–dissipation" condition $iG^{<}(\omega) = f_F(\omega - \mu)\, 2\,\mathrm{Im}G^R(\omega)$ (cf. Chapter 30 in Part II) enables one to test whether the system, after undergoing a nonequilibrium process, has eventually reached a steady state and whether the steady state is in consistent with thermal equilibrium. The main result of this work is that a sudden quench to the NFL (FL) leads to rapid (slow) thermalization, with drastic differences between the two cases. In addition, differences are also seen in the expected temperature of the putative thermal state at long times as estimated from the total energy (which is conserved after the quench), with the sudden quench leading to higher final temperatures. The thermalization behavior of the interacting model is reproduced in Fig. 32.7(b). Here, for $p \lesssim 3.0$, the energy is seen to rapidly approach the expected equilibrium value (arrowheads) after the quench, and the system gets equilibrated very quickly, while for $p \gtrsim 3.0$, the rate of equilibration slows down drastically, and the energy takes a much longer time to reach its equilibrium value.

32.7 Driven "Open" Quantum Systems, with Emphasis on Dissipation

We recall from Section 32.1 that an "open" quantum system is a quantum-mechanical system that interacts with an external quantum system, known as the environment (or reservoir, or bath). In general, these interactions significantly change the dynamics of the system and result in quantum *dissipation*, in such a way that (at least part of) the information contained in the system is lost to the environment. This is quite a common situation in nature because no quantum system is completely isolated from its surroundings. A full description of a quantum system then requires the inclusion of the environment, in which the system must be suitably "embedded," under the assumption that the entire system-plus-environment combination is a large closed system. Loss of energy to the environment is termed "quantum dissipation," while loss of coherence is termed "quantum decoherence."

The fact that every quantum system has some degree of openness implies that no quantum system can ever be in a pure state and has accordingly to be described by the formalism of density matrices. In this context, one looks for a reduced description, wherein the system's dynamics is considered explicitly and the reservoir (or bath) dynamics is described implicitly. For the combined system, the global Hamiltonian H is the sum of the system's Hamiltonian H_S, the bath Hamiltonian H_B, and the system-bath interaction H_{SB}. The state of the system can then be obtained from a partial trace over the combined system and bath, such that $\rho_S(t) = \mathrm{Tr}_B\{\rho(t)\}$. The "reduced" density matrix $\rho_S(t)$ is then the quantity of central interest in the description of open quantum systems.

A reasonable assumption, often used to make easier the solution of problems with a system–bath interaction, is that the system does not preserve memory of its previous states, in the sense that the state of the system at the next moment depends only on the current state of the system. This assumption is justified when the characteristic decay time of bath correlations is short compared with the relaxation time of the system coupled to the bath. Systems with this property are known as *Markov* systems. In addition, one usually assumes that the system–bath interaction is weak and that the system and bath are initially (i.e., at $t = t_0 = 0$ for simplicity, when their coupling is switched on) uncorrelated, such that $\rho(t=0) = \rho_S \otimes \rho_B$.

With these assumptions, one can show that the *Lindblad Master Equation* (LME) for $\rho_S(t)$ is obtained, in the form [184, 185]

$$\frac{d}{dt}\rho_S(t) = -i[H_S, \rho_S(t)] + \mathcal{L}(\rho_S(t)), \tag{32.1}$$

where H_S is the system's Hamiltonian introduced above and

$$\mathcal{L}(\rho_S(t)) = \sum_n \left(V_n \rho_S(t) V_n^\dagger - \frac{1}{2}\left(\rho_S(t) V_n^\dagger V_n + V_n^\dagger V_n \rho_S(t)\right)\right) \tag{32.2}$$

is the *Lindblad super-operator* which, by operating on the reduced density matrix $\rho_S(t)$, accounts for the dissipative part by describing implicitly the influence of the bath on the system through the (jump) operators V_n. The Markov property implies that the system and bath remain uncorrelated at all times, that is, that $\rho(t) = \rho_S(t) \otimes \rho_B(t)$ not only at $t = 0$ but also for $t > 0$. We postpone to Chapter 37 a brief generic derivation of the LME as given by Eqs. (32.1) and (32.2).

Giving due emphasis to the LME appears appropriate, owing to the increasing interest developing in the literature about its use for dealing with a variety of physical problems that involve dissipation [186, 187], which has also been extended to problems dealing with nonlinear parametric oscillators [188, 189].

More specifically, regarding the connection of the Schwinger–Keldysh field theory with dissipation, based on the early work of Refs. [153] and [155] mentioned in Section 32.3, in Ref. [190], the Schwinger–Keldysh field theory was exploited in the presence of dissipation (as related to dephasing processes). The essential idea was again that the effect of the leads and the phase-breaking processes can be represented as "boundary conditions" on the closed system, with the boundary conditions appearing as time-dependent self-energies that account for the degrees of freedom left out from the Hamiltonian of the closed system. Later on, Ref. [191] has further expanded on the work of the above references, by presenting an efficient method and a fast algorithm to calculate one-body observables in this context.

An updated version of this methodology was later proposed in Refs. [192] and [193] in the context of superconductors coupled to external baths (with the dynamics of superconductors treated within a mean-field decoupling). In particular, Ref. [192] was concerned with tr-ARPES (which was already mentioned for superconductors in Section 32.5) and explicitly considered the relaxation due to the coupling of the system with external baths. Later on, Ref. [193], although not concerned with tr-ARPES in superconductors, has further expanded on the approach of Ref. [192] by considering the time dependence of the superconducting order parameter after a quench of the coupling, when dissipation effects due to the coupling with a reservoir are included. In Ref. [193], it was further shown that excitation of the fermions balanced by heat loss due to thermal contact with a bath enables a stationary nonequilibrium state to exist. In Chapter 34, we shall consider this approach for fermionic superfluids in some detail.

In addition, an *explicit* connection between the Schwinger–Keldysh approach and the Lindblad equation was provided in Refs. [194] and [195], although with some relevant distinctions between the two references. In particular, Ref. [194] makes use of the Schwinger–Keldysh *functional integral* (specifically for bosons) in reverse, in the sense that it starts from the Lindblad equation and then builds up an equivalent Keldysh functional integral; while in Ref. [195], the Lindblad equation is derived from first principles in terms of the *operator* Schwinger–Keldysh approach within a simple (two-level) model coupled to a bath. In Chapter 35, we shall considerably expand on the issue of the connection between the approaches of the LME and of the Schwinger–Keldysh field theory.

It should, finally, be mentioned that open quantum systems that do not have the Markovian property are generally much more difficult to deal with. This is due to the fact that the next state of a non-Markovian system is determined by each of its previous states, a requirement that rapidly increases the *memory* requirements to compute the evolution of the system. One of the current methods for treating these systems is the "memory function formalism," which employs the so-called projection operators technique and leads to the Nakajima–Zwanzig equation [196, 197]. Although in the rest of Part III we shall not consider non-Markovian systems in detail, in Chapter 33 we shall nonetheless make a generic use of the Zwanzig projector operators technique in the time-dependent case, by separating the physical system of interest into a "relevant" part needed for the calculation of specific observables and a remaining "irrelevant" part, so as to deal formally with driven open quantum systems introduced in Section 32.3 that keep track of memory effects.

33

Driven Open Quantum Systems

In this chapter, we consider in detail an open quantum system, exemplified by a junction made up of a central region of finite size and of (at least two) connected terminals, with a time-dependent bias superposed on the terminals. For simplicity, in this chapter, we shall assume that the fermions in the terminals are noninteracting, while those in the central region are interacting. In particular, we shall be interested in calculating the time-dependent current flowing through this system using the Schwinger–Keldysh formalism developed in Part I for the normal phase.

Before proceeding along these lines, however, it is worth framing this problem in a more general context, by adapting the Zwanzig projection operators technique [197] to the present problem. This requires us to partition the Hilbert space of a generic many-body quantum system into a P subspace where the relevant physical processes take place and its complement Q (such that $P + Q = \mathbb{1}$ and $PQ = 0$), which may have influence on these processes. In this way, "memory" effects due to the transfer of information from P to Q (and vice versa) will arise. We recall that the Zwanzig P-Q partition was also utilized by Feshbach in the context of nuclear physics [105], leading also to the molecular Fano–Feshbach resonances mentioned in Chapter 26 in Part II in the context of the BCS–BEC crossover.

33.1 Memory Effects Arising from the P-Q Partition

Here, we apply the Zwanzig P-Q partition to the time development of a generic many-body wave function $\Psi(t)$ according to the Schrödinger equation

$$i \frac{\partial}{\partial t} \Psi(t) = \mathcal{H}(t)\, \Psi(t) \tag{33.1}$$

under the action of the time-dependent Hamiltonian $\mathcal{H}(t)$. We then write quite generally

$$\Psi(t) = (P + Q)\Psi(t) = P\Psi(t) + Q\Psi(t) = \Psi_{\mathrm{P}}(t) + \Psi_{\mathrm{Q}}(t)\,, \tag{33.2}$$

such that, by projecting Eq. (33.1) alternatively onto P and Q (under the assumption that P and Q do not depend on time), we obtain two coupled equations for $\Psi_\mathrm{P}(t)$ and $\Psi_\mathrm{Q}(t)$:

$$i\,\frac{\partial}{\partial t}\,P\,\Psi(t) = i\,\frac{\partial}{\partial t}\,\Psi_\mathrm{P}(t) = P\,\mathcal{H}(t)\,(P+Q)\,\Psi(t) = (P\,\mathcal{H}(t)\,P)\,P\,\Psi(t) \\ + (P\,\mathcal{H}(t)\,Q)\,Q\,\Psi(t) = \mathcal{H}_\mathrm{PP}(t)\,\Psi_\mathrm{P}(t) + \mathcal{H}_\mathrm{PQ}(t)\,\Psi_\mathrm{Q}(t), \tag{33.3}$$

and

$$i\,\frac{\partial}{\partial t}\,Q\,\Psi(t) = i\,\frac{\partial}{\partial t}\,\Psi_\mathrm{Q}(t) = Q\,\mathcal{H}(t)\,(P+Q)\,\Psi(t) = (Q\,\mathcal{H}(t)\,P)\,P\,\Psi(t) \\ + (Q\,\mathcal{H}(t)\,Q)\,Q\,\Psi(t) = \mathcal{H}_\mathrm{QP}(t)\,\Psi_\mathrm{P}(t) + \mathcal{H}_\mathrm{QQ}(t)\,\Psi_\mathrm{Q}(t)\,. \tag{33.4}$$

Here, we have used the properties $P^2 = P$ and $Q^2 = Q$ and defined

$$\mathcal{H}_\mathrm{PP}(t) = P\,\mathcal{H}(t)\,P,\;\; \mathcal{H}_\mathrm{PQ}(t) = P\,\mathcal{H}(t)\,Q,\;\; \mathcal{H}_\mathrm{QP}(t) = Q\,\mathcal{H}(t)\,P,\;\; \mathcal{H}_\mathrm{QQ}(t) = Q\,\mathcal{H}(t)\,Q. \tag{33.5}$$

Accordingly, the initial condition at time $t = t_0$ reads

$$\Psi(t_0) = \Psi_\mathrm{P}(t_0) + \Psi_\mathrm{Q}(t_0)\,, \tag{33.6}$$

whereby a simplification will occur in the following by setting $\Psi_\mathrm{Q}(t_0) = 0$.

In order to "fold" the effects of the Q subspace onto the P subspace, we solve Eq. (33.4) for $\Psi_\mathrm{Q}(t)$ and enter the result into Eq. (33.3). To this end, we introduce the time-evolution operator $U_\mathrm{QQ}(t, t_0)$ in the Q subspace, such that (cf. Eq. (2.18) of Part I)

$$i\frac{\partial}{\partial t}U_\mathrm{QQ}(t, t') = \mathcal{H}_\mathrm{QQ}(t)\,U_\mathrm{QQ}(t, t')\,. \tag{33.7}$$

The solution to Eq. (33.4) can then be written in the form

$$\boxed{\Psi_\mathrm{Q}(t) = -\,i\int_{t_0}^{t} dt'\;U_\mathrm{QQ}(t, t')\,\mathcal{H}_\mathrm{QP}(t')\,\Psi_\mathrm{P}(t') \;+\; U_\mathrm{QQ}(t, t_0)\,\Psi_\mathrm{Q}(t_0)\,,} \tag{33.8}$$

such that $\Psi_\mathrm{Q}(t_0) = U_\mathrm{QQ}(t_0, t_0)\,\Psi_\mathrm{Q}(t_0) = \Psi_\mathrm{Q}(t_0)$ as required.

Proof that the expression (33.8) solves Eq. (33.4).

Taking the time derivative of the expression (33.8), we obtain:

$$\begin{aligned}\frac{\partial}{\partial t}\Psi_{\mathrm{Q}}(t) &= -\,i\,U_{\mathrm{QQ}}(t,t)\,\mathcal{H}_{\mathrm{QP}}(t)\,\Psi_{\mathrm{P}}(t) - i\int_{t_0}^{t} dt'\,\frac{\partial}{\partial t}\,U_{\mathrm{QQ}}(t,t')\,\mathcal{H}_{\mathrm{QP}}(t')\,\Psi_{\mathrm{P}}(t')\\ &\quad+\frac{\partial}{\partial t}\,U_{\mathrm{QQ}}(t,t_0)\,\Psi_{\mathrm{Q}}(t_0) = -\,i\,\mathcal{H}_{\mathrm{QP}}(t)\,\Psi_{\mathrm{P}}(t)\\ &\quad-\,i\,\mathcal{H}_{\mathrm{QQ}}(t)\left[-i\int_{t_0}^{t} dt'\,U_{\mathrm{QQ}}(t,t')\,\mathcal{H}_{\mathrm{QP}}(t')\,\Psi_{\mathrm{P}}(t') + U_{\mathrm{QQ}}(t,t_0)\,\Psi_{\mathrm{Q}}(t_0)\right]\\ &= -\,i\,\mathcal{H}_{\mathrm{QP}}(t)\,\Psi_{\mathrm{P}}(t)\;-\;i\,\mathcal{H}_{\mathrm{QQ}}(t)\,\Psi_{\mathrm{Q}}(t)\,. \qquad \text{[QED]}\end{aligned} \tag{33.9}$$

We now enter the expression (33.8) into Eq. (33.3), to obtain:

$$\begin{aligned} i\,\frac{\partial}{\partial t}\,\Psi_{\mathrm{P}}(t) &= \mathcal{H}_{\mathrm{PP}}(t)\,\Psi_{\mathrm{P}}(t)\\ &\quad+\mathcal{H}_{\mathrm{PQ}}(t)\left[-i\int_{t_0}^{t} dt'\,U_{\mathrm{QQ}}(t,t')\,\mathcal{H}_{\mathrm{QP}}(t')\,\Psi_{\mathrm{P}}(t') + U_{\mathrm{QQ}}(t,t_0)\,\Psi_{\mathrm{Q}}(t_0)\right]. \end{aligned} \tag{33.10}$$

Finally, we introduce the "effective kernel"

$$\boxed{\mathcal{K}_{\mathrm{PP}}^{\mathrm{eff}}(t,t') \equiv -\,i\,\mathcal{H}_{\mathrm{PQ}}(t)\,U_{\mathrm{QQ}}(t,t')\,\mathcal{H}_{\mathrm{QP}}(t')} \tag{33.11}$$

and simplify Eq. (33.10) by setting $\Psi_{\mathrm{Q}}(t_0) = 0$, such that we obtain eventually for $t \geq t_0$:

$$\boxed{i\,\frac{\partial}{\partial t}\,\Psi_{\mathrm{P}}(t) = \mathcal{H}_{\mathrm{PP}}(t)\,\Psi_{\mathrm{P}}(t) + \int_{t_0}^{t} dt'\,\mathcal{K}_{\mathrm{PP}}^{\mathrm{eff}}(t,t')\,\Psi_{\mathrm{P}}(t')\,.} \tag{33.12}$$

This is an *effective* equation of motion for $\Psi_{\mathrm{P}}(t)$ only, which contains the "direct" effect of the dynamics of the P subspace through $\mathcal{H}_{\mathrm{PP}}(t)$ as well as the "indirect" effect of the dynamics of the Q subspace through its folding into the P subspace via the kernel $\mathcal{K}_{\mathrm{PP}}^{\mathrm{eff}}(t,t')$. Note how the expression (33.12) contains *memory effects* on the past history of $\Psi_{\mathrm{P}}(t)$, which result from the presence of the kernel $\mathcal{K}_{\mathrm{PP}}^{\mathrm{eff}}(t,t')$. In Section 33.2, we shall obtain contributions similar to those occurring on the right-hand side of Eq. (33.12) also in terms of the Schwinger–Keldysh Green's functions approach.

It is worth mentioning that an approach similar to that just outlined has been discussed in the context of the "damping theory" in Quantum Mechanics [198], as well as in the context of the "contraction of information" bearing on projected stochastic processes in Statistical Physics [199].

Remark: About the physical meaning of the "complement" subspace Q
Formally, the above P-Q partition applies quite generally to all physical cases treated in the present chapter and in Chapter 34. However, on physical grounds, one should distinguish between the cases when P and Q refer either to (i) different degrees of freedom (or states) of the *same* system or (ii) two different systems with their own quantum dynamics, which could even be made independent from each other. The first case applies, for example, to the Fano–Feshbach resonances mentioned in Chapter 26 (where the subspace P refers to the open channels and the subspace Q to the closed channels of a given molecular system – cf. Ref. [99]), as well as to the system considered later where a given particle can jump from the central region to the surrounding leads via a hopping Hamiltonian, similar to the treatment of the Josephson effect [107, 108]. The second case applies instead to the coupling of a "small" quantum system to a "large" reservoir (or bath) with a continuum spectrum, like that treated in Chapter 35, where the phonon bath has its own independent quantum dynamics.

33.2 Schwinger–Keldysh Description of System Plus Environment

With these premises, we now turn to the original problem of interest, where a central region of finite size is connected with two ("source" and "drain") terminals (or leads), although an additional number of terminals may as well be considered. This composite system is depicted schematically in Fig. 33.1.

We shall regard the central region as the (sub)system of interest (P) and the terminals as the environment (Q). In dealing with this composite system, we shall follow the treatment of Ref. [155] in the version reported in the more recent review article of Ref. [159], in which we shall incorporate our special emphasis on the P-Q partition utilized in Section 33.1.

To simplify the problem, the interparticle interaction will be assumed to be active only in the central region, while no interaction is active between a particle in the central region and a particle in the reservoirs as well as between particles in the reservoirs. (In particular, the particles we are considering are spin-$\frac{1}{2}$ fermions.) Accordingly, we write the total Hamiltonian in the form

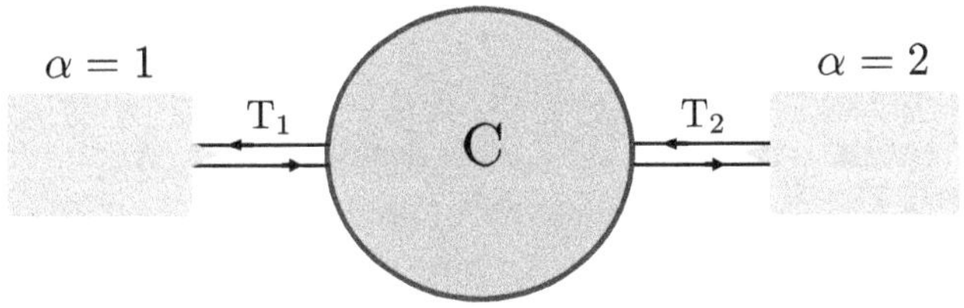

Figure 33.1 System composed of a central region (C) and two terminals (or reservoirs) labeled by $\alpha = 1$ (at left) and $\alpha = 2$ (at right) and connected with C via the links T_1 (at left) and T_2 (at right).

$$\mathcal{H}(t) = H_{\mathrm{C}}(t) + \sum_{\alpha=1}^{2} h_{\alpha}(t) + \sum_{\alpha=1}^{2} \left(T_{\alpha\mathrm{C}}(t) + \mathrm{h.c.}\right), \tag{33.13}$$

where:

(i)

$$H_{\mathrm{C}}(t) = \sum_{mn}\sum_{\sigma} d^{\dagger}_{m\sigma}\, h_{mn}(t)\, d_{n\sigma} + \frac{1}{2}\sum_{mn}\sum_{m'n'}\sum_{\sigma\sigma'} d^{\dagger}_{m\sigma}\, d^{\dagger}_{n\sigma'}\, v_{mnm'n'}\, d_{m'\sigma'}\, d_{n'\sigma} \tag{33.14}$$

is the Hamiltonian of the central (C) region, expressed in terms of basic functions specified by the integers $m, n, \cdots$, with spin label σ. Note that we have here considered the single-particle part of $H_{\mathrm{C}}(t)$ to depend on time through a generic time-dependent potential, while the two-particle part of $H_{\mathrm{C}}(t)$ has been assumed to be independent of time;

(ii)

$$h_{\alpha}(t) = \sum_{k\sigma} c^{\dagger}_{k\alpha\sigma}\, \varepsilon_{k\alpha}(t)\, c_{k\alpha\sigma} \qquad (\alpha = 1, 2) \tag{33.15}$$

is the single-particle Hamiltonian of the α-th reservoir (lead), whose levels $\varepsilon_{k\alpha}(t)$ are assumed to be time dependent (for instance, by taking $\varepsilon_{k\alpha}(t) = \varepsilon_{k\alpha} + V_{\alpha}(t)$);

(iii)

$$T_{\alpha\mathrm{C}}(t) = \sum_{mk}\sum_{\sigma} d^{\dagger}_{m\sigma}\, T^{(\alpha)}_{mk}(t)\, c_{k\alpha\sigma} \tag{33.16}$$

accounts for the tunneling between the α-th reservoir and the central region, with a time-dependent tunneling matrix $\mathbf{T}^{(\alpha)}(t)$.

The structure of the Hamiltonian (33.13)–(33.16) suggests us an analogy with the structure of Eqs. (33.3)–(33.5) in terms of the projector operators P and Q, by introducing "block matrices" in the subspaces identified by the labels m (for P) and $k\alpha$ (for Q) for a given spin σ. We then organize the single-particle parts of the Hamiltonian (33.13)–(33.16) in the block form reported in Fig. 33.2(a) (for given σ), where the correspondence of the blocks with the P and Q subspaces is indicated. Similarly, the single-particle Green's function can be decomposed in the block form reported in Fig. 33.2(b), where, unlike in the Hamiltonian matrix of Fig. 33.2(a), there now occur lead–lead coupling terms $G_{\alpha\alpha'}$ with $\alpha \neq \alpha'$ because different leads can be connected by hopping processes mediated by the central region. Finally, the self-energy takes the block form reported in Fig. 33.2(c) (for given σ), with nonvanishing contributions occurring only in the central region

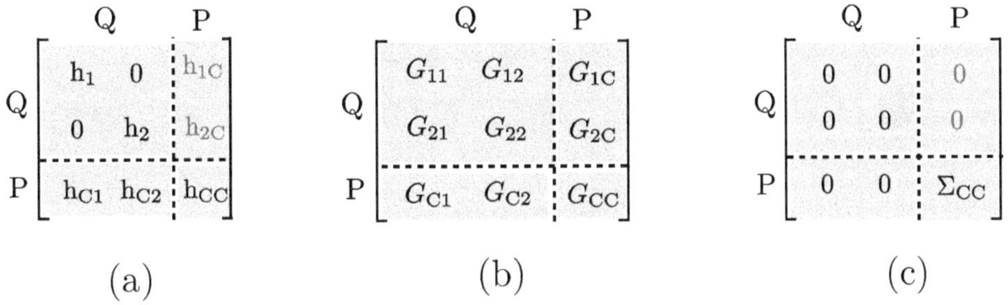

Figure 33.2 P-Q partition of the matrices representing (a) the single-particle part of the Hamiltonian, (b) the single-particle Green's functions, and (c) the self-energy. As in Fig. 33.1, the index C refers to the central region, while the indices (1, 2) refer, respectively, to the left and right reservoirs to which it is connected. For clarity, the dependence of all quantities on time variables is left implicit.

where the two-body interaction in the expression (33.14) is assumed to be active (thus taking into account that a diagrammatic expansion of the self-energy has to start and to end with an interaction line).

The block structure shown in the various panels of Fig. 33.2 can be exploited to project the Dyson equation (11.26) of Part I into the relevant blocks, retracing in this way what we did in Section 33.1 for the Schrödinger equation (33.1). In particular, for given σ, we obtain for the CC block

$$\left[i\frac{d}{dz} - h_{\mathrm{CC}}(z)\right] G_{\mathrm{CC}}(z,z') - \sum_{\alpha=1}^{2} h_{\mathrm{C}\alpha}(z)\, G_{\alpha\mathrm{C}}(z,z') - \int_{\gamma_{\mathrm{T}}} d\bar{z}\ \Sigma_{\mathrm{CC}}(z,\bar{z})\, G_{\mathrm{CC}}(\bar{z},z')$$
$$= \delta(z,z'), \tag{33.17}$$

where γ_{T} is the total contour of Fig. 8.1 of Part I. Note that Eq. (33.17) plays the role of Eq. (33.3) in the present context.

To express Eq. 33.17) in terms of G_{CC} only, we need an explicit expression of $G_{\alpha\mathrm{C}}$, which would play the role of Eq. (33.8) in the present context. To this end, it is sufficient to consider the αC blocks with $\alpha = (1,2)$, whereby

$$\left[i\frac{d}{dz} - h_{\alpha}(z)\right] G_{\alpha\mathrm{C}}(z,z') - h_{\alpha\mathrm{C}}(z)\, G_{\mathrm{CC}}(z,z') = 0\,. \tag{33.18}$$

This equation can be solved by considering the Green's function $g_{\alpha\alpha}(z,z')$ of an isolated lead α, which corresponds to the Hamiltonian with the matrix block $h_{\alpha}(z)$ in Fig. 33.2(a). This Green's function satisfies the equation (cf. Eq. (11.32) of Part I)

$$\left[i\frac{d}{dz} - h_\alpha(z)\right] g_{\alpha\alpha}(z,z') = \delta(z,z'), \tag{33.19}$$

whose inverse is given by (cf. Eq. (11.33) of Part I)

$$g^{-1}_{\alpha\alpha}(z,z') = \left[i\frac{d}{dz} - h_\alpha(z)\right] \delta(z,z'). \tag{33.20}$$

Accordingly, we first rewrite Eq. (33.18) in the form

$$\begin{aligned}\left[i\frac{d}{dz} - h_\alpha(z)\right] G_{\alpha C}(z,z') &= \int_{\gamma_T} d\bar{z} \left[i\frac{d}{dz} - h_\alpha(z)\right] \delta(z,\bar{z})\, G_{\alpha C}(\bar{z},z') \\ &\underset{[\text{Eq.}(33.20)]}{\underset{\uparrow}{=}} \int_{\gamma_T} d\bar{z}\, g^{-1}_{\alpha\alpha}(z,\bar{z})\, G_{\alpha C}(\bar{z},z') \underset{[\text{Eq.}(33.18)]}{\underset{\uparrow}{=}} h_{\alpha C}(z)\, G_{CC}(z,z'),\end{aligned} \tag{33.21}$$

and then multiply both sides of this equation from the left by $g_{\alpha\alpha}(\bar{\bar{z}},z)$ and integrate over z to obtain the required result

$$\begin{aligned}\int_{\gamma_T} dz\, d\bar{z}\, g_{\alpha\alpha}(\bar{\bar{z}},z)\, g^{-1}_{\alpha\alpha}(z,\bar{z})\, G_{\alpha C}(\bar{z},z') &= \int_{\gamma_T} d\bar{z}\, \delta(\bar{\bar{z}},\bar{z})\, G_{\alpha C}(\bar{z},z') \\ &= \boxed{G_{\alpha C}(\bar{\bar{z}},z') = \int_{\gamma_T} dz\, g_{\alpha\alpha}(\bar{\bar{z}},z)\, h_{\alpha C}(z)\, G_{CC}(z,z')}.\end{aligned} \tag{33.22}$$

Note here that:

(i) $g_{\alpha\alpha}(z,z')$ of Eq. (33.19) differs from $G^{-1}_{\alpha\alpha}(z,z')$ appearing in the QQ block of Fig. 33.2(b) since the latter satisfies the equation

$$\left[i\frac{d}{dz} - h_\alpha(z)\right] G_{\alpha\alpha}(z,z') - h_{\alpha C}(z)\, G_{C\alpha}(z,z') = 0; \tag{33.23}$$

(ii) The explicit expression of $g_{\alpha\alpha}(z,z')$ can be readily obtained as follows. According to the expression (33.15), the block matrices h_α appearing in the QQ block of Fig. 33.2(a) have only diagonal elements equal to $\varepsilon_{\alpha k}(z)$ for each σ. Correspondingly, $g_{\alpha\alpha}(z,z')$ has diagonal elements $g_{\alpha k}(z,z')$ given by

$$\boxed{g_{\alpha k}(z,z') = -i\left[\theta(z,z')\,(1 - f_F(\varepsilon_{\alpha k})) - \theta(z',z) f_F(\varepsilon_{\alpha k})\right] e^{-i\int_{z'}^{z} d\bar{z}\, \varepsilon_{\alpha k}(\bar{z})}}, \tag{33.24}$$

where $\varepsilon_{\alpha k} = \varepsilon_{\alpha k}(z = t_0)$ and $f_F(\varepsilon_{\alpha k})$ is the Fermi–Dirac distribution function corresponding to the initial equilibrium configuration at temperature T. To prove that this is the correct expression, we calculate its time derivative explicitly:

$$\frac{d}{dz}g_{\alpha k}(z,z') = -i\,[\delta(z,z')\,(1-f_F(\varepsilon_{\alpha k})) + \delta(z,z')f_F(\varepsilon_{\alpha k})]\,e^{-i\int_{z'}^{z}d\bar{z}\,\varepsilon_{\alpha k}(\bar{z})}$$

$$-i\,[\theta(z,z')\,(1-f_F(\varepsilon_{\alpha k}))-\theta(z',z)f_F(\varepsilon_{\alpha k})]\,(-i)\,\varepsilon_{\alpha k}(z)\,e^{-i\int_{z'}^{z}d\bar{z}\,\varepsilon_{\alpha k}(\bar{z})}$$

$$= -i\,\delta(z,z') - i\,\varepsilon_{\alpha k}(z)\,g_{\alpha k}(z,z'), \tag{33.25}$$

that is,

$$\left[i\frac{d}{dz} - \varepsilon_{\alpha k}(z)\right] g_{\alpha k}(z,z')) = \delta(z,z'). \qquad \text{[QED]} \tag{33.26}$$

(iii) The expression (33.24) satisfies the Kubo–Martin–Schwinger boundary condition (12.2) of Part I for fermions, in the form

$$g_{\alpha k}(t_0,z') = -\,g_{\alpha k}(t_0 - i\beta, z'). \tag{33.27}$$

This is because, according to Eq. (33.24),

$$g_{\alpha k}(t_0,z') = i f_F(\varepsilon_{\alpha k})\,e^{-i\int_{z'}^{t_0}d\bar{z}\,\varepsilon_{\alpha k}(\bar{z})}, \tag{33.28}$$

while

$$\begin{aligned} g_{\alpha k}(t_0 - i\beta, z') &= -\,i\,(1-f_F(\varepsilon_{\alpha k}))\,e^{-i\int_{z'}^{t_0-i\beta}d\bar{z}\,\varepsilon_{\alpha k}(\bar{z})} \\ &= -\,i\,(1-f_F(\varepsilon_{\alpha k}))\,e^{-\beta(\varepsilon_{\alpha k}-\mu)}e^{-i\int_{z'}^{t_0}d\bar{z}\,\varepsilon_{\alpha k}(\bar{z})} \\ &= -\,i f_F(\varepsilon_{\alpha k})\,e^{-i\int_{z'}^{t_0}d\bar{z}\,\varepsilon_{\alpha k}(\bar{z})} \underset{\text{[Eq.(33.28)]}}{\overset{\uparrow}{=}} -\,g_{\alpha k}(t_0,z'). \qquad \text{[QED]} \end{aligned} \tag{33.29}$$

Remark: Proof of the generic validity of the expression (33.24)

The form (33.24) can also be obtained for a generic noninteracting single-particle Green's function, directly from its definition

$$G_k^{(0)}(z,z') = -i\,\left\langle \mathcal{T}_\gamma\left\{e^{-i\int_\gamma d\bar{z}\,H_0(\bar{z})}\,c_k(z)\,c_k^\dagger(z')\right\}\right\rangle_0, \tag{33.30}$$

where γ is now the closed contour depicted in Fig. 8.1 of Part I, $H_0(\bar{z}) = \sum_k c_k^\dagger \varepsilon_k(\bar{z}) c_k$ is the time-dependent Hamiltonian of the noninteracting system, and $\langle\cdots\rangle_0$ is the thermal average with respect to the initial Hamiltonian $H_0(t_0) = \sum_k c_k^\dagger \varepsilon_k c_k$.

To prove this statement, we recall the structure of the time-evolution operator entering the expression (2.11) of Part I, as well as the identities utilized to prove the property (9.4) therein, and write accordingly

$$c_k(z)\,U_0(z,t_0) = U_0(z,t_0)\,c_k(z)\,e^{-i\int_{t_0}^{z}d\bar{z}\,\varepsilon_k(\bar{z})}, \tag{33.31}$$

$$c_k^\dagger(z)\,U_0(z,t_0) = U_0(z,t_0)\,c_k^\dagger(z)\,e^{i\int_{t_0}^{z}d\bar{z}\,\varepsilon_k(\bar{z})}, \tag{33.32}$$

where $U_0(z,z') = \mathcal{T}_\gamma\left\{e^{-i\int_{z'}^{z}d\bar{z}\,H_0(\bar{z})}\right\}$. We thus obtain for the expression (33.30):

(i) When $\boxed{z' > z}$ along the contour γ of Fig. 8.1 of Part I:

$$\begin{aligned}
i\, G_k^{(0)}(z,z') &= -\,\langle U_0(t_0,z')\, c_k^{\dagger}(z')\, U_0(z',z)\, \overbrace{c_k(z)\, U_0(z,t_0)}\rangle_0 \\
&\underset{\substack{\uparrow \\ [\text{Eq.(33.31)}]}}{=} -\,\langle U_0(t_0,z')\, \underbrace{c_k^{\dagger}(z')\, \overbrace{U_0(z',z)\, U_0(z,t_0)}}\, c_k(z)\rangle_0\, e^{-i\int_{t_0}^{z} d\bar{z}\, \varepsilon_k(\bar{z})} \\
&\underset{\substack{\uparrow \\ [\text{Eq.(33.32)}]}}{=} -\,\langle \overbrace{U_0(t_0,z')\, U_0(z',t_0)}\, c_k^{\dagger}(z')\, c_k(z)\rangle_0\, e^{i\int_{t_0}^{z'} d\bar{z}\, \varepsilon_k(\bar{z})}\, e^{-i\int_{t_0}^{z} d\bar{z}\, \varepsilon_k(\bar{z})} \\
&= -\,\langle c_k^{\dagger}(z')\, c_k(z)\rangle_0\, e^{-i\int_{z'}^{z} d\bar{z}\, \varepsilon_k(\bar{z})} = -f_F(\varepsilon_k)\, e^{-i\int_{z'}^{z} d\bar{z}\, \varepsilon_k(\bar{z})}\,; \qquad (33.33)
\end{aligned}$$

(ii) When $\boxed{z > z'}$ along the contour γ of Fig. 8.1 of Part I:

$$\begin{aligned}
i\, G_k^{(0)}(z,z') &= \langle U_0(t_0,z)\, c_k(z)\, U_0(z,z')\, \overbrace{c_k^{\dagger}(z')\, U_0(z',t_0)}\rangle_0 \\
&\underset{\substack{\uparrow \\ [\text{Eq.(33.32)}]}}{=} \langle U_0(t_0,z)\, \underbrace{c_k(z)\, \overbrace{U_0(z,z')\, U_0(z',t_0)}}\, c_k^{\dagger}(z')\rangle_0\, e^{i\int_{t_0}^{z'} d\bar{z}\, \varepsilon_k(\bar{z})} \\
&\underset{\substack{\uparrow \\ [\text{Eq.(33.31)}]}}{=} \langle \overbrace{U_0(t_0,z)\, U_0(z,t_0)}\, c_k(z)\, c_k^{\dagger}(z')\rangle_0\, e^{-i\int_{t_0}^{z} d\bar{z}\, \varepsilon_k(\bar{z})}\, e^{i\int_{t_0}^{z'} d\bar{z}\, \varepsilon_k(\bar{z})} \\
&= \langle c_k(z)\, c_k^{\dagger}(z')\rangle_0\, e^{-i\int_{z'}^{z} d\bar{z}\, \varepsilon_k(\bar{z})} = (1 - f_F(\varepsilon_k))\; e^{-i\int_{z'}^{z} d\bar{z}\, \varepsilon_k(\bar{z})}\,, \qquad (33.34)
\end{aligned}$$

where in the last line of both Eqs. (33.33) and (33.34), we have recalled that the variables z in $c_k(z)$ and z' in $c_k^{\dagger}(z')$ are only meant to specify the positions of these operators along the contour γ (cf. also Section 7.2 of Part I), such that $\langle c_k^{\dagger}(z')\, c_k(z)\rangle_0 \to \langle c_k^{\dagger}\, c_k\rangle_0 = f_F(\varepsilon_k)$ and $\langle c_k(z)\, c_k^{\dagger}(z')\rangle_0 \to \langle c_k\, c_k^{\dagger}\rangle_0 = \langle\big(1 - c_k^{\dagger}\, c_k\big)\rangle_0 = 1 - f_F(\varepsilon_k)$. Taken together, the expressions (33.33) and (33.34) are seen to reproduce the result (33.24), as anticipated. [QED]

Remark: The related lesser, greater, retarded, and advanced noninteracting Green's functions

From the expressions (33.30), (33.33), and (33.34), we can readily obtain the corresponding expressions for the lesser, greater, retarded, and advanced noninteracting Green's functions (which will be of use in the following):

(i) $\boxed{\text{Lesser Green's function:}}$ From the definition (13.5) of Part I, we obtain for $z' > z$

$$G_k^{(0)<}(t,t') = i f_F(\varepsilon_k)\, e^{-i\int_{t'}^{t} d\bar{t}\, \varepsilon_k(\bar{t})}\,; \qquad (33.35)$$

(ii) Greater Green's function: From the definition (13.6) of Part I, we obtain for $z > z'$

$$G_k^{(0)>}(t,t') = -\,i\,(1 - f_F(\varepsilon_k))\, e^{-\,i\int_{t'}^{t} d\bar{t}\,\varepsilon_k(\bar{t})}\,; \tag{33.36}$$

(iii) Retarded Green's function: The previous results (33.35) and (33.36) can be combined with the definition (30.9) of Part II, to yield

$$G_k^{(0)\mathrm{R}}(t,t') = -\,i\,\theta(t-t')\, e^{-\,i\int_{t'}^{t} d\bar{t}\,\varepsilon_k(\bar{t})}\,; \tag{33.37}$$

(iv) Advanced Green's function: Similarly, the results (33.35) and (33.36) can be combined with the definition (30.11) of Part II, to yield

$$G_k^{(0)\mathrm{A}}(t,t') = i\,\theta(t'-t)\, e^{-\,i\int_{t'}^{t} d\bar{t}\,\varepsilon_k(\bar{t})}\,. \tag{33.38}$$

Note that in the time-independent case, whereby $\int_{t'}^{t} d\bar{t}\,\varepsilon_k(\bar{t}) \to (t-t')\,\varepsilon_k$, Eqs. (33.35), (33.36), and (33.37)–(33.38), once Fourier transformed to frequency space, reduce to Eqs. (20.10), (20.8), and (20.6) of Part I, in the order.

At this point, we can enter the expression (33.22) into Eq. (33.17), to obtain

$$\left[i\frac{d}{dz} - h_{\mathrm{CC}}(z)\right] G_{\mathrm{CC}}(z,z') - \int_{\gamma_{\mathrm{T}}} d\bar{z} \sum_{\alpha=1}^{2} h_{\mathrm{C}\alpha}(z)\, g_{\alpha\alpha}(z,\bar{z})\, h_{\alpha\mathrm{C}}(\bar{z})\, G_{\mathrm{CC}}(\bar{z},z')$$
$$- \int_{\gamma_{\mathrm{T}}} d\bar{z}\, \Sigma_{\mathrm{CC}}(z,\bar{z})\, G_{\mathrm{CC}}(\bar{z},z') = \delta(z,z')\,. \tag{33.39}$$

In addition, we can identify the so-called *embedding self-energy* given by the expression

$$\boxed{\Sigma_{\mathrm{em}}(z,z') = \sum_{\alpha=1}^{2} h_{\mathrm{C}\alpha}(z)\, g_{\alpha\alpha}(z,z')\, h_{\alpha\mathrm{C}}(z')}\,, \tag{33.40}$$

such that Eq. (33.39) can be rewritten in the compact form

$$\left[i\frac{d}{dz} - h_{\mathrm{CC}}(z)\right] G_{\mathrm{CC}}(z,z') - \int_{\gamma_{\mathrm{T}}} d\bar{z}\, (\Sigma_{\mathrm{em}}(z,\bar{z}) + \Sigma_{\mathrm{CC}}(z,\bar{z}))\, G_{\mathrm{CC}}(\bar{z},z') = \delta(z,z')\,. \tag{33.41}$$

The analogy between the expressions (33.40) and (33.11), as well as between the equations (33.41) and (33.12), is now apparent. All in all, what we have effectively achieved is to "fold" the effect of the leads (environment) into an equation for the central region (subsystem) of interest.

33.3 Calculation of the Time-Dependent Current

In the context of the composite system depicted in Fig. 33.1 that we are considering, a physical quantity of interest is the *total current* flowing between (say) the left lead (with $\alpha = 1 \rightarrow L$) and the central region. Let $N_L = \sum_{k\sigma} c^\dagger_{kL\sigma} c_{kL\sigma}$ then be the total particle operator of the left lead, whose (real) time dependence is given by the average value over the initial distribution

$$N_L(t) = \langle U^\dagger(t, t_0)\, N_L\, U(t, t_0)\rangle\,. \tag{33.42}$$

The particle current flowing from the left lead to the central region is thus given by

$$I_L(t) = -\frac{dN_L(t)}{dt} = i\,\langle U^\dagger(t, t_0)\,[N_L, \mathcal{H}(t)]\,U(t, t_0)\rangle \tag{33.43}$$

with the total Hamiltonian $\mathcal{H}(t)$ given by Eq. (33.13). Since both $H_C(t)$ and $h_\alpha(t)$ commute with N_L, we are left with evaluating the commutator $[N_L, (T_{LC}(t) + \text{h.c.})]$, where

$$\begin{aligned}[N_L, T_{LC}(t)] &= \sum_m \sum_k \sum_\sigma d^\dagger_{m\sigma}\, T^{(L)}_{mk}(t)\, [c^\dagger_{kL\sigma} c_{kL\sigma}, c_{kL\sigma}] \\ &= -\sum_m \sum_k \sum_\sigma d^\dagger_{m\sigma}\, T^{(L)}_{mk}(t)\, c_{kL\sigma}, \end{aligned} \tag{33.44}$$

while

$$[N_L, T_{LC}(t)^\dagger] = -[N_L, T_{LC}(t)]^\dagger = \sum_m \sum_k \sum_\sigma c^\dagger_{kL\sigma}\, T^{(L)}_{mk}(t)^*\, d_{m\sigma}\,. \tag{33.45}$$

The expression (33.43) for $I_L(t)$ then becomes

$$\begin{aligned} I_L(t) &= -\,i \sum_m \sum_k \sum_\sigma \Big(T^{(L)}_{mk}(t)\, \langle U^\dagger(t, t_0)\, d^\dagger_{m\sigma}\, c_{kL\sigma}\, U(t, t_0)\rangle \\ &\quad - T^{(L)}_{mk}(t)^* \langle U^\dagger(t, t_0)\, c^\dagger_{kL\sigma}\, d_{m\sigma}\, U(t, t_0)\rangle \Big) \\ &= -\sum_m \sum_k \sum_\sigma \Big(T^{(L)}_{mk}(t)\, G^<_{kLm,\sigma}(t, t) - T^{(L)}_{mk}(t)^*\, G^<_{mkL,\sigma}(t, t) \Big), \end{aligned} \tag{33.46}$$

where we have used the definition (13.5) of Part I for the lesser (mixed) Green's functions. In addition, from the property (13.22) of Part I, which in the present case reads

$$G^<_{mkL,\sigma}(t, t)^* = -G^<_{kLm,\sigma}(t, t)\,, \tag{33.47}$$

the expression (33.46) becomes eventually:

$$
\begin{aligned}
I_L(t) &= -\sum_m \sum_k \sum_\sigma \left(T_{mk}^{(L)}(t)\, G_{kLm,\sigma}^{<}(t,t) + T_{mk}^{(L)}(t)^*\, G_{kLm,\sigma}^{<}(t,t)^* \right) \\
&= -2\,\mathfrak{Re}\left\{ \sum_m \sum_k \sum_\sigma T_{mk}^{(L)}(t)\, G_{kLm,\sigma}^{<}(t,t) \right\} . \qquad (33.48)
\end{aligned}
$$

There remains to make use of the result (33.22) for expressing Eq. (33.48) entirely in terms of the Green's functions of the central region. To this end, it is first convenient to cast Eq. (33.48) in matrix form according to the convention of Fig. 33.2(a) (and at the same time to eliminate the spin index by replacing $\sum_\sigma \to 2$, since in the present case no physical quantity depends explicitly on spin), thereby arriving at the compact expression

$$
I_L(t) = -4\,\mathfrak{Re}\left\{ \mathrm{Tr}_{\mathrm{C}}\left[h_{CL}(t)\, G_{LC}^{<}(t,t) \right] \right\} , \qquad (33.49)
$$

where Tr_{C} is the trace over the quantum numbers "m" of the central region, and matrix multiplication over the quantum numbers "k" of the L lead is implied. We then write with the help of Eq. (33.22) for generic time variables z and z' along the contour γ_{T}:

$$
\begin{aligned}
\mathrm{Tr}_{\mathrm{C}}\left[h_{CL}(z)\, G_{LC}(z,z') \right] &= \mathrm{Tr}_{\mathrm{C}}\left[\int_{\gamma_{\mathrm{T}}} d\bar{z}\, h_{CL}(z)\, g_L(z,\bar{z})\, h_{LC}(\bar{z})\, G_{\mathrm{CC}}(\bar{z},z') \right] \\
&= \mathrm{Tr}_{\mathrm{C}}\left[\int_{\gamma_{\mathrm{T}}} d\bar{z}\, \Sigma_{\mathrm{em}}^{(L)}(z,\bar{z})\, G_{\mathrm{CC}}(\bar{z},z') \right] , \qquad (33.50)
\end{aligned}
$$

where now

$$
\Sigma_{\mathrm{em}}^{(L)}(z,z') = h_{CL}(z)\, g_L(z,z')\, h_{LC}(z') \qquad (33.51)
$$

is the contribution from the left lead to the embedding self-energy (33.40). Taking also into account that

$$
\mathrm{Tr}_{\mathrm{C}}\left[h_{CL}(z)\, G_{LC}(z,z') \right]^{<} = \mathrm{Tr}_{\mathrm{C}}\left[h_{CL}(t)\, G_{LC}^{<}(t,t') \right] , \qquad (33.52)
$$

we are thus left with considering the lesser component of the convolution (33.50). This can be readily done by recalling the definition (14.8) and the result (14.25) of Part I, yielding

$$
\begin{aligned}
&\mathrm{Tr}_{\mathrm{C}}\left[\int_{\gamma_{\mathrm{T}}} d\bar{z}\, \Sigma_{\mathrm{em}}^{(L)}(z,\bar{z})\, G_{\mathrm{CC}}(\bar{z},z') \right]^{<}_{z=z'=t} \\
&= \mathrm{Tr}_{\mathrm{C}}\left[\int_{t_0}^{\infty} d\bar{t} \left(\Sigma_{\mathrm{em}}^{(L)\mathrm{R}}(t,\bar{t})\, G_{\mathrm{CC}}^{<}(\bar{t},t) + \Sigma_{\mathrm{em}}^{(L)<}(t,\bar{t})\, G_{\mathrm{CC}}^{\mathrm{A}}(\bar{t},t) \right) \right. \\
&\quad \left. -i \int_0^{\beta} d\bar{\tau}\, \Sigma_{\mathrm{em}}^{(L)\rceil}(t,\bar{\tau})\, G_{\mathrm{CC}}^{\lceil}(\bar{\tau},t) \right] , \qquad (33.53)
\end{aligned}
$$

where only the Green's functions of the central region appear eventually. Note that the last term on the right-hand side of Eq. (33.53) accounts for the dependence of the particle current on the "initial" preparation of the compound system, which gives rise to transient effects for t, not much later than t_0. However, transient effects can also originate from an abrupt time dependence of the external potential, even past the time when transient effects related to the initial preparation are washed out.

The first term on the right-hand side of Eq. (33.53) can be further manipulated as follows:

(i) The upper limit of the integral over $\bar{t}$ can be set equal to the external time t, owing to the presence therein of $\Sigma_{\text{em}}^{(L)\text{R}}(t,\bar{t})$ for which $t > \bar{t}$ and of $G_{\text{CC}}^{\text{A}}(\bar{t},t)$ for which $\bar{t} < t$.

(ii) Under these circumstances, we may subtract from the integrand of the integral over $\bar{t}$ the two terms

$$\Sigma_{\text{em}}^{(L)<}(t,\bar{t})\, G_{\text{CC}}^{\text{R}}(\bar{t},t) + \Sigma_{\text{em}}^{(L)\text{A}}(t,\bar{t})\, G_{\text{CC}}^{<}(\bar{t},t) \tag{33.54}$$

since both $G_{\text{CC}}^{\text{R}}(\bar{t},t)$ and $\Sigma_{\text{em}}^{(L)\text{A}}(t,\bar{t})$ would require $\bar{t} > t$, which is not compatible with the condition $t > \bar{t}$ of the integral itself. In this way, the first term on the right-hand side of Eq. (33.53) can be rewritten as follows:

$$\begin{aligned}
&\int_{t_0}^{\infty} d\bar{t}\, \Big(\Sigma_{\text{em}}^{(L)\text{R}}(t,\bar{t})\, G_{\text{CC}}^{<}(\bar{t},t) + \Sigma_{\text{em}}^{(L)<}(t,\bar{t})\, G_{\text{CC}}^{\text{A}}(\bar{t},t) \\
&\qquad - \Sigma_{\text{em}}^{(L)<}(t,\bar{t})\, G_{\text{CC}}^{\text{R}}(\bar{t},t) - \Sigma_{\text{em}}^{(L)\text{A}}(t,\bar{t})\, G_{\text{CC}}^{<}(\bar{t},t)\Big) \\
&= \int_{t_0}^{\infty} d\bar{t}\, \Big(\Sigma_{\text{em}}^{(L)<}(t,\bar{t}) \Big[G_{\text{CC}}^{\text{A}}(\bar{t},t) - G_{\text{CC}}^{\text{R}}(\bar{t},t)\Big] + \Big[\Sigma_{\text{em}}^{(L)\text{R}}(t,\bar{t}) - \Sigma_{\text{em}}^{(L)\text{A}}(t,\bar{t})\Big]\, G_{\text{CC}}^{<}(\bar{t},t)\Big) \\
&= i \int_{t_0}^{\infty} d\bar{t}\, \Big(\Sigma_{\text{em}}^{(L)<}(t,\bar{t})\, A_{\text{CC}}(\bar{t},t) - \Gamma_{\text{em}}^{(L)}(t,\bar{t})\, G_{\text{CC}}^{<}(\bar{t},t)\Big)\,,
\end{aligned} \tag{33.55}$$

where we have introduced the notation

$$A_{\text{CC}}(t,t') = i\left[G_{\text{CC}}^{\text{R}}(t,t') - G_{\text{CC}}^{\text{A}}(t,t')\right] \tag{33.56}$$

in analogy with the expression (30.49) of Part II for the single-particle spectral function (here projected into the central region), as well as its counterpart

$$\Gamma_{\text{em}}^{(L)}(t,t') = i\left[\Sigma_{\text{em}}^{(L)\text{R}}(t,t') - \Sigma_{\text{em}}^{(L)\text{A}}(t,t')\right] \tag{33.57}$$

for the spectral function of the embedding self-energy.

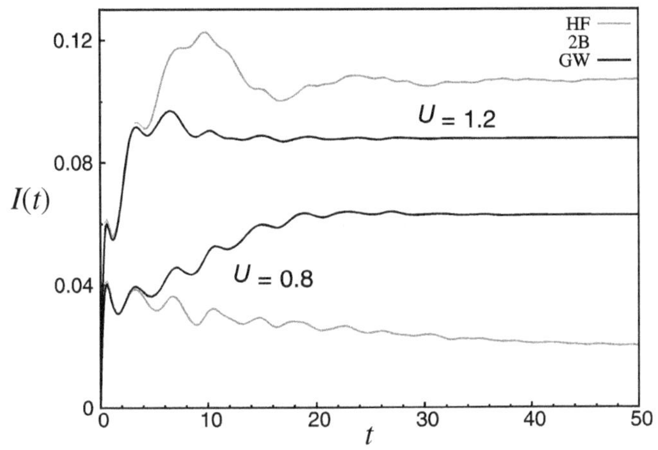

Figure 33.3 The transient current flowing into the right lead in a four-site molecular junction with long-range interactions is compared at different levels of self-energy approximations (Hartree–Fock (HF), second-order Born (2B), and GW) and for two values (0.8 and 1.2) of the bias $U = U_L = -U_R$, which is suddenly applied with a different sign to the left (L) and right (R) leads at time $t_0 = 0$. All quantities are in arbitrary units. [Reproduced from Fig. 6 of Ref. [200], with permission.]

In conclusion, the expression (33.49) for the current can be cast in its final form:

$$\boxed{\begin{aligned} I_L(t) = &-4\,\Re\mathrm{e}\left\{ i\,\mathrm{Tr}_C \left[\int_{t_0}^{\infty} d\bar{t} \left(\Sigma^{(L)<}_{\mathrm{em}}(t,\bar{t})\, A_{CC}(\bar{t},t) - \Gamma^{(L)}_{\mathrm{em}}(t,\bar{t})\, G^{<}_{CC}(\bar{t},t) \right) \right.\right. \\ &\left.\left. - \int_0^{\beta} d\bar{\tau}\, \Sigma^{(L)\rceil}_{\mathrm{em}}(t,\bar{\tau})\, G^{\lceil}_{CC}(\bar{\tau},t) \right] \right\} . \end{aligned}} \tag{33.58}$$

Examples of how the current flowing through the central region is affected, not only by transient effects related to sudden jumps in the potential applied to the leads but also by the interparticle interaction active in the central region, are shown in Fig. 33.3, where different levels of self-energy approximations (namely, Hartree–Fock (HF), second-order Born (2B), and GW) have been utilized.

In the expression (33.58) for the current, the following approximations can further be made:

(i) Let t_0 recede backward with respect to the time t at which the current is measured. In this case, the last term in Eq. (33.49) can be neglected because memory of the initial preparation at time t_0 is eventually washed out. Accordingly, we can take $t_0 \to -\infty$, such that the whole approach reduces to that obtained within the Keldysh formalism. This is what was done in the pioneering work of Ref. [155]. Examples of the time dependence of the current obtained in this way were already reported in Section 32.3, for the simpler case when the interparticle interaction in the central region is neglected.

(ii) In the limit $t \to +\infty$, provided the time-dependent external disturbance gets switched off at some finite time, the system is eventually expected to reach a *steady state*. In this case, the Green's functions and the self-energies become functions of the time difference $t - \bar{t}$ only and can accordingly be Fourier transformed to the real frequency ω. Under these circumstances, the steady-state value of the current obtained from the expression (33.58) reads:

$$\begin{aligned} I_L(t \to +\infty) = &- 4\,\mathfrak{Re}\left\{ i\,\mathrm{Tr_C}\left[\int_{-\infty}^{+\infty} d\bar{t}\,\left(\Sigma_{\mathrm{em}}^{(L)<}(t-\bar{t})\, A_{\mathrm{CC}}(\bar{t}-t) \right.\right.\right. \\ &\left.\left.\left. - \Gamma_{\mathrm{em}}^{(L)}(t-\bar{t})\, G_{\mathrm{CC}}^{<}(\bar{t}-t) \right) \right] \right\} \\ = &- 4\,\mathfrak{Re}\left\{ i\,\mathrm{Tr_C}\left[\int_{-\infty}^{+\infty} \frac{d\omega}{2\pi} \left(\Sigma_{\mathrm{em}}^{(L)<}(\omega)\, A_{\mathrm{CC}}(\omega) - \Gamma_{\mathrm{em}}^{(L)}(\omega)\, G_{\mathrm{CC}}^{<}(\omega) \right) \right] \right\}. \end{aligned} \tag{33.59}$$

This is what was done in the even earlier work of Ref. [154]. Note that, in the expression (33.59), $A_{\mathrm{CC}}(\omega)$ and $\Gamma_{\mathrm{em}}^{(L)}(\omega)$ are Hermitian matrices in C-space, while $G_{\mathrm{CC}}^{<}(\omega)$ and $\Sigma_{\mathrm{em}}^{(L)<}(\omega)$ are anti-Hermitian matrices, such that the quantity $I_L(t \to +\infty)$ is real as expected.

Note finally that in the derivations here considered, we have assumed that for $t < t_0$, the whole system (leads plus central region) is in thermodynamic equilibrium with unique temperature and chemical potential. This situation corresponds to the so-called *partition-free approach* of Ref. [201]. This is to be distinguished from the so-called *partitioned approach* of Ref. [153], whereby for $t < t_0$ the leads are disconnected from the central region, such that they are in separate thermodynamic equilibrium with different temperatures and chemical potentials.

33.4 Wide-Band-Limit Approximation

A simpler version of this problem corresponds to neglecting the interparticle interaction also in the central region, whereby $\Sigma_{\mathrm{CC}} = 0$. In this case, a full solution can be obtained with the additional assumptions that in Eq. (33.15) $\varepsilon_{k\alpha}(t) = \varepsilon_{k\alpha} + V_\alpha(t)$, where $V_\alpha(t)$ is suddenly switched on at time t_0 to a constant value and that the time dependence of $T_{mk}^{(L)}$ is altogether neglected [15]. In this context, it is also relevant to adopt the so-called *wide-band-limit approximation* (WBLA), whereby the lead density of state is considered featureless over the characteristic energy scale of the central region (mostly near the Fermi energy). Analytically, these approximations significantly simplify the form of embedding self-energy entering the Dyson equation (33.41) in terms of the following features.

Let's consider, in particular, the retarded component of the embedding self-energy (33.51). According to the convention of Eq. (33.16) and the result (33.37), we write for its matrix elements

$$\begin{aligned}\Sigma^{(L)\mathrm{R}}_{\mathrm{em}}(t,t')_{mn} &= \sum_k T^{(L)}_{mk}\, g^{(L)\mathrm{R}}_k(t,t')\, T^{(L)\,*}_{nk} = -i\,\theta(t-t') \sum_k T^{(L)}_{mk}\, e^{-i\int_{t'}^{t} d\bar{t}\,\varepsilon_{kL}(\bar{t})}\, T^{(L)\,*}_{nk} \\ &= -i\,\theta(t-t')\, e^{-i\int_{t'}^{t} d\bar{t}\, V_L(\bar{t})} \sum_k T^{(L)}_{mk}\, e^{-i\varepsilon_{kL}(t-t')}\, T^{(L)\,*}_{nk} \\ &\equiv e^{-i\psi_L(t,t')}\, \tilde{\Sigma}^{(L)\mathrm{R}}_{\mathrm{em}}(t-t')_{mn}. \end{aligned} \tag{33.60}$$

Here, we have defined

$$\psi_L(t,t') = \int_{t'}^{t} d\bar{t}\; V_L(\bar{t}), \tag{33.61}$$

which depends explicitly on the bias voltage, and

$$\begin{aligned}\tilde{\Sigma}^{(L)\mathrm{R}}_{\mathrm{em}}(t-t')_{mn} &= -i\,\theta(t-t') \sum_k T^{(L)}_{mk}\, e^{-i\varepsilon_{kL}(t-t')}\, T^{(L)\,*}_{nk} \\ &= \int_{-\infty}^{+\infty} \frac{d\omega}{2\pi}\, e^{-i\omega(t-t')}\, \tilde{\Sigma}^{(L)\mathrm{R}}_{\mathrm{em}}(\omega)_{mn}, \end{aligned} \tag{33.62}$$

where ($\eta = 0^+$)

$$\begin{aligned}\tilde{\Sigma}^{(L)\mathrm{R}}_{\mathrm{em}}(\omega)_{mn} &= -i \sum_k T^{(L)}_{mk} \int_0^{\infty} dt\, e^{i(\omega+i\eta-\varepsilon_{kL})\,t}\; T^{(L)\,*}_{nk} = \sum_k T^{(L)}_{mk}\, \frac{1}{\omega+i\eta-\varepsilon_{kL}}\, T^{(L)\,*}_{nk} \\ &= \int_{-\infty}^{+\infty} \frac{d\omega'}{2\pi}\, \frac{\tilde{\Gamma}^{(L)}_{\mathrm{em}}(\omega')_{mn}}{\omega-\omega'} - \frac{i}{2}\, \tilde{\Gamma}^{(L)}_{\mathrm{em}}(\omega)_{mn} \end{aligned} \tag{33.63}$$

with

$$\boxed{\tilde{\Gamma}^{(L)}_{\mathrm{em}}(\omega)_{mn} = 2\pi \sum_k T^{(L)}_{mk}\, \delta(\omega-\varepsilon_{kL})\, T^{(L)\,*}_{nk}}\,. \tag{33.64}$$

The two terms on the right-hand side of Eq. (33.63) correspond, respectively, to a *level shift* and a *level width*. According to the WBLA, one is allowed to neglect the ω-dependence of the expression (33.64), whereby $\tilde{\Gamma}^{(L)}_{\mathrm{em}}(\omega)_{mn} \to \tilde{\Gamma}^{(L)}_{\mathrm{em}}|_{mn}$ is a *constant*. Under these circumstances, the first term on the right-hand side of Eq. (33.63) vanishes. As a consequence, the expression (33.62) reduces to

$$\tilde{\Sigma}^{(L)\mathrm{R}}_{\mathrm{em}}(t-t')_{mn} \;\to\; -\frac{i}{2}\, \tilde{\Gamma}^{(L)}_{\mathrm{em}}|_{mn}\; \delta(t-t')\,, \tag{33.65}$$

such that the expression (33.60) in turn becomes

$$\Sigma_{\text{em}}^{(L)\text{R}}(t,t')_{mn} \to -\frac{i}{2}\,\tilde{\Gamma}_{\text{em}}^{(L)}|_{mn}\;\delta(t-t'), \tag{33.66}$$

which depends on $t - t'$ only and where memory effects are lost.

The efficiency of the WBLA compared to more sophisticated approaches has been explicitly emphasized in the context of transport in molecular junctions [202]. The WBLA will also be utilized in Chapter 34, in the context of superfluid fermions coupled with an environment.

34
Extension to Superfluid Fermi Systems

The effects of the coupling to the environment can manifest itself also in a superfluid Fermi system. In this chapter, we explicitly consider this case, by addressing the time-dependent behavior of the gap parameter following a sharp quench of the coupling parameter v_0 of the contact interaction. In this case, coupling the system to the environment is important for reaching equilibrium eventually. Several simplifying assumptions will be adopted for treating this problem in an as simple as possible way, following the lines of Ref. [193] where this problem was first examined.

34.1 A Number of Simplifying Assumptions

We consider the time evolution of a superfluid Fermi system, from an out-of-equilibrium regime which develops just after a quench of v_0, to an asymptotic thermal state, for reaching which the coupling to the environment proves important due to dissipation. To this end, the dynamics of the gap parameter has to be consistently determined in terms of the Schwinger–Keldysh Green's functions in the superfluid phase. As anticipated earlier, this problem will be dealt with in a minimal way by relying on the following *simplifying assumptions*:

(i) The superconducting dynamics is accounted for in terms of the s-wave mean-field decoupling as described in Chapter 23 of Part II.
(ii) The environment is assumed to be time translational invariant (i.e., no external potential is present), so that the phase factor $\psi_L(t, t')$ of Eq. (33.61) vanishes and the Green's functions (33.35)–(33.38) of the environment depend only on the time difference $t - t'$.
(iii) In this way, also the embedding self-energy (33.40), when specified to real times, depends only on the time difference $t - t'$.

(iv) A time-dependent protocol is established by a sudden quench of the interaction parameter, such that

$$v_0(t) = v_0^i\, \theta(-t) + v_0^f\, \theta(t) \tag{34.1}$$

with a jump $v_0^f - v_0^i$ at time $t = 0$ from the initial v_0^i to the final v_0^f values.

(v) The whole system (including the environment) is further assumed to be spatially homogeneous, such that the wave vector $\mathbf{k}$ is a good quantum number.

(vi) The WBLA that has led to Eq. (33.66) is also adopted, where now not only the frequency but also the wave-vector dependence is neglected.

In practice, to determine the time dependence of the (spatially homogeneous) gap parameter $\Delta(t)$, it is sufficient to determine the corresponding time dependence of the anomalous lesser Green's function at equal time

$$G^{<}(\mathbf{r}t, \mathbf{r}t)_{12} = \int \frac{d\mathbf{k}}{(2\pi)^3}\, G^{<}_{\mathbf{k}}(t)_{12} \tag{34.2}$$

according to Eq. (23.21) of Part II, where now we let $v_0 \to v_0(t)$ like in the expression (34.1), such that

$$\Delta(t) = -\, i\, v_0(t) \int \frac{d\mathbf{k}}{(2\pi)^3}\, G^{<}_{\mathbf{k}}(t)_{12}\,. \tag{34.3}$$

In the present context, it is convenient to determine the lesser Green's function $G^{<}(\mathbf{x}t, \mathbf{x}'t)$ (from which the required $G^{<}_{\mathbf{k}}(t)_{12}$ will eventually be obtained) by relying on the integral relation

$$\boxed{G^{<}(t, t') = \int_{-\infty}^{+\infty} dt_1 \int_{-\infty}^{+\infty} dt_2\, G^{\mathrm{R}}(t, t_1)\, \Sigma^{<}(t_1, t_2)\, G^{\mathrm{A}}(t_2, t'),} \tag{34.4}$$

where for simplicity the variables $\mathbf{x} = (\mathbf{r}, i)$ with Nambu index i, as well as the related integrals and summations, have not been explicitly indicated.

Proof of Eq. (34.4)

Apply the operator $\left[i\frac{\partial}{\partial t} - h(t)\right]$ to both sides of the expression (34.4) (where a matrix notation and the presence of the Pauli matrix τ_3 in $h(t)$ are implied, according to Eqs. (23.6)–(23.9) of Part II):

$$\left[i\frac{\partial}{\partial t} - h(t)\right] G^{<}(t, t') = \int_{-\infty}^{+\infty} dt_1 \int_{-\infty}^{+\infty} dt_2\, \overbrace{\left[i\frac{\partial}{\partial t} - h(t)\right] G^{\mathrm{R}}(t, t_1)}\, \Sigma^{<}(t_1, t_2)\, G^{\mathrm{A}}(t_2, t')$$

$$= \int_{-\infty}^{+\infty} dt_1 \int_{-\infty}^{+\infty} dt_2 \left\{\delta(t-t_1) + \int_{-\infty}^{+\infty} d\bar{t}\, \Sigma^{\mathrm{R}}(t,\bar{t})\, G^{\mathrm{R}}(\bar{t},t_1)\right\} \Sigma^{<}(t_1,t_2)\, G^{\mathrm{A}}(t_2,t')$$

$$= \int_{-\infty}^{+\infty} dt_2\, \Sigma^{<}(t,t_2)\, G^{\mathrm{A}}(t_2,t')$$

$$+ \int_{-\infty}^{+\infty} d\bar{t}\; \Sigma^{\mathrm{R}}(t,\bar{t}) \int_{-\infty}^{+\infty} dt_1 \int_{-\infty}^{+\infty} dt_2\; \overbrace{G^{\mathrm{R}}(\bar{t},t_1)\, \Sigma^{<}(t_1,t_2)\, G^{\mathrm{A}}(t_2,t')}$$

$$= \int_{-\infty}^{+\infty} dt_2\, \Sigma^{<}(t,t_2)\, G^{\mathrm{A}}(t_2,t') + \int_{-\infty}^{+\infty} d\bar{t}\; \Sigma^{\mathrm{R}}(t,\bar{t})\, G^{<}(\bar{t},t'), \tag{34.5}$$

where the equation of motion for $G^{\mathrm{R}}(t,t_1)$ has been utilized in the second line and the expression (34.4) has consistently been taken into account in the last line. The equation of motion (34.5) is thus seen to coincide with that of $G^{<}(t,t')$ (cf. Eq. (15.13) of Part I, once rephrased in Nambu space), *provided* one lets $t_0 \to -\infty$ therein and neglects the term containing $\Sigma^{\rceil} \star G^{\lceil}$. [QED]

Remark: About the initial conditions

Strictly speaking, in the expression (34.4) one could have kept the lower limits of the time integrals at a finite value t_0 and added the term (cf. Eq. (3.13) of Ref. [39])

$$\int d\mathbf{x}_2 \int d\mathbf{x}_3\, G^{\mathrm{R}}(\mathbf{x}t, \mathbf{x}_2 t_0)\, G^{<}(\mathbf{x}_2 t_0, \mathbf{x}_3 t_0^{+})\, G^{\mathrm{A}}(\mathbf{x}_3 t_0, \mathbf{x}'t'), \tag{34.6}$$

where we have restored the $\mathbf{x}$ variables for the present needs. This is because, when $t \to t_0^{+}$ and $t' \to t_0^{++}$, $G^{\mathrm{R}}(\mathbf{x}t_0^{+}, \mathbf{x}_2 t_0) = -i\delta(\mathbf{x},\mathbf{x}_2)$ and $G^{\mathrm{A}}(\mathbf{x}_3 t_0, \mathbf{x}'t_0^{++}) = i\delta(\mathbf{x}_3,\mathbf{x}')$ (cf. Eqs. (13.15) and (13.17) of Part I, respectively), such that in this limit the expression (34.6) accounts for the initial condition $G^{<}(\mathbf{x}t_0, \mathbf{x}'t_0^{+})$ for the lesser Green's function (provided this can somehow be determined in practice), while the expression (34.4) with $-\infty \to t_0$ in the time integrals would vanish in this limit. One can also readily verify that the presence of the additional term (34.6) does not modify the validity of the result (34.5).

In any case, in the present context it is appropriate to let $t_0 \to -\infty$ in the expression (34.4) and correspondingly to neglect the term (34.6), because for a dissipative system coupled to a heat bath the information on the initial state and the initial transient dynamics are expected to be essentially wiped out in a finite lapse of time. Accordingly, the Keldysh formalism is applicable to the equilibrium state established in the past with the initial value v_0^i of the coupling constant, sufficiently before the quench (34.1) is applied at $t = 0$.

34.2 The Lesser Green's Function and Its Ingredients

In the present problem, the single-particle self-energy contains contributions *both* from the mean-field (BCS) approximation Σ_{BCS} given by Eq. (23.10) of Part II (which is off-diagonal in the Nambu indices and contains only a singular retarded contribution) and from the embedding self-energy Σ_{em} given by Eq. (33.40) (which is diagonal in the Nambu indices since the tunneling Hamiltonian (33.16) does not admit spin flips). Accordingly, in the expression (34.4) we consider $\Sigma^{<}$ to contain only the embedding contribution Σ_{em}, while G^{R} and G^{A} contain the effects of both Σ_{BCS} and Σ_{em}. We thus need the explicit expressions of the lesser, retarded, and advanced components of Σ_{em} (diagonal in the Nambu indices). These can be obtained from the general expression (33.40), where $\alpha = 1$ because only a single reservoir is here considered.

Before doing that, however, it is convenient to frame the expressions (33.35)–(33.38) in the context of the Nambu spin indices, by taking into account the results (33.33) and (33.34). Upon recalling Eq. (22.1) of Part II, we obtain for the relevant cases to be utilized in Eq. (33.40):

$$g^{<}_{\mathbf{k}}(t,t')_{11} = i\,\langle c^{\dagger}_{\mathbf{k}\uparrow}(t')\, c_{\mathbf{k}\uparrow}(t)\rangle = i f_F(\varepsilon_{\mathbf{k}})\, e^{-i\varepsilon_{\mathbf{k}}(t-t')} \tag{34.7}$$

$$g^{<}_{\mathbf{k}}(t,t')_{22} = i\,\langle c_{\mathbf{k}\downarrow}(t')\, c^{\dagger}_{\mathbf{k}\downarrow}(t)\rangle = i\,(1-f_F(\varepsilon_{\mathbf{k}}))\; e^{i\varepsilon_{\mathbf{k}}(t-t')} = g^{<}_{\mathbf{k}}(t,t')_{11}\Big|_{\varepsilon_{\mathbf{k}}\to-\varepsilon_{\mathbf{k}}} \tag{34.8}$$

$$g^{>}_{\mathbf{k}}(t,t')_{11} = -\,i\,\langle c_{\mathbf{k}\uparrow}(t)\, c^{\dagger}_{\mathbf{k}\uparrow}(t')\rangle = -\,i\;(1-f_F(\varepsilon_{\mathbf{k}}))\;e^{-i\varepsilon_{\mathbf{k}}(t-t')} \tag{34.9}$$

$$g^{>}_{\mathbf{k}}(t,t')_{22} = -\,i\,\langle c^{\dagger}_{\mathbf{k}\downarrow}(t)\, c_{\mathbf{k}\downarrow}(t')\rangle = -\,i f_F(\varepsilon_{\mathbf{k}})\, e^{i\varepsilon_{\mathbf{k}}(t-t')} = g^{>}_{\mathbf{k}}(t,t')_{11}\Big|_{\varepsilon_{\mathbf{k}}\to-\varepsilon_{\mathbf{k}}} \tag{34.10}$$

since $f_F(-\varepsilon_{\mathbf{k}}) = 1 - f_F(\varepsilon_{\mathbf{k}})$, such that

$$g^{>}_{\mathbf{k}}(t,t')_{11} - g^{<}_{\mathbf{k}}(t,t')_{11} = -\,i\,e^{-i\varepsilon_{\mathbf{k}}(t-t')} \tag{34.11}$$

$$g^{>}_{\mathbf{k}}(t,t')_{22} - g^{<}_{\mathbf{k}}(t,t')_{22} = -\,i\,e^{i\varepsilon_{\mathbf{k}}(t-t')}\,. \tag{34.12}$$

The lesser component of Σ_{em} in Nambu space

Consider first element 11 of $\Sigma^{<}_{\text{em}}$. From Eqs. (33.40) and (34.7), we get

$$\Sigma^{<}_{\mathbf{k}}(t,t')_{11} = \sum_{\mathbf{k}'} T_{\mathbf{k}\mathbf{k}'}\, g^{<}_{\mathbf{k}'}(t,t')_{11}\, T^{*}_{\mathbf{k}\mathbf{k}'} = i\sum_{\mathbf{k}'} T_{\mathbf{k}\mathbf{k}'}\, f_F(\varepsilon_{\mathbf{k}'})\, e^{-i\varepsilon_{\mathbf{k}'}(t-t')}\, T^{*}_{\mathbf{k}\mathbf{k}'}, \tag{34.13}$$

where $\mathbf{k}$ refers to the system and $\mathbf{k}'$ the environment. Proceeding similar to the way in Section 33.4, we then obtain

$$\Sigma^{<}_{\mathbf{k}}(t,t')_{11} = \Sigma^{<}_{\mathbf{k}}(t-t')_{11} = \int_{-\infty}^{+\infty} \frac{d\omega}{2\pi}\, e^{-i\omega(t-t')}\, \Sigma^{<}_{\mathbf{k}}(\omega)_{11}, \tag{34.14}$$

where

$$\Sigma^<_{\mathbf{k}}(\omega)_{11} = \int_{-\infty}^{+\infty} dt\, e^{i\omega t}\, i \sum_{\mathbf{k}'} T_{\mathbf{k}\mathbf{k}'} f_F(\varepsilon_{\mathbf{k}'})\, e^{-i\varepsilon_{\mathbf{k}'} t}\, T^*_{\mathbf{k}\mathbf{k}'}$$
$$= 2\pi i \sum_{\mathbf{k}'} T_{\mathbf{k}\mathbf{k}'} f_F(\varepsilon_{\mathbf{k}'})\, \delta(\omega - \varepsilon_{\mathbf{k}'})\, T^*_{\mathbf{k}\mathbf{k}'} \,. \tag{34.15}$$

When substituted in Eq. (34.14), this result gives

$$\Sigma^<_{\mathbf{k}}(t-t')_{11} = \int_{-\infty}^{+\infty} \frac{d\omega}{2\pi}\, e^{-i\omega(t-t')}\, 2\pi i \sum_{\mathbf{k}'} T_{\mathbf{k}\mathbf{k}'} f_F(\varepsilon_{\mathbf{k}'})\, \delta(\omega - \varepsilon_{\mathbf{k}'})\, T^*_{\mathbf{k}\mathbf{k}'}$$
$$= i \int_{-\infty}^{+\infty} \frac{d\omega}{2\pi}\, e^{-i\omega(t-t')}\, f_F(\omega)\, \overbrace{2\pi \sum_{\mathbf{k}'} T_{\mathbf{k}\mathbf{k}'}\, \delta(\omega - \varepsilon_{\mathbf{k}'})\, T^*_{\mathbf{k}\mathbf{k}'}}$$
$$\underset{[\text{Eq.(33.64)}]}{\overset{\uparrow}{=}} i \int_{-\infty}^{+\infty} \frac{d\omega}{2\pi}\, e^{-i\omega(t-t')} f_F(\omega)\, \tilde{\Gamma}_{\mathbf{k}}(\omega) \underset{[\text{WBLA}]}{\overset{\uparrow}{\simeq}} i\gamma \int_{-\infty}^{+\infty} \frac{d\omega}{2\pi}\, e^{-i\omega(t-t'+i\delta)} f_F(\omega)$$
$$\underset{[T=0 \text{ and } \mu=0]}{\overset{\uparrow}{=}} i\gamma \int_{-\infty}^{0} \frac{d\omega}{2\pi}\, e^{-i\omega(t-t'+i\delta)} = -\frac{\gamma}{2\pi} \frac{1}{t-t'+i\delta}\,, \tag{34.16}$$

where $\delta = 0^+$ is needed to regularize the frequency integral after applying the WBLA replacement $\tilde{\Gamma}_{\mathbf{k}}(\omega) \to \gamma$, thereby neglecting not only the frequency but also the wave vector dependence of $\tilde{\Gamma}_{\mathbf{k}}(\omega)$.

In addition, the expression for $\Sigma^<_{\mathbf{k}}(t-t')_{22}$ can be readily obtained by taking into account the result (34.8), thereby letting $\varepsilon_{\mathbf{k}'} \to -\varepsilon_{\mathbf{k}'}$ in the first line of the expression (34.16) for $\Sigma^<_{\mathbf{k}}(t-t')_{11}$. Accordingly, we write

$$\Sigma^<_{\mathbf{k}}(t-t')_{22} = \int_{-\infty}^{+\infty} \frac{d\omega}{2\pi}\, e^{-i\omega(t-t')}\, 2\pi i \sum_{\mathbf{k}'} T_{\mathbf{k}\mathbf{k}'} f_F(-\varepsilon_{\mathbf{k}'})\, \delta(\omega + \varepsilon_{\mathbf{k}'})\, T^*_{\mathbf{k}\mathbf{k}'}$$
$$= i \int_{-\infty}^{+\infty} \frac{d\omega}{2\pi}\, e^{-i\omega(t-t')}\, f_F(\omega)\, \overbrace{2\pi \sum_{\mathbf{k}'} T_{\mathbf{k}\mathbf{k}'}\, \delta(\omega + \varepsilon_{\mathbf{k}'})\, T^*_{\mathbf{k}\mathbf{k}'}}$$
$$\underset{[\text{WBLA}]}{\overset{\uparrow}{\simeq}} i\gamma \int_{-\infty}^{+\infty} \frac{d\omega}{2\pi}\, e^{-i\omega(t-t'+i\delta)} f_F(\omega) = -\frac{\gamma}{2\pi} \frac{1}{t-t'+i\delta} = \Sigma^<_{\mathbf{k}}(t-t')_{11} \,. \tag{34.17}$$

The retarded component of Σ_{em} in Nambu space

From Eqs. (33.40) and (34.11), we get

$$
\begin{aligned}
\Sigma^{R}_{\mathbf{k}}(t,t')_{11} &= \sum_{\mathbf{k}'} T_{\mathbf{k}\mathbf{k}'}\, g^{R}_{\mathbf{k}'}(t,t')_{11}\, T^{*}_{\mathbf{k}\mathbf{k}'} = -\,i\,\theta(t-t') \sum_{\mathbf{k}'} T_{\mathbf{k}\mathbf{k}'}\, e^{-i\varepsilon_{\mathbf{k}'}(t-t')}\, T^{*}_{\mathbf{k}\mathbf{k}'} \\
&\underset{[\text{Eq.}(33.63)]}{\underset{\uparrow}{\simeq}} \int_{-\infty}^{+\infty} \frac{d\omega}{2\pi}\, e^{-i\omega(t-t')} \sum_{\mathbf{k}'} T_{\mathbf{k}\mathbf{k}'}\, \frac{1}{\omega + i\eta - \varepsilon_{\mathbf{k}'}}\, T^{*}_{\mathbf{k}\mathbf{k}'} \underset{[\text{WBLA}]}{\underset{\uparrow}{\simeq}} -\,i\,\frac{\gamma}{2}\,\delta(t-t')
\end{aligned}
\tag{34.18}
$$

by retracing what was done to arrive at Eq. (33.66) within the WBLA.

In addition, from the result (34.12) we obtain for element 22:

$$
\begin{aligned}
\Sigma^{R}_{\mathbf{k}}(t,t')_{22} &= \sum_{\mathbf{k}'} T_{\mathbf{k}\mathbf{k}'}\, g^{R}_{\mathbf{k}'}(t,t')_{22}\, T^{*}_{\mathbf{k}\mathbf{k}'} = -\,i\,\theta(t-t') \sum_{\mathbf{k}'} T_{\mathbf{k}\mathbf{k}'}\, e^{i\varepsilon_{\mathbf{k}'}(t-t')}\, T^{*}_{\mathbf{k}\mathbf{k}'} \\
&\underset{[\text{Eq.}(33.63)]}{\underset{\uparrow}{\simeq}} \int_{-\infty}^{+\infty} \frac{d\omega}{2\pi}\, e^{-i\omega(t-t')} \sum_{\mathbf{k}'} T_{\mathbf{k}\mathbf{k}'}\, \frac{1}{\omega + i\eta + \varepsilon_{\mathbf{k}'}}\, T^{*}_{\mathbf{k}\mathbf{k}'} \underset{[\text{WBLA}]}{\underset{\uparrow}{\simeq}} -\,i\,\frac{\gamma}{2}\,\delta(t-t')\,.
\end{aligned}
\tag{34.19}
$$

The advanced component of Σ_{em} in Nambu space

Both elements 11 and 22 of the advanced component of Σ_{em} can be readily obtained from the corresponding elements (34.18) and (34.19) of the retarded component by letting $i\eta \to -i\eta$ in those expressions, namely:

$$
\Sigma^{A}_{\mathbf{k}}(t,t')_{11} = \int_{-\infty}^{+\infty} \frac{d\omega}{2\pi}\, e^{-i\omega(t-t')} \sum_{\mathbf{k}'} T_{\mathbf{k}\mathbf{k}'}\, \frac{1}{\omega - i\eta - \varepsilon_{\mathbf{k}'}}\, T^{*}_{\mathbf{k}\mathbf{k}'} \underset{[\text{WBLA}]}{\underset{\uparrow}{\simeq}} i\,\frac{\gamma}{2}\,\delta(t-t') \tag{34.20}
$$

$$
\Sigma^{A}_{\mathbf{k}}(t,t')_{22} = \int_{-\infty}^{+\infty} \frac{d\omega}{2\pi}\, e^{-i\omega(t-t')} \sum_{\mathbf{k}'} T_{\mathbf{k}\mathbf{k}'}\, \frac{1}{\omega - i\eta + \varepsilon_{\mathbf{k}'}}\, T^{*}_{\mathbf{k}\mathbf{k}'} \underset{[\text{WBLA}]}{\underset{\uparrow}{\simeq}} i\,\frac{\gamma}{2}\,\delta(t-t')\,. \tag{34.21}
$$

The interacting retarded and advanced Green's functions in Nambu space

The explicit expressions of the retarded and advanced Green's functions are to be inserted in Eq. (34.4). To this end, we consider the differential Dyson equation for the retarded Green's function in Nambu representation, which contains both the off-diagonal mean-field self-energy (23.10) of Part II and the diagonal embedding self-energy (34.18)-(34.19) within the WBLA. With the notation of Eq. (23.9) of Part II, we then write for the equation satisfied by the 2×2 matrix $\overleftrightarrow{G}^{R}(\mathbf{r}t, \mathbf{r}'t')$:

$$
\begin{aligned}
&\begin{pmatrix} i\frac{\partial}{\partial t} - h(\mathbf{r},t) + \mu + i\frac{\gamma}{2} & -\Delta(\mathbf{r},t) \\ -\Delta(\mathbf{r},t)^{*} & i\frac{\partial}{\partial t} + h(\mathbf{r},t) - \mu + i\frac{\gamma}{2} \end{pmatrix} \overleftrightarrow{G}^{R}(\mathbf{r}t, \mathbf{r}'t') \\
&= \delta(\mathbf{r}-\mathbf{r}')\,\delta(t-t') \begin{pmatrix} 1 & 0 \\ 0 & 1 \end{pmatrix},
\end{aligned}
\tag{34.22}
$$

where we have kept the spatial dependence in the gap parameter as well as the time dependence in the single-particle Hamiltonian $h(\mathbf{r}, t)$ (from which we have also subtracted the thermodynamic chemical potential μ, consistently with the considerations made in Section 25.1 of Part II). The dependence of $\overleftrightarrow{G}^{\mathrm{R}}(\mathbf{r}t, \mathbf{r}'t')$ on γ can be readily disentangled from Eq. (34.22), by setting

$$\boxed{\overleftrightarrow{G}^{\mathrm{R}}(\mathbf{r}t, \mathbf{r}'t') = \overleftrightarrow{G}^{\mathrm{R}}_{\mathrm{mf}}(\mathbf{r}t, \mathbf{r}'t')\, e^{-\frac{\gamma}{2}(t-t')},} \tag{34.23}$$

where $\overleftrightarrow{G}^{\mathrm{R}}_{\mathrm{mf}}(\mathbf{r}t, \mathbf{r}'t')$ is the mean-field (mf) counterpart. This is because

$$i\frac{\partial}{\partial t}\overleftrightarrow{G}^{\mathrm{R}}(\mathbf{r}t, \mathbf{r}'t') = e^{-\frac{\gamma}{2}(t-t')}\left(i\frac{\partial}{\partial t} - i\frac{\gamma}{2}\right)\overleftrightarrow{G}^{\mathrm{R}}_{\mathrm{mf}}(\mathbf{r}t, \mathbf{r}'t'), \tag{34.24}$$

such that Eq. (34.22) becomes

$$\begin{pmatrix} i\frac{\partial}{\partial t} - h(\mathbf{r}, t) + \mu & -\Delta(\mathbf{r}, t) \\ -\Delta(\mathbf{r}, t)^* & i\frac{\partial}{\partial t} + h(\mathbf{r}, t) - \mu \end{pmatrix}\overleftrightarrow{G}^{\mathrm{R}}_{\mathrm{mf}}(\mathbf{r}t, \mathbf{r}'t') = \delta(\mathbf{r} - \mathbf{r}')\,\delta(t - t')\begin{pmatrix} 1 & 0 \\ 0 & 1 \end{pmatrix}, \tag{34.25}$$

which is the differential Dyson equation for the mean-field retarded Green's function $\overleftrightarrow{G}^{\mathrm{R}}_{\mathrm{mf}}(\mathbf{r}t, \mathbf{r}'t')$. Note that the expression (34.23) gives also the correct dependence on t', since

$$-i\frac{\partial}{\partial t'}\overleftrightarrow{G}^{\mathrm{R}}(\mathbf{r}t, \mathbf{r}'t') = e^{-\frac{\gamma}{2}(t-t')}\left(-i\frac{\partial}{\partial t'} - i\frac{\gamma}{2}\right)\overleftrightarrow{G}^{\mathrm{R}}_{\mathrm{mf}}(\mathbf{r}t, \mathbf{r}'t'). \tag{34.26}$$

In addition, owing to the sign difference between the expressions (34.18) and (34.19) on the one hand and (34.20) and (34.21) on the other hand (whereby $\gamma \to -\gamma$), the corresponding disentangling for the advanced Green's function reads

$$\boxed{\overleftrightarrow{G}^{\mathrm{A}}(\mathbf{r}t, \mathbf{r}'t') = \overleftrightarrow{G}^{\mathrm{A}}_{\mathrm{mf}}(\mathbf{r}t, \mathbf{r}'t')\, e^{\frac{\gamma}{2}(t-t')}.} \tag{34.27}$$

Finally, with the results (34.16) and (34.17) for the lesser embedding self-energy, (34.23) for the retarded Green's function, and (34.27) for the advanced Green's function, the expression (34.4) for the lesser Green's function, once implemented for a spatially homogeneous system, becomes:

$$\begin{aligned}\overleftrightarrow{G}^{<}_{\mathbf{k}}(t, t') &= \int_{-\infty}^{+\infty} dt_1 \int_{-\infty}^{+\infty} dt_2\, \overleftrightarrow{G}^{\mathrm{R}}_{\mathrm{mf},\mathbf{k}}(t, t_1)\, e^{-\frac{\gamma}{2}(t-t_1)} \\ &\quad \times \left(-\frac{\gamma}{2\pi}\frac{1}{t_1 - t_2 + i\delta}\right)\begin{pmatrix} 1 & 0 \\ 0 & 1 \end{pmatrix}\overleftrightarrow{G}^{\mathrm{A}}_{\mathrm{mf},\mathbf{k}}(t_2, t')\, e^{\frac{\gamma}{2}(t_2-t')} \\ &= -\frac{\gamma}{2\pi}e^{-\frac{\gamma}{2}(t+t')}\int_{-\infty}^{+\infty} dt_1 \int_{-\infty}^{+\infty} dt_2\, \overleftrightarrow{G}^{\mathrm{R}}_{\mathrm{mf},\mathbf{k}}(t, t_1)\frac{e^{\frac{\gamma}{2}(t_1+t_2)}}{t_1 - t_2 + i\delta}\overleftrightarrow{G}^{\mathrm{A}}_{\mathrm{mf},\mathbf{k}}(t_2, t').\end{aligned} \tag{34.28}$$

34.3 An Equation of Motion with a Memory Kernel

At this point, a nontrivial equation of motion can be obtained for $\overleftrightarrow{G}^{<}_{\mathbf{k}}(t,t)$, whose element 12 is needed to get the time dependence of the gap parameter from Eq. (34.3), by setting $t=t'$ in the expression (34.28) and then applying the operator $\begin{pmatrix} i\frac{\partial}{\partial t} & 0 \\ 0 & i\frac{\partial}{\partial t} \end{pmatrix}$ to both sides of the resulting expression. By noting that $\begin{pmatrix} i\frac{\partial}{\partial t} & 0 \\ 0 & i\frac{\partial}{\partial t} \end{pmatrix} \overleftrightarrow{G}^{<}_{\mathbf{k}}(t,t) = i\frac{\partial \overleftrightarrow{G}^{<}_{\mathbf{k}}(t,t)}{\partial t}$, and recalling the equation of motion (34.25) for the retarded mean-field Green's function $\overleftrightarrow{G}^{\mathrm{R}}_{\mathrm{mf}}(\mathbf{r}t,\mathbf{r}_1 t_1)$ (plus the corresponding equation for the advanced mean-field Green's function $\overleftrightarrow{G}^{\mathrm{A}}_{\mathrm{mf},\mathbf{k}}(\mathbf{r}_2 t_2,\mathbf{r}' t)$), we obtain eventually:

$$
\begin{aligned}
i\frac{\partial}{\partial t}\overleftrightarrow{G}^{<}_{\mathbf{k}}(t,t) + i\gamma\,\overleftrightarrow{G}^{<}_{\mathbf{k}}(t,t) &= -\frac{\gamma e^{-\gamma t}}{2\pi}\int_{-\infty}^{+\infty} dt_1 \int_{-\infty}^{+\infty} dt_2 \frac{e^{\frac{\gamma}{2}(t_1+t_2)}}{t_1-t_2+i\delta} \\
&\times \left\{ i\frac{\partial \overleftrightarrow{G}^{\mathrm{R}}_{\mathrm{mf},\mathbf{k}}(t,t_1)}{\partial t}\,\overleftrightarrow{G}^{\mathrm{A}}_{\mathrm{mf},\mathbf{k}}(t_2,t) + \overleftrightarrow{G}^{\mathrm{R}}_{\mathrm{mf},\mathbf{k}}(t,t_1)\, i\frac{\partial \overleftrightarrow{G}^{\mathrm{A}}_{\mathrm{mf},\mathbf{k}}(t_2,t)}{\partial t} \right\} \\
&= -\frac{\gamma e^{-\gamma t}}{2\pi}\int_{-\infty}^{+\infty} dt_1 \int_{-\infty}^{+\infty} dt_2 \frac{e^{\frac{\gamma}{2}(t_1+t_2)}}{t_1-t_2+i\delta} \\
&\times \left\{ \begin{pmatrix} \xi_{\mathbf{k}} & \Delta(t) \\ \Delta(t)^* & -\xi_{\mathbf{k}} \end{pmatrix} \overleftrightarrow{G}^{\mathrm{R}}_{\mathrm{mf},\mathbf{k}}(t,t_1)\,\overleftrightarrow{G}^{\mathrm{A}}_{\mathrm{mf},\mathbf{k}}(t_2,t) + \delta(t-t_1)\,\overleftrightarrow{G}^{\mathrm{A}}_{\mathrm{mf},\mathbf{k}}(t_2,t) \right. \\
&\left. - \overleftrightarrow{G}^{\mathrm{R}}_{\mathrm{mf},\mathbf{k}}(t,t_1)\,\overleftrightarrow{G}^{\mathrm{A}}_{\mathrm{mf},\mathbf{k}}(t_2,t) \begin{pmatrix} \xi_{\mathbf{k}} & \Delta(t) \\ \Delta(t)^* & -\xi_{\mathbf{k}} \end{pmatrix} - \delta(t_2-t)\,\overleftrightarrow{G}^{\mathrm{R}}_{\mathrm{mf},\mathbf{k}}(t,t_1) \right\},
\end{aligned}
\tag{34.29}
$$

where the equation of motion of $\overleftrightarrow{G}^{\mathrm{A}}_{\mathrm{mf},\mathbf{k}}(t',t)$ has been obtained from that of $\overleftrightarrow{G}^{\mathrm{R}}_{\mathrm{mf},\mathbf{k}}(t,t')$ by exploiting the symmetry property $\overleftrightarrow{G}^{\mathrm{A}}_{\mathrm{mf},\mathbf{k}}(t',t) = \overleftrightarrow{G}^{\mathrm{R}}_{\mathrm{mf},\mathbf{k}}(t,t')^{\dagger}$ (cf. Eq. (13.25) of Part I), with the Hermitian conjugate here applying only to the matrices in Nambu indices.

Equation (34.29) can, finally, be rewritten in the compact form:

$$
\boxed{
\begin{aligned}
i\frac{\partial}{\partial t}\overleftrightarrow{G}^{<}_{\mathbf{k}}(t,t) &= -i\gamma\,\overleftrightarrow{G}^{<}_{\mathbf{k}}(t,t) + \left[\begin{pmatrix} \xi_{\mathbf{k}} & \Delta(t) \\ \Delta(t)^* & -\xi_{\mathbf{k}} \end{pmatrix}, \overleftrightarrow{G}^{<}_{\mathbf{k}}(t,t) \right] \\
&- \frac{\gamma}{2\pi}\int_{-\infty}^{+\infty} d\bar{t}\, e^{-\frac{\gamma}{2}(t-\bar{t})} \left(\frac{\overleftrightarrow{G}^{\mathrm{R}}_{\mathrm{mf},\mathbf{k}}(t,\bar{t})}{t-\bar{t}-i\delta} + \frac{\overleftrightarrow{G}^{\mathrm{A}}_{\mathrm{mf},\mathbf{k}}(\bar{t},t)}{t-\bar{t}+i\delta} \right)
\end{aligned}
}
\tag{34.30}
$$

where the integration variable $\bar{t}$ in the last term extends up to t. This equation has to be solved numerically in a self-consistent way together with the definition (34.3)

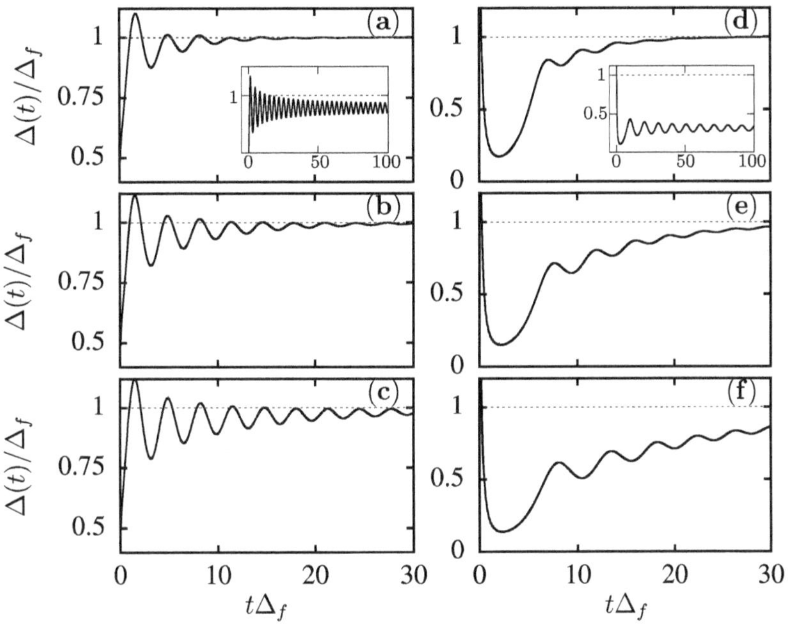

Figure 34.1 The time dependence of $\Delta(t)$ is shown for two values of Δ_0/Δ_f (0.4 for panels (a)–(c) and 4.0 for panels (d)–(f)) and three values of γ/Δ_f $(0.2, 0.1,$ and 0.05, from top to bottom). [Reproduced from Fig. 1 of Ref. [193], with permission.]

for $\Delta(t)$ and the equation of motion for $G^{\mathrm{R}}_{\mathrm{mf},\mathbf{k}}(t, t')$, which was utilized to get the right-hand side of Eq. (34.29). Note that the equation of motion (34.30) contains a *memory kernel*, which originates from the exponential factor of the retarded (34.23) and advanced (34.27) Green's functions. This kernel vanishes when $\gamma \to 0$.

Equations (34.25) and (34.30) were solved numerically in Ref. [193], where the quantum quench was parametrized not by the change of the interaction constant like in Eq. (34.1), but rather by the ratio Δ_0/Δ_f, where Δ_0 and Δ_f are the equilibrium values of the gap parameter corresponding to v_0^i and v_0^f in the expression (34.1), respectively. Numerical results for the ensuing time dependence of $\Delta(t)$ are shown in Fig. 34.1. Note here that larger values of the quantum quench make it harder for the system to reach the equilibrium value, as well as the presence of persistent oscillations that are more efficiently damped by the coupling to the bath for increasing γ.

Finally, it should be mentioned that the physical problem considered in this chapter is somewhat intermediate between what can be experimentally realized in condensed matter and in cold atom physics. This is because in a condensed-matter sample, it is hard to modify in a time-dependent way the interparticle coupling which is responsible for the superconducting behavior, as it was assumed in Eq. (34.1), while the sample is connected in some way with an external bath. In this respect, in the pump-probe experiments that explore metastable photo-induced superconductivity as mentioned in Sections 32.4 and 32.5, the time-dependent pump acts on the sample as a single-particle external potential and does not affect the interparticle coupling directly. In ultra-cold Fermi gases, on the other hand, the interparticle coupling (as represented by the dimensionless parameter $(k_F a_F)^{-1}$ mentioned in Chapter 26 of Part II) can be experimentally varied in a time-dependent way [24, 25], but the gas cloud itself is almost ideally isolated from the external world with no system-bath coupling. In addition, for an ultra-cold Fermi gas, treating the effects of the attractive interparticle interaction by the mean-field (BCS) decoupling considered in Eq. (34.22) is known not to be appropriate, especially when approaching the unitary regime [20]. This beyond-mean-field scenario may as well apply to nonconventional superconductors [102], provided that what corresponds to the unitary regime in these systems would be judiciously identified [103]. Work on time-dependent many-body problems along these lines is awaited at the time of writing.

35

Connection between the Schwinger–Keldysh Closed-Contour Approach and the Lindblad Master Equation

As anticipated in Section 32.7, there is an increasing interest in the literature in the Lindblad Master equation (LME) for the many-body density matrix, which describes the transient dynamics in *dissipative* systems subject to fast driving fields. The relevance of the LME also stems from the fact that it preserves the trace and positivity of the many-body density matrix [185, 186]. In this chapter, we shall explore to what extent the closed-contour Schwinger–Keldysh approach and the LME are connected. To this end, as a general framework we shall explicitly consider (and suitably expand on) the approach of Ref. [195], where the connection with the Schwinger–Keldysh closed-contour approach does not involve the full machinery of the Green's functions method. Rather, it refers directly to the time evolution of the many-body density matrix, since this contains a *forward* evolution operator $U(t, t_0)$ from the reference time t_0 to the measuring time t and a *backward* evolution operator $U(t_0, t)$ from t back to t_0. In this respect, it is worth mentioning the more recent contribution by Ref. [203], where the original Schwinger–Keldysh idea is reformulated to accommodate for Lindbladian time evolution, extending in this way the Kadanoff–Baym equations and generalizing the diagrammatic perturbation theory for many-body Lindblad operators.

The above general methods, based on either the Master Equations for the many-body density matrix or the Kadanoff–Baym equations for the Green's functions, are alternative (or complementary, depending on the point of view) to each other, in the sense that they put *different emphasis* on the operators on which the time-dependent Hamiltonian $\mathcal{H}(t)$ of Eq. (2.1) of Part I acts. Correspondingly, the expression (4.14) of Part I for the time-dependent average of an operator $A(t)$ (possibly, with its own parametric time dependence) can be cast in either one of the following forms:

$$\langle A(t)\rangle \underset{\substack{\uparrow\\ \text{[either]}}}{=} \mathrm{Tr}\{\rho(t_0)\, A_{\mathrm{H}}(t)\} = \mathrm{Tr}\{\rho(t_0)\, U(t_0,t)A(t)U(t,t_0)\}$$

$$= \mathrm{Tr}\{U(t,t_0)\rho(t_0)\, U(t_0,t)A(t)\} \tag{35.1}$$

$$\underset{\substack{\uparrow\\ \text{[or]}}}{=} \mathrm{Tr}\{\rho(t)\, A(t)\}\,, \tag{35.2}$$

where the cyclic property of the trace has been used and $\rho(t) = U(t,t_0)\rho(t_0)\, U(t_0,t)$ stands for the time development of the density matrix starting from its initial configuration $\rho = \rho(t_0)$. In the Green's functions method, it is the operator $A_{\mathrm{H}}(t)$ (and eventually the field operators) to carry the time dependence, while in the Master Equations method, it is the density matrix $\rho(t)$ to do the job.

Note that in the expression (35.1), we have kept disjoint the forward and backward time evolution, thereby avoiding to fold them into a single closed contour as we did in Eq. (6.6) of Part I in terms of the contour time-ordering operator, which is instead essential to the Schwinger–Keldysh approach.

Remark: One time argument in the density matrix versus two time arguments in the Green's functions

The transferring to the density matrix $\rho(t)$ of the time argument t associated with the evolution operator $U(t,t_0)$, like in Eq. (35.2), is possible because the operator $A_{\mathrm{H}}(t)$ therein contains only *one* time argument. Correspondingly, the single-particle Green's function given by Eq. (7.12) of Part I with *two* time arguments cannot be dealt with in this way. However, physical quantities of interest, such as the time-dependent particle density (cf. Eq. (18.3) of Part I) and the time-dependent current density (cf. Eq. (18.4) of Part I), depend on one time argument only such that they require knowledge of the lesser single-particle Green's function $G^{<}(t,t^{+})$ at time t. This single-time dependence could as well be dealt with like in Eq. (35.2). In this respect, a second time variable is introduced in the correlation function by the time dependence of an external field in the linear regime, like in the retarded density–density correlation function given by Eq. (18.10) of Part I. Accordingly, when dealing explicitly with correlation functions within linear response theory, it is more convenient to attribute the time dependence to the field operators and to regard the density matrix as independent of time [2].

To set up the generic framework we are leading to, a schematic derivation of the LME will be provided in Section 37.1, where it will be organized through clearly identified steps. Here, we shall initially consider a more elaborate, and somewhat more general, approach to the problem, which will then be specified and implemented for a two-level system coupled to a phonon bath, yielding in this way a simplified version of the LME. For the sake of clearness, reference to the subsequent steps identified in Section 37.1 will purposely be given along the way.

35.1 A Few Useful Preliminaries

Let A_S be a generic operator acting on the open (sub)system S which interacts with a reservoir (or bath) B and ρ the density matrix for the total system (including the reservoir or bath). In this case, the following identity holds

$$\langle A_S \rangle = \mathrm{Tr}\{\rho\, A_S\} = \mathrm{Tr}_S\{\rho_S\, A_S\}\,, \tag{35.3}$$

where

$$\rho_S = \mathrm{Tr}_B\{\rho\} \tag{35.4}$$

is the *reduced density matrix* of the open subsystem S. In Eq. (35.3), Tr_S denotes the partial trace over the Hilbert space of the open subsystem S, while in Eq. (35.4), Tr_B denotes the partial trace over the Hilbert space of the reservoir (or bath) B. The reduced density matrix ρ_S is the quantity of central interest in the description of open quantum systems. In addition, according to Eq. (35.4), the time development of ρ_S is related to the time development of the total density matrix ρ as follows:

$$\rho_S(t) = \mathrm{Tr}_B\{\rho(t)\} = \mathrm{Tr}_B\{U(t,t_0)\,\rho(t_0)\,U(t_0,t)\}\,, \tag{35.5}$$

where Eqs. (35.1) and (35.2) have been taken into account.

Proof of Eqs. (35.3) and (35.4)

Let $\{|m_S\rangle\}$ and $\{|m_B\rangle\}$ be complete sets of states for the subsystem S and the reservoir (or bath) B, respectively, and $\{w_{m_S m_B}\}$ the weights associated with the product states $\{|m_S\rangle|m_B\rangle\}$ in the total density matrix ρ, such that

$$\begin{aligned}
\langle A_S\rangle &= \mathrm{Tr}\,\{\rho A_S\} = \sum_{m_S m_B} \langle m_S|\langle m_B|\rho\, A_S|m_S\rangle|m_B\rangle \\
&= \sum_{m_S m_B} \langle m_S|\langle m_B| \overbrace{\sum_{n_S n_B} |n_S\rangle|n_B\rangle w_{n_S n_B}\langle n_S|\langle n_B|}^{=\rho}\, A_S|m_S\rangle|m_B\rangle \\
&= \sum_{n_S n_B} \langle n_S|\langle n_B|A_S|n_S\rangle|n_B\rangle\, w_{n_S n_B} = \sum_{n_S}\langle n_S|A_S|n_S\rangle \sum_{n_B} w_{n_S n_B},
\end{aligned} \tag{35.6}$$

where the orthonormality property of the states $\{|m_S\rangle\}$ and $\{|m_B\rangle\}$ has been utilized. This expression can be further manipulated as follows:

$$\begin{aligned}
\langle A_S\rangle &= \sum_{n_S}\langle n_S|A_S|n_S\rangle \sum_{n_B} w_{n_S n_B} = \sum_{n_S}\langle n_S|A_S\; \overbrace{\sum_{m_S} |m_S\rangle \sum_{n_B} w_{m_S n_B}\langle m_S|}^{=\rho_S}\;|n_S\rangle \\
&= \sum_{n_S}\langle n_S|A_S\,\rho_S|n_S\rangle = \mathrm{Tr}_S\{A_S\,\rho_S\}\,,
\end{aligned} \tag{35.7}$$

which coincides with Eq. (35.3), where we have introduced the notation:

$$\rho_\mathrm{S} = \sum_{m_\mathrm{S}} |m_\mathrm{S}\rangle \sum_{n_\mathrm{B}} w_{m_\mathrm{S} n_\mathrm{B}} \langle m_\mathrm{S}| = \sum_{n_\mathrm{B}} \langle n_\mathrm{B}| \overbrace{\sum_{m_\mathrm{S} m_\mathrm{B}} |m_\mathrm{S}\rangle |m_\mathrm{B}\rangle w_{m_\mathrm{S} m_\mathrm{B}} \langle m_\mathrm{S}| \langle m_\mathrm{B}|}^{=\rho} |n_\mathrm{B}\rangle$$
$$= \sum_{n_\mathrm{B}} \langle n_\mathrm{B}|\rho|n_\mathrm{B}\rangle = \mathrm{Tr}_\mathrm{B}\{\rho\} \,, \tag{35.8}$$

which coincides with Eq. (35.4). [QED]
[The results (35.3) and (35.4) will also be utilized in Step # 2 of Section 37.1].

In the above expressions, $\{|m_\mathrm{S}\rangle\}$ and $\{|m_\mathrm{B}\rangle\}$ are not necessarily eigenstates of the Hamiltonians H_S of the subsystem S and H_B of the reservoir (bath) B, respectively. We now introduce an additional complete set of states $\{|\bar{k}\rangle\}$ of the subsystem S, which are eigenstates of the Hamiltonian H_S of the subsystem S and do not necessarily coincide with the states $\{|m_\mathrm{S}\rangle\}$ of the previous set. Out of this new set $\{|\bar{k}\rangle\}$, we then pick up two states $|k\rangle$ and $|k'\rangle$, with which we form the operator $|k\rangle\langle k'|$ that acts on the Hilbert space of the subsystem S. We then associate $|k\rangle\langle k'|$ with the operator A_S entering the expressions (35.3) and (35.4) and write accordingly in the place of Eqs. (35.7) and (35.8):

$$\rho_\mathrm{S}^{k'k} \equiv \mathrm{Tr}\{\rho |k\rangle\langle k'|\} = \sum_{m_\mathrm{S} m_\mathrm{B}} \langle m_\mathrm{S}|\langle m_\mathrm{B}|\rho\, |k\rangle\langle k'|m_\mathrm{S}\rangle|m_\mathrm{B}\rangle$$
$$= \sum_{m_\mathrm{S} m_\mathrm{B}} \langle m_\mathrm{S}|\langle m_\mathrm{B}| \overbrace{\sum_{n_\mathrm{S} n_\mathrm{B}} |n_\mathrm{S}\rangle|n_\mathrm{B}\rangle w_{n_\mathrm{S} n_\mathrm{B}} \langle n_\mathrm{S}|\langle n_\mathrm{B}|}^{=\rho}\, |k\rangle\langle k'|m_\mathrm{S}\rangle|m_\mathrm{B}\rangle$$
$$= \sum_{n_\mathrm{S}} \langle n_\mathrm{S}|k\rangle\langle k'|n_\mathrm{S}\rangle \sum_{n_\mathrm{B}} w_{n_\mathrm{S} n_\mathrm{B}} \underset{\substack{\uparrow \\ [\mathrm{Eq.}(35.8)]}}{=} \langle k'|\rho_\mathrm{S}|k\rangle \tag{35.9}$$

as well as

$$\rho_\mathrm{S}^{k'k} = \sum_{n_\mathrm{S}} \langle n_\mathrm{S}|k\rangle\langle k'|n_\mathrm{S}\rangle \sum_{n_\mathrm{B}} w_{n_\mathrm{S} n_\mathrm{B}} = \sum_{n_\mathrm{S}} \langle n_\mathrm{S}|k\rangle\langle k'| \overbrace{\sum_{m_\mathrm{S}} |m_\mathrm{S}\rangle \sum_{n_\mathrm{B}} w_{m_\mathrm{S} n_\mathrm{B}} \langle m_\mathrm{S}|}^{=\rho_\mathrm{S}}\, |n_\mathrm{S}\rangle$$
$$= \mathrm{Tr}_\mathrm{S}\{|k\rangle\langle k'|\rho_\mathrm{S}\} \,. \tag{35.10}$$

These results, in turn, imply that

$$\rho_\mathrm{S} = \sum_{k'k} |k'\rangle\langle k'|\rho_\mathrm{S}|k\rangle\langle k| = \sum_{k'k} |k'\rangle \rho_\mathrm{S}^{k'k} \langle k| \,, \tag{35.11}$$

from which we conclude that knowledge of the matrix elements $\rho_\mathrm{S}^{k'k}$ (with their time dependence) suffices to know the reduced density matrix ρ_S (with its time dependence). In Section 35.2, we shall thus focus on finding an approximate expression for the time dependence of the matrix elements $\rho_S^{k'k}$.

35.2 Time Evolution of the Reduced Density Matrix in the Interaction Picture

Let H_S be the Hamiltonian of the subsystem S and H_B the Hamiltonian of the environment (or bath) B, which we assume to be both time independent, while the time-dependent interaction Hamiltonian $H_{SB}(t)$ between the subsystem and the bath starts being effective at time t_0. Let also $H_0 = H_S + H_B$ be the total unperturbed Hamiltonian for the closed combined system plus environment, to which there corresponds the evolution operator $U_0(t, t_0) = e^{-iH_0(t-t_0)}$. From the first line of Eq. (35.9), we then write

$$\begin{aligned} \rho_S^{k'k}(t) &= \mathrm{Tr}\{\rho(t)|k\rangle\langle k'|\} = \mathrm{Tr}\Big\{U_0(t,t_0)\, U_0^\dagger(t,t_0)\, \rho(t)\, U_0(t,t_0)\, U_0^\dagger(t,t_0)|k\rangle\langle k'|\Big\} \\ &= \mathrm{Tr}\Big\{\Big(U_0^\dagger(t,t_0)\, \rho(t)\, U_0(t,t_0)\Big)\Big(U_0^\dagger(t,t_0)\, |k\rangle\langle k'|\, U_0(t,t_0)\Big)\Big\} \ , \end{aligned} \tag{35.12}$$

where we introduce the quantity

$$P_{k'k}(t) \equiv U_0^\dagger(t,t_0)\, |k\rangle\langle k'|\, U_0(t,t_0) \tag{35.13}$$

and write with the help of Eq. (35.2) and of Eq. (3.25) of Part I

$$\begin{aligned} U_0^\dagger(t,t_0)\, \rho(t)\, U_0(t,t_0) &= U_0^\dagger(t,t_0)\, U(t,t_0)\, \rho(t_0)\, U^\dagger(t,t_0)\, U_0(t,t_0) \\ &= \tilde{W}(t,t_0)\, \rho(t_0)\, \tilde{W}^\dagger(t,t_0) \equiv \rho_I(t) \end{aligned} \tag{35.14}$$

for the density matrix operator in the interaction picture. [A similar definition will be used in Step # 1 of Section 37.1.] To alleviate the notation, in the following we shall omit the tilde over the symbols W in Eq. (35.14) and $H_I^{SB}(t)$ introduced below just after Eq. (35.22).

In this way, the expression (35.12) can eventually be cast in the form:

$$\begin{aligned} \rho_S^{k'k}(t) &= \mathrm{Tr}\{\rho_I(t)\, P_{k'k}(t)\} = \mathrm{Tr}\big\{W(t,t_0)\, \rho(t_0)\, W^\dagger(t,t_0)\, P_{k'k}(t)\big\} \\ &= \mathrm{Tr}\big\{\rho(t_0)\, W^\dagger(t,t_0)\, P_{k'k}(t)\, W(t,t_0)\big\} \ . \end{aligned} \tag{35.15}$$

The complete transformation of $\rho_S^{k'k}(t)$ to the interaction picture will be considered later in Section 35.7.

35.3 Assumption of Initial Factorization

We assume that the subsystem and the bath are not correlated with each other, at least up to the reference time t_0 when they begin to interact with each other. On physical grounds, this should be the case not only when the subsystem and the bath have not interacted at times earlier than t_0 but also when the correlations between the subsystem and the bath are short lived. Accordingly, we set

$$\rho(t_0) = \rho_{\rm B}(t_0) \otimes \rho_{\rm S}(t_0), \tag{35.16}$$

where owing to Eq. (35.11)

$$\rho_{\rm S}(t_0) = \sum_{\bar{k}'\bar{k}} |\bar{k}'\rangle\, \rho_{\rm S}^{\bar{k}'\bar{k}}(t_0)\, \langle\bar{k}|\,. \tag{35.17}$$

[This kind of assumption will be made also in Step # 3 of Section 37.1.]

Remark: A complete decoupling between the subsystem and the bath
A complete decoupling between subsystem and bath occurs when $w_{m_{\rm S}m_{\rm B}} \to w_{m_{\rm S}}\, w_{m_{\rm B}}$. In this case,

$$\rho = \sum_{m_{\rm S}m_{\rm B}} |m_{\rm S}\rangle|m_{\rm B}\rangle w_{m_{\rm S}m_{\rm B}} \langle m_{\rm S}|\langle m_{\rm B}| \to \left(\sum_{m_{\rm S}} |m_{\rm S}\rangle w_{m_{\rm S}} \langle m_{\rm S}|\right)\left(\sum_{m_{\rm B}} |m_{\rm B}\rangle w_{m_{\rm B}} \langle m_{\rm B}|\right) = \bar{\rho}_{\rm S}\, \rho_{\rm B}, \tag{35.18}$$

where $\rho_{\rm B} = \sum_{m_{\rm B}} |m_{\rm B}\rangle w_{m_{\rm B}} \langle m_{\rm B}|$ is the density matrix of the reservoir (or bath) and $\bar{\rho}_{\rm S}$ coincides with the reduced density matrix of Eq. (35.8) since

$$\rho_{\rm S} = \sum_{m_{\rm S}} |m_{\rm S}\rangle \sum_{m_{\rm B}} w_{m_{\rm S}m_{\rm B}} \langle m_{\rm S}| \to \sum_{m_{\rm B}} w_{m_{\rm B}} \sum_{m_{\rm S}} |m_{\rm S}\rangle w_{m_{\rm S}} \langle m_{\rm S}| = \left(\sum_{m_{\rm B}} w_{m_{\rm B}}\right) \bar{\rho}_{\rm S} = \bar{\rho}_{\rm S}\,. \tag{35.19}$$

The result (35.16) is thus recovered.

With the assumption (35.16) and taking into account the expression (35.17), Eq. (35.15) becomes

$$\begin{aligned}
\rho_{\rm S}^{k'k}(t) &= {\rm Tr}\left\{\rho_{\rm B}(t_0) \sum_{\bar{k}'\bar{k}} |\bar{k}'\rangle\, \rho_{\rm S}^{\bar{k}'\bar{k}}(t_0)\, \langle\bar{k}|\, W^\dagger(t,t_0)\, P_{k'k}(t)\, W(t,t_0)\right\} \\
&= \sum_{\bar{k}''} \langle\bar{k}''|\, {\rm Tr_B}\left\{\rho_{\rm B}(t_0) \sum_{\bar{k}'\bar{k}} |\bar{k}'\rangle\, \rho_{\rm S}^{\bar{k}'\bar{k}}(t_0)\, \langle\bar{k}|\, W^\dagger(t,t_0)\, P_{k'k}(t)\, W(t,t_0)\right\} |\bar{k}''\rangle \\
&= \sum_{\bar{k}'\bar{k}} \rho_{\rm S}^{\bar{k}'\bar{k}}(t_0)\, {\rm Tr_B}\left\{\rho_{\rm B}(t_0)\, \langle\bar{k}|\, W^\dagger(t,t_0)\, P_{k'k}(t)\, W(t,t_0)|\bar{k}'\rangle\right\} \\
&= \sum_{\bar{k}'\bar{k}} \rho_{\rm S}^{\bar{k}'\bar{k}}(t_0)\, \Pi_{\bar{k}'\bar{k}\to k'k}(t_0,t),
\end{aligned} \tag{35.20}$$

where in the last line we have defined

$$\Pi_{\bar{k}'\bar{k}\to k'k}(t_0,t) \equiv {\rm Tr_B}\left\{\rho_{\rm B}(t_0)\, \langle\bar{k}|\, W^\dagger(t,t_0)\, P_{k'k}(t)\, W(t,t_0)|\bar{k}'\rangle\right\}\,. \tag{35.21}$$

35.4 Assumptions of Weak Coupling

There now remains to obtain an explicit expression for the quantity (35.21). With the assumption that the coupling between the subsystem and the reservoir (or bath) is weak, this task can be performed with perturbation theory in terms of the interaction Hamiltonian $H_{SB}(t)$. From Eq. (3.25) of Part I, we then write

$$W(t,t_0) = \mathcal{T}\left\{e^{-i\int_{t_0}^{t} d\bar{t}\, H_I^{SB}(\bar{t})}\right\} \tag{35.22}$$

with $H_I^{SB}(t) = e^{i(H_S+H_B)(t-t_0)} H_{SB}(t)\, e^{-i(H_S+H_B)(t-t_0)}$, as well as

$$W(t,t_0) = \mathbb{1} - i\int_{t_0}^{t} d\bar{t}\, H_I^{SB}(\bar{t})\, W(\bar{t},t_0), \tag{35.23}$$

where $W^\dagger(t,t_0)W(t,t_0) = \mathbb{1}$. [Recall that, for simplicity, we have here dropped the tilde over the symbols W and $H_I(t)$ that instead appear in Eq. (3.25) of Part I.] We can then make use of the expression (35.23) to expand $W(t,t_0)$ in powers of $H_{SB}(t)$ as follows:

$$\begin{aligned} W(t,t_0) &= \mathbb{1} - i\int_{t_0}^{t} dt_1\, H_I^{SB}(t_1)\, W(t_1,t_0) \\ &= \mathbb{1} - i\int_{t_0}^{t} dt_1\, H_I^{SB}(t_1)\left(\mathbb{1} - i\int_{t_0}^{t_1} dt_2\, H_I^{SB}(t_2)\, W(t_2,t_0)\right) \\ &= \mathbb{1} - i\int_{t_0}^{t} dt_1\, H_I^{SB}(t_1) + i^2\int_{t_0}^{t} dt_1\int_{t_0}^{t_1} dt_2\, H_I^{SB}(t_1)\, H_I^{SB}(t_2) + \cdots \end{aligned} \tag{35.24}$$

such that

$$W^\dagger(t,t_0) = \mathbb{1} + i\int_{t_0}^{t} dt_1\, H_I^{SB}(t_1) + i^2\int_{t_0}^{t} dt_1\int_{t_0}^{t_1} dt_2\, H_I^{SB}(t_2)\, H_I^{SB}(t_1) + \cdots . \tag{35.25}$$

Up to second order in $H_{SB}(t)$, the expression (35.21) then becomes:

$$\begin{aligned} &\Pi_{\bar{k}'\bar{k}\to k'k}(t_0,t) \\ &\simeq \langle\bar{k}|\,\mathrm{Tr}_B\Bigg\{\rho_B(t_0)\left(\mathbb{1} + i\int_{t_0}^{t} dt_1\, H_I^{SB}(t_1) + i^2\int_{t_0}^{t} dt_1\int_{t_0}^{t_1} dt_2\, H_I^{SB}(t_2)\, H_I^{SB}(t_1)\right) \\ &\times P_{k'k}(t)\left(\mathbb{1} - i\int_{t_0}^{t} d\bar{t}_1\, H_I^{SB}(\bar{t}_1) + i^2\int_{t_0}^{t} d\bar{t}_1\int_{t_0}^{\bar{t}_1} d\bar{t}_2\, H_I^{SB}(\bar{t}_1)\, H_I^{SB}(\bar{t}_2)\right)\Bigg\}|\bar{k}'\rangle \\ &\simeq \langle\bar{k}|\,\mathrm{Tr}_B\{\rho_B(t_0)\, P_{k'k}(t)\}|\bar{k}'\rangle \qquad \left[\text{term } \textcircled{1} \times \textcircled{1}\right] \end{aligned}$$

$$+ \langle \bar{k}| \, \mathrm{Tr_B} \left\{ \rho_B(t_0) \int_{t_0}^{t} dt_1 \, H_I^{SB}(t_1) \, P_{k'k}(t) \int_{t_0}^{t} d\bar{t}_1 \, H_I^{SB}(\bar{t}_1) \right\} |\bar{k}'\rangle \qquad \left[\text{term} \, ② \times ②\right]$$

$$- \langle \bar{k}| \, \mathrm{Tr_B} \left\{ \rho_B(t_0) \int_{t_0}^{t} dt_1 \int_{t_0}^{t_1} dt_2 \, H_I^{SB}(t_2) \, H_I^{SB}(t_1) \, P_{k'k}(t) \right\} |\bar{k}'\rangle \qquad \left[\text{term} \, ③ \times ①\right]$$

$$- \langle \bar{k}| \, \mathrm{Tr_B} \left\{ \rho_B(t_0) \, P_{k'k}(t) \int_{t_0}^{t} d\bar{t}_1 \int_{t_0}^{\bar{t}_1} d\bar{t}_2 \, H_I^{SB}(\bar{t}_1) \, H_I^{SB}(\bar{t}_2) \right\} |\bar{k}'\rangle \qquad \left[\text{term} \, ① \times ③\right] . \tag{35.26}$$

Here, the numbering ⓘ × ⓙ of the various terms corresponds to that of (i) the three terms in the expression (35.25) for $W^{\dagger}(t, t_0)$ as far as the index ⓘ is concerned and (ii) the three terms in the expression (35.24) for $W(t, t_0)$ as far as the index ⓙ is concerned. Note accordingly that on the right-hand side of Eq. (35.26), we have omitted the two linear terms ① × ② and ② × ①, consistently with the assumption (37.9) used in Step # 4 of Section 37.1 (cf. the end of Section 35.5, where also these terms will be explicitly calculated for completeness).

Remark: No need here for the assumption of persistent factorization
In Section 37.1, the assumption of weak coupling will appear implicitly in the assumption of a "persistent" factorization made in Step # 5 therein. There, this assumption results in the presence of only two powers of $H_{SB}(t)$ appearing in the approximate equation of motion (37.13) for the reduced density matrix, with no way, however, for including higher-order terms. In the present approach, on the other hand, we rely on the less restrictive assumption of an "initial" factorization (35.16), which (at least in principle) allows us to consider any power of $H_{SB}(t)$ in the equation of motion for the reduced density matrix. Nevertheless, in Eq. (35.26), we have considered only terms up to second order in $H_{SB}(t)$, because our aim here is to eventually recover the form of the Lindblad equation, which is already beginning to take shape from the presence of the retained terms.

35.5 Analysis up to Second Order

We pass now to explicitly calculate the contribution of the four terms reported on the right-hand side of Eq. (35.26). To this end, it is further convenient to assume that:

(i) The states $\{|k\rangle\}$, which appear in the above expressions, are as anticipated eigenstates of the subsystem Hamiltonian H_S with eigenvalues $\{\epsilon_k\}$, such that

$$H_S \, |k\rangle = |k\rangle \, \epsilon_k \, . \tag{35.27}$$

(ii) The states $\{|m_{\rm B}\rangle\}$ entering the density matrix of the reservoir (or bath) $\rho_{\rm B}(t_0) \leftrightarrow \rho_{\rm B} = \sum_{m_{\rm B}} |m_{\rm B}\rangle w_{m_{\rm B}} \langle m_{\rm B}|$ (with $\mathrm{Tr_B}\{\rho_{\rm B}(t_0)\} = \sum_{m_{\rm B}} w_{m_{\rm B}} = 1$), which also appear in the above expressions, are eigenstates of the reservoir (or bath) Hamiltonian $H_{\rm B}$, such that $[\rho_{\rm B}(t_0), H_{\rm B}] = 0$.

(iii) The time-dependent behavior of the interaction Hamiltonian between the subsystem and the reservoir (or bath) is taken of the simple form $H_{\rm SB}(t) = \theta(t - t_0)\, H_{\rm SB}$ (cf. also Step # 3 of Section 37.1).

Term ① × ①

$$\begin{aligned}
\langle \bar{k}|\, \mathrm{Tr_B}\, \{\rho_{\rm B}(t_0)\, P_{k'k}(t)\}\, |\bar{k}'\rangle &\underset{[\text{Eq.}(35.13)]}{\overset{}{=}} \langle \bar{k}|\, \mathrm{Tr_B}\, \left\{\rho_{\rm B}(t_0)\, U_0^{\dagger}(t, t_0)\, |k\rangle\langle k'|\, U_0(t, t_0)\right\} |\bar{k}'\rangle \\
&= \langle \bar{k}|\, \mathrm{Tr_B}\, \left\{e^{-iH_{\rm B}(t-t_0)}\, \rho_{\rm B}(t_0)\, e^{iH_{\rm B}(t-t_0)}\, e^{iH_{\rm S}(t-t_0)} |k\rangle\langle k'|\, e^{-iH_{\rm S}(t-t_0)})\right\} |\bar{k}'\rangle \\
&= \mathrm{Tr_B}\, \left\{\rho_{\rm B}(t_0)\, \langle \bar{k}|\, e^{iH_{\rm S}(t-t_0)} |k\rangle\langle k'|\, e^{-iH_{\rm S}(t-t_0)})\, |\bar{k}'\rangle\right\} \\
&= \langle \bar{k}|\, e^{iH_{\rm S}(t-t_0)} |k\rangle\langle k'|\, e^{-iH_{\rm S}(t-t_0)})\, |\bar{k}'\rangle \\
&= \delta_{\bar{k}k}\, \delta_{\bar{k}'k'}\, e^{i(\epsilon_k - \epsilon_{k'})(t-t_0)} \equiv \Pi^0_{\bar{k}'\bar{k} \to k'k}(t_0, t),
\end{aligned} \tag{35.28}$$

where the suffix "0" signifies that the interaction Hamiltonian $H_{\rm SB}(t)$ has no effect on this term.

Term ② × ②

$$\begin{aligned}
&\langle \bar{k}|\, \mathrm{Tr_B}\, \left\{\rho_{\rm B}(t_0) \int_{t_0}^{t} dt_1\, H_{\rm I}^{\rm SB}(t_1)\, P_{k'k}(t) \int_{t_0}^{t} d\bar{t}_1\, H_{\rm I}^{\rm SB}(\bar{t}_1)\right\} |\bar{k}'\rangle \\
&= \int_{t_0}^{t} dt_1 \int_{t_0}^{t_1} d\bar{t}_1\, \langle \bar{k}|\, \mathrm{Tr_B} \left\{\rho_{\rm B}(t_0) \left(H_{\rm I}^{\rm SB}(t_1)\, P_{k'k}(t)\, H_{\rm I}^{\rm SB}(\bar{t}_1) + H_{\rm I}^{\rm SB}(\bar{t}_1)\, P_{k'k}(t)\, H_{\rm I}^{\rm SB}(t_1)\right)\right\} |\bar{k}'\rangle \\
&= \int_{t_0}^{t} dt_1 \int_{t_0}^{t_1} d\bar{t}_1 \left(e^{i\epsilon_{\bar{k}}(t_1-t_0)}\, e^{-i\epsilon_k(t_1-t)}\, e^{i\epsilon_{k'}(\bar{t}_1-t)}\, e^{-i\epsilon_{\bar{k}'}(\bar{t}_1-t_0)} \right. \\
&\quad \left. \times \mathrm{Tr_B} \left\{\rho_{\rm B}(t_0)\, \langle \bar{k}| e^{iH_{\rm B}(t_1-t_0)} H_{\rm SB} e^{-iH_{\rm B}(t_1-t_0)} |k\rangle\langle k'| e^{iH_{\rm B}(t_2-t_0)} H_{\rm SB} e^{-iH_{\rm B}(t_2-t_0)} |\bar{k}'\rangle\right\} + t_1 \rightleftarrows \bar{t}_1 \right) \\
&= \sum_{k_1 k_1'} \sum_{k_2 k_2'} \int_{t_0}^{t} dt_1 \int_{t_0}^{t_1} d\bar{t}_1\, \Pi^0_{\bar{k}'\bar{k} \to k_2' k_2}(t_0, \bar{t}_1)
\end{aligned}$$

$$\times \Bigg(\mathrm{Tr_B} \left\{ \rho_\mathrm{B}(t_0) \langle k_2 | e^{iH_\mathrm{B}(t_1-t_0)} H_\mathrm{SB} e^{-iH_\mathrm{B}(t_1-t_0)} | k_1 \rangle \langle k_1' | e^{iH_\mathrm{B}(\bar{t}_1-t_0)} H_\mathrm{SB} e^{-iH_\mathrm{B}(\bar{t}_1-t_0)} | k_2' \rangle \right\}$$

$$\times e^{-i(\epsilon_{k_1'}-\epsilon_{k_2})(t_1-\bar{t}_1)}$$

$$+ \mathrm{Tr_B} \left\{ \rho_\mathrm{B}(t_0) \langle k_2 | e^{iH_\mathrm{B}(\bar{t}_1-t_0)} H_\mathrm{SB} e^{-iH_\mathrm{B}(\bar{t}_1-t_0)} | k_1 \rangle \langle k_1' | e^{iH_\mathrm{B}(t_1-t_0)} H_\mathrm{SB} e^{-iH_\mathrm{B}(t_1-t_0)} | k_2' \rangle \right\}$$

$$\times e^{-i(\epsilon_{k_2'}-\epsilon_{k_1})(t_1-\bar{t}_1)} \Bigg) \Pi^0_{k_1'k_1 \to k'k}(t_1, t), \tag{35.29}$$

where we have highlighted the presence of the quantity Π^0 introduced in the last line of Eq. (35.28). The expression within parentheses on the right-hand side of Eq. (35.29) will be identified as the contribution of the term ② × ② to the *self-energy* $\Sigma_{k_2'k_2 \to k_1'k_1}(t_2, t_1)$ of relevance in the present context.

Term ③ × ①

$$- \langle \bar{k} | \mathrm{Tr_B} \left\{ \rho_\mathrm{B}(t_0) \int_{t_0}^{t} dt_1 \int_{t_0}^{t_1} dt_2 \, H_\mathrm{I}^\mathrm{SB}(t_2) \, H_\mathrm{I}^\mathrm{SB}(t_1) P_{k'k}(t) \right\} | \bar{k}' \rangle$$

$$= \sum_{k_1 k_1'} \sum_{k_2 k_2'} \int_{t_0}^{t} dt_1 \int_{t_0}^{t_1} dt_2 \, \Pi^0_{\bar{k}'\bar{k} \to k_2' k_2}(t_0, t_2)$$

$$\times \Bigg((-) \mathrm{Tr_B} \left\{ \rho_\mathrm{B}(t_0) \langle k_2 | e^{iH_\mathrm{B}(t_2-t_0)} H_\mathrm{SB} e^{-iH_\mathrm{B}(t_2-t_0)} e^{-iH_\mathrm{S}(t_2-t_1)} \right.$$

$$\left. \times e^{iH_\mathrm{B}(t_1-t_0)} H_\mathrm{SB} e^{-iH_\mathrm{B}(t_1-t_0)} | k_1 \rangle \right\} \delta_{k_1' k_2'} e^{-i\epsilon_{k_1'}(t_1-t_2)} \Bigg) \Pi^0_{k_1' k_1 \to k'k}(t_1, t), \tag{35.30}$$

where the expression within parentheses on the right-hand side of Eq. (35.30) will correspond to the contribution of the term ③ × ① to the *self-energy* $\Sigma_{k_2'k_2 \to k_1'k_1}(t_2, t_1)$ mentioned earlier.

Term ① × ③

$$- \langle \bar{k} | \mathrm{Tr_B} \left\{ \rho_\mathrm{B}(t_0) \, P_{k'k}(t) \int_{t_0}^{t} d\bar{t}_1 \int_{t_0}^{\bar{t}_1} d\bar{t}_2 \, H_\mathrm{I}^\mathrm{SB}(\bar{t}_1) H_\mathrm{I}^\mathrm{SB}(\bar{t}_2) \right\} | \bar{k}' \rangle$$

$$= \sum_{k_1 k_1'} \sum_{k_2 k_2'} \int_{t_0}^{t} d\bar{t}_1 \int_{t_0}^{\bar{t}_1} d\bar{t}_2 \, \Pi^0_{\bar{k}'\bar{k} \to k_2' k_2}(t_0, \bar{t}_2) \Bigg((-) \delta_{k_1 k_2} \, e^{-i\epsilon_{k_2}(\bar{t}_2 - \bar{t}_1)}$$

$$\times \operatorname{Tr}_B \Big\{ \rho_B(t_0) \langle k_1' | e^{iH_B(\bar{t}_1 - t_0)} H_{SB} e^{-iH_B(\bar{t}_1 - t_0)} e^{-iH_S(\bar{t}_1 - \bar{t}_2)}$$

$$\times e^{iH_B(\bar{t}_2 - t_0)} H_{SB} e^{-iH_B(\bar{t}_2 - t_0)} | k_2' \rangle \Big\} \Bigg) \Pi^0_{k_1' k_1 \to k' k}(\bar{t}_1, t), \tag{35.31}$$

where the expression within parentheses on the right-hand side of Eq. (35.31) will correspond to the contribution of the term ①×③ to the *self-energy* $\Sigma_{k_2' k_2 \to k_1' k_1}(t_2, t_1)$.

Remark: Contribution of the terms ①×② and ②×①

For completeness, we explicitly calculate also the contributions of the terms ①× ② and ②×①, that were disregarded on the right-hand side of Eq. (35.26) in line with the assumption (37.9) of Section 37.1.

Term ①×②

$$- i \langle \bar{k} | \operatorname{Tr}_B \left\{ \rho_B(t_0) P_{k'k}(t) \int_{t_0}^{t} d\bar{t}_1 H_I^{SB}(\bar{t}_1) \right\} | \bar{k}' \rangle = -i \delta_{k\bar{k}} e^{i(\epsilon_{\bar{k}} - \epsilon_{k'})(t - t_0)} \int_{t_0}^{t} d\bar{t}_1 e^{i(\epsilon_{k'} - \epsilon_{\bar{k}'})(\bar{t}_1 - t_0)}$$

$$\times \langle k' | \operatorname{Tr}_B \left\{ \rho_B(t_0) e^{iH_B(\bar{t}_1 - t_0)} H_{SB} e^{-iH_B(\bar{t}_1 - t_0)} \right\} | \bar{k}' \rangle, \tag{35.32}$$

which vanishes in accordance with the assumption (37.9), as anticipated. A similar analysis holds for the other linear term ②×①.

35.6 A Diagrammatic Representation

To better appreciate the structure of the four contributions (35.28)–(35.31), it is convenient to illustrate them in a *diagrammatic representation*, as shown in the four panels of Fig. 35.1, respectively. Here, conventions and symbols are as follows:

(i) The block representing the term ①×① corresponds to the quantity $\Pi^0_{k'\bar{k} \to k'k}(t_0, t)$ introduced in Eq. (35.28). Here, lines with arrows stand for time propagation in terms of the subsystem Hamiltonian H_S, connecting two of its eigenstates (indicated by bras and kets at the ends).

(ii) The block Π^0 appears also at the left and right sides of the remaining terms ②×②, ③×①, and ①×③. In these terms, the central blocks contain twice the interaction Hamiltonian H_{SB} between the subsystem and the reservoir (or bath). In the central blocks, the connection between two factors H_{SB} is indicated by a dashed line, with an arrow pointing from the second to the first time index. The presence of the evolution operator with H_S is indicated when necessary.

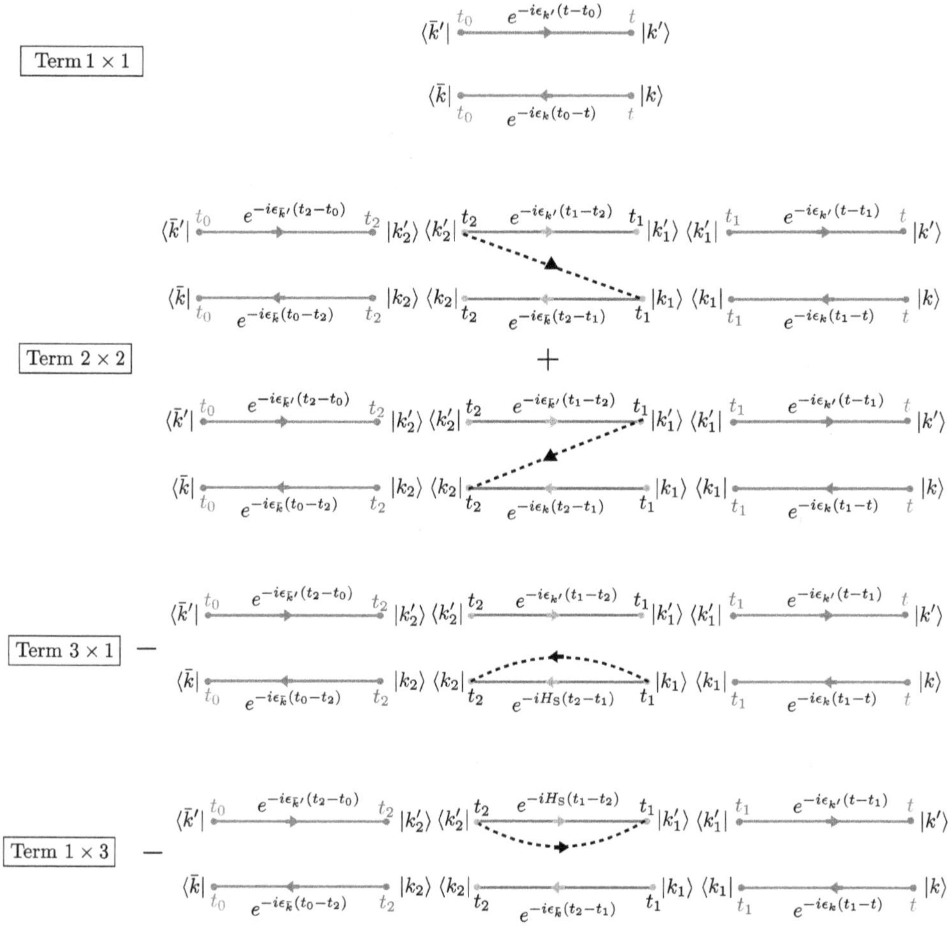

Figure 35.1 Diagrammatic representation of the four contributions (35.28)–(35.31). The full account of conventions and symbols is given in the text.

(iii) In all terms, the upper lines propagate *forward* in time from t_0 up to t, while the lower lines propagate *backwards* in time from t back to t_0. It is in this sense that the Schwinger–Keldysh closed-contour approach and the Lindblad equation are connected.

35.7 Extension to Higher Orders and Dyson-Like Equation of Motion for the Reduced Density Matrix

The analysis carried out thus far, at second order in the interaction Hamiltonian H_{SB} between the subsystem and the bath, can, in principle, be extended to higher orders. In the process, one ends up with contractions involving strings of the operators $H_{\rm SB}$, which can be handled by Wick's theorem in the following way.

A generic term of the expansion of the expression (35.21) in powers of $H_{\rm SB}(t)$ contains the factor

$$i^{N_{\rm L}}(-i)^{N_{\rm R}}{\rm Tr}_{\rm B}\left\{\rho_{\rm B}(t_0)\langle\bar{k}|H_{\rm I}^{\rm SB}(t_1)\cdots H_{\rm I}^{\rm SB}(t_{N_{\rm L}})|k\rangle \right. \\ \left. \times\,\langle k'|H_{\rm I}^{\rm SB}(t_{N_{\rm L}+1})\cdots H_{\rm I}^{\rm SB}(t_{N_{\rm L}+N_{\rm R}})|\bar{k}'\rangle\right\}, \tag{35.33}$$

with $N_{\rm L}$ ($N_{\rm R}$) terms at the left (right) of $P_{k'k}(t)$ (and with a further phase factor $e^{i(\epsilon_k-\epsilon_{k'})(t-t_0)}$ that originates from $P_{k'k}(t)$). Suppose now that $H_{\rm SB}$ has the "linear" form

$$H_{\rm SB}=\sum_i A_iB_i, \tag{35.34}$$

where the Hermitian operators $\{A_i\}$ act on the subsystem and $\{B_i\}$ on the bath. [This assumption will be made also in Step # 4 of Section 37.1.] A generic matrix element of $H_{\rm I}^{\rm SB}(t)$ has then the form:

$$\begin{aligned}\langle k_\gamma|H_{\rm I}^{\rm SB}(t)|k_{\gamma'}\rangle &= \langle k_\gamma|e^{i(H_{\rm S}+H_{\rm B})(t-t_0)}H_{\rm SB}e^{-i(H_{\rm S}+H_{\rm B})(t-t_0)}|k_{\gamma'}\rangle \\ &= e^{i(\epsilon_{k_\gamma}-\epsilon_{k_{\gamma'}})(t-t_0)}\langle k_\gamma|e^{iH_{\rm B}(t-t_0)}H_{\rm SB}e^{-iH_{\rm B}(t-t_0)}|k_{\gamma'}\rangle \\ &= e^{i(\epsilon_{k_\gamma}-\epsilon_{k_{\gamma'}})(t-t_0)}\sum_i\langle k_\gamma|A_i|k_{\gamma'}\rangle e^{iH_{\rm B}(t-t_0)}B_ie^{-iH_{\rm B}(t-t_0)} \\ &= \sum_i\mathcal{F}_i^{\gamma\gamma'}(t)B_{\rm I}^i(t),\end{aligned} \tag{35.35}$$

where we have defined $\mathcal{F}_i^{\gamma\gamma'}(t)=e^{i(\epsilon_{k_\gamma}-\epsilon_{k_{\gamma'}})(t-t_0)}\langle k_\gamma|A_i|k_{\gamma'}\rangle$ and $B_{\rm I}^i(t)=e^{iH_{\rm B}(t-t_0)}$ $B_i\,e^{-iH_{\rm B}(t-t_0)}$. The expression (35.33) then contains a bunch of contractions of the type

$${\rm Tr}_{\rm B}\left\{\rho_{\rm B}(t_0)\,B_{\rm I}^{i_1}(t_1)\,B_{\rm I}^{i_2}(t_2)\,\cdots\,B_{\rm I}^{i_{N_{\rm L}}}(t_{N_{\rm L}})\,\cdots\,B_{\rm I}^{i_{N_{\rm L}+N_{\rm R}}}(t_{N_{\rm L}+N_{\rm R}})\right\}. \tag{35.36}$$

With the further assumption that B_i are bosonic-like operators, the Wick's theorem for pairwise contractions discussed in Chapter 9 in Part I can be applied to the expression (35.36). Owing to the property (37.9) of Step # 4 of Section 37.1, only terms (35.36) with even $N_{\rm L}+N_{\rm R}$ are nonvanishing. In addition, no contraction can join the external vertices at times t_0 and t in the expression (35.21) for $\Pi(t_0,t)$.

Accordingly, at any successive order in perturbation theory, two vertices $H_{\rm SB}$ of the type (35.34), plus two evolution operators U_0 and two evolution operators $U_0^\dagger$, are added to the diagrammatic expression of Π. All the ensuing terms can then be formally re-summed by introducing a suitable self-energy operator Σ, in the form

$$\Pi = \Pi^0 + \Pi^0 \Sigma \Pi^0 + \Pi^0 \Sigma \Pi^0 \Sigma \Pi^0 + \cdots = \Pi^0 + \left(\Pi_0 + \Pi^0 \Sigma \Pi^0 + \cdots\right) \Sigma \Pi^0$$
$$= \Pi^0 + \Pi \Sigma \Pi^0 \tag{35.37}$$

with Π^0 defined in Eq. (35.28). In the above expression, matrix multiplication entails sums over the internal bra and ket symbols like in Fig. 35.1 plus integrations over the internal time variables, which are assumed to be suitably nested into one another as explicitly shown in the expressions (35.29)–(35.31). In the present context, there is no point in deriving in detail the topological structure of the diagrams for Σ entering Eq. (35.37) as well as the associated diagrammatic rules, since we shall here limit to consider explicitly only the expressions (35.29)–(35.31) up to second order in H_{SB}.

With these premises, we can go back to the expression (35.20) for the reduced density matrix and cast it in the following meaningful form:

$$\rho_{\mathrm{S}}^{k'k}(t) = \sum_{\bar{k}'\bar{k}} \rho_{\mathrm{S}}^{\bar{k}'\bar{k}}(t_0)\Pi_{\bar{k}'\bar{k}\to k'k}(t_0, t) = \sum_{\bar{k}'\bar{k}} \rho_{\mathrm{S}}^{\bar{k}'\bar{k}}(t_0)\Bigg(\Pi^0_{\bar{k}'\bar{k}\to k'k}(t_0, t)$$
$$+ \sum_{k_2'k_2}\sum_{k_1'k_1}\int_{t_0}^{t} dt_1 \int_{t_0}^{t_1} dt_2 \Pi_{\bar{k}'\bar{k}\to k_2'k_2}(t_0, t_2)\Sigma_{k_2'k_2\to k_1'k_1}(t_2, t_1)\Pi^0_{k_1'k_1\to k'k}(t_1, t)\Bigg)$$
$$= \rho_{\mathrm{S}}^{k'k}(t_0)e^{-i(\epsilon_{k'}-\epsilon_k)(t-t_0)} + \sum_{k_2'k_2}\int_{t_0}^{t} dt_1 \int_{t_0}^{t_1} dt_2 \rho_{\mathrm{S}}^{k_2'k_2}(t_2)$$
$$\times \Sigma_{k_2'k_2\to k'k}(t_2, t_1)e^{-i(\epsilon_{k'}-\epsilon_k)(t-t_1)}. \tag{35.38}$$

At this point, we can complete the transformation of the reduced density matrix to the interaction picture as anticipated at the end of Section 35.2, by defining

$$\rho_{\mathrm{I}}^{k'k}(t) \equiv e^{i(\epsilon_{k'}-\epsilon_k)(t-t_0)}\, \rho_{\mathrm{S}}^{k'k}(t) \tag{35.39}$$

such that Eq. (35.38) becomes

$$\rho_{\mathrm{I}}^{k'k}(t) = \rho_{\mathrm{I}}^{k'k}(t_0) + \sum_{k_2'k_2}\int_{t_0}^{t} dt_1 \int_{t_0}^{t_1} dt_2\, \rho_{\mathrm{I}}^{k_2'k_2}(t_2)\, \Sigma^{\mathrm{I}}_{k_2'k_2\to k'k}(t_2, t_1), \tag{35.40}$$

where

$$\Sigma^{\mathrm{I}}_{k_2'k_2\to k_1'k_1}(t_2, t_1) \equiv e^{-i(\epsilon_{k_2'}-\epsilon_{k_2})(t_2-t_0)}\, \Sigma_{k_2'k_2\to k_1'k_1}(t_2, t_1)\, e^{i(\epsilon_{k_1'}-\epsilon_{k_1})(t_1-t_0)}. \tag{35.41}$$

The equation of motion for $\rho_{\mathrm{I}}^{k'k}(t)$ is eventually obtained upon taking the derivative of both sides of Eq. (35.40) with respect to t, yielding the formally exact result

$$\boxed{\frac{\partial}{\partial t}\rho_{\mathrm{I}}^{k'k}(t) = \sum_{k_2'k_2}\int_{t_0}^{t} dt_2\, \rho_{\mathrm{I}}^{k_2'k_2}(t_2)\, \Sigma^{\mathrm{I}}_{k_2'k_2\to k'k}(t_2,t)\,.} \tag{35.42}$$

Note that this equation contains "memory effects" about the past history of $\rho_{\mathrm{I}}^{k_2'k_2}(t)$, whose knowledge, however, will not be required in the context of the Lindblad equation. We shall return to this issue in Section 35.10.

35.8 The Case of a Two-Level System Embedded in a Phonon Bath

To proceed further, we select a particularly simple form of the Hamiltonians H_{S}, H_{B}, and H_{SB}. We assume that the subsystem is a *two-level system* coupled to a bath of phonons via a linear coupling like in Eq. (35.34). Accordingly, we take for the two-level system

$$H_{\mathrm{S}} = \frac{1}{2}\,\omega_0\,\sigma_z = \begin{pmatrix} \frac{\omega_0}{2} & 0 \\ 0 & -\frac{\omega_0}{2} \end{pmatrix}, \tag{35.43}$$

where $\sigma_z = \begin{pmatrix} 1 & 0 \\ 0 & -1 \end{pmatrix}$ is a Pauli matrix and

$$|k=1\rangle = \begin{pmatrix} 0 \\ 1 \end{pmatrix}, \quad \epsilon_{k=1} = -\frac{\omega_0}{2}, \quad |k=2\rangle = \begin{pmatrix} 1 \\ 0 \end{pmatrix}, \quad \epsilon_{k=2} = \frac{\omega_0}{2}, \tag{35.44}$$

with reference to Eq. (35.27), such that $\epsilon_{k=2} - \epsilon_{k=1} = \omega_0 > 0$. In addition, for the dynamics of the phonon bath, we take

$$H_{\mathrm{B}} = \sum_{\{j\}} \omega_j\, b_j^{\dagger}\, b_j, \tag{35.45}$$

where b_j are bosonic operators and the frequencies $\{\omega_j\}$ span a *continuum spectrum*. Finally, for the interaction Hamiltonian, we take

$$H_{\mathrm{SB}} = \sigma_+ \sum_j \beta_j\, b_j + \sigma_- \sum_j \beta_j\, b_j^{\dagger}, \tag{35.46}$$

where β_j are real quantities for all j and

$$\sigma_+ = \frac{1}{2}\left(\sigma_x + i\sigma_y\right) = \begin{pmatrix} 0 & 1 \\ 0 & 0 \end{pmatrix}, \quad \sigma_- = \frac{1}{2}\left(\sigma_x - i\sigma_y\right) = \begin{pmatrix} 0 & 0 \\ 1 & 0 \end{pmatrix} = \sigma_+^{\dagger} \tag{35.47}$$

with σ_x and σ_y Pauli matrices, such that

$$\sigma_+ \begin{pmatrix} 1 \\ 0 \end{pmatrix} = 0\,, \quad \sigma_+ \begin{pmatrix} 0 \\ 1 \end{pmatrix} = \begin{pmatrix} 1 \\ 0 \end{pmatrix}, \quad \sigma_- \begin{pmatrix} 1 \\ 0 \end{pmatrix} = \begin{pmatrix} 0 \\ 1 \end{pmatrix}, \quad \sigma_- \begin{pmatrix} 0 \\ 1 \end{pmatrix} = 0\,. \tag{35.48}$$

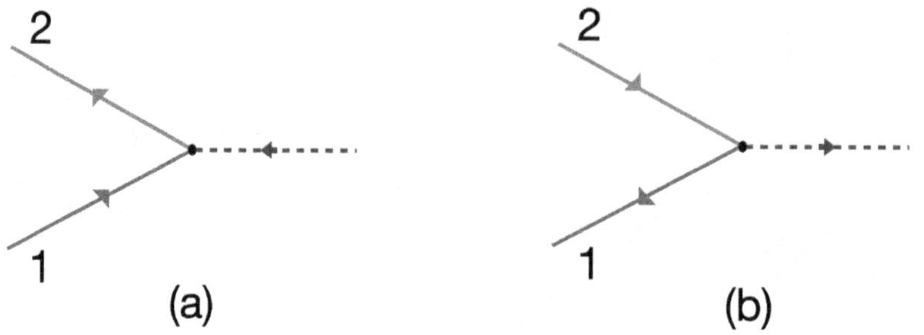

Figure 35.2 A two-level system either (a) absorbs a phonon (correspondingly jumping from the lower to the upper level) or (b) emits a phonon (correspondingly jumping from the upper to the lower level).

The two terms of Eq. (35.46) are represented pictorially in Fig. 35.2, where the *destruction* of a phonon makes the system pass from the lover level $\epsilon_{k=1}$ to the upper level $\epsilon_{k=2}$, while, vice versa, the creation of a phonon makes the system pass from the upper level $\epsilon_{k=2}$ to the lover level $\epsilon_{k=1}$.

With the help of the results utilized when proving Eq. (9.4) of Part I, we can also write:

$$\begin{aligned} e^{iH_{\mathrm{B}}(t-t_0)}\, H_{\mathrm{SB}}\, e^{-iH_{\mathrm{B}}(t-t_0)} &\rightarrow \sigma_+ \sum_j \beta_j\, e^{i\sum_{j'} \omega_{j'} b^\dagger_{j'} b_{j'} (t-t_0)}\, b_j\, e^{-i\sum_{j'} \omega_{j'} b^\dagger_{j'} b_{j'} (t-t_0)} + \text{h.c.} \\ &= \sigma_+ \sum_j \beta_j e^{-i\omega_j (t-t_0)}\, b_j + \text{h.c.}, \end{aligned} \tag{35.49}$$

which will be useful when calculating the expressions (35.29)–(35.31) explicitly.

Remark: Canonical form of the interaction Hamiltonian (35.46)
The interaction Hamiltonian (35.46) can be cast in the alternative form

$$H_{\mathrm{SB}} = \sigma_x \sum_j \beta_j \left(\frac{b_j^\dagger + b_j}{2}\right) + \sigma_y \sum_j \beta_j \left(\frac{b_j^\dagger - b_j}{2i}\right), \tag{35.50}$$

which complies with the expression (35.34) where both A_i and B_i are Hermitian operators. Nevertheless, below we will keep the form (35.46) because it is more useful for practical calculations.

Within this simple model, we now calculate the contributions of the terms ② × ②, ③ × ①, and ① × ③ to the self-energy (35.41) entering the equation of motion (35.42). In each case, only the nonvanishing matrix elements of the matrix (35.41) that result from the properties (35.48) will be reported below. In addition, from now on we shall set $t_0 = 0$ for simplicity.

Contribution to the self-energy (35.41) from the term ② × ②

$(k_2' = 2, k_2 = 2) \longrightarrow (k_1' = 1, k_1 = 1)$

$$\Sigma^{\mathrm{I}}_{k_2'=2k_2=2\to k_1'=1k_1=1}(t_2, t_1) = \sum_j \beta_j^2 \mathrm{Tr_B}\left\{\rho_{\mathrm{B}}(t_0=0) b_j b_j^\dagger\right\}$$
$$\times \langle k_1'=1|\sigma_-|k_2'=2\rangle(\downarrow)\langle k_2=2|\sigma_+|k_1=1\rangle\left(e^{i(\epsilon_2-\epsilon_1-\omega_j)(t_1-t_2)}+e^{-i(\epsilon_2-\epsilon_1-\omega_j)(t_1-t_2)}\right); \tag{35.51}$$

$(k_2' = 1, k_2 = 1) \longrightarrow (k_1' = 2, k_1 = 2)$

$$\Sigma^{\mathrm{I}}_{k_2'=1k_2=1\to k_1'=2k_1=2}(t_2, t_1) = \sum_j \beta_j^2 \mathrm{Tr_B}\left\{\rho_{\mathrm{B}}(t_0=0) b_j^\dagger b_j\right\}$$
$$\times \langle k_1'=2|\sigma_+|k_2'=1\rangle(\downarrow)\langle k_2=1|\sigma_-|k_1=2\rangle\left(e^{i(\epsilon_2-\epsilon_1-\omega_j)(t_1-t_2)}+e^{-i(\epsilon_2-\epsilon_1-\omega_j)(t_1-t_2)}\right). \tag{35.52}$$

Contribution to the self-energy (35.41) from the term ③ × ①

$(k_2' = 1, k_2 = 1) \longrightarrow (k_1' = 1, k_1 = 1)$

$$\Sigma^{\mathrm{I}}_{k_2'=1\,k_2=1\,\to k_1'=1\,k_1=1}(t_2, t_1) = -\,\delta_{k_1k_2}\,\delta_{k_1'k_2'}\sum_j \beta_j^2\,\mathrm{Tr_B}\left\{\rho_{\mathrm{B}}(t_0=0)\,b_j^\dagger b_j\right\}$$
$$\times(\downarrow)\,\langle k_2=1|\sigma_-|2\rangle\,\langle 2|\sigma_+|k_1=1\rangle\,e^{i(\epsilon_2-\epsilon_1-\omega_j)(t_1-t_2)}\,; \tag{35.53}$$

$(k_2' = 1, k_2 = 2) \longrightarrow (k_1' = 1, k_1 = 2)$

$$\Sigma^{\mathrm{I}}_{k_2'=1\,k_2=2\,\to k_1'=1\,k_1=2}(t_2, t_1) = -\,\delta_{k_1k_2}\,\delta_{k_1'k_2'}\sum_j \beta_j^2\,\mathrm{Tr_B}\left\{\rho_{\mathrm{B}}(t_0=0)\,b_j b_j^\dagger\right\}$$
$$\times(\downarrow)\,\langle k_2=2|\sigma_+|1\rangle\,\langle 1|\sigma_-|k_1=2\rangle\,e^{-i(\epsilon_2-\epsilon_1-\omega_j)(t_1-t_2)}\,; \tag{35.54}$$

$(k_2'=2, k_2=1) \longrightarrow (k_1'=2, k_1=1)$

$$\Sigma^{\mathrm{I}}_{k_2'=2\,k_2=1\,\to\,k_1'=2\,k_1=1}(t_2,t_1) = -\,\delta_{k_1k_2}\,\delta_{k_1'k_2'}\sum_j \beta_j^2\,\mathrm{Tr_B}\left\{\rho_{\mathrm{B}}(t_0=0)\,b_j^\dagger b_j\right\}$$
$$\times\,(\downarrow)\,\langle k_2=1|\sigma_-|2\rangle\,\langle 2|\sigma_+|k_1=1\rangle\,e^{i(\epsilon_2-\epsilon_1-\omega_j)(t_1-t_2)}\,; \tag{35.55}$$

$(k_2'=2, k_2=2) \longrightarrow (k_1'=2, k_1=2)$

$$\Sigma^{\mathrm{I}}_{k_2'=2\,k_2=2\,\to\,k_1'=2\,k_1=2}(t_2,t_1) = -\,\delta_{k_1k_2}\,\delta_{k_1'k_2'}\sum_j \beta_j^2\,\mathrm{Tr_B}\left\{\rho_{\mathrm{B}}(t_0=0)\,b_j b_j^\dagger\right\}$$
$$\times\,(\downarrow)\,\langle k_2=2|\sigma_+|1\rangle\,\langle 1|\sigma_-|k_1=2\rangle\,e^{-i(\epsilon_2-\epsilon_1-\omega_j)(t_1-t_2)}\,. \tag{35.56}$$

Contribution to the self-energy (35.41) from the term ① × ③

$(k_2'=1, k_2=1) \longrightarrow (k_1'=1, k_1=1)$

$$\Sigma^{\mathrm{I}}_{k_2'=1\,k_2=1\,\to\,k_1'=1\,k_1=1}(t_2,t_1) = -\,\delta_{k_1k_2}\,\delta_{k_1'k_2'}\sum_j \beta_j^2\,\mathrm{Tr_B}\left\{\rho_{\mathrm{B}}(t_0=0)\,b_j^\dagger b_j\right\}$$
$$\times\,\langle k_1'=1|\sigma_-|2\rangle\,\langle 2|\sigma_+|k_2'=1\rangle\,(\downarrow)\,e^{-i(\epsilon_2-\epsilon_1-\omega_j)(t_1-t_2)}\,; \tag{35.57}$$

$(k_2'=1, k_2=2) \longrightarrow (k_1'=1, k_1=2)$

$$\Sigma^{\mathrm{I}}_{k_2'=1\,k_2=2\,\to\,k_1'=1\,k_1=2}(t_2,t_1) = -\,\delta_{k_1k_2}\,\delta_{k_1'k_2'}\sum_j \beta_j^2\,\mathrm{Tr_B}\left\{\rho_{\mathrm{B}}(t_0=0)\,b_j^\dagger b_j\right\}$$
$$\times\,\langle k_1'=1|\sigma_-|2\rangle\,\langle 2|\sigma_+|k_2'=1\rangle\,(\downarrow)\,e^{-i(\epsilon_2-\epsilon_1-\omega_j)(t_1-t_2)}\,; \tag{35.58}$$

$(k_2'=2, k_2=1) \longrightarrow (k_1'=2, k_1=1)$

$$\Sigma^{\mathrm{I}}_{k_2'=2\,k_2=1\,\to\,k_1'=2\,k_1=1}(t_2,t_1) = -\,\delta_{k_1k_2}\,\delta_{k_1'k_2'}\sum_j \beta_j^2\,\mathrm{Tr_B}\left\{\rho_{\mathrm{B}}(t_0=0)\,b_j b_j^\dagger\right\}$$
$$\times\,\langle k_1'=2|\sigma_+|1\rangle\,\langle 1|\sigma_-|k_2'=2\rangle\,(\downarrow)\,e^{i(\epsilon_2-\epsilon_1-\omega_j)(t_1-t_2)}\,; \tag{35.59}$$

$$(k_2' = 2, k_2 = 2) \longrightarrow (k_1' = 2, k_1 = 2)$$

$$\Sigma^{\mathrm{I}}_{k_2'=2\,k_2=2 \to k_1'=2\,k_1=2}(t_2, t_1) = -\,\delta_{k_1 k_2}\,\delta_{k_1' k_2'} \sum_j \beta_j^2\,\mathrm{Tr_B}\left\{\rho_{\mathrm{B}}(t_0 = 0)\,b_j b_j^\dagger\right\}$$
$$\times \langle k_1' = 2|\sigma_+|1\rangle\,\langle 1|\sigma_-|k_2' = 2\rangle\,(\downarrow)\,e^{i(\epsilon_2 - \epsilon_1 - \omega_j)(t_1 - t_2)}\,. \qquad (35.60)$$

In the above expressions, we have:

(i) Adopted the short-hand notation $\epsilon_{k=1} = \epsilon_1$ and $\epsilon_{k=2} = \epsilon_2$.
(ii) Taken into account the property

$$\mathrm{Tr_B}\{\rho_{\mathrm{B}}(t_0 = 0)\,b_j b_j\} = \mathrm{Tr_B}\left\{\rho_{\mathrm{B}}(t_0 = 0)\,b_j^\dagger b_j^\dagger\right\} = 0 \qquad (35.61)$$

in line with the assumption (37.9) of Section 37.1, and highlighted the presence of the thermal averages

$$\mathrm{Tr_B}\left\{\rho_{\mathrm{B}}(t_0 = 0)\,b_j^\dagger b_j\right\} = n_{\mathrm{th}}(\omega_j) = \frac{1}{e^{\beta\omega_j} - 1} \qquad (35.62)$$

$$\mathrm{Tr_B}\left\{\rho_{\mathrm{B}}(t_0 = 0)\,b_j b_j^\dagger\right\} = \mathrm{Tr_B}\left\{\rho_{\mathrm{B}}(t_0 = 0)\left(b_j^\dagger b_j + 1\right)\right\} = n_{\mathrm{th}}(\omega_j) + 1 \qquad (35.63)$$

where $n_{\mathrm{th}}(\omega_j)$ is the *bath thermal distribution* at temperature T with $\beta = (k_B T)^{-1}$. If we further assume that the coupling coefficients $\{\beta_j\}$ depend on the index j only through the phonon frequency ω_j, we can write

$$\sum_{\{j\}} f_j \to \sum_{\{j\}} f(\omega_j) = \sum_{\{j\}} \int_0^\infty d\omega\, f(\omega)\,\delta(\omega - \omega_j)$$
$$= \int_0^\infty d\omega\, f(\omega) \sum_{\{j\}} \delta(\omega - \omega_j) = \int_0^\infty d\omega\, f(\omega)\,\nu(\omega), \qquad (35.64)$$

where f_j is either $\beta_j^2\,n_{\mathrm{th}}(\omega_j)$ or $\beta_j^2\left(n_{\mathrm{th}}(\omega_j) + 1\right)$. In the expression (35.64), we have assumed that $\{\omega_j \geq 0\}$ and introduced a frequency decomposition in terms of the *bath density of states* (cf. also Step # 7 of Section 37.1),

$$\nu(\omega) = \sum_{\{j\}} \delta(\omega - \omega_j)\,. \qquad (35.65)$$

(iii) Inserted a down arrow ($\downarrow$) in the places where the density matrix $\rho_{\mathrm{I}}^{k_2' k_2}(t_2)$ is going to be located when the above matrix elements are transferred into the equation of motion (35.42). This result implies that the contributions of the terms ② × ②, ③ × ①, and ① × ③ will give rise, respectively, to the first, second, and third terms on the right-hand side of the Lindblad equation given by Eqs. (32.1)–(32.2) in Section 32.7 (or, else, by Eq. (37.41) in Section 37.1), also with the correct relative signs and 2:1 weights.

(iv) For the sake of formal manipulations, left explicit the matrix elements of the *jump operators* $\sigma_\pm$ (which in the present case represent the operators V_n of Eqs. (32.1)–(32.2), or else the operators $A_r(\omega)$ of Eq. (37.41) in Section 37.1), although all the relevant matrix elements of $\sigma_\pm$ are actually unity owing to Eq. (35.48).

Note finally that, when entering the results (35.51)–(35.60) for $\Sigma^{\mathrm{I}}_{k_2'k_2\to k'k}(t_2,t)$ into the Dyson-like equation of motion (35.42) for $\rho_{\mathrm{I}}^{k'k}(t)$, "memory effects" about the past history contained in $\rho_{\mathrm{I}}^{k_2'k_2}(t_2)$ for times $t_0 \le t_2 \le t$ have still to be taken into account, as we have pointed out near the end of Section 35.7. In Section 35.10, we will discuss the additional assumptions required to get rid of these memory effects, retrieving in this way the Lindblad equation in its canonical form. Before doing that, however, due to the rather simple structure of the results (35.51)–(35.60), in Section 35.9 we will briefly consider the way Eq. (35.42) can formally be solved even when taking into account memory effects.

35.9 Solving for the Equation of Motion with Memory Effects

The dynamics of the two-level system coupled to a phonon bath that we are considering is simple enough that memory effects can be fully taken into account. To this end, it is convenient to resort to the Laplace transform, whose properties relevant in the present context are summarized as follows.

Let $f(t)$ be a function defined for all real numbers $t \ge 0$ and s a complex variable with $\Re\{s\} \ge 0$. Its Laplace transform, f_s is then given by (cf., e.g., Chapter 29 of Ref. [204]):

$$f_s = \int_0^\infty dt\, e^{-st} f(t)\,. \tag{35.66}$$

Let's consider, in particular, the following cases:

$$\begin{aligned}
&\text{(i) } f(t) = e^{-i\omega t} && \Longrightarrow && f_s = \int_0^\infty dt\, e^{-st} e^{-i\omega t} = \tfrac{1}{s+i\omega}\ ;\\
&\text{(ii) } f(t) = g(t)\, e^{-i\omega t} && \Longrightarrow && f_s = \int_0^\infty dt\, e^{-st} g(t)\, e^{-i\omega t} = g_{s+i\omega}\ ;\\
&\text{(iii) } f(t) = \tfrac{dg(t)}{dt} && \Longrightarrow && f_s = \int_0^\infty dt\, e^{-st}\, \tfrac{dg(t)}{dt} = s\, g_s - g(t=0)\ ;\\
&\text{(iv) } f(t) = \int_0^t dt'\, g(t') && \Longrightarrow && f_s = \tfrac{1}{s}\, g_s\ ;
\end{aligned}$$

with the assumption that all integrals extending up to ∞ are convergent.

With these provisions, we now take the Laplace transform of both sides of the equation of motion (35.42) with $t_0 = 0$, where the matrix elements of the self-energy $\Sigma^{\mathrm{I}}_{k_2'k_2\to k'k}(t_2,t)$ are given by the expressions (35.51)–(35.60). Owing to the above properties of the Laplace transform, for the left-hand side of the equation of motion we obtain

$$\int_0^\infty dt\, e^{-st} \frac{\partial}{\partial t}\rho_{\rm I}^{k'k}(t) = s\,\rho_{\rm I}^{k'k}(s) - \rho_{\rm I}^{k'k}(t=0)\,, \tag{35.67}$$

while, with reference to the expressions (35.51)–(35.60), the right-hand side of the equation of motion contains the factors

$$\int_0^\infty dt\, e^{-st} \int_0^t dt'\, \rho_{\rm I}^{k_2'k_2}(t')\, e^{\pm i\Omega(t-t')} = \frac{\rho_{\rm I}^{k_2'k_2}(s)}{s \mp i\Omega}, \tag{35.68}$$

where $\Omega = \epsilon_2 - \epsilon_1 - \omega = \omega_0 - \omega$. We further introduce the notation

$$\frac{1}{s \pm i\Omega} = \frac{s \mp i\Omega}{s^2 + \Omega^2} = \pi\, \mathcal{L}(\Omega|s) \mp i\, \frac{\Omega}{s^2 + \Omega^2}, \tag{35.69}$$

where

$$\mathcal{L}(\Omega|s) = \frac{1}{\pi}\, \frac{s}{s^2 + \Omega^2} \tag{35.70}$$

is a Lorentzian function of width s in the variable Ω, such that

$$\frac{1}{s + i\Omega} + \frac{1}{s - i\Omega} = 2\pi\, \mathcal{L}(\Omega|s)\,. \tag{35.71}$$

We then obtain for the various matrix elements of the equation of motion:

$$\begin{aligned} s\,\rho_{\rm I}^{11}(s) - \rho_{\rm I}^{11}(t=0) &= \int_0^\infty d\omega\, \nu(\omega)\, \beta(\omega)^2 (n_{\rm th}(\omega) + 1)\, 2\pi\, \mathcal{L}(\omega_0 - \omega|s)\, \rho_{\rm I}^{22}(s) \\ &\quad - \int_0^\infty d\omega\, \nu(\omega)\, \beta(\omega)^2\, n_{\rm th}(\omega)\, 2\pi\, \mathcal{L}(\omega_0 - \omega|s)\, \rho_{\rm I}^{11}(s) \end{aligned} \tag{35.72}$$

$$s\,\rho_{\rm I}^{12}(s) - \rho_{\rm I}^{12}(t=0) = -\int_0^\infty d\omega\, \nu(\omega)\, \beta(\omega)^2 (2n_{\rm th}(\omega) + 1)\, \frac{1}{s + i(\omega_0 - \omega)}\, \rho_{\rm I}^{12}(s) \tag{35.73}$$

$$s\,\rho_{\rm I}^{21}(s) - \rho_{\rm I}^{21}(t=0) = -\int_0^\infty d\omega\, \nu(\omega)\, \beta(\omega)^2 (2n_{\rm th}(\omega) + 1)\, \frac{1}{s - i(\omega_0 - \omega)}\, \rho_{\rm I}^{21}(s) \tag{35.74}$$

$$\begin{aligned} s\,\rho_{\rm I}^{22}(s) - \rho_{\rm I}^{22}(t=0) &= \int_0^\infty d\omega\, \nu(\omega)\, \beta(\omega)^2\, n_{\rm th}(\omega)\, 2\pi\, \mathcal{L}(\omega_0 - \omega|s)\, \rho_{\rm I}^{11}(s) \\ &\quad - \int_0^\infty d\omega\, \nu(\omega)\, \beta(\omega)^2 (n_{\rm th}(\omega) + 1)\, 2\pi\, \mathcal{L}(\omega_0 - \omega|s)\, \rho_{\rm I}^{22}(s)\,. \end{aligned} \tag{35.75}$$

For given value of s, this represents a set of four linear equations in the four variables $\rho_{\rm I}^{ij}(s)$ with $i, j = 1, 2$ and known terms $\rho_{\rm I}^{ij}(t=0)$, which can be readily solved over a suitable interval of values of s. Once the functions $\rho_{\mathrm{I}}^{ij}(s)$ are obtained in this way, the functions $\rho_{\mathrm{I}}^{ij}(t)$ of real time t can then be obtained numerically by Laplace transform methods [205].

In particular, if the quantities $\rho_{\mathrm{I}}^{ij}(t)$ vary "sufficiently" slowly in time, we can limit to consider only sufficiently small values of s, such that we may approximate

$$\lim_{s\to 0} \mathcal{L}(\omega_0 - \omega|s) = \delta(\omega - \omega_0) \quad \text{and} \quad \lim_{s\to 0} \frac{\omega_0 - \omega}{s^2 + (\omega_0 - \omega)^2} = \mathcal{P}\left(\frac{1}{\omega_0 - \omega}\right), \tag{35.76}$$

where $\mathcal{P}$ denotes principal value. In this case, the coefficients of Eqs. (35.72)–(35.75) simplify to the form

$$\int_0^\infty d\omega\, \nu(\omega)\, \beta(\omega)^2\, (n_{\mathrm{th}}(\omega) + 1)\, 2\pi\, \mathcal{L}(\omega_0 - \omega|s)$$
$$\to 2\pi\, \nu(\omega_0)\, \beta(\omega_0)^2\, (n_{\mathrm{th}}(\omega_0) + 1) \equiv \gamma_+ \tag{35.77}$$

$$\int_0^\infty d\omega\, \nu(\omega)\, \beta(\omega)^2\, n_{\mathrm{th}}(\omega)\, 2\pi\, \mathcal{L}(\omega_0 - \omega|s) \;\to\; 2\pi\, \nu(\omega_0)\, \beta(\omega_0)^2\, n_{\mathrm{th}}(\omega_0) \equiv \gamma_- \tag{35.78}$$

$$\int_0^\infty d\omega\, \nu(\omega)\, \beta(\omega)^2\, (2n_{\mathrm{th}}(\omega) + 1)\, \frac{1}{s \pm i(\omega_0 - \omega)} \;\to\; \left(\frac{\gamma_+ + \gamma_-}{2}\right) \mp i S, \tag{35.79}$$

where we have introduced the further short-hand notation

$$S \equiv \int_0^\infty d\omega\, \nu(\omega)\, \beta(\omega)^2 (2n_{\mathrm{th}}(\omega) + 1)\, \mathcal{P}\left(\frac{1}{\omega_0 - \omega}\right). \tag{35.80}$$

By utilizing these approximate solutions in Eqs. (35.72)–(35.75) and transforming the resulting expressions back to real times, the elements $\rho_{\mathrm{I}}^{k'k}(t)$ entering both sides of the equation of motion (35.42) appear to be taken at the *same time* t. This result is consistent with the Markov approximation, signifying that with the approximation (35.76) memory effects have been lost along the way (as it was implicitly implied by the initial assumption of a slow time variation of $\rho_{\mathrm{I}}^{ij}(t)$). This is actually what is needed for retrieving the Lindblad equation (as it is done in Step # 6 of Section 37.1).

35.10 Retrieving the Lindblad Equation

To better appreciate the physical circumstances under which memory effects can safely be dismissed in the equation of motion (35.42) with the self-energy given by the expressions (35.51)–(35.60), it is convenient to go back to the original time representation of the equation of motion and analyze the relevant terms on its right-hand side along the following lines.

As a typical example, let's consider the first term of the expression (35.51), which gives the following contribution to the right-hand side of the equation of motion (35.42):

$$
\begin{aligned}
&\int_0^t dt_2\, \rho_{\rm I}^{22}(t_2) \sum_j \beta_j^2\, {\rm Tr}_{\rm B}\left\{\rho_{\rm B}(t_0=0)\, b_j b_j^\dagger\right\}\, e^{i(\epsilon_2-\epsilon_1-\omega_j)(t-t_2)} && \text{(i)}\\
&= \int_0^t dt_2\, \rho_{\rm I}^{22}(t_2)\, e^{i\omega_0(t-t_2)} \underbrace{\sum_j \beta_j^2\, {\rm Tr}_{\rm B}\left\{\rho_{\rm B}(t_0=0)\, b_{j{\rm H}}(t)\, b_{j{\rm H}}^\dagger(t_2)\right\}}_{\text{short range in } (t-t_2)} && \text{(ii)}\\
&\simeq \rho_{\rm I}^{22}(t) \int_0^t dt_2 \int_0^\infty d\omega\, \nu(\omega)\, \beta(\omega)^2 (n_{\rm th}(\omega)+1)\, e^{i(\omega_0-\omega)(t-t_2)} && \text{(iii)}\\
&= \rho_{\rm I}^{22}(t) \int_0^t dt' \int_0^\infty d\omega\, \nu(\omega)\, \beta(\omega)^2 (n_{\rm th}(\omega)+1)\, e^{i(\omega_0-\omega)t'} && \text{(iv)}\\
&\simeq \rho_{\rm I}^{22}(t) \int_0^\infty dt' \int_0^\infty d\omega\, \nu(\omega)\, \beta(\omega)^2 (n_{\rm th}(\omega)+1)\, e^{i(\omega_0-\omega+i\eta)t'} && \text{(v)}\\
&= \rho_{\rm I}^{22}(t) \int_0^\infty d\omega\, \nu(\omega)\, \beta(\omega)^2 (n_{\rm th}(\omega)+1) \int_0^\infty dt'\, e^{i(\omega_0-\omega+i\eta)t'} && \text{(vi)}\\
&= \rho_{\rm I}^{22}(t) \int_0^\infty d\omega\, \nu(\omega)\, \beta(\omega)^2 (n_{\rm th}(\omega)+1) \left(\pi\, \delta(\omega_0-\omega) + i\,\mathcal{P}\left(\frac{1}{\omega_0-\omega}\right)\right). && \text{(vii)}
\end{aligned}
\tag{35.81}
$$

In Eq. (35.81), we notice the following:

(i) The first line corresponds to the original expression which keeps memory of the past history of $\rho_{\rm I}^{22}(t_2)$ for times $t_2 \le t$.

(ii) The second line highlights the presence of the bath correlation function, averaged over all bath modes with weights β_j^2, where $b_{j{\rm H}}(t) = e^{iH_{\rm B}t} b_j e^{-iH_{\rm B}t}$ is the Heisenberg representation of the bath operator b_j. This correlation function is regarded to be of short range in the time difference $(t-t_2)$, with respect to the time variations of the density matrix $\rho_{\rm I}^{22}(t_2)$ itself. Physically, this is related to the fact that the spectrum $\{\omega_j\}$ form a *continuum* (cf. Step # 8 of Section 37.1).

(iii) The third line implements the consequences of the above assumptions, by replacing $\rho_{\rm I}^{22}(t_2) \to \rho_{\rm I}^{22}(t)$ and introducing the integral over the continuous frequency ω as we did in Eq. (35.64). This corresponds to the so-called Markov approximation discussed in Step # 6 of Section 37.1.

(iv) The fourth line changes the time integration variable from t_2 to the relative time $t' = t - t_2$. In this way, the external time t appears only in the upper limit of the time integral.

(v) The fifth line lets the upper limit of the time integral extend up to ∞, on the basis of the fact that the bath correlation function is of short range in t'. At the same time, to avoid introducing spurious divergencies, the integral over t' is regularized by letting $\omega_0 \rightarrow \omega_0 + i\eta$ in the exponential factor with $\eta = 0^+$.

(vi) The sixth line simply interchanges the integral over t' with the integral over ω.

(vii) The seventh line calculates the integral over t' explicitly, yielding the "on-the-energy-shell" term $\delta(\omega_0 - \omega)$ and the "off-the-energy-shell" term $\mathcal{P}\left(\frac{1}{\omega_0-\omega}\right)$ with respect to the system excitation energy ω_0. Similar to the configuration interaction of a discrete state embedded in a continuum [104], here the first term is responsible for the decay of the system excitation into the bath continuum, while the second term produces a shift of the subsystem excitation energy ω_0 caused by interactions with virtual phonons, which is in this sense analogous to the Lamb shift (cf. Step # 9 of Section 37.1).

The above analysis can be repeated for the various contributions to the equation of motion (35.42) which originate from the expressions (35.51)–(35.60) of the self-energy. The end result is the following set of four time-dependent equations:

$$\frac{\partial}{\partial t}\rho_{\rm I}^{11}(t) = \gamma_+\, \rho_{\rm I}^{22}(t) \;-\; \gamma_-\, \rho_{\rm I}^{11}(t) \tag{35.82}$$

$$\frac{\partial}{\partial t}\rho_{\rm I}^{12}(t) = -\left(\frac{\gamma_+ + \gamma_-}{2}\right)\rho_{\rm I}^{12}(t) \;+\; i\,S\,\rho_{\rm I}^{12}(t) \tag{35.83}$$

$$\frac{\partial}{\partial t}\rho_{\rm I}^{21}(t) = -\left(\frac{\gamma_+ + \gamma_-}{2}\right)\rho_{\rm I}^{21}(t) \;-\; i\,S\,\rho_{\rm I}^{21}(t) \tag{35.84}$$

$$\frac{\partial}{\partial t}\rho_{\rm I}^{22}(t) = \gamma_-\, \rho_{\rm I}^{11}(t) \;-\; \gamma_+\, \rho_{\rm I}^{22}(t), \tag{35.85}$$

where γ_+, γ_-, and S are given by Eqs. (35.77), (35.78), and (35.80), respectively. It can be readily verified that these equations correspond to the k'–k matrix elements of the LME with jump operators $\sigma_\pm$, of the form:

$$\boxed{\begin{cases} \frac{\partial}{\partial t}\rho_{\rm I}(t) = \gamma_+\sigma_-\rho_{\rm I}(t)\sigma_+ + \gamma_-\sigma_+\rho_{\rm I}(t)\sigma_- - \frac{1}{2}\gamma_+\rho_{\rm I}(t)\sigma_+\sigma_- - \frac{1}{2}\gamma_-\rho_{\rm I}(t)\sigma_-\sigma_+ \\ \qquad -\frac{1}{2}\gamma_+\sigma_+\sigma_-\rho_{\rm I}(t) - \frac{1}{2}\gamma_-\sigma_-\sigma_+\rho_{\rm I}(t) - i\left[H^{\rm LS}, \rho_{\rm I}(t)\right]. \end{cases}} \tag{35.86}$$

Note that $H^{\rm LS}$ is diagonal in this basis with $H^{\rm LS}_{22} - H^{\rm LS}_{11} = S$, such that $H^{\rm LS}$ affects only the off-diagonal elements (35.83) and (35.84) (cf. also Step # 10 of Section 37.1).

35.11 A Simple Solution

To a first approximation, we may neglect the quantity S given by Eq. (35.80), with the argument that the frequency integrand therein has opposite signs on the two sides of ω_0, which is assumed to lie well within the phonon spectral range. Recalling further the definitions (35.77) for γ_+ and (35.78) for γ_-, we may there set $\gamma_+ = (N_0 + 1)\,\gamma_0$ and $\gamma_- = N_0\,\gamma_0$, where $\gamma_0 = 2\,\pi\nu(\omega_0)\beta(\omega_0)^2$ and $N_0 = n_{\text{th}}(\omega_0)$ is the number of phonons at the transition frequency ω_0, such that $\gamma_+ + \gamma_- = (2N_0 + 1)\gamma_0 \equiv \gamma$.

Next we note that the four equations (35.82)–(35.85) can be reduced to three only since

$$\frac{\partial}{\partial t}\left(\rho_{\mathrm{I}}^{11}(t) + \rho_{\mathrm{I}}^{22}(t)\right) = 0\,, \tag{35.87}$$

which implies that $\rho_{\mathrm{I}}^{11}(t) + \rho_{\mathrm{I}}^{22}(t) = \rho_{\mathrm{I}}^{11}(t=0) + \rho_{\mathrm{I}}^{22}(t=0) = 1$. We then consider the equation for the difference $f(t) = \rho_{\mathrm{I}}^{11}(t) - \rho_{\mathrm{I}}^{22}(t)$, which reads

$$\begin{aligned}\frac{\partial}{\partial t}f(t) &= \frac{\partial}{\partial t}\left(\rho_{\mathrm{I}}^{11}(t) - \rho_{\mathrm{I}}^{22}(t)\right) = 2\gamma_+\rho_{\mathrm{I}}^{22}(t) - 2\gamma_-\rho_{\mathrm{I}}^{11}(t)\\ &= (2N_0 + 2)\gamma_0\rho_{\mathrm{I}}^{22}(t) - 2N_0\gamma_0\rho_{\mathrm{I}}^{11}(t) = (2N_0 + 1)\gamma_0\left(\rho_{\mathrm{I}}^{22}(t) - \rho_{\mathrm{I}}^{11}(t)\right)\\ &\quad + \gamma_0\left(\rho_{\mathrm{I}}^{11}(t) + \rho_{\mathrm{I}}^{22}(t)\right) = -\gamma\left(\rho_{\mathrm{I}}^{11}(t) - \rho_{\mathrm{I}}^{22}(t)\right) + \gamma_0\\ &= -\gamma f(t) + \gamma_0.\end{aligned} \tag{35.88}$$

The function $f(t)$ has thus the stationary solution $f_{\mathrm{s}} \equiv \frac{\gamma_0}{\gamma} = \frac{1}{2\,N_0+1}$. In addition, the remaining off-diagonal equations (35.83) and (35.84) read simply $\frac{\partial}{\partial t}\rho_{\mathrm{I}}^{12}(t) = -\frac{\gamma}{2}\,\rho_{\mathrm{I}}^{12}(t)$ and $\frac{\partial}{\partial t}\rho_{\mathrm{I}}^{21}(t) = -\frac{\gamma}{2}\,\rho_{\mathrm{I}}^{21}(t)$, yielding an exponential decay for $\rho_{\mathrm{I}}^{12}(t)$ and $\rho_{\mathrm{I}}^{21}(t)$ with half the coefficient γ of Eq. (35.88).

Suppose now for definiteness that the subsystem was initially prepared in the excited state $|2\rangle$, such that $f(t=0) = -1$. The corresponding solution to Eq. (35.88) then reads

$$f(t) = -\,e^{-\gamma t}(f_{\mathrm{s}} + 1) + f_{\mathrm{s}}\,, \tag{35.89}$$

such that $\lim_{t\to\infty} f(t) = f_{\mathrm{s}}$. This gives for the diagonal matrix elements of $\rho_{\mathrm{I}}(t)$:

$$\rho_{\mathrm{I}}^{11}(t) = \frac{1 + f(t)}{2} \xrightarrow{t\to\infty} \frac{1 + f_{\mathrm{s}}}{2} = \frac{\gamma + \gamma_0}{2\gamma} = \frac{N_0 + 1}{2N_0 + 1} \tag{35.90}$$

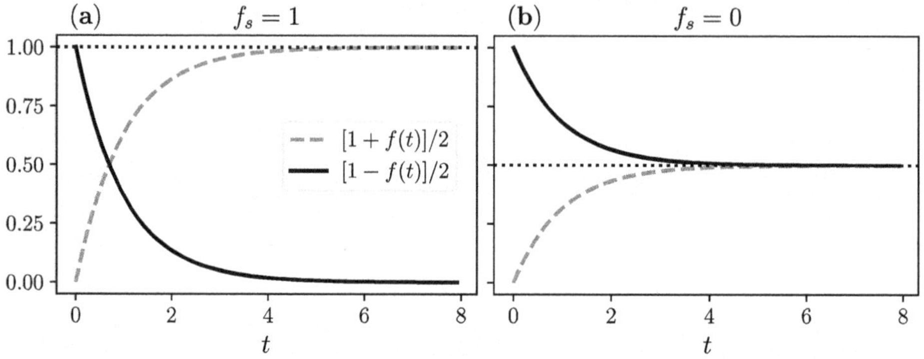

Figure 35.3 The diagonal elements $\rho_{\mathrm{I}}^{11}(t)$ (dashed lines) and $\rho_{\mathrm{I}}^{22}(t)$ (full lines), given by Eqs. (35.90) and (35.91), respectively, where $f(t)$ is given by the expression (35.89) with $\gamma = 1$, are plotted as function of t (a) at zero temperature when $f_{\mathrm{s}} = 1$ and (b) at high temperature when $f_{\mathrm{s}} \simeq 0$.

$$\rho_{\mathrm{I}}^{22}(t) = \frac{1-f(t)}{2} \xrightarrow{t\to\infty} \frac{1-f_{\mathrm{s}}}{2} = \frac{\gamma-\gamma_0}{2\gamma} = \frac{N_0}{2N_0+1}\,. \tag{35.91}$$

In particular:

(i) At zero temperature whereby $N_0 = 0$, $f_{\mathrm{s}} = 1$ such that $\lim_{t\to\infty} \rho_{\mathrm{I}}^{11}(t) = 1$ and $\lim_{t\to\infty} \rho_{\mathrm{I}}^{22}(t) = 0$;
(ii) At finite temperature with $N_0 \gg 1$, $f_{\mathrm{s}} \simeq 0$ such that $\lim_{t\to\infty} \rho_{\mathrm{I}}^{11}(t) \simeq \frac{1}{2}$ and $\lim_{t\to\infty} \rho_{\mathrm{I}}^{22}(t) \simeq \frac{1}{2}$.

These two opposite situations are represented graphically in Fig. 35.3, in panel (a) at zero temperature when $f_{\mathrm{s}} = 1$ and in panel (b) at high temperature when $f_{\mathrm{s}} \simeq 0$. Note that in the first case, the system tends asymptotically to populate the lower energy state $|1\rangle$, while in the second case, an asymptotic equilibrium is reached between the populations of the energy states $|1\rangle$ and $|2\rangle$ owing to the effect of the thermal bath.

A variant of the above calculation was provided in Ref. [206], where to the subsystem diagonal Hamiltonian H_{S} given by Eq. (35.43) an off-diagonal term of the form $-\frac{\Omega_{\mathrm{R}}}{2}\,\sigma_x$ is added (Ω_{R} being the Rabi frequency), with the consequence that the eigenvectors $|\pm\rangle$ of the total subsystem Hamiltonian are now linear combinations of those of H_{S} given by Eq. (35.43). This feature, in turn, reflects itself also in the matrix elements of the density matrix $\rho_{\mathrm{I}}(t)$. In addition, in Ref. [206], only the zero-temperature case was considered, such that $N_0 = 0$ with $\gamma_- = 0$ and $\gamma_+ \neq 0$. Figure 35.4 shows the ensuing numerical results for the time-dependent populations of the ground $|-\rangle$ and excited $|+\rangle$ states for various values of the Rabi frequency Ω_{R}, with the same initial conditions utilized in Fig. 35.3. It is seen that

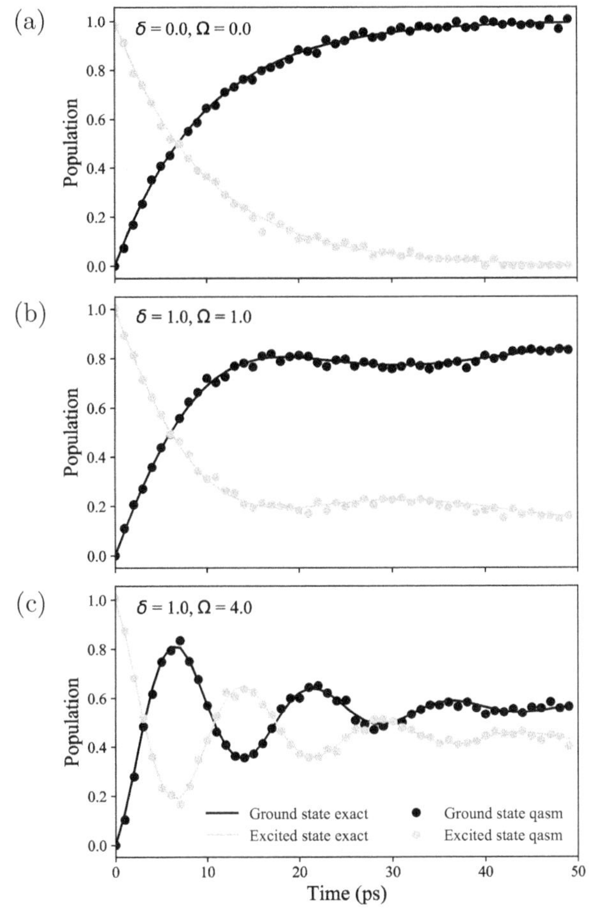

Figure 35.4 (a)–(c) The populations of the ground and excited states of the two-level Hamiltonian $-\frac{1}{2}(\delta\,\sigma_z + \Omega\,\sigma_x)$ (with the mapping $\delta \leftrightarrow -\omega_0$ and $\Omega \leftrightarrow \Omega_R$ into our conventions) are shown for various values of the Rabi frequency Ω when $\gamma_+ = 1$ and $\gamma_- = 0$ in the Lindblad equation. [Reproduced from Fig. 1 of Ref. [206], under the Creative Commons Attribution 4.0 International license (https://creativecommons.org/licenses/by/4.0/).]

increasing values of Ω_R not only lead to larger oscillations but also make the two populations to equilibrate asymptotically like in the thermal case of Fig. 35.3(b) since $|\pm\rangle \rightarrow (|k=1\rangle \mp |k=2\rangle)/\sqrt{2}$ when $\Omega_R \gg \omega_0$.

Note that in all the above cases, the solution of the LME (35.86) has resulted in an *irreversible* evolution toward equilibrium, as it would have been expected on physical grounds for an open quantum system coupled to an infinitely large environment (or bath) with a continuum spectrum of frequencies. This irreversibility thus selects a particular arrow of time, namely, from $t=0$ to $t=\infty$, evidencing the fact that the Lindblad equation as it stands is not time-reversal symmetric. In this context, it is interesting to mention a recent formal analysis reported in Ref. [207], which has traced the origin of the irreversibility of the Lindblad equation in the approximation made in the fifth line of Eq. (35.81) (cf. also Step # 6 of Section 37.1), where the upper limit of the time interval was let to go to ∞ at an early stage of the derivation of the Lindblad equation itself.

36

State-of-the-Art Numerical Methods

For quantum many-body systems at equilibrium, diagrammatic methods are nowadays routinely implemented with powerful numerical techniques. In particular, within the zero-temperature formalism [32], one should mention the GW approximation (cf. Section 17.4 of Part I for its time-dependent version) for predicting electronic single-particle excitations in chemical compounds and materials [53], with the associated Bethe–Salpeter equation (cf. Section 17.5 of Part I for its time-dependent version) for two-particle excited state calculations [208, 209]. Or, within the Matsubara formalism with the ensuing analytic continuation from imaginary to real frequencies, one should mention the t-matrix approximation (cf. Chapter 16 of Part I in the normal phase and Chapter 24 of Part II in the superfluid phase for its time-dependent versions) for systems like ultracold gases with short-range interparticle interaction, even at temperatures comparable with the Fermi temperature [48].

For quantum many-body systems out of equilibrium, on the other hand, the main difficulties one encounters in practice when implementing the solution of the Kadanoff–Baym (KB) equations stem from the fact that, in this case, the single-particle Green's functions $G(t, t')$ depend separately on two times t and t' (and not on their difference $t - t'$ like in the equilibrium case). Added to this double-time structure (which is rather expensive as far as both computing time and storage memory are concerned) is the proliferation of the components of the KB equations to be simultaneously solved (which are now four, instead of two of the equilibrium case – cf. also Table 30.1 of Part II).

In this chapter, a brief survey will first be given about the so-called time-stepping procedure and predictor-corrector scheme for solving the KB equations with two time variables, and then, the "generalized Kadanoff–Baym ansatz" (GKBA) will be introduced, which aims at somewhat simplifying the solution of the KB equations themselves.

36.1 The Time-Stepping Procedure for the KB Equations

In nonequilibrium situations, discretizing (or, digitalizing) the time variables transforms the real-time KB equations into a linear system of equations, which can be solved by means of standard operations such as matrix multiplications and inversions. In this respect, a useful algorithm for solving numerically the KB equations is provided by the *time-stepping procedure* introduced in Ref. [210], whose numerical performance can further be improved by supplementing it with the *predictor-corrector scheme* (cf., e.g., section S1 of Ref. [182]).

Here, we provide a short summary of this combined "procedure-plus-scheme," which is only meant to serve as a concise introduction to the topic. For a more extensive treatment, the interested reader may consult the survey on the state-of-the-art numerical calculations for solving the KB equations, which can be found in Ref. [159] (cf., in particular, section 6 therein). In addition, one may mention the description of the open-source software package NESSi given in Ref. [211], which allows one to perform many-body dynamics simulations based specifically on the Green's functions for the Schwinger–Keldysh closed-contour approach.

36.1.1 General Considerations

The differential equations to be solved have the generic structure

$$\frac{df(t)}{dt} = g(t), \tag{36.1}$$

where $g(t)$ is a known function and $f(t)$ is the function to be determined. Suppose that $f(t_0)$ at the initial time t_0 is known. Then, at a later time $t = t_0 + \delta t$ with δt sufficiently small, we can approximate

$$f(t_0 + \delta t) = f(t_0) + \delta t\, g(t_0)\,. \tag{36.2}$$

We can go on in this way to further later times $t_0 + 2\delta t$, $t_0 + 3\delta t, \cdots, t_0 + (n+1)\delta t$ with time steps of length δt and write accordingly

$$f(t_0 + (n+1)\delta t) = f(t_0 + n\delta t) + \delta t\, g(t_0 + n\delta t)\,. \tag{36.3}$$

This *time-stepping procedure* can be further improved by using, for example, an average value like

$$\frac{1}{2}\left[g(t_0 + n\delta t) + g(t_0 + (n+1)\delta t)\right] \tag{36.4}$$

in the place of $g(t_0 + n\delta t)$ on the right-hand side of Eq. (36.3), to connect with the final value $f(t_0 + (n+1)\delta t)$. This represents a simplified form of the *predictor-corrector scheme*, which is discussed more extensively in section S1 of Ref. [182].

We now pass to discuss the numerical solution of the KB equations, using the aforementioned time-stepping technique (supplemented when necessary by the predictor-corrector scheme).

36.1.2 Selection of the Equations to Be Solved Numerically

For the present purposes, it is convenient to reconsider the choice made in Chapter 15 of Part I for the set of four equations to be solved numerically. Here, in the place of Eq. (15.13) of Part I for $G^<$, we shall make use of Eq. (15.14) of Part I for $G^>$ (where the time derivative is taken with respect to the first time variable t) and replace Eq. (15.19) of Part I for G^{R} by the equation for $G^<$ with the time derivative now taken with respect to the second time variable t', namely,

$$\boxed{\begin{aligned}&\left[-i\frac{\partial}{\partial t'} - h(t')\right] G^<(t,t') - \int_{t_0}^{+\infty} d\bar{t}\; G^{\mathrm{R}}(t,\bar{t})\,\Sigma^<(\bar{t},t') - \int_{t_0}^{+\infty} d\bar{t}\; G^<(t,\bar{t})\,\Sigma^{\mathrm{A}}(\bar{t},t')\\ &\qquad + i\int_0^{\beta} d\bar{\tau}\; G^{\rceil}(t,\bar{\tau})\,\Sigma^{\lceil}(\bar{\tau},t') = 0\end{aligned}} \tag{36.5}$$

in accordance with the structure of Eq. (11.28) of Part I. Note that, here like in Chapter 15 of Part I, the spatial and spin variables have been suppressed for simplicity. [In the superfluid phase, on the other hand, the structure of the KB equations in each of the two time variables follows from Eqs. (31.5) and (31.6) of Part II in the Nambu representation.]

In the following, we shall also need the equation for the *time-diagonal* lesser Green's function $G^<(t,t)$, which is obtained by setting $t' = t^+$ in both Eq. (15.13) of Part I and Eq. (36.5) above and then subtracting the ensuing expressions:

$$\begin{aligned}&\left[i\frac{\partial}{\partial t} - h(t)\right] G^<(t,t')|_{t'=t^+} + \left[i\frac{\partial}{\partial t'} + h(t')\right] G^<(t,t')|_{t'=t^+}\\ &\quad - \int_{t_0}^{+\infty} d\bar{t}\left(\Sigma^{\mathrm{R}}(t,\bar{t})\,G^<(\bar{t},t) - G^{\mathrm{R}}(t,\bar{t})\,\Sigma^<(\bar{t},t)\right)\\ &\quad - \int_{t_0}^{+\infty} d\bar{t}\left(\Sigma^<(t,\bar{t})\,G^{\mathrm{A}}(\bar{t},t) - G^<(t,\bar{t})\,\Sigma^{\mathrm{A}}(\bar{t},t)\right)\\ &\quad + i\int_0^{\beta} d\bar{\tau}\left(\Sigma^{\rceil}(t,\bar{\tau})\,G^{\lceil}(\bar{\tau},t) - G^{\rceil}(t,\bar{\tau})\,\Sigma^{\lceil}(\bar{\tau},t)\right) = 0\,.\end{aligned} \tag{36.6}$$

Remark: About the diagonal time derivative

In the expression (36.6), we can write

$$\left(\frac{\partial}{\partial t} + \frac{\partial}{\partial t'}\right) G^<(t,t')|_{t'=t^+} = \frac{d}{dt} G^<(t,t)\,. \tag{36.7}$$

This result follows from the property of a generic function $f(x,y)$ of two variables

$$\left(\frac{\partial}{\partial x}+\frac{\partial}{\partial y}\right)f(x,y)|_{x=y}=\frac{d}{dx}f(x,x)\,, \tag{36.8}$$

which can be proved as follows (with $\epsilon = 0^+$):

$$\begin{aligned}\frac{d}{dx}f(x,x)=\left.\frac{f(x+\epsilon,y+\epsilon)-f(x,y)}{\epsilon}\right|_{x=y} &= \left.\frac{f(x,y)+\epsilon\frac{\partial}{\partial x}f(x,y)+\epsilon\frac{\partial}{\partial y}f(x,y)-f(x,y)}{\epsilon}\right|_{x=y}\\ &= \left[\frac{\partial f(x,y)}{\partial x}+\frac{\partial f(x,y)}{\partial y}\right]_{x=y}\,.\end{aligned} \tag{36.9}$$

Once $G^<(\mathbf{x}t,\mathbf{x}'t)$ is known (where $\mathbf{x}=(\mathbf{r},\sigma)$ stands for the restored space and spin variables), $G^>(\mathbf{x}t,\mathbf{x}'t)$ can be obtained directly from the definitions (13.5) and (13.6) of Part I, whereby for fermions

$$G^<(\mathbf{x}t,\mathbf{x}'t)-G^>(\mathbf{x}t,\mathbf{x}'t)=i\,\langle\,\psi_{\mathrm{H}}^{\dagger}(\mathbf{x}'t)\,\psi_{\mathrm{H}}(\mathbf{x}t)\,\rangle+i\,\langle\,\psi_{\mathrm{H}}(\mathbf{x}t)\,\psi_{\mathrm{H}}^{\dagger}(\mathbf{x}'t)\,\rangle=i\,\delta(\mathbf{x}-\mathbf{x}') \tag{36.10}$$

owing to the anti-commutation relations (2.2) of the field operators.

36.1.3 Combining Solving for $G^>(t,t')$ When $t>t'$ and for $G^<(t,t')$ When $t<t'$

With these premises, we can limit ourselves to calculating numerically $G^>$ for $t>t'$ and $G^<$ for $t<t'$ when both t and t' are larger than t_0, with *initial conditions* specified at the initial time t_0 when the time-dependent perturbation begins to act on the system. Both $G^>$ and $G^<$ can then be obtained over the whole t-t' plane by exploiting the symmetry properties (13.22) and (13.23) of Part I. Recall also that the presence of an integral over the imaginary time τ in Eqs. (36.5) and (36.6) implies that we are assuming that at time t_0 the system was in thermal equilibrium. Otherwise, when different initial conditions are assumed, the integral over the imaginary time is omitted. Later on, we shall discuss how to obtain $G^{\rceil}$ and $G^{\lceil}$, which also enter Eqs. (36.5) and (36.6).

In any event, we shall consider the "positive quadrant" of the t-t' plane for $t>t_0$ and $t'>t_0$ sketched in Fig. 36.1, where we have superimposed a square lattice of points with the same time step δt along both directions. To determine both $G^>$ and $G^<$ over this whole quadrant, we then proceed through the following *steps* with reference to Fig. 36.1:

Step # 1 Starting from the initial point (t_0,t_0), move along the horizontal (t) axis by solving Eq. (15.14) of Part I for $G^>$ using the time-stepping technique of Subsection 36.1.1 with time step δt. In this way, propagate the solution up to a maximum time $t_{\max}$ of interest.

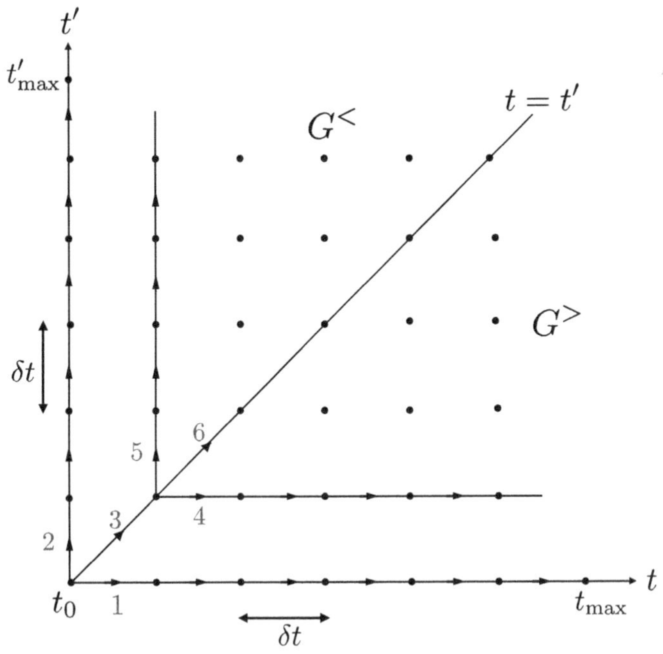

Figure 36.1 The "positive quadrant" of the t-t' plane for $t_0 < t < t_{max}$ and $t_0 < t' < t'_{max}$ is subdivided into a grid of square cells of width δt, over which the equations (15.14) of Part I for $G^>$ and (36.5)–(36.6) for $G^<$ are suitably solved using the time-stepping technique, following the steps indicated in the text.

Step # 2 Starting again from the initial point (t_0, t_0), move along the vertical (t') axis by solving Eq. (36.5) for $G^<$ using the time-stepping technique with time step δt. In this way, propagate the solution up to a maximum time t'_{max} of interest.

Step # 3 Starting once more from the initial point (t_0, t_0), move along the diagonal $t = t'$ by one time step of length $\sqrt{2}\,\delta t$ by solving Eq. (36.6) for $G^<$ and then use the identity (36.10) (possibly projected onto a suitable basis) to determine also $G^>$ at times $(t_0 + \sqrt{2}\,\delta t, t_0 + \sqrt{2}\,\delta t)$.

Step # 4 Go back to Eq. (15.14) of Part I and use the time-stepping technique with time step δt to solve it with a fixed second argument $t' = t_0 + \delta t$ along a line parallel to the horizontal axis.

Step # 5 Go back to Eq. (36.5) and use the time-stepping technique with time step δt to solve it with a fixed first argument $t = t_0 + \delta t$ along a line parallel to the vertical axis.

Step # 6 Go back to Eq. (36.6) for $G^<$ and move one step further along the diagonal, from $(t_0 + \sqrt{2}\,\delta t, t_0 + \sqrt{2}\,\delta t)$ to $(t_0 + 2\sqrt{2}\,\delta t, t_0 + 2\sqrt{2}\,\delta t)$, to determine $G^<$ at this final point, and then use the identity (36.10) to determine $G^>$ correspondingly.

This procedure can be repeated over and over again to determine $G^>$ on the whole half positive quadrant with $t \geq t'$ and $G^<$ on the whole half positive quadrant with $t' \geq t$. At this point, we can invoke the symmetry properties (13.22) and (13.23) of Part I, to obtain $G^>$ and $G^<$ in the corresponding specular positive quadrants, as we have anticipated. Knowledge of $G^>$ and $G^<$ (as well as of $G^\rceil$) suffices to obtain all components of the contour single-particle Green's function (cf. also Table 30.1 of Part II).

36.1.4 Solving for $G^\rceil(t, \tau)$

We now go back to the KB equations that determine the initial equilibrium condition. They are Eq. (15.2) of Part I to determine G^{M} and Eq. (15.11) of Part I to determine $G^\rceil$, while $G^\lceil$ needed in Eq. (15.13) of Part I to determine $G^<$ can be obtained from $G^\rceil$ using the symmetry property (13.28) of Part I. Whereas Eq. (15.2) of Part I for G^{M} can be solved in Matsubara frequency space (a task that may anyhow require considerable numerical efforts, especially when one plans to achieve a fully self-consistent solution), Eq. (15.11) of Part I for $G^\rceil$ requires itself the use of the time-stepping technique in the real-time coordinate t for given imaginary time τ. The time-stepping procedure for propagating $G^\rceil(t, \tau)$ out of $G^\rceil(t_0, \tau)$ in terms of the differential equation (15.11) of Part I works along vertical lines in the τ-t plane, as sketched in Fig. 36.2.

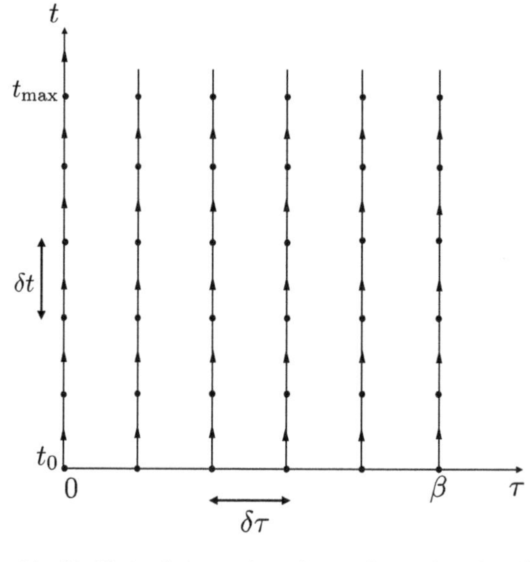

Figure 36.2 A grid of cells in the τ-t plane is used to solve the differential equation for $G^\rceil(t, \tau)$ with the time-stepping procedure, which starts from $G^\rceil(t_0, \tau)$ for given τ and propagates it vertically with time steps δt up to a maximum time t_{max}. Steps of width $\delta\tau$ are used along the horizontal imaginary time axis to connect $\tau = 0$ with $\tau = \beta$ (where $\beta = (k_B T)^{-1}$ is the inverse temperature).

As it was already mentioned in Section 15.1 of Part I, in this case, the initial condition reads (Fig. 36.3)

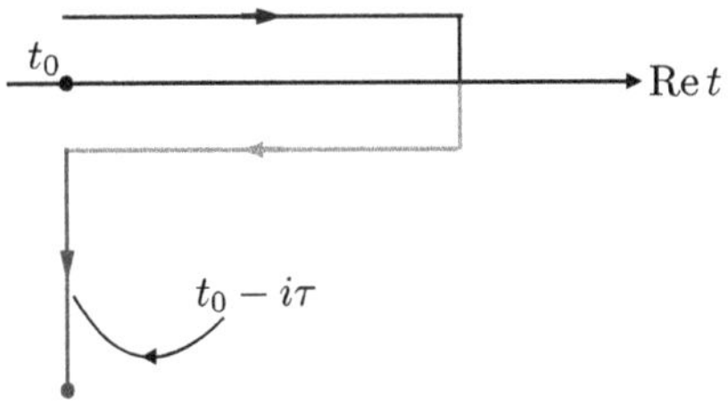

Figure 36.3 Graphical view of Eq. (36.11).

$$\boxed{G^{\rceil}(t_0, \tau) = G^{\mathrm{M}}(\tau = 0, \tau)}\,, \qquad (36.11)$$

which can be seen to hold by comparing Eq. (13.7) of Part I in the form

$$G^{\rceil}(t_0, \tau) = i\,\langle\, \psi^{\dagger}_{\mathrm{H}}(\mathbf{x}' t_0 - i\tau)\, \psi_{\mathrm{H}}(\mathbf{x} t_0)\,\rangle \qquad (36.12)$$

with Eq. (13.11) of Part I in the form

$$G^{\mathrm{M}}(0, \tau) = i\,\langle\, \psi^{\dagger}_{\mathrm{H}}(\mathbf{x}' t_0 - i\tau)\, \psi_{\mathrm{H}}(\mathbf{x} t_0)\,\rangle\,. \qquad (36.13)$$

Here and in Subsection 36.1.5, only the fermionic case is considered for definiteness and the space–spin variables $\mathbf{x}$ are restored when needed.

36.1.5 About the Initial Conditions for $G^<$ and $G^>$

Finally, the initial conditions for $G^<$ and $G^>$, to be utilized in the time-stepping procedure sketched in Fig. 36.1 when the system is initially in thermodynamic equilibrium, are as follows (Figs. 36.4–36.5):

(i)

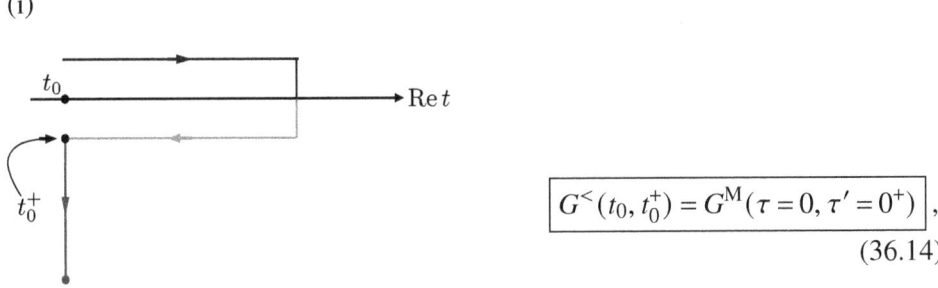

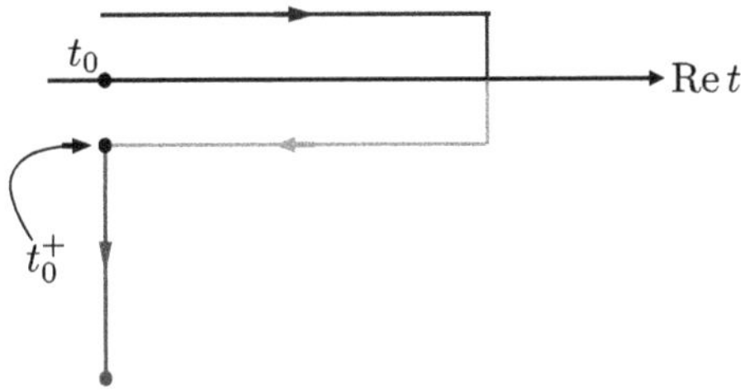

Figure 36.4 Graphical view of Eq. (36.14).

$$\boxed{G^<(t_0, t_0^+) = G^{\mathrm{M}}(\tau = 0, \tau' = 0^+)}\,, \qquad (36.14)$$

which can be seen to hold by comparing Eq. (13.5) of Part I in the form

$$G^{<}(t_0, t_0^{+}) = i \langle \psi_{\mathrm{H}}^{\dagger}(\mathbf{x}' t_0^{+})\, \psi_{\mathrm{H}}(\mathbf{x} t_0) \rangle \tag{36.15}$$

with Eq. (13.11) of Part I in the form

$$G^{\mathrm{M}}(\tau = 0, \tau' = 0^{+}) = i \langle \psi_{\mathrm{H}}^{\dagger}(\mathbf{x}' t_0^{+})\, \psi_{\mathrm{H}}(\mathbf{x} t_0) \rangle . \tag{36.16}$$

(ii)

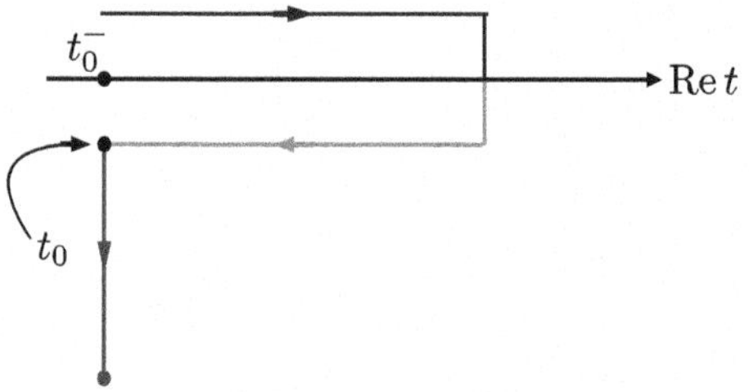

Figure 36.5 Graphical view of Eq. (36.17).

$$\boxed{G^{>}(t_0, t_0^{-}) = G^{\mathrm{M}}(\tau = 0, \tau' = 0^{-})}, \tag{36.17}$$

which can be seen to hold by comparing Eq. (13.6) of Part I in the form

$$G^{>}(t_0, t_0^{-}) = -\, i \langle \psi_{\mathrm{H}}(\mathbf{x} t_0)\, \psi_{\mathrm{H}}^{\dagger}(\mathbf{x}' t_0^{-}) \rangle \tag{36.18}$$

with Eq. (13.11) of Part I in the form

$$G^{\mathrm{M}}(\tau = 0, \tau' = 0^{-}) = -\, i \langle \psi_{\mathrm{H}}(\mathbf{x} t_0)\, \psi_{\mathrm{H}}^{\dagger}(\mathbf{x}' t_0^{-}) \rangle . \tag{36.19}$$

Both initial conditions (36.14) and (36.17) were already mentioned in section 15.1 of Part I.

36.2 The Generalized Kadanoff-Baym Ansatz

The computational scaling of the time propagation of the KB equations described in Section 36.1 is rather demanding since it increases with the number of time steps cube. This feature puts considerable limitations to direct applications of the KB equations for physical problems of interest. To overcome this restriction, a few approximations have been proposed over the years, starting from the original formulation of the KB equations themselves in Ref. [5].

36.2.1 A Glimpse of the Original Kadanoff-Baym Ansatz

Specifically, on page 139 of their pioneering work of Ref. [5], Kadanoff and Baym introduced what was later referred to as the Kadanoff–Baym ansatz (KBA), with the purpose of making closed the equation of motion for the one-particle density matrix $\rho(t) = -iG^{<}(t,t)$ [212]. The final aim was to construct in this way a quantum Boltzmann equation starting from the KB equations for $G^{<}$ and $G^{>}$. Physically, the central idea behind the KBA is that the spectral properties of the system are assumed to vary not as strongly with time as the population dynamics. Equivalently, this type of "decoherence" assumption considers the quasiparticle lifetime to be relatively long compared with the average collision time, a picture that is locally correct only in the presence of external fields, which also vary smoothly in space and time. The KBA can then be viewed as a nonequilibrium extension of the fluctuation–dissipation theorem to locally equilibrated states, suitably not very far from equilibrium. We shall provide more details about the original KBA in Section 37.2.

36.2.2 Need for Improving on the Kadanoff-Baym Ansatz

One challenging problem with the original KBA is that it does not satisfy the requirement of causal time evolution. Overcoming this shortcoming was first considered in Ref. [212], where the so-called generalized Kadanoff-Baym ansatz (GKBA) was originally proposed. An additional important feature of the GKBA is that it reduces the dynamics of the two-time single-particle Green's function $G(t,t')$ to a propagation along the time diagonal $t = t'$, from which the two-time Green's function $G(t,t')$ is then reconstructed. In this way, the scaling in the number of time steps is brought down from cubic to quadratic. In addition, in closed systems, the GKBA can be exactly reformulated in terms of a coupled set of ordinary differential equations (ODE), which further yields a speed-up of another order of magnitude to linear scaling [213]. An extension of the ODE formulation to open systems was later provided in Ref. [214].

36.2.3 The Structure of the Generalized Kadanoff-Baym Ansatz

A well-organized summary, about how the GKBA can be derived from first principles and under what kind of approximations, can be found in Ref. [38], where a mean-field (HF) form is further utilized for the advanced and retarded Green's functions (for this reason, the method is referred to as HF-GKBA). Additional useful descriptions about the GKBA can be found in Refs. [215] and [159]. Here, for completeness, we provide a concise yet detailed derivation of the GKBA and point out explicitly the approximations on which it is based.

In Section 36.1, we have seen that, to implement the time-stepping procedure described therein, one needs also the time-diagonal lesser Green's function $G^{<}(t,t)$, whose equation of motion is given by Eq. (36.6) supplemented by the result (36.7). Knowledge of $G^<(\mathbf{x}t, \mathbf{x}'t)$ sufficed to obtain the local number density (cf. Eq. (18.3) of Part I, for fermions per spin component)

$$n(\mathbf{x}, t) = -\, i\, G^<(\mathbf{x}t, \mathbf{x}t), \tag{36.20}$$

as well as the local current density (cf. Eq. (18.4) of Part I, again for fermions per spin component)

$$\mathbf{j}(\mathbf{x}, t) = -\frac{1}{2m}\,(\nabla_{\mathbf{r}} - \nabla_{\mathbf{r}'})\, G^<(\mathbf{r}\sigma t, \mathbf{r}'\sigma t)\Big|_{\mathbf{r}'=\mathbf{r}}\,. \tag{36.21}$$

The problem with the equation of motion (36.6), however, is that it is not "closed," in the sense that it depends on the time-off-diagonal lesser and greater Green's functions (also through the retarded G^{R} and advanced G^{A} Green's functions). The *generalized Kadanoff–Baym ansatz* (to be obtained below) aims at simplifying this equation, by writing approximately

$$\boxed{G^{\gtrless}(t, t') \simeq -\, G^{\mathrm{R}}(t, t')\, n^{\gtrless}(t') \,+\, n^{\gtrless}(t)\, G^{\mathrm{A}}(t, t')\,,} \tag{36.22}$$

such that one is able to reconstruct the time-off-diagonal $G^{\gtrless}(t, t')$ from the diagonal quantities $n^{\gtrless}(t)$ and $n^{\gtrless}(t')$. In the expression (36.22):

(i) Spatial ($\mathbf{r}$) and spin (σ) indices have been suppressed for simplicity, such that a matrix convention is implied for these indices;
(ii) For fermions,

$$n^<(t) \;\longleftrightarrow\; n^<(\mathbf{x}, \mathbf{x}'; t) \;=\; -\, i\, G^<(\mathbf{x}t, \mathbf{x}'t^+) \tag{36.23}$$

$$n^>(t) \;\longleftrightarrow\; n^<(\mathbf{x}, \mathbf{x}'; t) - \delta(\mathbf{x} - \mathbf{x}') \;=\; -\, i\, G^>(\mathbf{x}t, \mathbf{x}'t^-)\,; \tag{36.24}$$

(iii) To fulfill the program (thus making the expression (36.22) useful in practice), a further approximation has to be introduced for the retarded and advanced Green's functions.

The approximate expression (36.22) can be obtained through the following line of reasoning.

The route to Eq. (36.22)

We begin by using the property $\theta(t_1 - t_2) + \theta(t_2 - t_1) = 1$ of the Heaviside unit step function in real time, so as to split

$$\begin{aligned} G^{\gtrless}(t_1, t_2) &= (\theta(t_1 - t_2) + \theta(t_2 - t_1))\, G^{\gtrless}(t_1, t_2) \\ &= \theta(t_1 - t_2)\, G^{\gtrless}(t_1, t_2) + \theta(t_2 - t_1)\, G^{\gtrless}(t_1, t_2) \\ &\equiv \mathcal{R}^{\gtrless}(t_1, t_2) - \mathcal{A}^{\gtrless}(t_1, t_2)\,, \end{aligned} \tag{36.25}$$

where it is understood that $t_1 \geq t_0$ and $t_2 \geq t_0$. The two terms on the right-hand side of this expression can be manipulated separately as follows:

Manipulating the term $\mathcal{R}^{\gtrless}(t_1, t_2)$ We make use of the equation of motion (15.19) of Part I for G^{R} to manipulate the expression of $\mathcal{R}^{\gtrless}(t_1, t_2) = \theta(t_1 - t_2)\, G^{\gtrless}(t_1, t_2)$ in the following way:

$$
\begin{aligned}
\mathcal{R}^{\gtrless}(t_1, t_2) &= \int_{t_0}^{\infty} d\bar{t}_3 \delta(t_1 - \bar{t}_3) \mathcal{R}^{\gtrless}(\bar{t}_3, t_2) = \int_{t_0}^{\infty} d\bar{t}_1 d\bar{t}_3 G^{\mathrm{R}}(t_1, \bar{t}_1) G^{\mathrm{R}^{-1}}(\bar{t}_1, \bar{t}_3) \mathcal{R}^{\gtrless}(\bar{t}_3, t_2) \\
&= \int_{t_0}^{\infty} d\bar{t}_1 d\bar{t}_3 d\bar{t}_4 G^{\mathrm{R}}(t_1, \bar{t}_1) \delta(\bar{t}_1 - \bar{t}_4) G^{\mathrm{R}^{-1}}(\bar{t}_4, \bar{t}_3) \mathcal{R}^{\gtrless}(\bar{t}_3, t_2) \\
&\underset{[\text{Eq.(15.19)}]}{\uparrow{=}} \int_{t_0}^{\infty} d\bar{t}_1 d\bar{t}_3 d\bar{t}_4 G^{\mathrm{R}}(t_1, \bar{t}_1) \left\{ \left[i\frac{\partial}{\partial \bar{t}_1} - h(\bar{t}_1) \right] G^{\mathrm{R}}(\bar{t}_1, \bar{t}_4) \right. \\
&\quad \left. - \int_{t_0}^{\infty} d\bar{t}_2 \Sigma^{\mathrm{R}}(\bar{t}_1, \bar{t}_2) G^{\mathrm{R}}(\bar{t}_2, \bar{t}_4) \right\} G^{\mathrm{R}^{-1}}(\bar{t}_4, \bar{t}_3) \mathcal{R}^{\gtrless}(\bar{t}_3, t_2) \\
&= \int_{t_0}^{\infty} d\bar{t}_1 d\bar{t}_2 d\bar{t}_3 G^{\mathrm{R}}(t_1, \bar{t}_1) \left\{ \delta(\bar{t}_1 - \bar{t}_2) \left[i\frac{\partial}{\partial \bar{t}_2} - h(\bar{t}_2) \right] - \Sigma^{\mathrm{R}}(\bar{t}_1, \bar{t}_2) \right\} \\
&\quad \times \underbrace{\int_{t_0}^{\infty} d\bar{t}_4 G^{\mathrm{R}}(\bar{t}_2, \bar{t}_4) G^{\mathrm{R}^{-1}}(\bar{t}_4, \bar{t}_3)}_{=\delta(\bar{t}_2 - \bar{t}_3)} \mathcal{R}^{\gtrless}(\bar{t}_3, t_2) \\
&= \int_{t_0}^{\infty} d\bar{t}_1 G^{\mathrm{R}}(t_1, \bar{t}_1) \int_{t_0}^{\infty} d\bar{t}_2 \left\{ \delta(\bar{t}_1 - \bar{t}_2) \left[i\frac{\partial}{\partial \bar{t}_2} - h(\bar{t}_2) \right] - \Sigma^{\mathrm{R}}(\bar{t}_1, \bar{t}_2) \right\} \\
&\quad \times \mathcal{R}^{\gtrless}(\bar{t}_2, t_2).
\end{aligned} \tag{36.26}
$$

This expression can be further manipulated by using the definition (36.25) of $\mathcal{R}^{\gtrless}(\bar{t}_2, t_2)$, to rewrite the integral over $\bar{t}_2$ therein in the form:

$$
\begin{aligned}
&\int_{t_0}^{\infty} d\bar{t}_2 \left\{ \delta(\bar{t}_1 - \bar{t}_2) \left[i\frac{\partial}{\partial \bar{t}_2} - h(\bar{t}_2) \right] - \Sigma^{\mathrm{R}}(\bar{t}_1, \bar{t}_2) \right\} \mathcal{R}^{\gtrless}(\bar{t}_2, t_2) \\
&= \int_{t_0}^{\infty} d\bar{t}_2 \left\{ \delta(\bar{t}_1 - \bar{t}_2) \left[i\frac{\partial}{\partial \bar{t}_2} - h(\bar{t}_2) \right] - \Sigma^{\mathrm{R}}(\bar{t}_1, \bar{t}_2) \right\} \theta(\bar{t}_2 - t_2)\, G^{\gtrless}(\bar{t}_2, t_2) \\
&= \int_{t_0}^{\infty} d\bar{t}_2 \left\{ \delta(\bar{t}_1 - \bar{t}_2)\, i \left(\delta(\bar{t}_2 - t_2)\, G^{\gtrless}(\bar{t}_2, t_2) + \theta(\bar{t}_2 - t_2) \frac{\partial}{\partial \bar{t}_2} G^{\gtrless}(\bar{t}_2, t_2) \right) \right. \\
&\quad \left. - \left(\delta(\bar{t}_1 - \bar{t}_2)\, h(\bar{t}_2) + \Sigma^{\mathrm{R}}(\bar{t}_1, \bar{t}_2) \right) \theta(\bar{t}_2 - t_2)\, G^{\gtrless}(\bar{t}_2, t_2) \right\} \\
&= i\, \delta(\bar{t}_1 - t_2)\, G^{\gtrless}(\bar{t}_1, t_2) + \theta(\bar{t}_1 - t_2) \left[i\frac{\partial}{\partial \bar{t}_1} - h(\bar{t}_1) \right] G^{\gtrless}(\bar{t}_1, t_2)
\end{aligned}
$$

$$
- \int_{t_0}^{\infty} d\bar{t}_2\, \theta(\bar{t}_2 - t_2)\, \Sigma^{\mathrm{R}}(\bar{t}_1, \bar{t}_2)\, G^{\gtrless}(\bar{t}_2, t_2)
$$

$$
\underset{\substack{\uparrow \\ \text{[Eqs.(15.13) and (15.14) of Part I]}}}{=} i\,\delta(\bar{t}_1 - t_2)\, G^{\gtrless}(\bar{t}_1, t_2) - \int_{t_0}^{\infty} d\bar{t}\, \theta(\bar{t} - t_2)\, \underbrace{\Sigma^{\mathrm{R}}(\bar{t}_1, \bar{t})\, G^{\gtrless}(\bar{t}, t_2)}_{1^{st}}
$$

$$
+ \theta(\bar{t}_1 - t_2) \left\{ \int_{t_0}^{+\infty} d\bar{t} \left(\underbrace{\Sigma^{\gtrless}(\bar{t}_1, \bar{t})\, G^{\mathrm{A}}(\bar{t}, t_2)}_{2^{nd}} + \underbrace{\Sigma^{\mathrm{R}}(\bar{t}_1, \bar{t})\, G^{\gtrless}(\bar{t}, t_2)}_{3^{rd}} \right) \right.
$$

$$
\left. - i \int_0^{\beta} d\bar{\tau}\, \Sigma^{\rceil}(\bar{t}_1, \bar{\tau})\, G^{\lceil}(\bar{\tau}, t_2) \right\}
$$

$$
\underset{\substack{\uparrow \\ \text{[cf. Fig.36.3]}}}{=} i\delta(\bar{t}_1 - t_2) G^{\gtrless}(\bar{t}_1, t_2) - i\theta(\bar{t}_1 - t_2) \int_0^{\beta} d\bar{\tau} \Sigma^{\rceil}(\bar{t}_1, \bar{\tau}) G^{\lceil}(\bar{\tau}, t_2)
$$

$$
+ \theta(\bar{t}_1 - t_2) \int_{t_0}^{t_2} d\bar{t} \left(\Sigma^{\gtrless}(\bar{t}_1, \bar{t}) G^{\mathrm{A}}(\bar{t}, t_2) + \Sigma^{\mathrm{R}}(\bar{t}_1, \bar{t}) G^{\gtrless}(\bar{t}, t_2) \right), \tag{36.27}
$$

where we have made explicit the integration domains of the three integrals over $\bar{t}$, as shown in Fig. 36.6.

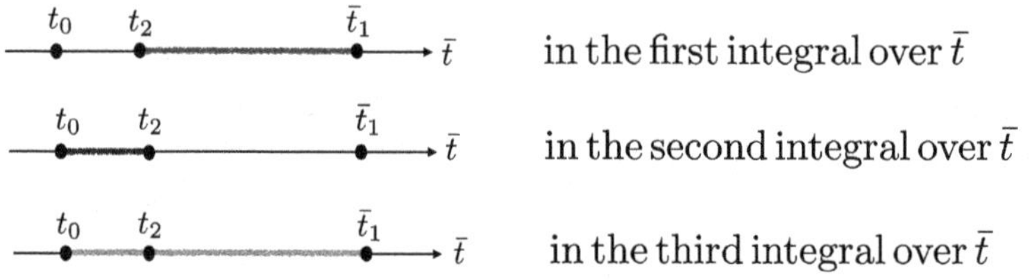

Figure 36.6 Integration domains (thick lines) for the three integrals indicated in the next-to-last line of Eq. (36.27).

Entering the result (36.27) into the expression (36.26), we obtain eventually:

$$
\mathcal{R}^{\gtrless}(t_1, t_2) = iG^{\mathrm{R}}(t_1, t_2) G^{\gtrless}(t_2, t_2) - i \int_{t_2}^{t_1} d\bar{t}_1 G^{\mathrm{R}}(t_1, \bar{t}_1) \int_0^{\beta} d\bar{\tau} \Sigma^{\rceil}(\bar{t}_1, \bar{\tau}) G^{\lceil}(\bar{\tau}, t_2)
$$

$$
+ \int_{t_2}^{t_1} d\bar{t}_1 \int_{t_0}^{t_2} d\bar{t} G^{\mathrm{R}}(t_1, \bar{t}_1) \left(\Sigma^{\gtrless}(\bar{t}_1, \bar{t}) G^{\mathrm{A}}(\bar{t}, t_2) + \Sigma^{\mathrm{R}}(\bar{t}_1, \bar{t}) G^{\gtrless}(\bar{t}, t_2) \right), \tag{36.28}
$$

where (cf. Eqs. (36.23) and (36.24))

$$
iG^{<}(t_2, t_2) \longleftrightarrow iG^{<}(\mathbf{x}t_2, \mathbf{x}'t_2^{+}) = -n^{<}(\mathbf{x}, \mathbf{x}'; t_2), \tag{36.29}
$$

$$iG^>(t_2,t_2) \longleftrightarrow iG^>(\mathbf{x}t_2,\mathbf{x}'t_2^-) = \delta(\mathbf{x}-\mathbf{x}') - n^<(\mathbf{x},\mathbf{x}';t_2) = -n^>(\mathbf{x},\mathbf{x}';t_2). \tag{36.30}$$

Manipulating the term $\mathcal{A}^{\gtrless}(t_1,t_2)$ Proceeding similarly to what we have done earlier for the term $\mathcal{R}^{\gtrless}(t_1,t_2)$, for $\mathcal{A}^{\gtrless}(t_1,t_2) = -\theta(t_2-t_1)G^{\gtrless}(t_1,t_2)$, we now use the equation of motion for G^A to write:

$$\mathcal{A}^{\gtrless}(t_1,t_2) = \int_{t_0}^{\infty} d\bar{t}_3 \mathcal{A}^{\gtrless}(t_1,\bar{t}_3)\delta(\bar{t}_3-t_2) = \int_{t_0}^{\infty} d\bar{t}_1 d\bar{t}_3 \mathcal{A}^{\gtrless}(t_1,\bar{t}_3) G^{A^{-1}}(\bar{t}_3,\bar{t}_1) G^A(\bar{t}_1,t_2)$$

$$= \int_{t_0}^{\infty} d\bar{t}_1 d\bar{t}_3 d\bar{t}_4 \mathcal{A}^{\gtrless}(t_1,\bar{t}_4)\delta(\bar{t}_4-\bar{t}_3) G^{A^{-1}}(\bar{t}_3,\bar{t}_1) G^A(\bar{t}_1,t_2)$$

$$\underset{\substack{\uparrow \\ [\text{Eq.}(15.22) \text{ of Part I}]}}{=} \int_{t_0}^{\infty} d\bar{t}_1 d\bar{t}_3 d\bar{t}_4 \mathcal{A}^{\gtrless}(t_1,\bar{t}_4) \left\{ \left[i\frac{\partial}{\partial \bar{t}_4} - h(\bar{t}_4) \right] G^A(\bar{t}_4,\bar{t}_3) \right.$$

$$\left. - \int_{t_0}^{\infty} d\bar{t}_2 \Sigma^A(\bar{t}_4,\bar{t}_2) G^A(\bar{t}_2,\bar{t}_3) \right\} G^{A^{-1}}(\bar{t}_3,\bar{t}_1) G^A(\bar{t}_1,t_2)$$

$$= \int_{t_0}^{\infty} d\bar{t}_1 d\bar{t}_2 d\bar{t}_4 \mathcal{A}^{\gtrless}(t_1,\bar{t}_4) \left\{ \delta(\bar{t}_4-\bar{t}_2) \left[i\frac{\partial}{\partial \bar{t}_2} - h(\bar{t}_2) \right] - \Sigma^A(\bar{t}_4,\bar{t}_2) \right\}$$

$$\times \underbrace{\int_{t_0}^{\infty} d\bar{t}_3 G^A(\bar{t}_2,\bar{t}_3) G^{A^{-1}}(\bar{t}_3,\bar{t}_1)}_{=\delta(\bar{t}_2-\bar{t}_1)} G^A(\bar{t}_1,t_2)$$

$$= \int_{t_0}^{\infty} d\bar{t}_2 d\bar{t}_4 \mathcal{A}^{\gtrless}(t_1,\bar{t}_4) \left\{ \delta(\bar{t}_4-\bar{t}_2) \left[i\frac{\partial}{\partial \bar{t}_2} - h(\bar{t}_2) \right] - \Sigma^A(\bar{t}_4,\bar{t}_2) \right\} G^A(\bar{t}_2,t_2)$$

$$= \int_{t_0}^{\infty} d\bar{t}_2 \mathcal{A}^{\gtrless}(t_1,\bar{t}_2) \left[i\frac{\partial}{\partial \bar{t}_2} - h(\bar{t}_2) \right] G^A(\bar{t}_2,t_2)$$

$$- \int_{t_0}^{\infty} d\bar{t}_2 d\bar{t}_4 \mathcal{A}^{\gtrless}(t_1,\bar{t}_4) \Sigma^A(\bar{t}_4,\bar{t}_2) G^A(\bar{t}_2,t_2)$$

$$\underset{\substack{\uparrow \\ [\text{by parts}]}}{=} \int_{t_0}^{\infty} d\bar{t}_2 \left(\left[-i\frac{\partial}{\partial \bar{t}_2} - h(\bar{t}_2) \right] \mathcal{A}^{\gtrless}(t_1,\bar{t}_2) \right) G^A(\bar{t}_2,t_2)$$

$$- \int_{t_0}^{\infty} d\bar{t}_2 d\bar{t}_4 \mathcal{A}^{\gtrless}(t_1,\bar{t}_4) \Sigma^A(\bar{t}_4,\bar{t}_2) G^A(\bar{t}_2,t_2), \tag{36.31}$$

where the boundary terms do not contribute to the integral by parts since

(i) for $\bar{t}_2 = t_0 \implies \mathcal{A}^{\gtrless}(t_1,\bar{t}_2=t_0) \propto \theta(t_0-t_1) = 0$,
(ii) for $\bar{t}_2 = \infty \implies G^A(\bar{t}_2=\infty,t_2) \propto \theta(t_2-\infty) = 0$.

The first term on the right-hand side of Eq. (36.31) then yields:

$$
\begin{aligned}
&\int_{t_0}^{\infty} d\bar{t}_2 \left(\left[-i\frac{\partial}{\partial \bar{t}_2} - h(\bar{t}_2) \right] \mathcal{A}^{\gtrless}(t_1, \bar{t}_2) \right) G^{\mathrm{A}}(\bar{t}_2, t_2) \\
&= - \int_{t_0}^{\infty} d\bar{t}_2 \left(\left[-i\frac{\partial}{\partial \bar{t}_2} - h(\bar{t}_2) \right] \theta(\bar{t}_2 - t_1) G^{\gtrless}(t_1, \bar{t}_2) \right) G^{\mathrm{A}}(\bar{t}_2, t_2) \\
&= \int_{t_0}^{\infty} d\bar{t}_2 \left\{ i\delta(\bar{t}_2 - t_1) G^{\gtrless}(t_1, \bar{t}_2) - \theta(\bar{t}_2 - t_1) \left[-i\frac{\partial}{\partial \bar{t}_2} - h(\bar{t}_2) \right] G^{\gtrless}(t_1, \bar{t}_2) \right\} G^{\mathrm{A}}(\bar{t}_2, t_2) \\
&\underset{[\text{Eq.}(36.5)]}{\overset{\uparrow}{=}} iG^{\gtrless}(t_1, t_1) G^{\mathrm{A}}(t_1, t_2) - \int_{t_0}^{\infty} d\bar{t}_2 \theta(\bar{t}_2 - t_1) \left\{ \int_{t_0}^{\infty} d\bar{t} G^{\gtrless}(t_1, \bar{t}) \Sigma^{\mathrm{A}}(\bar{t}, \bar{t}_2) \right. \\
&\quad \left. + \int_{t_0}^{\infty} d\bar{t} G^{\mathrm{R}}(t_1, \bar{t}) \Sigma^{\gtrless}(\bar{t}, \bar{t}_2) - i \int_0^{\beta} d\bar{\tau} G^{\rceil}(t_1, \bar{\tau}) \Sigma^{\lceil}(\bar{\tau}, \bar{t}_2) \right\} G^{\mathrm{A}}(\bar{t}_2, t_2).
\end{aligned}
\tag{36.32}
$$

Entering this result into Eq. (36.31), we obtain

$$
\begin{aligned}
\mathcal{A}^{\gtrless}(t_1, t_2) &= i\, G^{\gtrless}(t_1, t_1)\, G^{\mathrm{A}}(t_1, t_2) - \int_{t_1}^{t_2} d\bar{t}_2 \int_{t_0}^{\bar{t}_2} d\bar{t}\; G^{\gtrless}(t_1, \bar{t})\, \Sigma^{\mathrm{A}}(\bar{t}, \bar{t}_2)\, G^{\mathrm{A}}(\bar{t}_2, t_2) \\
&\quad - \int_{t_1}^{t_2} d\bar{t}_2 \int_{t_0}^{t_1} d\bar{t}\; G^{\mathrm{R}}(t_1, \bar{t})\, \Sigma^{\gtrless}(\bar{t}, \bar{t}_2)\, G^{\mathrm{A}}(\bar{t}_2, t_2) \\
&\quad + \int_{t_0}^{t_2} d\bar{t}_2 \int_{t_1}^{\bar{t}_2} d\bar{t}\; G^{\gtrless}(t_1, \bar{t})\, \Sigma^{\mathrm{A}}(\bar{t}, \bar{t}_2)\, G^{\mathrm{A}}(\bar{t}_2, t_2) \\
&\quad + i \int_{t_1}^{t_2} d\bar{t}_2 \int_0^{\beta} d\bar{\tau}\; G^{\rceil}(t_1, \bar{\tau})\, \Sigma^{\lceil}(\bar{\tau}, \bar{t}_2)\, G^{\mathrm{A}}(\bar{t}_2, t_2)\,,
\end{aligned}
\tag{36.33}
$$

where the first and third integrals on the right-hand side can be combined together as illustrated graphically in Fig. 36.7, to yield[1]

$$
\begin{aligned}
\mathcal{A}^{\gtrless}(t_1, t_2) &= iG^{\gtrless}(t_1, t_1) G^{\mathrm{A}}(t_1, t_2) + i \int_{t_1}^{t_2} d\bar{t}_2 \int_0^{\beta} d\bar{\tau} G^{\rceil}(t_1, \bar{\tau}) \Sigma^{\lceil}(\bar{\tau}, \bar{t}_2) G^{\mathrm{A}}(\bar{t}_2, t_2) \\
&\quad - \int_{t_1}^{t_2} d\bar{t}_2 \int_{t_0}^{t_1} d\bar{t} \left(G^{\gtrless}(t_1, \bar{t}) \Sigma^{\mathrm{A}}(\bar{t}, \bar{t}_2) + G^{\mathrm{R}}(t_1, \bar{t}) \Sigma^{\gtrless}(\bar{t}, \bar{t}_2) \right) G^{\mathrm{A}}(\bar{t}_2, t_2).
\end{aligned}
\tag{36.34}
$$

The full expression for $G^{\gtrless}(t_1, t_2)$ is eventually obtained by subtracting the result (36.34) for $\mathcal{A}^{\gtrless}(t_1, t_2)$ from the result (36.28) for $\mathcal{R}^{\gtrless}(t_1, t_2)$, according to the definition (36.25).

[1] Note that the singular parts of Σ^{R} and Σ^{A} do not contribute, respectively, to Eqs. (36.28) for $\mathcal{R}^{\gtrless}$ and (36.34) for $\mathcal{A}^{\gtrless}$.

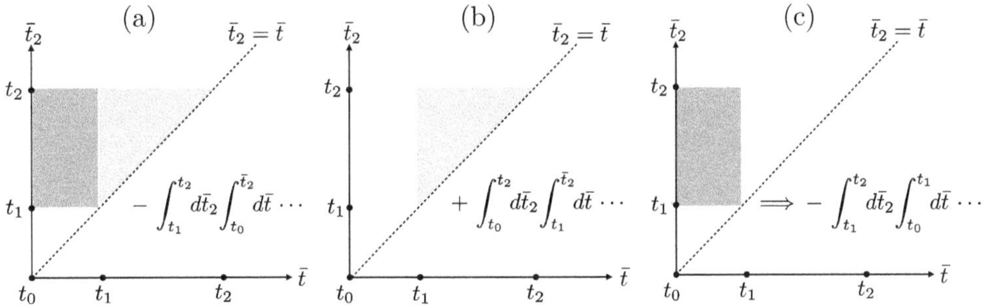

Figure 36.7 The integration domains for the (a) first and (b) third integral on the right-hand side of Eq. (36.33) combine together to yield (c) the integration domain utilized in Eq. (36.34).

All manipulations thus far are exact. At this point, the GKBA consists of neglecting *all* integral terms in Eqs. (36.28) and (36.34), in such a way that the approximate expression (36.22) is obtained by taking into account the identities (36.23) and (36.24) [38, 212]. [QED]

The plausibility of the physical conditions that have led to the simplified expression (36.22) was originally discussed in Ref. [212]. From a physical point of view, what the GKBA achieves is to decouple the "spectral information" contained in G^{R} and G^{A} from the "population dynamics" contained in $n^{\gtrless}$. To this end, the spectral properties of the system are assumed to vary slowly in time with respect to the population dynamics. That is to say, the quasi-particle lifetime is assumed to be relatively long compared with the average collision time, in line with the original KBA mentioned in Subsection 36.2.1.

To make the use of the GKBA relatively handy, *a further simplifying assumption* is adopted for the retarded and advanced Green's functions therein, by writing (cf. Eqs. (33.37) and (33.38))

$$G^{\mathrm{R}}(\mathbf{x}t, \mathbf{x}'t') = -i\theta(t - t')\langle \mathbf{x}|e^{-i\int_{t'}^{t} d\bar{t} h_{\mathrm{eff}}(\bar{t})}|\mathbf{x}'\rangle, \tag{36.35}$$

$$G^{\mathrm{A}}(\mathbf{x}t, \mathbf{x}'t') = i\theta(t' - t)\langle \mathbf{x}|e^{-i\int_{t'}^{t} d\bar{t} h_{\mathrm{eff}}(\bar{t})}|\mathbf{x}'\rangle, \tag{36.36}$$

where $h_{\mathrm{eff}}(t)$ is a suitably chosen single-particle Hamiltonian, which depends on the problem at hand. In this way, $G^{\gtrless}$ in Eq. (36.22) depend solely on the functional form of $n^{\gtrless}$.

In addition, the GKBA satisfies the following *properties*.

Property #1: The GKBA is internally consistent with the identities (36.23) and (36.24) in the appropriate limits:

(i) When $t' = t^+$, $G^{\mathrm{R}}(t, t^+) = 0$ and

$$G^{\mathrm{A}}(t, t^+) = G^<(t, t) - G^>(t, t) \longleftrightarrow i\left\langle\left(\psi_{\mathrm{H}}(\mathbf{x}t)\psi^\dagger_{\mathrm{H}}(\mathbf{x}'t) + \psi^\dagger_{\mathrm{H}}(\mathbf{x}'t)\psi_{\mathrm{H}}(\mathbf{x}t)\right)\right\rangle$$
$$= i\delta(\mathbf{x} - \mathbf{x}'), \tag{36.37}$$

such that from the GKBA expression (36.22), we obtain

$$G^<(\mathbf{x}t, \mathbf{x}'t^+) = i\int d\mathbf{x}''\, n^<(\mathbf{x}, \mathbf{x}''; t)\, \delta(\mathbf{x}'' - \mathbf{x}') = i\, n^<(\mathbf{x}, \mathbf{x}'; t), \tag{36.38}$$

and Eq. (36.23) is recovered;

(ii) When $t' = t^-$, $G^{\mathrm{A}}(t, t^-) = 0$ and

$$G^{\mathrm{R}}(t, t^-) = G^>(t, t) - G^<(t, t) \longleftrightarrow -i\,\delta(\mathbf{x} - \mathbf{x}'), \tag{36.39}$$

such that from the GKBA expression (36.22), we obtain

$$G^>(\mathbf{x}t, \mathbf{x}'t^-) = i\int d\mathbf{x}''\, \delta(\mathbf{x} - \mathbf{x}'')\, n^>(\mathbf{x}'', \mathbf{x}'; t) = i\, n^>(\mathbf{x}, \mathbf{x}'; t), \tag{36.40}$$

and Eq. (36.24) is as well recovered.

Property #2: The GKBA satisfies the spectral identity (13.19) of Part I. This is because

$$\begin{aligned}
G^>(\mathbf{x}t, \mathbf{x}'t') - G^<(\mathbf{x}t, \mathbf{x}'t') &\underset{\substack{\uparrow\\ [\mathrm{GKBA}]}}{=} \int d\mathbf{x}''\Big(-G^{\mathrm{R}}(\mathbf{x}t, \mathbf{x}''t')n^>(\mathbf{x}'', \mathbf{x}'; t')\\
&+ n^>(\mathbf{x}, \mathbf{x}''; t)G^{\mathrm{A}}(\mathbf{x}''t, \mathbf{x}'t') + G^{\mathrm{R}}(\mathbf{x}t, \mathbf{x}''t')n^<(\mathbf{x}'', \mathbf{x}'; t')\\
&- n^<(\mathbf{x}, \mathbf{x}''; t)G^{\mathrm{A}}(\mathbf{x}''t, \mathbf{x}'t')\Big) = \int d\mathbf{x}''\Big[G^{\mathrm{R}}(\mathbf{x}t, \mathbf{x}''t')\Big(n^<(\mathbf{x}'', \mathbf{x}'; t') - n^>(\mathbf{x}'', \mathbf{x}'; t')\Big)\\
&- \Big(n^<(\mathbf{x}, \mathbf{x}''; t') - n^>(\mathbf{x}, \mathbf{x}''; t')\Big)G^{\mathrm{A}}(\mathbf{x}''t, \mathbf{x}'t')\Big] \underset{\substack{\uparrow\\ [\mathrm{Eq.}(36.24)]}}{=} G^{\mathrm{R}}(\mathbf{x}t, \mathbf{x}'t') - G^{\mathrm{A}}(\mathbf{x}t, \mathbf{x}'t').
\end{aligned} \tag{36.41}$$

Property #3: The GKBA obeys particle number conservation.

This is because, from the relations (36.20) and (36.21), the continuity equation for the number density can be expressed only in terms of $G^<(\mathbf{x}t, \mathbf{x}'t^+)$, which is just a quantity that the GKBA is able to approximate in the form (36.38).

Property #4: The GKBA exhibits an inherently causal structure.

This is because, in the expression (36.22), $G^{\mathrm{R}}(t, t')$ propagates from t' to $t > t'$ the information contained in $n^{\gtrless}(t')$, while $G^{\mathrm{A}}(t, t')$ propagates from t to $t' > t$ the information contained in $n^{\gtrless}(t)$.

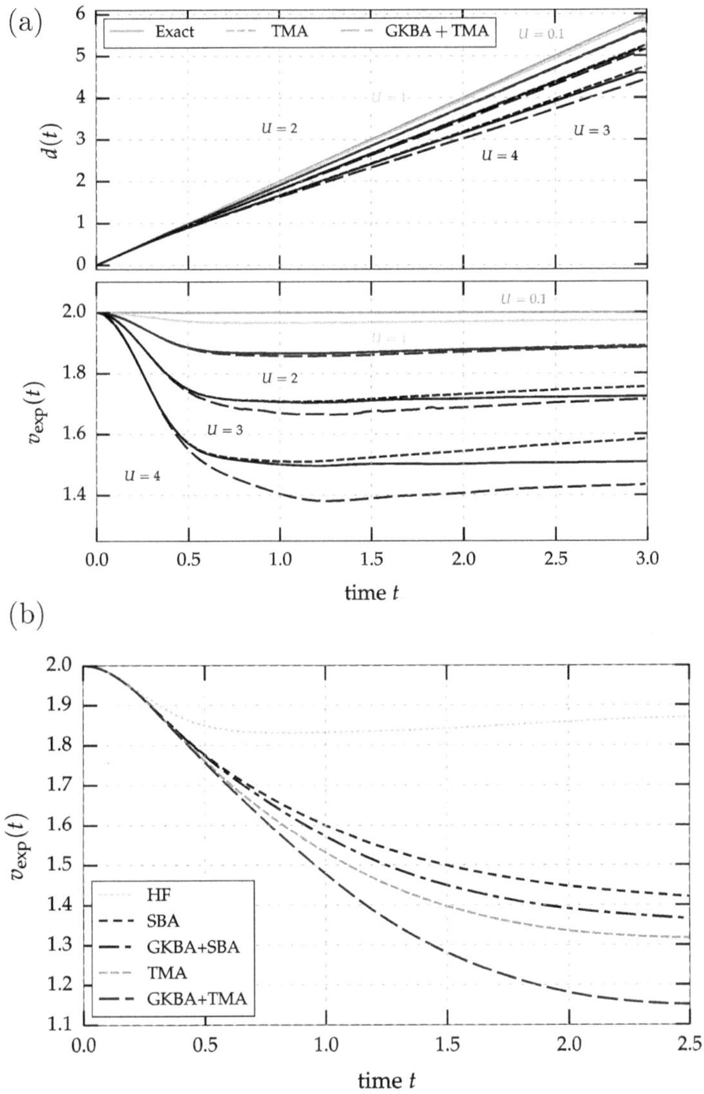

Figure 36.8 (a) Time-dependent diameter (upper panel) and expansion velocity (lower panel) for two fermions in a 2D lattice. [Reproduced from Fig. 33 of Ref. [38], with permission.] (b) Expansion velocity of fermions in a 2D lattice for different approximations. [Reproduced from Fig. 36 of Ref. [38], with permission.]

Remark: A drawback of the GKBA

The GKBA is not able to approximate the right ($G^{\rceil}$) and left ($G^{\lceil}$) Keldysh components, which are needed, in particular, in the equation of motion (36.6) for the time-diagonal Green's function $G^{<}$. Apart from calculating these additional Keldysh components independently (in terms of the equation of motion (15.11) of Part I, together with the symmetry property (13.28) therein), a possible alternative solution to deal with initial correlations within the GKBA has been proposed in Ref. [216].

In practice, the GKBA has been utilized in numerical calculations and tested against alternative approximations (as well as with exact results when available). Figure 36.8 shows a few examples of these comparisons, for a 2D system of N fermions embedded in a square lattice and mutually interacting in terms of a Hubbard model with a local repulsion U [38]. The system is initially prepared in a circular configuration and then let to expand from time $t_0 = 0$ on. In the upper panels, both the time-dependent diameter of the expanding cloud and the corresponding expansion velocity are shown when $N = 2$ for several values of U, as obtained within the t-matrix approximation (TMA) and when the GKBA is applied to the TMA (GKBA + TMA), and further compared with the exact result, which is available in this case. In the lower panel, the expansion velocity is shown for $N = 58$ and $U = 2.5$, as obtained within the second Born approximation (SBA) and the TMA, to both of which the GKBA is also applied (GKBA + SBA and GKBA + TMA). [The result of the Hartree–Fock (HF) approximation is also shown for comparison.] In all cases, the GKBA is seen to perform reasonably well with respect to the corresponding reference approximation. A further comparison among the results obtained from the full two-time calculations and the GKBA can be found, for example, in Ref. [159] (cf. also the references therein).

Finally, it is worth mentioning that the GKBA has sometimes been utilized in conjunction with the wide-band-limit approximation (WBLA) discussed in Section 33.4, which considerably simplified the numerical calculations and, in some cases, allows one to close the equations of motion [159]. However, limitations of this combined GKBA+WBLA approach arise when considering heat flow in addition to particle flow, as it was pointed out in Ref. [217].

37

Miscellany and Addenda to Part III

Similar to what we have done at the end of Parts I and II, here, we add the detailed treatment of a few issues that complement some of the topics treated in Chapters 35 and 36 of Part III.

37.1 A Schematic Derivation of the Lindblad Master Equation

It was pointed out in Chapter 35 that the Schwinger–Keldysh closed-contour approach and the Lindblad equation for Markovian systems can be connected to each other as far as the occurrence of a closed time contour is concerned in both cases. Here, for completeness, we briefly recall how the Lindblad equation can be derived from first principles in the context of the Quantum Master Equation and its Markov approximation. This will also help the reader in retracing and better identifying the essential steps made and approximations adopted in the course of the (somewhat more general) derivation presented in Chapter 35. A fuller account of the derivation and properties of the Lindblad equation can be found in Refs. [185, 186, 218].

The starting point is the expression (35.1)–(35.2) for the time-dependent average of an operator $A(t)$, which can be expressed in terms of the time-dependent density matrix $\rho(t) = U(t, t_0)\rho\, U^{\dagger}(t, t_0)$ that evolves in time out of an initial configuration $\rho = \rho(t_0)$. Here, $U(t, t_0)$ is the time-evolution operator introduced in Section 2.2 of Part I, which depends on the total Hamiltonian $\mathcal{H}(t)$.

The quantum system we are considering is globally closed and can be partitioned into the subsystem S of interest with Hamiltonian H_{S} and a surrounding environment (or heat bath) B with Hamiltonian H_{B}, which are both assumed to be independent of time, while the Hamiltonian $H_{\mathrm{SB}}(t)$ that describes the SB interaction between the subsystem and the bath may as well depend on time. Accordingly, the state of the subsystem S will change as a consequence of its internal dynamics and of the interaction with the environment B. This situation is sketched pictorially in Fig. 37.1.

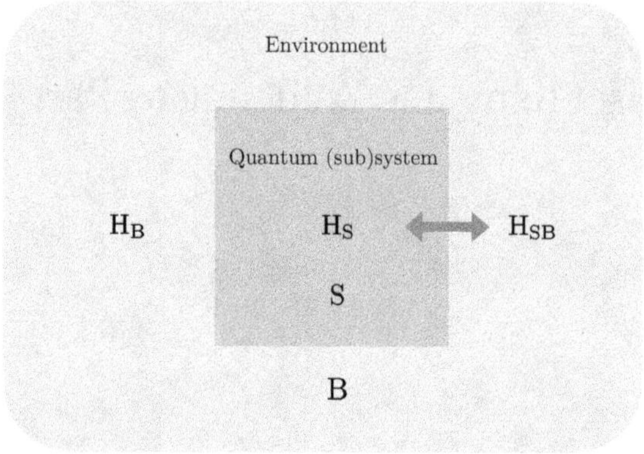

Figure 37.1 The quantum subsystem S coupled with the bath B.

Since all observations of interest refer to the subsystem S, we shall be concerned with deriving the equation of motion of the relevant density matrix on which these observations depend, arriving eventually at the Lindblad Master equation. To this end, we proceed by relying on a number of assumptions and related approximations, which we organize in subsequent steps as follows:

Step # 1: The interaction picture With the perspective that the action of $H_{SB}(t)$ can be considered "weak" in some sense, it is convenient to revert from the outset to the "interaction picture," as defined in Section 3.2 of Part I. We thus write

$$\rho_I(t) = W(t, t_0)\, \rho(t_0)\, W^\dagger(t, t_0), \tag{37.1}$$

where $W(t, t_0) = e^{iH_0 t}\, U(t, t_0)\, e^{-iH_0 t_0}$, such that

$$i \frac{\partial}{\partial t} W(t, t_0) = H_I(t)\, W(t, t_0)\,, \tag{37.2}$$

t_0 being the time at which the subsystem and the bath begin to interact with each other, in such a way that $[\rho(t_0), H_0] = 0$, where $H_0 = H_S + H_B$ in the present context. In addition,

$$H_I(t) \leftrightarrow H_I^{SB}(t) = e^{iH_0 t}\, H_{SB}(t)\, e^{-iH_0 t}\,. \tag{37.3}$$

In this way, we can rewrite $\langle A(t)\rangle = \mathrm{Tr}\{\rho_I(t)\, A_I(t)\}$, with $A_I(t) = e^{iH_0 t}\, A(t)\, e^{-iH_0 t}$.

The equation of motion of $\rho_I(t)$ then reads:

$$\begin{aligned}\frac{d}{dt}\rho_I(t) &= \left(\frac{\partial}{\partial t} W(t, t_0)\right) \rho(t_0)\, W^\dagger(t, t_0) + W(t, t_0)\, \rho(t_0) \left(\frac{\partial}{\partial t} W^\dagger(t, t_0)\right) \\ &= -\,i\, H_I(t)\, W(t, t_0)\, \rho(t_0)\, W^\dagger(t, t_0) + i\, W(t, t_0)\, \rho(t_0)\, W^\dagger(t, t_0)\, H_I(t) \\ &= -\,i\,[H_I(t), \rho_I(t)]\,.\end{aligned} \tag{37.4}$$

This equation can be integrated, to yield

$$\rho_{\rm I}(t) = \rho_{\rm I}(t_0) - i \int_{t_0}^{t} d\bar{t}\ [H_{\rm I}(\bar{t}), \rho_{\rm I}(\bar{t})]\ , \tag{37.5}$$

which can then be inserted back into the equation of motion (37.4), to obtain

$$\frac{d}{dt}\rho_{\rm I}(t) = -\,i\,[H_{\rm I}(t), \rho_{\rm I}(t_0)] - \int_{t_0}^{t} d\bar{t}\,[H_{\rm I}(t), [H_{\rm I}(\bar{t}), \rho_{\rm I}(\bar{t})]]\ . \tag{37.6}$$

From now on, we shall set $t_0 = 0$ without loss of generality.

Step # 2: Projection onto the S subspace As all observations of interest refer to the subsystem S, it is convenient to "trace out" at the outset the bath degrees of freedom, thereby effectively projecting into the S subspace. For a generic operator $A_{\rm S}$ acting only on the subsystem S, we can then write $\langle A_{\rm S}\rangle = {\rm Tr}\,\{\rho\, A_{\rm S}\} = {\rm Tr_S}\{\rho_{\rm S}\, A_{\rm S}\}$, where $\rho_{\rm S} = {\rm Tr_B}\,\{\rho\}$ is the *reduced density matrix*, as it was shown in Section 35.1.

We are thus led to take the trace of Eq. (37.6) over the bath, obtaining in this way the equation of motion of the reduced density matrix:

$$\frac{d}{dt}\rho_{\rm I}^{\rm S}(t) = -\,i\,{\rm Tr_B}\,\{[H_{\rm I}(t), \rho_{\rm I}(t_0)]\} - \int_{0}^{t} d\bar{t}\ {\rm Tr_B}\,\{[H_{\rm I}(t), [H_{\rm I}(\bar{t}), \rho_{\rm I}(\bar{t})]]\}\ . \tag{37.7}$$

The first term on the right-hand side of Eq. (37.7) can be proved to vanish, on the basis of the assumptions discussed in the following two steps.

Step # 3: Assumption of initial factorization We shall assume that, at the initial time $t_0 = 0$ before they start to interact, the subsystem S and its environment B are uncorrelated from each other, such that we set

$$\rho(t=0) = \rho_{\rm B}(t=0) \otimes \rho_{\rm S}(t=0)\ . \tag{37.8}$$

This would be the case if the system and the environment had not interacted at previous times, such that $H_{\rm SB}(t) = \theta(t - t_0)\, H_{\rm SB}$. Obviously, a more complicated time-dependent behavior of $H_{\rm SB}(t)$ for $t > t_0$ can occur.

Step # 4: Assumption of linear interaction We shall also assume that

$${\rm Tr_B}\left\{\rho_{\rm B}(t=0)\, H_{\rm I}^{\rm SB}(t)\right\} = 0\ . \tag{37.9}$$

This can be seen as a consequence [185] of taking the Hamiltonian that couples the subsystem to the environment in the form

$$H_{\rm SB} = \sum_{\alpha} A_\alpha\, B_\alpha, \tag{37.10}$$

where $A_\alpha^\dagger = A_\alpha$ refers to the subsystem and $B_\alpha^\dagger = B_\alpha$ refers to the environment. This is because for all α $\mathrm{Tr_B}\,\{\rho_\mathrm{B}(t=0) B_\alpha\} = 0$, as it is the case for a normal system in the absence of condensation.

With the assumptions (37.8) and (37.9), the first term on the right-hand side of Eq. (37.7) becomes:

$$\begin{aligned}
\mathrm{Tr_B}\,\{[H_\mathrm{I}(t), \rho_\mathrm{I}(t=0)]\} &\underset{[\mathrm{Eq.(37.8)}]}{\overset{}{=}} \mathrm{Tr_B}\,\{[H_\mathrm{I}(t), \rho_\mathrm{B}(t=0)\,\rho_\mathrm{S}(t=0)]\} \\
&= \mathrm{Tr_B}\,\{H_\mathrm{I}(t)\,\rho_\mathrm{B}(t=0)\,\rho_\mathrm{S}(t=0)\} - \mathrm{Tr_B}\{\rho_\mathrm{B}(t=0)\,\rho_\mathrm{S}(t=0)\,H_\mathrm{I}(t)\} \\
&= \mathrm{Tr_B}\,\{H_\mathrm{I}(t)\,\rho_\mathrm{B}(t=0)\}\rho_\mathrm{S}(t=0) - \rho_\mathrm{S}(t=0)\,\mathrm{Tr_B}\{\rho_\mathrm{B}(t=0)\,H_\mathrm{I}(t)\} \underset{[\mathrm{Eq.(37.9)}]}{=} 0. \qquad (37.11)
\end{aligned}$$

In this way, we are left only with the second term on the right-hand side of Eq. (37.7).

Step # 5: Assumption of persistent factorization $\leftrightarrow$ weak coupling To proceed further, we also extend the assumption of factorization (37.8) to later times $t > t_0 = 0$, by writing approximately

$$\rho_\mathrm{I}(t) \simeq \rho_\mathrm{B}(t=0) \otimes \rho_\mathrm{I}^\mathrm{S}(t)\,. \qquad (37.12)$$

This corresponds to assuming that the coupling between the subsystem and the environment remains weak enough, such that the environment state specified by ρ_B is always decoupled from the subsystem state and does not change in time, even after the environment and the subsystem are put in contact.

With the assumption of persistent factorization (37.12), the equation of motion (37.7) becomes

$$\frac{d}{dt}\rho_\mathrm{I}^\mathrm{S}(t) = -\int_0^t d\bar{t}\;\mathrm{Tr_B}\left\{\left[H_\mathrm{I}(t), \left[H_\mathrm{I}(\bar{t}), \rho_\mathrm{I}^\mathrm{S}(\bar{t})\,\rho_\mathrm{B}(t=0)\right]\right]\right\}\,, \qquad (37.13)$$

which is a closed integro-differential equation for the reduced density matrix $\rho_\mathrm{I}^\mathrm{S}(t)$ in the interaction representation. Note how Eq. (37.13) contains "memory effects" about the dynamics of the subsystem for $0 \le \bar{t} < t$.

Step # 6: The Markov approximation Equation (37.13) can be further simplified by adopting the so-called *Markov approximation*, whereby the time development of the state of the subsystem at time t depends only on the present state and not on the past history. Physically, this amounts to assuming that the characteristic decay time τ_B of the environment correlations is short compared with the relaxation time τ_S of the subsystem, owing to the weakness of the interaction between the subsystem and the environment. Note in this context that this decay is possible only when the energy spectrum of the environment is continuous. Mathematically, the above assumption amounts to replacing $\rho_{\rm I}^{\rm S}(\bar{t}) \rightarrow \rho_{\rm I}^{\rm S}(t)$ in the integrand on the right-hand side of Eq. (37.13), thus yielding

$$\frac{d}{dt}\rho_{\rm I}^{\rm S}(t) = -\int_0^t d\bar{t}\ {\rm Tr_B}\left\{\left[H_{\rm I}(t), \left[H_{\rm I}(\bar{t}), \rho_{\rm I}^{\rm S}(t)\,\rho_{\rm B}(t=0)\right]\right]\right\}\,. \tag{37.14}$$

We may also change the integration variable by setting $\tau = t - \bar{t}$, such that Eq. (37.14) becomes

$$\frac{d}{dt}\rho_{\rm I}^{\rm S}(t) = -\int_0^t d\tau\ {\rm Tr_B}\left\{\left[H_{\rm I}(t), \left[H_{\rm I}(t-\tau), \rho_{\rm I}^{\rm S}(t)\,\rho_{\rm B}(t=0)\right]\right]\right\}\,. \tag{37.15}$$

We may further exploit the above assumption about the fast decay of the environment correlations and consider the kernel in the above integral to decay fast enough, such that the upper limit of the integration can be extended to infinity with no change in its outcome. This yields

$$\frac{d}{dt}\rho_{\rm I}^{\rm S}(t) = -\int_0^\infty d\tau\ {\rm Tr_B}\left\{\left[H_{\rm I}(t), \left[H_{\rm I}(t-\tau), \rho_{\rm I}^{\rm S}(t)\,\rho_{\rm B}(t=0)\right]\right]\right\}\,, \tag{37.16}$$

which is known as the *Markovian Quantum Master Equation*.

Step # 7: A frequency decomposition To connect with the energy spectrum of the bath, it is convenient to introduce a frequency decomposition as follows:

Let $\{|\phi_\varepsilon\rangle\}$ be the set of eigenfunctions of the subsystem Hamiltonian $H_{\rm S}$, each associated with an eigenvalue ε. Let further Π_ε be the projection operator belonging to the eigenvalue ε, such that

$$\sum_\varepsilon \Pi_\varepsilon = \mathbb{1} \tag{37.17}$$

is the identity operator. For each operator A_α of the subsystem entering the decomposition (37.10), we can then write

$$\begin{aligned} A_\alpha &= \left(\sum_\varepsilon \Pi_\varepsilon\right) A_\alpha \left(\sum_{\varepsilon'} \Pi_{\varepsilon'}\right) = \sum_\varepsilon \sum_{\varepsilon'} \Pi_\varepsilon\, A_\alpha\, \Pi_{\varepsilon'} = \sum_\omega \sum_{\{\varepsilon'=\epsilon+\omega\}} \Pi_\varepsilon\, A_\alpha\, \Pi_{\varepsilon'} \\ &= \sum_\omega A_\alpha(\omega), \end{aligned} \tag{37.18}$$

where we have rearranged the sums over ε and ε', as indicated schematically in Fig. 37.2 and defined

$$A_\alpha(\omega) \equiv \sum_{\{\varepsilon'=\varepsilon+\omega\}} \Pi_\varepsilon\, A_\alpha\, \Pi_{\varepsilon'}\,. \tag{37.19}$$

From this definition, it follows that

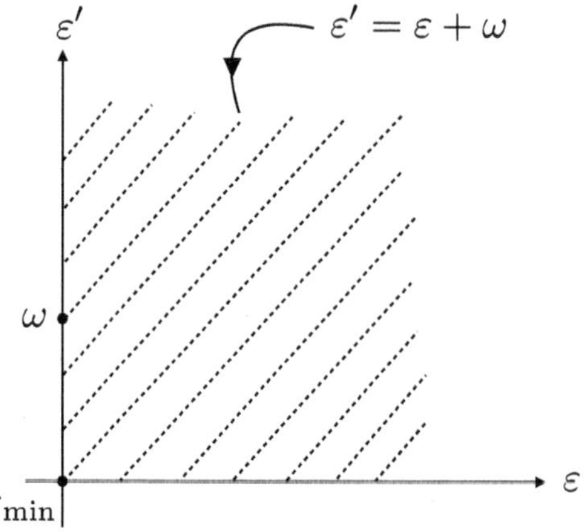

Figure 37.2 Rearranging the sums over ε and ε'. For given ω, $\varepsilon' \geq \varepsilon_{\rm min}$ is associated with $\varepsilon \geq \varepsilon_{\rm min}$ according to $\varepsilon' = \varepsilon + \omega$, where $\varepsilon_{\rm min}$ is the minimum eigenvalue of the subsystem Hamiltonian $H_{\rm S}$.

$$H_{\rm S} A_\alpha(\omega) = H_{\rm S} \sum_{\{\varepsilon'=\epsilon+\omega\}} \Pi_\varepsilon\, A_\alpha\, \Pi_{\varepsilon'} = \varepsilon A_\alpha(\omega), \tag{37.20}$$

$$A_\alpha(\omega) H_{\rm S} = \sum_{\{\varepsilon'=\epsilon+\omega\}} \Pi_\varepsilon\, A_\alpha\, \Pi_{\varepsilon'}\, H_{\rm S} = \varepsilon' A_\alpha(\omega)\,, \tag{37.21}$$

from which

$$H_{\rm S} A_\alpha(\omega) - A_\alpha(\omega) H_{\rm S} = (\varepsilon - \varepsilon')\, A_\alpha(\omega) = -\,\omega\, A_\alpha(\omega)\,. \tag{37.22}$$

This leads to the result

$$e^{iH_{\rm S}t} A_\alpha(\omega)\, e^{-iH_{\rm S}t} = e^{-i\omega t} A_\alpha(\omega)\,, \tag{37.23}$$

which can be proved as follows. Recall the Baker–Campbell–Hausdorff identity [219]

$$e^{\mathcal{S}}\, \mathcal{O}\, e^{-\mathcal{S}} = \mathcal{O} + [\mathcal{S}, \mathcal{O}] + \frac{1}{2!}\, [\mathcal{S}, [\mathcal{S}, \mathcal{O}]] + \frac{1}{3!}\, [\mathcal{S}, [\mathcal{S}, [\mathcal{S}, \mathcal{O}]]] + \cdots\,, \tag{37.24}$$

which holds for any pair of operators $\mathcal{S}$ and $\mathcal{O}$. In our case, $\mathcal{S} \leftrightarrow iH_{\rm S}t$ and $\mathcal{O} \leftrightarrow A_\alpha(\omega)$, with $[\mathcal{S}, \mathcal{O}] \leftrightarrow i\,t\,[H_{\rm S}, A_\alpha(\omega)] = -\,i\,\omega\,t\,A_\alpha(\omega)$ according to Eq. (37.22). We thus obtain

$$\begin{aligned} e^{iH_{\rm S}t} A_\alpha(\omega)\, e^{-iH_{\rm S}t} &= A_\alpha(\omega) \;-\; i\,\omega\,t\,A_\alpha(\omega) \;+\; \frac{1}{2!}\,(-\,i\,\omega\,t)^2\, A_\alpha(\omega) \\ &\quad + \frac{1}{3!}\,(-\,i\,\omega\,t)^3\, A_\alpha(\omega) \;+\; \cdots = e^{-i\omega t} A_\alpha(\omega)\,. \end{aligned} \tag{37.25}$$

In addition, from the definition (37.19), we obtain

$$A_\alpha^\dagger(\omega) \equiv \sum_{\{\varepsilon'=\varepsilon+\omega\}} \Pi_{\varepsilon'} A_\alpha \Pi_\varepsilon = \sum_{\{\varepsilon=\varepsilon'-\omega\}} \Pi_{\varepsilon'} A_\alpha \Pi_\varepsilon = A_\alpha(-\omega)\,. \tag{37.26}$$

With these premises, the interaction Hamiltonian (37.3), with $H^{SB}(t)$ given by Eq. (37.10), can be written in the following form:

$$\begin{aligned} H_\mathrm{I}(t) \leftrightarrow H_\mathrm{I}^{SB}(t) &= e^{i(H_\mathrm{S}+H_\mathrm{B})t} \sum_\alpha A_\alpha B_\alpha e^{-i(H_\mathrm{S}+H_\mathrm{B})t} \\ &= \sum_\alpha \left(e^{iH_\mathrm{S}t} A_\alpha e^{-iH_\mathrm{S}t}\right)\left(e^{iH_\mathrm{B}t} B_\alpha e^{-iH_\mathrm{B}t}\right) \\ &\underset{[\text{Eq.}(37.18)]}{\overset{\uparrow}{=}} \sum_\alpha \sum_\omega \left(e^{iH_\mathrm{S}t} A_\alpha(\omega) e^{-iH_\mathrm{S}t}\right)\left(e^{iH_\mathrm{B}t} B_\alpha e^{-iH_\mathrm{B}t}\right) \\ &\underset{[\text{Eq.}(37.25)]}{\overset{\uparrow}{=}} \sum_\alpha \sum_\omega e^{-i\omega t} A_\alpha(\omega) B_\alpha(t), \end{aligned} \tag{37.27}$$

where

$$B_\alpha(t) = e^{iH_\mathrm{B}t} B_\alpha e^{-iH_\mathrm{B}t} \tag{37.28}$$

in the interaction picture for the environment. In terms of the result (37.27), the Markovian equation (37.16) can be manipulated as follows:

$$\begin{aligned} \frac{d}{dt}\rho_\mathrm{I}^\mathrm{S}(t) &= -\int_0^\infty d\tau\, \mathrm{Tr}_\mathrm{B}\Big\{H_\mathrm{I}(t)\left(H_\mathrm{I}(t-\tau)\rho_\mathrm{I}^\mathrm{S}(t)\,\rho_\mathrm{B}(t=0) - \rho_\mathrm{I}^\mathrm{S}(t)\,\rho_\mathrm{B}(t=0) H_\mathrm{I}(t-\tau)\right) \\ &\quad - \left(H_\mathrm{I}(t-\tau)\,\rho_\mathrm{I}^\mathrm{S}(t)\,\rho_\mathrm{B}(t=0) - \rho_\mathrm{I}^\mathrm{S}(t)\rho_\mathrm{B}(t=0)\, H_\mathrm{I}(t-\tau)\right) H_\mathrm{I}(t)\Big\} \\ &\underset{[[\rho_\mathrm{I}^\mathrm{S}(t),\,\rho_\mathrm{B}(t=0)]=0]}{\overset{\uparrow}{=}} \int_0^\infty d\tau\, \mathrm{Tr}_\mathrm{B}\Big\{H_\mathrm{I}(t-\tau)\,\rho_\mathrm{I}^\mathrm{S}(t)\rho_\mathrm{B}(t=0)\, H_\mathrm{I}(t) \\ &\quad - H_\mathrm{I}(t) H_\mathrm{I}(t-\tau)\rho_\mathrm{I}^\mathrm{S}(t)\,\rho_\mathrm{B}(t=0)\Big\} + \text{h.c.} \\ &\underset{[\text{Eq. }(37.27)]}{\overset{\uparrow}{=}} \sum_{\alpha\omega}\sum_{\alpha'\omega'} e^{i(\omega'-\omega)t} \int_0^\infty d\tau\, e^{i\omega\tau}\, \mathrm{Tr}_\mathrm{B}\left\{B_{\alpha'}^\dagger(t)\, B_\alpha(t-\tau)\,\rho_\mathrm{B}(t=0)\right\} \\ &\quad \times \left(A_\alpha(\omega)\,\rho_\mathrm{I}^\mathrm{S}(t)\, A_{\alpha'}^\dagger(\omega') - A_{\alpha'}^\dagger(\omega')\, A_\alpha(\omega)\,\rho_\mathrm{I}^\mathrm{S}(t)\right) + \text{h.c.} \\ &= \sum_{\alpha\omega}\sum_{\alpha'\omega'} e^{i(\omega'-\omega)t}\Gamma_{\alpha'\alpha}(\omega)\left(A_\alpha(\omega)\rho_\mathrm{I}^\mathrm{S}(t)A_{\alpha'}^\dagger(\omega') - A_{\alpha'}^\dagger(\omega')A_\alpha(\omega)\rho_\mathrm{I}^\mathrm{S}(t)\right) \\ &\quad + \text{h.c.}, \end{aligned} \tag{37.29}$$

where h.c. stands for the Hermitian conjugate and we have introduced the one-sided Fourier transform of the reservoir correlation functions

$$\Gamma_{\alpha'\alpha}(\omega) = \int_0^\infty d\tau\, e^{i\omega\tau}\, \mathrm{Tr}_\mathrm{B}\left\{B^\dagger_{\alpha'}(t)\, B_\alpha(t-\tau)\, \rho_\mathrm{B}(t=0)\right\}$$
$$\rightarrow \int_0^\infty d\tau\, e^{i\omega\tau}\, \mathrm{Tr}_\mathrm{B}\left\{B^\dagger_{\alpha'}(\tau)\, B_\alpha(0)\, \rho_\mathrm{B}(t=0)\right\} . \qquad (37.30)$$

Note that in the last line we have dropped the dependence on the time t because we have assumed that the reservoir correlation functions are homogeneous in time.

Step # 8: The rotating wave approximation A further simplification of Eq. (37.29) stems by adopting the so-called rotating wave approximation (RWA) (cf. Ref. [220] for a recent review about RWA). On physical grounds, this simplification is related to the basic condition underlying the Markov approximation, according to which the reservoir correlation functions decay over a time $\tau_\mathrm{B} \ll \tau_\mathrm{S}$. This condition translates into Eq. (37.29) by noting that terms for which $|\omega - \omega'|\tau_\mathrm{S} \gg 1$ may be neglected, since they oscillate very rapidly during the characteristic relaxation time τ_S of the subsystem over which $\rho_\mathrm{I}^\mathrm{S}(t)$ varies appreciably (as shown pictorially in Fig. 37.3). As a consequence, to a first approximation, only terms with $\omega = \omega'$ will be retained in Eq. (37.29) according to the RWA, thus yielding the result:

$$\frac{d}{dt}\rho_\mathrm{I}^\mathrm{S}(t) = \sum_\omega \sum_{\alpha\,\alpha'} \Gamma_{\alpha'\alpha}(\omega) \left(A_\alpha(\omega)\, \rho_\mathrm{I}^\mathrm{S}(t)\, A^\dagger_{\alpha'}(\omega) - A^\dagger_{\alpha'}(\omega)\, A_\alpha(\omega)\, \rho_\mathrm{I}^\mathrm{S}(t)\right) + \mathrm{h.c.}\,. \qquad (37.31)$$

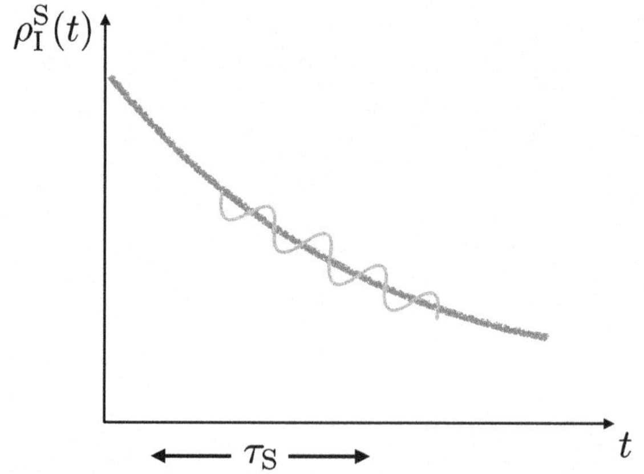

Figure 37.3 For the sake of comparison, fast oscillations with $\omega \neq \omega'$ in Eq. (37.29) are superimposed on the slower overall variation of $\rho_\mathrm{I}^\mathrm{S}(t)$ over the time scale τ_S.

Step # 9: A further cosmetic A further manipulation can be made on the expression (37.31), by writing

$$\Gamma_{\alpha\beta}(\omega) = \frac{1}{2}\Big(\Gamma_{\alpha\beta}(\omega) + \Gamma^*_{\beta\alpha}(\omega)\Big) + i\,\frac{1}{2i}\Big(\Gamma_{\alpha\beta}(\omega) - \Gamma^*_{\beta\alpha}(\omega)\Big) \equiv \gamma_{\alpha\beta}(\omega) + i\,S_{\alpha\beta}(\omega), \tag{37.32}$$

where

$$\begin{aligned} 2\,\gamma_{\alpha\beta}(\omega) = \Gamma_{\alpha\beta}(\omega) + \Gamma^*_{\beta\alpha}(\omega) &\underset{[\text{Eq.37.30}]}{\underset{\uparrow}{=}} \int_0^\infty d\tau\, e^{i\omega\tau}\, \langle B^\dagger_\alpha(\tau)\, B_\beta(0)\rangle_B \\ &+ \int_0^\infty d\tau\, e^{-i\omega\tau}\, \langle B^\dagger_\beta(\tau)\, B_\alpha(0)\rangle^*_B \\ &= \int_0^\infty d\tau\, e^{i\omega\tau}\, \langle B^\dagger_\alpha(\tau)\, B_\beta(0)\rangle_B - \int_0^{-\infty} d\tau\, e^{i\omega\tau}\, \langle B^\dagger_\alpha(0)\, B_\beta(-\tau))\rangle_B \\ &= \int_{-\infty}^\infty d\tau\, e^{i\omega\tau}\, \langle B^\dagger_\alpha(\tau)\, B_\beta(0)\rangle_B \end{aligned} \tag{37.33}$$

is now the full Fourier transform of the reservoir correlation functions, and

$$S_{\alpha\beta}(\omega) = \frac{1}{2i}\Big(\Gamma_{\alpha\beta}(\omega) - \Gamma^*_{\beta\alpha}(\omega)\Big)\,. \tag{37.34}$$

Note that both $\gamma_{\alpha\beta}(\omega)^* = \gamma_{\beta\alpha}(\omega)$ and $S_{\alpha\beta}(\omega)^* = S_{\beta\alpha}(\omega)$ are Hermitian matrices.

With these definitions, Eq. (37.31) becomes eventually:

$$\begin{aligned} \frac{d}{dt}\rho^S_I(t) = \sum_\omega \sum_{\alpha\beta} \gamma_{\alpha\beta}(\omega)\Big(&2\,A_\beta(\omega)\,\rho^S_I(t)\,A^\dagger_\alpha(\omega) - \rho^S_I(t)\,A^\dagger_\alpha(\omega)\,A_\beta(\omega) \\ &- A^\dagger_\alpha(\omega)\,A_\beta(\omega)\,\rho^S_I(t)\Big) \\ - i\sum_\omega \sum_{\alpha\beta} S_{\alpha\beta}(\omega)\Big(&A^\dagger_\alpha(\omega)\,A_\beta(\omega)\,\rho^S_I(t) - \rho^S_I(t)\,A^\dagger_\alpha(\omega)\,A_\beta(\omega)\Big)\,. \end{aligned} \tag{37.35}$$

The last term in this expression can be rewritten in the form of $(-i)$ times the commutator $\left[H^{\text{LS}}, \rho^S_I(t)\right]$ where

$$H^{\text{LS}} = \sum_\omega \sum_{\alpha\beta} A^\dagger_\alpha(\omega)\, S_{\alpha\beta}(\omega)\, A_\beta(\omega) \tag{37.36}$$

with $(H^{\mathrm{LS}})^{\dagger} = H^{\mathrm{LS}}$ is the so-called *Lamb shift Hamiltonian* since it leads to a Lamb-type renormalization [221] of the unperturbed energy levels of the subsystem as induced by the coupling between the subsystem and the environment.

Step # 10: The Lindblad equation, eventually Since the matrix $\gamma_{\alpha\beta}(\omega)$ given by the expression (37.33) is Hermitian for given ω, it can be diagonalized with real eigenvalues. We thus write

$$\gamma_{\alpha\beta}(\omega) = \sum_r \mathcal{U}_{\alpha r}(\omega)\, \tilde{\gamma}_r(\omega)\, \mathcal{U}^{\dagger}_{r\beta}(\omega) , \tag{37.37}$$

such that in Eq. (37.31)

$$\sum_{\alpha\beta} \gamma_{\alpha\beta}(\omega)\, A_{\beta}(\omega)\, \rho^{\mathrm{S}}_{\mathrm{I}}(t)\, A^{\dagger}_{\alpha}(\omega) = \sum_r \tilde{\gamma}_r(\omega)\, A_r(\omega)\, \rho^{\mathrm{S}}_{\mathrm{I}}(t)\, A^{\dagger}_r(\omega), \tag{37.38}$$

$$\sum_{\alpha\beta} \gamma_{\alpha\beta}(\omega)\, A^{\dagger}_{\alpha}(\omega)\, A_{\beta}(\omega) = \sum_r \tilde{\gamma}_r(\omega)\, A^{\dagger}_r(\omega)\, A_r(\omega) , \tag{37.39}$$

where we have defined

$$A_r(\omega) = \sum_{\alpha} \mathcal{U}^{\dagger}_{r\alpha}(\omega)\, A_{\alpha}(\omega) \quad \text{and} \quad A^{\dagger}_r(\omega) = \sum_{\alpha} A^{\dagger}_{\alpha}(\omega)\mathcal{U}_{\alpha r}(\omega) . \tag{37.40}$$

In this way, Eq. (37.35) acquires the final form of the *Lindblad Master Equation* (LME):

$$\frac{d}{dt}\rho^{\mathrm{S}}_{\mathrm{I}}(t) = \sum_{\omega}\sum_r \tilde{\gamma}_r(\omega) \Big(2\, A_r(\omega)\, \rho^{\mathrm{S}}_{\mathrm{I}}(t)\, A^{\dagger}_r(\omega) - \rho^{\mathrm{S}}_{\mathrm{I}}(t)\, A^{\dagger}_r(\omega)\, A_r(\omega)$$
$$-A^{\dagger}_r(\omega)\, A_r(\omega)\, \rho^{\mathrm{S}}_{\mathrm{I}}(t)\Big) - i\left[H^{\mathrm{LS}}, \rho^{\mathrm{S}}_{\mathrm{I}}(t)\right] . \tag{37.41}$$

This result can be compared with Eqs. (32.1) and (32.2), where we now identify $n \leftrightarrow (\omega, r)$ and $V_n \leftrightarrow \sqrt{2\, \tilde{\gamma}_r(\omega)}\, A_r(\omega)$ for the jump operators (provided the eigenvalues $\tilde{\gamma}_r(\omega)$ are all positive), as well as $H_{\mathrm{S}} \leftrightarrow H^{\mathrm{LS}}$ since we are now working in the interaction picture.

Proof that $\tilde{\gamma}_r(\omega) \geq 0$

We begin by rewriting the last line of Eq. (37.27) in terms of the matrix $\mathcal{U}$, such that

$$\sum_{\alpha} A_{\alpha}(\omega) B_{\alpha}(t) = \sum_r \left(\sum_{\alpha} \mathcal{U}^{\dagger}_{r\alpha}(\omega)\, A_{\alpha}(\omega)\right) \left(\sum_{\beta} B_{\beta}(t)\, \mathcal{U}_{\beta r}(\omega)\right)$$
$$\underset{[\text{Eq.}(37.40)]}{\overset{\uparrow}{=}} \sum_r A_r(\omega) B_r(t; \omega) , \tag{37.42}$$

where we have identified $B_r(t;\omega) = \sum_\beta B_\beta(t)\,\mathcal{U}_{\beta r}(\omega)$. Accordingly, from Eq. (37.37) and the definition (37.33), we obtain

$$\begin{aligned}
2\,\tilde{\gamma}_r(\omega) &= \sum_{\alpha\beta} \mathcal{U}^\dagger_{r\alpha}(\omega)\,2\,\gamma_{\alpha\beta}(\omega)\,\mathcal{U}_{\beta r}(\omega) \\
&= \int_{-\infty}^{\infty} d\tau\, e^{i\omega\tau} \left\langle \left(\sum_\alpha \mathcal{U}^\dagger_{r\alpha}(\omega)\,B^\dagger_\alpha(\tau)\right)\left(\sum_\beta B_\beta(0)\,\mathcal{U}_{\beta r}(\omega)\right)\right\rangle_B \\
&= \int_{-\infty}^{\infty} d\tau\, e^{i\omega\tau} \langle B^\dagger_r(\tau;\omega)\,B_r(0;\omega)\rangle_B \\
&= \int_{-\infty}^{\infty} d\tau\, e^{i\omega\tau} \sum_{n_B} w_{n_B} \langle n_B| e^{iH_B\tau}\,B^\dagger_r(0;\omega)\,e^{-iH_B\tau}\,B_r(0;\omega)|n_B\rangle \\
&= \sum_{m_B n_B} w_{n_B} \int_{-\infty}^{\infty} d\tau\, e^{i(\omega+E_{n_B}-E_{m_B})\tau} \langle n_B|B^\dagger_r(0;\omega)|m_B\rangle\langle m_B|B_r(0;\omega)|n_B\rangle \\
&= 2\pi \sum_{m_B n_B} w_{n_B}\,\delta(\omega + E_{n_B} - E_{m_B})\,|\langle n_B|B^\dagger_r(0;\omega)|m_B\rangle|^2 \geq 0\,,
\end{aligned} \tag{37.43}$$

where the ensemble average over the "bath" with weights $\{w_{n_B}\}$ is made explicit. [QED]

Finally, we note that the LME (37.41) preserves in time the value of the trace of the density matrix $\rho^{\mathrm{S}}_{\mathrm{I}}(t)$. This is proved by taking the trace $\mathrm{Tr_S}$ of both sides of Eq. (37.41) over the Hilbert space of the subsystem, such that $\mathrm{Tr_S}\{\frac{d}{dt}\rho^{\mathrm{S}}_{\mathrm{I}}(t)\} = \frac{d}{dt}\mathrm{Tr_S}\{\rho^{\mathrm{S}}_{\mathrm{I}}(t)\} = 0$, where the cyclic property of the trace has been exploited.

37.2 The Original Kadanoff–Baym Ansatz

In Subsection 36.2.1, we have mentioned the original version of the Kadanoff–Baym ansatz (KBA), which was introduced by Kadanoff and Baym themselves in their pioneering work on the time-dependent many-body problem [5]. In that work, the approximate form for $G^<$ (given by Eq. (11.1) therein) was stated essentially like an *ansatz* for the case of a "weak" nonequilibrium situation. A way to arrive at the KBA in a more systematic way was later described in section II-C of Ref. [212], where the GKBA was also first introduced (for a more extended treatment, see also Ref. [222]). Here, we will take up that treatment, by specifying the physical assumptions leading to the KBA and further mentioning the limitations of this approximation.

We start from the first of Eq. (30.39) of Part II valid for fermions at equilibrium, expressed for the case of a homogeneous system, in the form

$$G^<(\mathbf{k},\omega) = -f_F(\omega)\,\left[G^{\mathrm{R}}(\mathbf{k},\omega) - G^{\mathrm{A}}(\mathbf{k},\omega)\right]\,, \tag{37.44}$$

where $\mathbf{k}$ is a wave vector, $f_F(\omega) = \left(e^{\beta(\omega-\mu)} + 1\right)^{-1}$ is the Fermi–Dirac distribution function, and the spin indices have been suppressed for simplicity. By recalling further the definition (30.46) of Part II, which we now write in the form

$$A(\mathbf{k}, \omega) = i \left[G^{\mathrm{R}}(\mathbf{k}, \omega) - G^{\mathrm{A}}(\mathbf{k}, \omega)\right] \tag{37.45}$$

valid for a homogeneous system, Eq. (37.44) becomes

$$G^{<}(\mathbf{k}, \omega) = f_F(\omega)\, i\, A(\mathbf{k}, \omega)\,. \tag{37.46}$$

In the weak scattering limit, we may approximate

$$A(\mathbf{k}, \omega) \simeq 2\pi\, \delta(\omega - \varepsilon(\mathbf{k})), \tag{37.47}$$

where $\varepsilon(\mathbf{k})$ is the quasi-particle dispersion (for which the sum rule (30.56) of Part II is satisfied). The expression (37.46) then becomes approximately

$$G^{<}(\mathbf{k}, \omega) \simeq f_F(\varepsilon(\mathbf{k}))\, i\, A(\mathbf{k}, \omega)\,. \tag{37.48}$$

In addition, Eq. (36.20) reads in the present context

$$n^{<}(\mathbf{k}) = -\, i \int_{-\infty}^{+\infty} \frac{d\omega}{2\pi}\, G^{<}(\mathbf{k}, \omega) \simeq f_F(\varepsilon(\mathbf{k})) \int_{-\infty}^{+\infty} \frac{d\omega}{2\pi}\, A(\mathbf{k}, \omega) \underset{\substack{\uparrow \\ \text{[Eq.(37.47)]}}}{=} f_F(\varepsilon(\mathbf{k}))\,, \tag{37.49}$$

such that Eq. (37.48) becomes eventually

$$G^{<}(\mathbf{k}, \omega) \simeq n^{<}(\mathbf{k})\, i\, A(\mathbf{k}, \omega)\,. \tag{37.50}$$

Kadanoff and Baym assumed that this relation remains *locally valid in space and time* for the case of a "weak" nonequilibrium situation, thereby arriving at their KBA in the form

$$\boxed{G^{<}(\mathbf{k}, \omega|\mathbf{R}, T) \simeq n^{<}(\mathbf{k}|\mathbf{R}, T))\, i\, A(\mathbf{k}, \omega|\mathbf{R}, T)),} \tag{37.51}$$

where $\mathbf{R} = (\mathbf{r} + \mathbf{r}')/2$, $T = (t + t')/2$, $\mathbf{k}$ is the wave vector Fourier conjugated to $(\mathbf{r} - \mathbf{r}')$, and ω is the frequency Fourier conjugated to $(t - t')$. In terms of the original space $(\mathbf{r}, \mathbf{r}')$ and time (t, t') coordinates, the KBA (37.51) reads

$$G^{<}(\mathbf{x}t, \mathbf{x}'t') \simeq -\, n^{<}(\mathbf{x}, \mathbf{x}'; T) \left[G^{\mathrm{R}}(\mathbf{x}t, \mathbf{x}'t') - G^{\mathrm{A}}(\mathbf{x}t, \mathbf{x}'t')\right], \tag{37.52}$$

where the spin indices have now been restored. Apart from the different convention $g^{<} = -iG^{<}$, the expression (37.52) above coincides with Eq. (36) of Ref. [212]. Note finally that, contrary to the GKBA result (36.22), the KBA result (37.52) violates the requirement of a causal time evolution, to the extent that $n^{<}$ therein depends only on the midpoint time T [222].

Bibliography

[1] J. Schwinger, *Brownian motion of a quantum oscillator*, J. Math. Phys. **2**, 407 (1961).
[2] L. V. Keldysh, *Diagram technique for non-equilibrium processes*, Sov. Phys. JETP **20**, 1018 (1965) [Zh. Eksp. Teor. Fiz. (U.S.S.R.) **47**, 1515 (1964)].
[3] G. Baym and L. P. Kadanoff, *Conservation laws and correlation functions*, Phys. Rev. **124**, 287 (1961).
[4] G. Baym, *Self-consistent approximations in many-body systems*, Phys. Rev. **127**, 1391 (1962).
[5] L. P. Kadanoff and G. Baym, *Quantum Statistical Mechanics* (W. A. Benjamin, Menlo Park, 1962).
[6] A. A. Abrikosov, L. P. Gorkov, and I. E. Dzyaloshinski, *Methods of Quantum Field Theory in Statistical Physics* (Dover, New York, 1963).
[7] A. L. Fetter and J. D. Walecka, *Quantum Theory of Many-Particle Systems* (McGraw-Hill, New York, 1971).
[8] A. Altland and B. Simons, *Condensed Matter Field Theory* (Cambridge University Press, Cambridge, 2013).
[9] G. Rickayzen, *Green's Functions and Condensed Matter* (Academic Press, London, 1980).
[10] G. D. Mahan, *Many-Particle Physics* (Plenum Press, New York, 1981).
[11] J. W. Negele and H. Orland, *Quantum Many-Particle Systems* (Addison-Wesley, Reading, 1988).
[12] P. Nozières, *Theory of Interacting Fermi Systems* (W. A. Benjamin, New York, 1964).
[13] J. R. Schrieffer, *Theory of Superconductivity* (W. A. Benjamin, New York, 1964).
[14] J. Rammer, *Quantum Field Theory of Non-Equilibrium States* (Cambridge University Press, Cambridge, 2009).
[15] G. Stefanucci and R. van Leeuwen, *Nonequilibrium Many Body Theory of Quantum Systems* (Cambridge University Press, Cambridge, 2013).
[16] A. Kamenev, *Field Theory of Non-Equilibrium Systems* (Cambridge University Press, Cambridge, 2013).
[17] M. Pourfath, *The Non-Equilibrium Green's Function Method for Nanoscale Device Simulation* (Springer-Verlag, Wien, 2014).
[18] M. Schultze, M. Fieß, N. Karpowicz, J. Gagnon, M. Korbman, M. Hofstetter, S. Neppl, A. L. Cavalieri, Y. Komninos, Th. Mercouris, C. A. Nicolaides, R. Pazourek, S. Nagele, J. Feist, J. Burgdörfer, A. M. Azzeer, R. Ernstorfer, R. Kienberger, U. Kleineberg, E. Goulielmakis, F. Krausz, and V. S. Yakovlev, *Delay in photoemission*, Science **328**, 1658 (2010).

[19] M. Nisoli, P. Decleva, F. Calegari, A. Palacios, and F. Martín, *Attosecond electron dynamics in molecules*, Chem. Rev. **117**, 10760 (2017).

[20] G. Calvanese Strinati, P. Pieri, G. Röpke, P. Schuck, and M. Urban, *The BCS–BEC crossover: From ultra-cold Fermi gases to nuclear systems*, Phys. Rep. **738**, 1 (2018).

[21] D. Fausti, R. I. Tobey, N. Dean, S. Kaiser, A. Dienst, M. C. Hoffmann, S. Pyon, T. Takayama, H. Takagi, and A. Cavalleri, *Light-induced superconductivity in a stripe-ordered cuprate*, Science **331**, 189 (2011).

[22] M. Mitrano, A. Cantaluppi, D. Nicoletti, S. Kaiser, A. Perucchi, S. Lupi, P. Di Pietro, D. Pontiroli, M. Riccò, S. R. Clark, D. Jaksch, and A. Cavalleri, *Possible light-induced superconductivity in* K_3C_{60} *at high temperature*, Nature **530**, 461 (2016).

[23] M. Buzzi, D. Nicoletti, M. Fechner, N. Tancogne-Dejean, M. A. Sentef, A. Georges, T. Biesner, E. Uykur, M. Dressel, A. Henderson, T. Siegrist, J. A. Schlueter, K. Miyagawa, K. Kanoda, M.-S. Nam, A. Ardavan, J. Coulthard, J. Tindall, F. Schlawin, D. Jaksch, and A. Cavalleri, *Photomolecular high-temperature superconductivity*, Phys. Rev. X **10**, 031028 (2020).

[24] P. Dyke, A. Hogan, I. Herrera, C. C. N. Kuhn, S. Hoinka, and C. J. Vale, *Dynamics of a Fermi gas quenched to unitarity*, Phys. Rev. Lett. **127**, 100405 (2021).

[25] M. Breyer, D. Eberz, A. Kell, and M. Köhl, *Quenching a Fermi superfluid across the BEC-BCS crossover*, SciPost Phys. **18**, 053 (2025).

[26] E. Fermi, *Quantum theory of radiation*, Rev. Mod. Phys. **4**, 87 (1932).

[27] J. Rammer and H. Smith, *Quantum field-theoretical methods in transport theory of metals*, Rev. Mod. Phys. **58**, 323 (1986).

[28] L. I. Schiff, *Quantum Mechanics* (McGraw-Hill, New York, 1968).

[29] R. P. Feynman, *An operator calculus having applications in quantum electrodynamics*, Phys. Rev. **54**, 108 (1951).

[30] U. Fano, *Description of states in quantum mechanics by density matrix and operator techniques*, Rev. Mod. Phys. **29**, 74 (1957).

[31] K. Huang, *Statistical Mechanics* (John Wiley & Sons, New York, 1963).

[32] G. Strinati, *Application of the Green's functions method to the study of the optical properties of semiconductors*, La Rivista del Nuovo Cimento **11**, (12) 1 (1998).

[33] M. Gell-Mann and F. Low, *Bound states in quantum field theory*, Phys. Rev. **84**, 350 (1951).

[34] R. A. Craig, *Perturbation expansion for real-time Green's functions*, J. Math. Phys. **9**, 605 (1968).

[35] T. Matsubara, *A New approach to quantum-statistical mechanics*, Prog. Theor. Phys. **14**, 351 (1955).

[36] O. V. Konstantinov and V. I. Perel', *A diagram technique for evaluation transport quantities*, Sov. Phys. JETP **12**, 142 (1961) [Zh. Eksp. Teor. Fiz. (U.S.S.R.) **39**, 197 (1960)].

[37] V. Špička, A. Kalvová, and B. Velický, *Dynamics of mesoscopic systems: Non-equilibrium Green's functions approach*, Physica E **42**, 525 (2010).

[38] N. Schlünzen and M. Bonitz, *Nonequilibrium Green functions approach to strongly correlated fermions in lattice systems*, Contrib. Plasma Phys. **56**, 5 (2016).

[39] P. Danielewicz, *Quantum theory of nonequilibrium processes*, Ann. Phys. **152**, 239 (1984).

[40] H. Aoki, N. Tsuji, M. Eckstein, M. Kollar, T. Oka, and P. Werner, *Nonequilibrium dynamical mean-field theory and its applications*, Rev. Mod. Phys. **86**, 779 (2014).

[41] P. C. Hohenberg and P. C. Martin, *Microscopic theory of superfluid helium*, Ann. Phys. **34**, 291 (1965).
[42] D. Forster, *Hydrodynamic Fluctuations, Broken Symmetry, and Correlation Functions* (W. A. Benjamin, Reading, 1975).
[43] N. N. Bogoliubov, *On the theory of superfluidity*, J. Phys. (U.S.S.R.) **11**, 23 (1947).
[44] N. N. Boboliubov, *A new method in the theory of superconductivity. I* , Sov. Phys. JETP **34**, 41 (1958) [Zh. Eksp. Teor. Fiz. (U.S.S.R.) **34**, 58 (1958)].
[45] J. G. Valatin, *Comments on the theory of superconductivity*, Il Nuovo Cimento **7**, 843 (1958).
[46] L. P. Gor'kov, *On the energy spectrum of superconductors*, Sov. Phys. JETP **34**, 505 (1958) [Zh. Eksp. Teor. Fiz. (U.S.S.R.) **34**, 735 (1958)].
[47] D. C. Langreth and J. W. Wilkins, *Theory of spin resonance in dilute magnetic alloys*, Phys. Rev. B **6**, 3189 (1972).
[48] M. Pini, P. Pieri, and G. Calvanese Strinati, *Fermi gas throughout the BCS-BEC crossover: Comparative study of t-matrix approaches with various degrees of self-consistency*, Phys. Rev. B **99**, 094502 (2019).
[49] L. Hedin, *New method for calculating the one-particle Green's function with application to the electron-gas problem*, Phys. Rev. **139**, A796 (1965).
[50] G. Strinati, H. J. Mattausch, and W. Hanke, *Dynamical correlation effects on the quasiparticle Bloch states of a covalent crystal*, Phys. Rev. Lett. **45**, 290 (1980).
[51] G. Strinati, H. J. Mattausch, and W. Hanke, *Dynamical aspects of correlation corrections in a covalent crystal*, Phys. Rev. B **25**, 2867 (1982).
[52] X. Leng, F. Jin, M. Wei, and Y. Ma, *GW method and Bethe–Salpeter equation for calculating electronic excitations*, WIREs Comput. Mol. Sci. **6**, 532 (2016).
[53] D. Golze, M. Dvorak, and P. Rinke, *The GW compendium: A practical guide to theoretical photoemission spectroscopy*, Front. Chem. **7**, 377 (2019).
[54] V. M. Galitskii, *The energy spectrum of a non-ideal Fermi gas*, Sov. Phys. JETP **7**, 104 (1958) [Zh. Eksp. Teor. Fiz. **34**, 151 (1958)].
[55] L. P. Gor'kov and T. M. Melik-Barkhudarov, *Contribution to the theory of superfluidity in an imperfect Fermi gas*, Sov. Phys. JETP **13**, 1018 (1961) [Zh. Eksp. Teor. Fiz. **40**, 1452 (1961)].
[56] L. Pisani, A. Perali, P. Pieri, and G. Calvanese Strinati, *Entanglement between pairing and screening in the Gorkov-Melik-Barkhudarov correction to the critical temperature throughout the BCS-BEC crossover*, Phys. Rev. B **97**, 014528 (2018).
[57] L. Pisani, P. Pieri, and G. Calvanese Strinati, *Gap equation with pairing correlations beyond the mean-field approximation and its equivalence to a Hugenholtz-Pines condition for fermion pairs*, Phys. Rev. B **98**, 104507 (2018).
[58] P. F. Loos and P. Romaniello, *Static and dynamic Bethe–Salpeter equations in the T-matrix approximation*, J. Chem. Phys. **156**, 164101 (2022).
[59] J. Schwinger, *On the Green's functions of quantized fields. I* , Proc. Natl. Acad. Sci. U.S.A. **37**, 452 (1951).
[60] F. J. Dyson, *The S matrix in quantum electrodynamics*, Phys. Rev. **75**, 1736 (1949).
[61] E. Perfetto, Y. Pavlyukh, and G. Stefanucci, *Real-time GW: Toward an ab-initio description of the ultrafast carrier and exciton dynamics in two-dimensional materials*, Phys. Rev. Lett. **128**, 016801 (2022).
[62] J.-P. Joost, N. Schlünzen, H. Ohldag, M. Bonitz, F. Lackner, and I. Březinová, *Dynamically screened ladder approximation: Simultaneous treatment of strong electronic correlations and dynamical screening out of equilibrium*, Phys. Rev. B **105**, 165155 (2022).

[63] F. Palestini and G. Calvanese Strinati, *Temperature dependence of the pair coherence and healing lengths for a fermionic superfluid throughout the BCS-BEC crossover*, Phys. Rev. B **89**, 224508 (2014).

[64] L. Pisani, P. Pieri, and G. Calvanese Strinati, *Spatial emergence of off-diagonal long-range order throughout the BCS-BEC crossover*, Phys. Rev. B **105**, 054505 (2022).

[65] E. E. Salpeter and H. A. Bethe, *A relativistic equation for bound-state problems*, Phys. Rev. **84**, 1232 (1951).

[66] T. Hansen and T. Pullerits, *Nonlinear response theory on the Keldysh contour*, J. Phys. B: At. Mol. Opt. Phys. **45**, 154014 (2012).

[67] C. Kittel, *Quantum Theory of Solids* (John Wiley & Sons, New York, 1963).

[68] O. Goulko, A. S. Mishchenko, L. Pollet, N. Prokofèv, and B. Svistunov, *Numerical analytic continuation: Answers to well-posed questions*, Phys. Rev. B **95**, 014102 (2017).

[69] P. M. Chaikin and T. C. Lubensky, *Principles of Condensed Matter Physics* (Cambridge University Press, Cambridge, 1997), Chapter 4.

[70] J. Bardeen, L. N. Cooper, and J. R. Schrieffer, *Theory of superconductivity*, Phys. Rev. **108**, 1175 (1957).

[71] R. B. Laughlin, *Anomalous quantum Hall effect: An incompressible quantum fluid with fractionally charged excitations*, Phys. Rev. Lett. **50**, 1395 (1983).

[72] B. H. Bransden and C. J. Joachain, *Physics of Atoms and Molecules* (Longman Scientific & Technical, Harlow, 1994), Chapter 7.

[73] D. R. Hartree, *The wave mechanics of an atom with a non-Coulomb central field. Part II. Some results and discussion*, Proc. Cambridge Phil. Soc. **24**, 111 (1928).

[74] A. Dalgarno and G. A. Victor, *The time-dependent-coupled Hartree-Fock approximation*, Proc. Royal Soc. **291**, 1425 (1966).

[75] P. Bonche, S. Koonin, and J. W. Negele, *One-dimensional nuclear dynamics in the time-dependent Hartree-Fock approximation*, Phys. Rev. C **13**, 1226 (1976).

[76] C. N. Yang, *Concept of off-diagonal long-range order and the quantum phases of liquid He and of superconductors*, Rev. Mod. Phys. **34**, 694 (1962).

[77] P. W. Anderson, *More is different*, Science **177**, 393 (1972).

[78] Y. Nambu and G. Jona-Lasinio, *Dynamical model of elementary particles based on an analogy with superconductivity. I,* Phys. Rev. **122**, 345 (1961).

[79] Y. Nambu, *Quasi-particles and gauge invariance in the theory of superconductivity*, Phys. Rev. **117**, 648 (1960).

[80] G. Calvanese Strinati, *BCS and Eliashberg Theories*, Proceedings of the International School of Physics "Enrico Fermi" Course CXXXVI, G. Iadonisi, J. R. Schrieffer, and M. L. Chiofalo, Eds. (IOS Press, Amsterdam, 1998).

[81] R. Haussmann, *Crossover from BCS superconductivity to Bose-Einstein condensation: A self-consistent theory*, Z. Phys. B: Condens. Matter **91**, 291 (1993).

[82] R. Haussmann, *Properties of a Fermi liquid at the superfluid transition in the crossover region between BCS superconductivity and Bose-Einstein condensation*, Phys. Rev. B **49**, 12975 (1994).

[83] N. Andrenacci, P. Pieri, and G. Calvanese Strinati, *Evolution from BCS superconductivity to Bose-Einstein condensation: Current correlation function in the broken symmetry phase*, Phys. Rev. B **68**, 144507 (2003).

[84] P. Pieri, L. Pisani, and G. Calvanese Strinati, *BCS-BEC crossover at finite temperature in the broken-symmetry phase*, Phys. Rev. B **70**, 094508 (2004).

[85] R. Haussmann, M. Punk, and W. Zwerger, *Spectral functions and rf response of ultracold fermionic atoms*, Phys. Rev. A **80**, 063612 (2009).

[86] P. G. De Gennes, *Superconductivity of Metals and Alloys* (W. A. Benjamin, New York, 1966).
[87] L. S. Rodberg and R. M. Thaler, *Introduction to the Quantum Theory of Scattering* (Academic Press, New York, 1967).
[88] J.-X. Zhu, *Bogoliubov-de Gennes Method and Its Applications*, Lecture Notes in Physics Vol. **924** (Springer International Publishing, Cham, Switzerland, 2016).
[89] P. Pieri and G. Calvanese Strinati, *Derivation of the Gross-Pitaevskii equation for condensed bosons from the Bogoliubov–de Gennes equations for superfluid fermions*, Phys. Rev. Lett. **91**, 030401 (2003).
[90] S. Simonucci, P. Pieri, and G. Calvanese Strinati, *Temperature dependence of a vortex in a superfluid Fermi gas*, Phys. Rev. B **87**, 214507 (2013).
[91] D. R. Tilley and J. Tilley, *Superfluidity and Superconductivity* (Adam Hilger, Bristol, 1986).
[92] M. R. Schafroth, S. T. Butler, and J. M. Blatt, *Quasi-chemical equilibrium model to superconductivity*, Helv. Phys. Acta **30**, 93 (1957).
[93] L. V. Keldysh and Y. V. Kopaev, *Possible instability of the semimetallic state toward Coulomb interaction*, Sov. Phys. Solid State **6**, 2219 (1965) [Fiz. Tverd. Tela (Leningrad) **6**, 2791 (1964)].
[94] V. N. Popov, *Theory of a Bose gas produced by bound states of Fermi particles*, Sov. Phys. JETP **23**, 1034 (1966) [Zh. Eksp. Teor. Fiz. **50**, 1550 (1966)].
[95] D. M. Eagles, *Possible pairing without superconductivity at low carrier concentrations in bulk and thin-film superconducting semiconductors*, Phys. Rev. **186**, 456 (1969).
[96] A. J. Leggett, *Diatomic molecules and Cooper pairs*, in A. Pekalski and R. Przystawa, Eds., Modern Trends in the Theory of Condensed Matter, Lecture Notes in Physics (Springer-Verlag, Berlin, 1980), Vol. **115**, p. 13.
[97] P. Nozières and S. Schmitt-Rink, *Bose condensation in an attractive fermion gas: From weak to strong coupling superconductivity*, J. Low Temp. Phys. **59**, 195 (1985).
[98] L. Pitaevskii and S. Stringari, *Bose-Einstein Condensation* (Clarendon Press, Oxford, 2003).
[99] C. J. Pethick and H. Smith, *Bose-Einstein Condensation in Dilute Gases* (Cambridge University Press, Cambridge, 2008).
[100] S. Giorgini, L. P. Pitaevskii, and S. Stringari, *Theory of ultracold atomic Fermi gases*, Rev. Mod. Phys. **80**, 1215 (2008).
[101] W. Zwerger, Ed., *The BCS-BEC Crossover and the Unitary Fermi Gas*, Lecture Notes in Physics Vol. **836** (Springer-Verlag, Berlin, 2012).
[102] Q. Chen, Z. Wang, R. Boyack, and K. Levin, *When superconductivity crosses over: From BCS to BEC*, Rev. Mod. Phys. **96**, 025002 (2024).
[103] L. Pisani, A. G. Moshe, P. Pieri, G. Calvanese Strinati, and G. Deutscher, *Tunneling spectra of unconventional quasi-two-dimensional superconductors*, Phys. Rev. B **110**, L100506 (2024).
[104] U. Fano, *Effects of configuration interaction on intensities and phase shifts*, Phys. Rev. **124**, 1866 (1961).
[105] H. Feshbach, *A unified theory of nuclear reactions. II*, Ann. Phys. **19**, 287 (1962).
[106] C. Chin, R. Grimm, P. Julienne, and E. Tiesinga, *Feshbach resonances in ultracold gases*, Rev. Mod. Phys. **82**, 1225 (2010).
[107] B. D. Josephson, *Possible new effects in superconductive tunnelling*, Phys. Lett. **1**, 251 (1962).

[108] A. Barone and G. Paternò, *Physics and Applications of the Josephson Effect* (J. Wiley & Sons, New York, 1982).

[109] A. Spuntarelli, P. Pieri, and G. Calvanese Strinati, *Solution of the Bogoliubov-de Gennes equations at zero temperature throughout the BCS-BEC crossover: Josephson and related effects*, Phys. Rep. **488**, 111 (2010).

[110] W. J. Kwon, G. Del Pace, R. Panza, M. Inguscio, W. Zwerger, M. Zaccanti, F. Scazza, and G. Roati, *Strongly correlated superfluid order parameters from dc Josephson supercurrents*, Science **369**, 84 (2020).

[111] G. Del Pace, W. J. Kwon, M. Zaccanti, G. Roati, and F. Scazza, *Tunneling transport of unitary fermions across the superfluid transition*, Phys. Rev. Lett. **126**, 055301 (2021).

[112] G. Calvanese Strinati, *A survey on the crossover from BCS superconductivity to Bose-Einstein condensation*, Phys. Essays **13**, 427 (2000).

[113] M. Marini, F. Pistolesi, and G. Calvanese Strinati, *Evolution from BCS superconductivity to Bose condensation: Analytic results for the crossover in three dimensions*, Eur. Phys. J. B **1**, 151 (1998).

[114] F. Pistolesi and G. Calvanese Strinati, *Evolution from BCS superconductivity to Bose condensation: Role of the parameter $k_F\xi$*, Phys. Rev. B **49**, 6356 (1994).

[115] A. Perali, P. Pieri, G. Calvanese Strinati, and C. Castellani, *Pseudogap and spectral function from superconducting fluctuations to the bosonic limit*, Phys. Rev. B **66**, 024510 (2002).

[116] P. Pieri and G. Calvanese Strinati, *Strong-coupling limit in the evolution from BCS superconductivity to Bose-Einstein condensation*, Phys. Rev. B **61**, 15370 (2000).

[117] M. Pini, P. Pieri, and G. Calvanese Strinati, *Evolution of an attractive polarized Fermi gas: From a Fermi liquid of polarons to a non-Fermi liquid at the Fulde-Ferrell-Larkin-Ovchinnikov quantum critical point*, Phys. Rev. B **107**, 054505 (2023).

[118] A. Perali, P. Pieri, L. Pisani, and G. Calvanese Strinati, *BCS-BEC crossover at finite temperature for superfluid trapped Fermi atoms*, Phys. Rev. Lett. **92**, 220404 (2004).

[119] M. Pini, P Pieri, M. Jäger, J. Hecker Denschlag, and G. Calvanese Strinati, *Pair correlations in the normal phase of an attractive Fermi gas*, New J. Phys. **22**, 083008 (2020).

[120] G. Calvanese Strinati, *Recent advances in the theory of the BCS–BEC crossover for fermionic superfluidity*, Physica C: Supercond. Appl. **614**, 1354377 (2023).

[121] P. Dyke, S. Musolino, H. Kurkjian, D. J. M. Ahmed-Braun, A. Pennings, I. Herrera, S. Hoinka, S. J. J. M. F. Kokkelmans, V. E. Colussi, and C. J. Vale, *Higgs oscillations in a unitary Fermi superfluid*, Phys. Rev. Lett. **132**, 223402 (2024).

[122] F. Pistolesi and G. Calvanese Strinati, *Evolution from BCS superconductivity to Bose condensation: Calculation of the zero-temperature phase coherence length*, Phys. Rev. B **53**, 15168 (1996).

[123] L. P. Gor'kov, *Microscopic derivation of the Ginzburg-Landau equations in the theory of superconductivity*, Sov. Phys. JETP **36**, 1364 (1959) [Zh. Eksp. Teor. Fiz. **36**, 1918 (1959)].

[124] M. A. Baranov and D. S. Petrov, *Critical temperature and Ginzburg-Landau equation for a trapped Fermi gas*, Phys. Rev. A **58**, R801 (1998).

[125] S. Simonucci and G. Calvanese Strinati, *Equation for the superfluid gap obtained by coarse graining the Bogoliubov-deGennes equations throughout the BCS-BEC crossover*, Phys. Rev. B **89**, 054511 (2014).

[126] D. S. Petrov, C. Salomon, and G. V. Shlyapnikov, *Scattering properties of weakly bound dimers of fermionic atoms*, Phys. Rev. A **71**, 012708 (2005).

[127] I. V. Brodsky, M. Y. Kagan, A. V. Klaptsov, R. Combescot, and X. Leyronas, *Exact diagrammatic approach for dimer-dimer scattering and bound states of three and four resonantly interacting particles*, Phys. Rev. A **73**, 032724 (2006).

[128] F. Dalfovo, S. Giorgini, L. P. Pitaevskii, and S. Stringari, *Theory of Bose-Einstein condensation in trapped gases*, Rev. Mod. Phys. **71**, 463 (1999).

[129] S. Simonucci, P. Pieri, and G. Calvanese Strinati, *Vortex arrays in neutral trapped Fermi gases through the BCS-BEC crossover*, Nat. Phys. **11**, 941 (2015).

[130] S. Simonucci and G. Calvanese Strinati, *Nonlocal equation for the superconducting gap parameter*, Phys. Rev. B **96**, 054502 (2017).

[131] V. Piselli, S. Simonucci, and G. Calvanese Strinati, *Optimizing the proximity effect along the BCS side of the BCS-BEC crossover*, Phys. Rev. B **98**, 144508 (2018).

[132] V. Piselli, S. Simonucci, and G. Calvanese Strinati, *Josephson effect at finite temperature along the BCS-BEC crossover*, Phys. Rev. B **102**, 144517 (2020).

[133] L. Pisani, V. Piselli, and G. Calvanese Strinati, *Inclusion of pairing fluctuations in the differential equation for the gap parameter for superfluid fermions in the presence of nontrivial spatial constraints*, Phys. Rev. B **108**, 214503 (2023).

[134] V. Piselli, L. Pisani, and G. Calvanese Strinati, *Josephson current flowing through a nontrivial geometry: Role of pairing fluctuations across the BCS-BEC crossover*, Phys. Rev. B **108**, 214504 (2023).

[135] L. Pisani, V. Piselli, and G. Calvanese Strinati, *Critical current throughout the BCS-BEC crossover with the inclusion of pairing fluctuations*, Phys. Rev. A **109**, 033306 (2024).

[136] This property can be traced to a theorem for integrals depending on a complex parameter, as discussed in V. I. Smirnov, *A Course of Higher Mathematics*, Vol. **3**, Part 2 (Pergamon Press, Oxford, 1964), Sec. 70.

[137] E. Abrahams and T. Tsuneto, *Time variation of the Ginzburg-Landau order parameter*, Phys. Rev. **152**, 416 (1966).

[138] P. W. Anderson, *Coherent excited states in the theory of superconductivity: Gauge invariance and the Meissner effect*, Phys. Rev. **110**, 827 (1958).

[139] K. Xhani, E. Neri, L. Galantucci, F. Scazza, A. Burchianti, K.-L. Lee, C. F. Barenghi, A. Trombettoni, M. Inguscio, M. Zaccanti, G. Roati, and N. P. Proukakis, *Critical transport and vortex dynamics in a thin atomic Josephson junction*, Phys. Rev. Lett. **124**, 045301 (2020).

[140] W. Magnus, F. Oberhettinger, and R. P. Soni, *Formulas and Theorems for the Special Functions of Mathematical Physics* (Springer-Verlag, New York, 1966).

[141] C. Di Castro and W. Young, *Density-matrix methods and time dependence of order parameter in superconductors*, Nuovo Cimento B **62**, 273 (1969).

[142] C. Xue, Q.-Y. Wang, H.-X. Ren, A. He, and A. V. Silhanek, *Case studies on time-dependent Ginzburg-Landau simulations for superconducting applications*, Electromagn. Sci. **2**, 0060121 (2024), and references therein.

[143] H. J. Vidberg and J. W. Serene, *Solving the Eliashberg equations by means of N-point Padé approximants*, J. Low Temp. Phys. **29**, 179 (1977).

[144] X. Dong. E. Gull, and H. U. R. Strand, *Excitations and spectra from equilibrium real-time Green's functions*, Phys. Rev. B **106**, 125153 (2022).

[145] C. H. Johansen, B. Frank, and J. Lang, *Spectral functions of the strongly interacting three-dimensional Fermi gas*, Phys. Rev. A **109**, 023324 (2024).

[146] T. Enss, *Particle and pair spectra for strongly correlated Fermi gases: A real-frequency solver*, Phys. Rev. A **109**, 023325 (2024).

[147] E. Dizer, J. Horak, and J. M. Pawlowski, *Spectral properties and observables in ultracold Fermi gases*, Phys. Rev. A **109**, 063311 (2024).

[148] I. S. Gradshteyn and I. M. Ryzhik, *Tables of Integrals, Series, and Products* (Academic Press, San Diego, 1980).

[149] M. Ossiander, K. Golyari, K. Scharl, L. Lehnert, F. Siegrist, J. P. Bürger, D. Zimin, J. A. Gessner, M. Weidman, I. Floss, V. Smejkal , S. Donsa, C. Lemell, F. Libisch, N. Karpowicz, J. Burgdörfer, F. Krausz, and M. Schultze, *The speed limit of optoelectronics*, Nat. Comm. **13**, 1620 (2022).

[150] J. Lloyd-Hughes, P. M. Oppeneer, T. Pereira dos Santos, A. Schleife, S. Meng, M. A. Sentef, M. Ruggenthaler, A. Rubio, I. Radu, M. Murnane, X. Shi, H. Kapteyn, B. Stadtmüller, K. M. Dani, F. H. da Jornada, E. Prinz , M. Aeschlimann, R. L. Milot, M. Burdanova, J. Boland, T. Cocker, and F. Hegmann, *The 2021 ultrafast spectroscopic probes of condensed matter roadmap*, J. Phys. Condens. Matter **33**, 353001 (2021).

[151] P. Bocchieri and A. Loinger, *Quantum recurrence theorem*, Phys. Rev. **107**, 337 (1957).

[152] U. Schneider, L. Hackermüller, J. P. Ronzheimer, S. Will, S. Braun, T. Best, I. Bloch, E. Demler, S. Mandt, D. Rasch, and A. Rosch, *Fermionic transport in a homogeneous Hubbard model: Out-of-equilibrium dynamics with ultracold atoms*, Nat. Phys. **8**, 213 (2012).

[153] C. Caroli, R. Combescot, P. Nozières, and D. Saint-James, *Direct calculation of the tunneling current*, J. Phys. C Solid St. Phys. **4**, 916 (1971).

[154] Y. Meir and N. S. Wingreen, *Landauer formula for the current through an interacting electron region*, Phys. Rev. Lett. **68**, 2512 (1992).

[155] A. P. Jauho, N. S. Wingreen, and Y. Meir, *Time-dependent transport in interacting and noninteracting resonant-tunneling systems*, Phys. Rev. B **50**, 5528 (1994).

[156] M. Ridley, A. MacKinnon, and L. Kantorovich, *Current through a multilead nanojunction in response to an arbitrary time-dependent bias*, Phys. Rev. B **91**, 125433 (2015).

[157] R. Tuovinen, R. van Leeuwen, E. Perfetto, and G. Stefanucci, *Time-dependent Landauer Büttiker formula for transient dynamics*, J. Phys. Conf. Ser. **427**, 012014 (2013).

[158] S. Latini, E. Perfetto, A.-M. Uimonen, R. van Leeuwen, and G. Stefanucci, *Charge dynamics in molecular junctions: Nonequilibrium Green's function approach made fast*, Phys. Rev. B **89**, 075306 (2014).

[159] M. Ridley, N. W. Talarico, D. Karlsson, N. Lo Gullo, and R. Tuovinen, *A many-body approach to transport in quantum systems: From the transient regime to the stationary state*, J. Phys. A Math. Theor. **55**, 273001 (2022), and references therein.

[160] C. Pellegrini, A. Marinelli, and S. Reiche, *The physics of x-ray free-electron lasers*, Rev. Mod. Phys. **88**, 015006 (2016).

[161] C. Caroli, D. Lederer-Rozenblatt, B. Roulet, and D. Saint-James, *Inelastic effects in photoemission: Microscopic formulation and qualitative discussion*, Phys. Rev. B **8**, 4552 (1973).

[162] M. Feidt, S. Mathias, and M. Aeschlimann, *Development of an analytical simulation framework for angle-resolved photoemission spectra*, Phys. Rev. Mater. **3**, 123801 (2019).

[163] J. K. Freericks, H. R. Krishnamurthy, and Th. Pruschke, *Theoretical description of time-resolved photoemission spectroscopy: Application to pump-probe experiments*, Phys. Rev. Lett. **102**, 136401 (2009).

[164] J. Braun, R. Rausch, M. Potthoff, J. Minár, and H. Ebert, *One-step theory of pump-probe photoemission*, Phys. Rev. B **91**, 035119 (2015).

[165] J. K. Freericks, H. R. Krishnamurthy, M. A. Sentef, and T. P. Devereaux, *Gauge invariance in the theoretical description of time-resolved angle-resolved pump/probe photoemission spectroscopy*, Phys. Scr. **T165**, 014012 (2015).

[166] R. Bertoncini and A. P. Jauho, *Gauge-invariant formulation of the intracollisional field effect including collisional broadening*, Phys. Rev. B **44**, 3655 (1991).

[167] E. Perfetto, D. Sangalli, A. Marini, and G. Stefanucci, *First-principles approach to excitons in time-resolved and angle-resolved photoemission spectra*, Phys. Rev. B **94**, 245303 (2016).

[168] Y.-H. Chan, D. Y. Qiu, F. H. da Jornada, and S. G. Louie, *Giant self-driven exciton-Floquet signatures in time-resolved photoemission spectroscopy of MoS2 from time-dependent GW approach*, arXiv:2302.01719.

[169] V. Gosetti, J. Cervantes-Villanueva, S. Mor, D. Sangalli, A. García-Cristóbal, A. Molina-Sánchez, V. F. Agekyan, M. Tuniz, D. Puntel, W. Bronsch, F. Cilento, and S. Pagliara, *Unveiling exciton formation: Exploring the early stages in time, energy and momentum domain*, arXiv:2412.02507.

[170] T. C. Rossi, L. Qiao, C. P. Dykstra, R. Rodrigues Pela, R. Gnewkow, R. F. Wallick, J. H. Burke, E. Nicholas, A-M. March, G. Doumy, D. B. Buchholz, C. Deparis, J. Zuñiga-Pérez, M. Weise, K. Ellmer, M. Fondell, C. Draxl, and R. M. van der Veen, *Ultrafast dynamic Coulomb screening of X-ray core excitons in photoexcited semiconductors*, arXiv:2412.01945.

[171] G. Strinati, *Dynamical shift and broadening of core excitons in semiconductors*, Phys. Rev. Lett. **49**, 1519 (1982).

[172] G. Strinati, *Effects of dynamical screening on resonances at inner-shell thresholds in semiconductors*, Phys. Rev. B **29**, 5718 (1984).

[173] M. Budden,T. Gebert, M. Buzzi, G. Jotzu, E. Wang, T. Matsuyama, G. Meier, Y. Laplace, D. Pontiroli, M. Riccò, F. Schlawin, D. Jaksch, and A. Cavalleri, *Evidence for metastable photo-induced superconductivity in K_3C_{60}*, Nat. Phys. **17**, 611 (2021).

[174] E. Rowe, B. Yuan, M. Buzzi, G. Jotzu, Y. Zhu, M. Fechner, M. Först, B. Liu, D. Pontiroli, M. Riccò, and A. Cavalleri, *Resonant enhancement for photo-induced superconductivity in K_3C_{60}*, Nat. Phys. **19**, 1821 (2023).

[175] A. F. Kemper, M. A. Sentef, B. Moritz, J. K. Freericks, and T. P. Devereaux, *Direct observation of Higgs mode oscillations in the pump-probe photoemission spectra of electron-phonon mediated superconductors*, Phys. Rev. B **92**, 224517 (2015).

[176] A. F. Kemper, M. A. Sentef, B. Moritz, T. P. Devereaux, and J. K. Freericks, *Review of the theoretical description of time-resolved angle-resolved photoemission spectroscopy in electron-phonon mediated superconductors*, Ann. Phys. (Berlin) **529**, 1600235 (2017).

[177] J. P. Revelle, A. Kumar, and A. F. Kemper, *Theory of time-resolved optical conductivity of superconductors: Comparing two methods for its evaluation*, Condens. Matter **4**, 79 (2019).

[178] J. Eisert, M. Friesdorf, and C. Gogolin, *Quantum many-body systems out of equilibrium*, Nat. Phys. **11**, 124 (2015).

[179] C. Eigen, J. A. P. Glidden, R. Lopes, E. A. Cornell, R. P. Smith, and Z. Hadzibabic, *Universal prethermal dynamics of Bose gases quenched to unitarity*, Nature **563**, 221 (2018).

[180] N. Lo Gullo and L. Dell'Anna, *Self-consistent Keldysh approach to quenches in the weakly interacting Bose-Hubbard model*, Phys. Rev. B **94**, 184308 (2016).

[181] A. Haldar, P. Haldar, S. Bera, I. Mandal, and S. Banerjee, *Quench, thermalization, and residual entropy across a non-Fermi liquid to Fermi liquid transition*, Phys. Rev. Res. **2**, 013307 (2020).

[182] Supplemental Material for Ref. [181].

[183] S. Banerjee and E. Altman, *Solvable model for a dynamical quantum phase transition from fast to slow scrambling*, Phys. Rev. B **95**, 134302 (2017).

[184] G. Lindblad, *On the generators of quantum dynamical semigroups*, Commun. Math. Phys. **48**, 119 (1976).

[185] D. Manzano, *A short introduction to the Lindblad master equation*, AIP Advances **10**, 025106 (2020).

[186] H.-P. Breuer and F. Petruccione, *The Theory of Open Quantum Systems* (Oxford University Press, Oxford, 2003).

[187] G. T. Landi, D. Poletti, and G. Schaller, *Non-equilibrium boundary-driven quantum systems: Models, methods, and properties*, Rev. Mod. Phys. **94**, 045006 (2022).

[188] H. Goto, *Quantum computation based on quantum adiabatic bifurcations of Kerr-nonlinear parametric oscillators*, J. Phys. Soc. Jpn. **88**, 061015 (2019).

[189] M. Calvanese Strinati and C. Conti, *Non-Gaussianity in the quantum parametric oscillator*, Phys. Rev. A **109**, 063519 (2024).

[190] H. M. Pastawski, *Classical and quantum transport from generalized Landauer-Büttiker equations. II. Time-dependent resonant tunneling*, Phys. Rev. B **46**, 4053 (1992).

[191] L. Arrachea, *Exact Green's function renormalization approach to spectral properties of open quantum systems driven by harmonically time-dependent fields*, Phys. Rev. B **75**, 035319 (2007).

[192] T. Xu, T. Morimoto, A. Lanzara, and J. E. Moore, *Efficient prediction of time- and angle-resolved photoemission spectroscopy measurements on a nonequilibrium BCS superconductor*, Phys. Rev. B **99**, 035117 (2019).

[193] H. P. Ojeda Collado, G. Usaj, J. Lorenzana, and C. A. Balseiro, *Fate of dynamical phases of a BCS superconductor beyond the dissipationless regime*, Phys. Rev. B **99**, 174509 (2019).

[194] L. M. Sieberer, M. Buchhold, and S. Diehl, *Keldysh field theory for driven open quantum systems*, Rep. Prog. Phys. **79**, 096001 (2016).

[195] C. Müller and T. M. Stace, *Deriving Lindblad master equations with Keldysh diagrams: Correlated gain and loss in higher order perturbation theory*, Phys. Rev. B **95**, 013847 (2017).

[196] S. Nakajima, *On quantum theory of transport phenomena: Steady diffusion*, Progr. Theor. Phys. **20**, 948 (1958).

[197] R. Zwanzig, *Ensemble method in the theory of irreversibility*, J. Chem. Phys. **33**, 1338 (1960).

[198] A. Messiah, *Quantum Mechanics* (Dover Publications, Mineola, 2014), Chapter XXI, Sec. 13.

[199] R. Kubo, M. Toda, and N. Hashitsume, *Statistical Physics II. Nonequilibrium Statistical Mechanics* (Springer-Verlag, Heidelberg, 1995), Sec. 2.5.

[200] P. Myöhänen, A. Stan, G. Stefanucci, and R. van Leeuwen, *Kadanoff-Baym approach to quantum transport through interacting nanoscale systems: From the transient to the steady-state regime*, Phys. Rev. B **80**, 115107 (2009).

[201] M. Cini, *Time-dependent approach to electron transport through junctions: General theory and simple applications*, Phys. Rev. B **22**, 5887 (1980).

[202] C. J. O. Verzijl, J. S. Seldenthuis, and J. M. Thijssen, *Applicability of the wide-band limit in DFT-based molecular transport calculations*, J. Chem. Phys. **138**, 094102 (2013).

[203] G. Stefanucci, *Kadanoff-Baym equations for interacting systems with dissipative Lindbladian dynamics*, Phys. Rev. Lett. **133**, 066901 (2024).

[204] M. Abramowitz and I. A. Stegun, *Handbook of Mathematical Functions* (National Bureau of Standards, Washington DC, 1972).

[205] A. M. Cohen, *Numerical Methods for Laplace Transform Inversion* (Springer, New York, 2010).

[206] A. W. Schlimgen, K. Head-Marsden, L. M. Sager, P. Narang, and D. A. Mazziotti, *Quantum simulation of the Lindblad equation using a unitary decomposition of operators*, Phys. Rev. Res. **4**, 023216 (2022).

[207] T. Guff, C. U. Shastry, and A. Rocco, *Emergence of opposing arrows of time in open quantum systems*, Sci. Rep. **15**, 3658 (2025).

[208] X. Leng, F. Jin, M. Wei, and Y. Ma, *GW method and Bethe-Salpeter equation for calculating electronic excitations*, Comput. Mol. Sci. **6**, 532 (2016).

[209] J. Authier and P.-F. Loos, *Dynamical kernels for optical excitations*, J. Chem. Phys. **153**, 184105 (2020).

[210] A. Stan, N. E. Dahlen, and R. van Leeuwen, *Time propagation of the Kadanoff–Baym equations for inhomogeneous systems*, J. Chem. Phys. **130**, 224101 (2009).

[211] M. Schüler, Denis Golež, Y. Murakami, N. Bittner, A. Herrmann, H. U. R. Strand, P. Werner, and M. Eckstein, *NESSi: The Non-Equilibrium Systems Simulation package*, Comp. Phys. Comm. **257**, 107484 (2020).

[212] P. Lipavský, V. Špička, and B. Velický, *Generalized Kadanoff-Baym ansatz for deriving quantum transport equations*, Phys. Rev. B **34**, 6933 (1986).

[213] J-P. Joost, N. Schlünzen, and M. Bonitz, *G1-G2 scheme: Dramatic acceleration of nonequilibrium Green functions simulations within the Hartree-Fock generalized Kadanoff-Baym ansatz*, Phys. Rev. B **101**, 245101 (2020).

[214] R. Tuovinen, Y. Pavlyukh, E. Perfetto, and G. Stefanucci, *Time-linear quantum transport simulations with correlated nonequilibrium Green's functions*, Phys. Rev. Lett. **130**, 246301 (2023).

[215] S. Hermanns, K. Balzer, and M. Bonitz, *The non-equilibrium Green function approach to inhomogeneous quantum many-body systems using the generalized Kadanoff–Baym ansatz*, Phys. Scripta **T151**, 014036 (2012).

[216] D. Karlsson, R. van Leeuwen, E. Perfetto, and G. Stefanucci, *The generalized Kadanoff-Baym ansatz with initial correlations*, Phys. Rev. B **98**, 115148 (2018).

[217] Y. Pavlyukh and R. Tuovinen, *Open system dynamics in linear-time beyond the wide-band limit*, arXiv:2502.04855.

[218] H-P. Breuer, E-M. Laine, J. Piilo, and B. Vacchini, *Colloquium: Non-Markovian dynamics in open quantum systems*, Rev. Mod. Phys. **88**, 021002 (2016).

[219] J. R. Schrieffer and P. A. Wolff, *Relation between the Anderson and Kondo Hamiltonians*, Phys. Rev. **149**, 491 (1966).

[220] D. Burgarth, P. Facchi, R. Hillier, and M. Ligabò, *Taming the rotating wave approximation*, Quantum **8**, 1262 (2024).

[221] W. E. Lamb, Jr. and R. C. Retherford, *Fine structure of the hydrogen atom by a microwave method*, Phys. Rev. **72**, 241 (1947).

[222] V. Špička, B. Velický, and A. Kalvová, *Long and short time quantum dynamics: I. Between Green's functions and transport equations*, Physica E **29**, 154 (2005).

Index

For EU product safety concerns, contact us at Calle de José Abascal, 56–1°, 28003 Madrid, Spain or eugpsr@cambridge.org.

www.ingramcontent.com/pod-product-compliance
Lightning Source LLC
LaVergne TN
LVHW081516060526
838200LV00005B/194